CELEBRATING
50 YEARS
Texas A&M University Press
publishing since 1974

THE OTHER SIDE
OF NOWHERE

Kathie and Ed Cox Jr. Books on Conservation Leadership

SPONSORED BY

THE MEADOWS CENTER
FOR WATER AND THE ENVIRONMENT
TEXAS STATE UNIVERSITY

Andrew Sansom, General Editor

THE OTHER SIDE OF NOWHERE

Exploring Big Bend Ranch State Park and Its Flora

ROY MOREY

Foreword by David Riskind

TEXAS A&M UNIVERSITY PRESS / COLLEGE STATION

Copyright © 2024 by Roy Morey
All rights reserved
First edition

♾ This paper meets the requirements of ANSI/NISO Z39.48–1992 (Permanence of Paper). Binding materials have been chosen for durability. Manufactured in China through Martin Book Management

Library of Congress Cataloging-in-Publication Data

Names: Morey, Roy, author. | Riskind, David H., writer of foreword. | Sansom, Andrew, writer of foreword.
Title: The other side of nowhere : exploring Big Bend Ranch State Park and its flora / Roy Morey ; foreword by Andrew Sansom.
Other titles: Kathie and Ed Cox Jr. books on conservation leadership.
Description: First edition. | College Station : Texas A&M University Press, [2024] | Series: Kathie and Ed Cox Jr. books on conservation leadership | Includes bibliographical references and index.
Identifiers: LCCN 2023024283 | ISBN 9781648431067 (hardcover) | ISBN 9781648431074 (ebook)
Subjects: LCSH: Plants—Texas—Big Bend Ranch State Park—Identification. | Big Bend Ranch State Park (Tex.)—Guidebooks. | BISAC: NATURE / Plants / General | NATURE / Reference | LCGFT: Guidebooks.
Classification: LCC QK188 .M674 2024 | DDC 581.9764/4—dc23/eng/ 20230705 LC record available at https://lccn.loc.gov/2023024283

Book Design by Kristie Lee

First frontispiece: *Blue Horsehead Nebula in Scorpius* by C. Peter Armstrong, Fort Davis, Texas astrophotographer. Big Bend Ranch State Park is part of the Greater Big Bend International Dark Sky Reserve, which spans over 15,000 square miles in Texas and Mexico, and the dark skies here are a phenomenal park attraction.

A list of titles in this series is available at the end of the book.

CONTENTS

Series Editor's Foreword, by Andrew Sansom	vii
Foreword, by David H. Riskind	ix
Preface	xi
Acknowledgments	xiii
Introduction	1

A TOUR OF BIG BEND RANCH STATE PARK BY PHYSIOGRAPHIC REGION — 9

Physiographic Region 1: Rio Grande Corridor	13
Physiographic Region 2: Alamito Creek–Terneros Creek Lowlands	29
Physiographic Region 3: Bofecillos Mountains: Highlands	33
Physiographic Region 4: Bofecillos Mountains: Llano	48
Physiographic Region 5: Solitario	57
Physiographic Region 6: Fresno Canyon–Contrabando Lowlands	73
Physiographic Region 7: Cienega	83

A FIELD GUIDE TO THE FLORA — 87

Appendix A. Plants on Texas Parks and Wildlife Department Rare, Threatened, and Endangered Species list	379
Appendix B. Other Plants of Possible Conservation Concern in Texas	381
Appendix C. Additions or Revisions to Floral Inventory of Big Bend Ranch	385
Glossary of Botanical and Geological Terms	393
Glossary of Spanish Terms	403
A Note on Sources	407
Literature Cited	411
Index	421

SERIES FOREWORD

I first heard about the remarkable place now known as Big Bend Ranch State Park in 1971 when a man named Robert L. Armstrong was elected Texas Land Commissioner. I was a junior aide in the Department of Interior in Washington when I learned of Armstrong's election. I was so excited that an environmentalist had been named to the office that I called him and offered to help him in Washington in any way I could.

Two months later, Armstrong showed up in my office. At that time he was pushing for acquisition by the State of Texas of a vast property in far West Texas owned by oilman Robert O. Anderson and known as the Big Bend Ranch. Though the ranch has many miles of Rio Grande river frontage, an entire volcanic crater, and two of the highest waterfalls in Texas, all he could talk about was the stars. I will never forget his rapturous description of the night sky as seen from the wilderness of this most remarkable place. The other side of nowhere.

Morey, a retired official of various state water agencies, including the comptroller's office and the Texas Attorney General's office, has here created the most comprehensive guidebook ever produced for those who wish to explore the land now known as Big Bend Ranch State Park. After leaving state government, the author migrated to the Trans-Pecos Region and dedicated his life to studying, photographing, and interpreting it.

After returning to Texas from Washington, I joined the staff of The Nature Conservancy in Texas and visited the ranch for the first time. It was actually the first time I had visited the Big Bend region and it took my breath away. By this time, Armstrong had retired from the General Land Office and served on the board of The Nature Conservancy in Texas where he continued his efforts to preserve the "other Big Bend," described here by Morey in sweeping and informative detail.

By 1988, Bob Armstrong had been appointed to the Texas Parks and Wildlife Commission by Governor Mark White, and the ranch was added to the Texas State System as the Big Bend Ranch State Natural Area, which became the largest state park in Texas and the third largest in the United States.

In 2008, while Bob Armstrong was still living, we celebrated the twentieth anniversary of the purchase of Big Bend Ranch State Park at the site. At the time, my grandson, Alex, was a five-year-old kindergartner from New Jersey, and we took him to the park for the festivities. When he returned to New Jersey, his teacher assigned the class to select an animal that migrates, tell where it would migrate to, and why.

Alex chose an eagle and wrote that he would migrate to Texas so he could go out to Big Bend and see the stars.

On these pages, Roy Morey gives us the dark sky at Big Bend Ranch and much more.

—**Andrew Sansom**
General Editor, Kathie and Ed Cox Jr. Books on Conservation Leadership

FOREWORD

Not widely known or appreciated, the Big Bend lies in an area of Texas subject to very early plant exploration. Presidio del Norte, and the great canyons of the Rio Grande below Presidio, were a nexus explored by botanists, surgeon-naturalists, geologists, and surveyors under the direction of Major W. H. Emory of the United States and Mexican Boundary Survey, Charles Parry and John Bigelow among them. A Bigelow collection on August 10, 1852, is the type of *Yucca thompsoniana*. The location "Mountains, Bufatello … ," a volcanic district near Presidio del Norte, is a misspelling of the Bofecillos Mountains. Collected farther downstream was *Vachellia (Acacia) schottii* by Parry.

Valery Havard, as part of a military expedition led by Major W. K. Livermore, visited the area on two occasions, May 1881 and September 1883. He recounted collections of *Eriogonum suffruticosum*, spoke of the vast grasslands of the Bofecillos Plateau, noted a collection of the rare grass of desert wetlands *Imperata*, and recounted the lush growth of grapes along an entrenched stream that defined the eastern limits of the Bofecillos—"Grapevine Canyon." Clearly, Havard was referring to Fresno Canyon as no other terrain feature fits into the landscape context he described.

Roy Morey was attracted to the Big Bend relatively recently, in 1986. Shortly after he was drawn into the complementary and intense pursuits of exploration, botany, photography, and authorship. By 2008 he had published *Little Big Bend* (LBB) and along the way becoming an expert at desert plant identification and photography. In the process, Roy explored more of Big Bend Ranch on foot, both spatially and temporally, than most anyone.

I "met" Roy as a reviewer of his LBB manuscript for Texas Tech University Press. At the time, as director for the Natural Resources Program for Texas State Parks, I was spearheading baseline assessment and resource planning for the recently acquired Big Bend Ranch State Park. I had been involved in several capacities, all resource related, for Big Bend Ranch since the early 1970s. Based on his efforts on the national park, I knew Roy was the one person with the unique skills, dedication, mettle, and perseverance to examine in detail the rich and diverse plant life and daunting landscape of the state park. I recruited him and arranged for permitting and some accommodation for special access to the 300,000-acre property. With the publication of *Little Big Bend*, Roy rededicated much of his time to exploring Big Bend Ranch and to learning the terrain, temperament, and sense of the place. He studied the literature and availed himself of the botanical mentors, students, and herbarium at Sul Ross.

Meanwhile, we continued resource assessment and synthesis of natural history and ecological works done previously. Collections by previous modern botanists, including Hinckley, Warnock, Johnston, Powell, Butterwick, Clark, Worthington, Poole, Webster and Westlund, Carr, Hardy, and others, were studied in detail. Using these data and his now-considerable field sense, Roy doubled down and searched Big Bend Ranch intensively through drought, monsoon, and virtually every season and then repeated the process over multiple years. Roy also sampled diverse terrains and substrate types as he mapped in detail through the geological investigations of C. D. Henry and colleagues.

Based on what we know about the Chihuahuan Desert, especially its ephemeral nature, the difficulty and diversity of terrain, and the overall richness of the Chihuahuan Desert biota, new discoveries and new plant distribution records were to be expected. Roy, a consummate and micro-detail-focused plant hunter, did not disappoint. This catalog of superb, artful plant portraits and carefully researched and documented accounts testifies to his success. Roy's work brings full circle the botanical exploration begun by Bigelow and Parry and carried through in modern times by colleagues who preceded Roy in documenting the rich and diverse flora of Big Bend Ranch. I hope you will appreciate and respect the Herculean effort by Roy Morey to bring forth these pages for you to enjoy.

—**David H. Riskind**
Former director, Natural Resources Program,
Texas Parks and Wildlife Department

PREFACE

I purchased my first camera for a trip to Chile in the early 1980s and produced some pleasing travel photos and keepsakes. Back in the United States, my interest soon focused on landscape photography and travel to national parks and other escapes. Of all the parks, Big Bend National Park made the greatest impression, not for the impressive landscape views but for the incredible park diversity, especially plant diversity. A very average landscape photographer, my interest quickly shifted to close-up photography as a tool to learn about the plants. I fell in love with the park, the plants, and the discovery process, which culminated in my publication of *Little Big Bend: Common, Uncommon, and Rare Plants of Big Bend National Park* in 2008.

In 2009, I made a timely move from Terlingua to Lajitas, 17 miles down the road, to continue my work at Big Bend Ranch State Park. For many years, Big Bend Ranch had been an afterthought—remote, out-of-the-way, with no paved roads, and without backcountry access. However, some locals and retired desert hermits were fans and spread the word about Big Bend Ranch as a truly magical, wilder, and more adventuresome outlet for wanderers like me. When the park fully opened in 2009, I was present at the ceremony.

My first step was to learn more about the park's human and natural history, natural resources, and plant communities. In the process of exploring Big Bend Ranch's 238 miles of trails, 70 miles of 4WD roads, and countless miles cross country, I developed a deep respect for and personal connection to the park—its isolation, wildness, hidden treasures, and subtle, understated beauty.

After scouting plant locations and targeting plants to seek out, my approach was simple: find and photograph a plant, record the location and habitat, and note plant features—stems, leaves, flower parts, and fruits. When needed for identification purposes, I would take a specimen with photos to Dr. Powell, at the A. Michael Powell Herbarium of Sul Ross State University.

This began a longer learning process. As my curiosity grew, I sought more details on a plant's geographic range, habitat, morphology, and human uses (for food, medicine, and such). I became intrigued with a plant's struggle to survive in this hostile environment, its coping mechanisms, and its protections from predators. I was enamored with the plant's botanical history, the early botanists who first collected and described the plant, and their history. This all seemed to be a part of the story.

My photographic approach is straightforward. I view the plant as the subject and the art, and myself as an apprentice craftsman, learning on the job, bringing some knowledge and tools to the subject. My focus is the natural beauty of the plant, often best illustrated in fine, close-up detail of small plant parts.

My journey from travel snapshots to grand landscape views to close-up photography to the appreciation of plants is now thirty-five years old, with twenty-three years at Big Bend National Park and twelve years at Big Bend Ranch. Big Bend is now ingrained in my mental make-up. Absence brings uneasiness, and presence renewed awe. Perhaps even more time is needed in the Big Bend—it takes time to become a part of this world.

An Important Note on Touring and Hiking the Park

The parkwide and physiographic region maps in this book offer only a "lay-of-the-land" overview of this immense park. One or more additional resources are essential in navigating Big Bend Ranch. These are listed in the following paragraph and can be obtained through the park's website (https://tpwd.texas.gov/state-parks/big-bend-ranch) and at the park (Barton

Warnock Visitor Center, east entrance; Fort Leaton State Historic Site, west entrance; Sauceda Ranger Station, interior).

The *Discovery Topographic Map* displays the roads, trails, park features, and park locations discussed herein. An even more detailed two-sided, waterproof topographic map, the *Exploration Map*, is not available online but can be purchased at park headquarters. Visitors must travel through private land and private inholdings within the park to reach the interior, and these maps clearly delineate those boundaries. A critical resource for negotiating Big Bend Ranch's difficult jeep routes, *Roads to Nowhere*, is a guide to the unmaintained 4 × 4 high-clearance roads, with equipment recommendations and safety admonitions. Using the *Campsite Guide*, campers can choose their campsite from descriptions and photographs of fifty-two campsite locations accessible by vehicle. An online *Interactive Map* displays Big Bend Ranch campsite, lodging, and permit locations. Numerous other topographic maps, trail maps, and interpretive guides are available.

The "Literature Cited" section of this book draws attention to resources used by the author in drafting this work, including park guides, geological and archaeological studies, natural area surveys, and other area studies. Many of these works should prove useful to prospective visitors and can enrich their Big Bend Ranch experience. An example: for those coming to view the park's incredible geological diversity, Chris Henry's *Geology of Big Bend Ranch State Park, Texas* (1998) contains a detailed geology map of Big Bend Ranch and describes the geology of each physiographic region.

For those needing a refresher, the Glossary of Botanical and Geological Terms should prove a welcome reference. For those curious about the meaning of Spanish place names or other Spanish terms, the Glossary of Spanish Terms has some answers.

ACKNOWLEDGMENTS

David H. Riskind, recently retired director of the Natural Resources Program for Texas State Parks, probably knows more about the ins and outs of Big Bend Ranch State Park, the human history and the trails, than any person living. David served as the principal reviewer of my 2008 work on the plants of Big Bend National Park, *Little Big Bend*. Immediately after that book was published, I relocated from Terlingua to Lajitas to photograph the plants of Big Bend Ranch State Park and ultimately write *The Other Side of Nowhere*. David's knowledge and love of Big Bend Ranch State Park led me there, and for years he acted as a distant trail guide from his Austin office at TPWD. Later, David guided me through various drafts of this work and led me to Texas A&M University Press.

In the same way that David Riskind served as my park guide, Dr. A. Michael Powell, professor emeritus of biology and director and curator of the A. Michael Powell Herbarium at Sul Ross State University, served as my guide to the plants. Dr. Powell encouraged my plant photography and my slowly developing botanical knowledge, politely corrected my errors in plant identification, and gave me tips from his own experiences in the field at Big Bend Ranch. This work would be much less without him.

Many individuals—rangers, staffers, and volunteers at Big Bend Ranch—assisted me in so many ways at Barton Warnock Visitor Center in Lajitas and at Sauceda (the backcountry headquarters) that I am unable to remember all the names after a decade of hiking the trails. I received assistance from a long series of park superintendents—Rod Treviso, Tony Gallego, David Long, Laird Considine, and Nathan Gold in Lajitas, and Barrett Durst and Karl Flocke at Sauceda. I benefited especially from David Long's experience with park trails and plants, and from his friendship. I remember Barrett most for his enthusiastic encouragement of my work, his love of the park, and his late evening rescue when my SUV broke down at the end of Las Burras 4WD Road. Karl graciously reviewed an early draft of my book and helped to improve it.

Blaine R. Hall, former Big Bend Ranch Interpretive Ranger and trainer for the Texas Master Naturalists and other educational groups, introduced me to park geology, became a parttime hiking companion, and later reviewed an early book draft. David Dodder, assistant superintendent at Sauceda, and his wife Vi became my friends. I relied on David's knowledge of backcountry trails and road conditions, and he and Vi graciously reviewed an early book draft.

Over the years, I became a friend and hiking companion of fellow photographer Gary Nored, a park volunteer at Sauceda and later Lajitas. Gary is well known to many park visitors and writes the popular blog An Eye for Texas (aneyefortexas.wordpress.com). Gary is the author of two works familiar to many visitors, *Big Bend Ranch State Park Road Guide* (Nored 2014a), and *Campsites of the Big Bend Ranch State Park* (Nored 2014b). Roy Saffel, a volunteer and later park ranger at Sauceda, became a friend and frequent hiking companion. We explored the very worst 4WD roads together.

Tim Gibbs, park archaeologist and coauthor of *Trails through Time*, an archaeological survey of selected park trail corridors, advised me on archaeological and rock art sites. He and Amber Harrison, park ranger in charge of trail maintenance and planning, both offered detailed advice on trails and trail conditions.

I first met Luis Hernandez, park ranger at Lajitas, when he was assigned to help me set up an exhibit of my photos at the Barton Warnock Visitor Center in Lajitas. He was generous with his firsthand knowledge of park locations and assisted me in various ways over the years.

Sylvia Aguilar, office manager at Sauceda, was a Big Bend Ranch fixture during my time there and a permanent anchor through many staff changes. I am deeply grateful for her assistance, support, and always welcoming manner.

In my early visits to Sauceda, I became friends with Edmundo Lujan, a jack-of-all-trades park ranger who seemed to do a little something for everyone. Edmundo, the most instinctive auto repairman I have ever met, rescued my vehicle on occasion and became an infrequent hiking companion.

For years, visitors could pay for hot meals at the Sauceda bunkhouse, and the meals were welcome rewards after long hikes. The first cook I met was Cindy Steinback-Martin, who soon left for greener pastures out of state. I promised her a copy of my book, but I doubt she thought it would take this long. A second cook and longtime area resident Sandra Cabezuela offered good food, warmth, and hospitality.

Victor Rivera, another good cook who specialized in barbecue, is also a backcountry ranger who keeps an eye out for plants. Victor, *mi amigo*, discovered two of the four known park sites of Havard's agave, a towering century plant rare at Big Bend Ranch, and he quickly conveyed their approximate locations to me.

After most of my field work was complete, Dr. Cathryn Hoyt, formerly at the Chihuahuan Desert Research Institute and now at Big Bend National Park, painstakingly reviewed an early draft of the 281 plant descriptions, and her comments led to improvements in both content and style. Jeff Keeling in Alpine, an expert on the flora of the Davis Mountains, and Gary Freeman in Fort Davis, professor emeritus of integrative biology, University of Texas at Austin, also kindly reviewed an early draft and offered helpful suggestions.

Dr. James Henrickson, visiting scholar at the Plant Resources Center and an expert on Chihuahuan Desert flora, graciously permitted me the use of an unpublished copy of *The Flora of the Chihuahuan Desert Region*, edited by himself and Marshall Johnston. This work proved invaluable in my efforts.

Toward the end of my work, two TPWD employees reviewed later drafts of the book and furnished assistance critical to its completion. In her draft review, Anna Strong, the TPWD's Rare Species Botanist, focused on my treatment of rare plants and suggested ways of discussing these plants while also protecting them from illegal collection. Price Rumbelow, Region 1 Habitat Conservation Specialist, with only a little help from me, prepared the parkwide and physiographic region maps so essential to any "tour" of the park. I am deeply grateful for his efforts.

THE OTHER SIDE
OF NOWHERE

INTRODUCTION

Big Bend Ranch State Park

Acquired by the state of Texas in 1988 and first opened to the public as Big Bend State Natural Area in 1991, Big Bend Ranch State Park (BBR) now encompasses 492 square miles of the incomparable Chihuahuan Desert (CD). Extending roughly 29 miles from east to west, and 22 miles from north to south, the park represents nearly half the total acreage of the state park system.

The park lies within the southern Big Bend of Trans-Pecos Texas, in southeast Presidio County and the southwest corner of Brewster County. The Rio Grande, the US-Mexico border in Texas, forms most of the park's southern boundary, from near Lajitas on the southeast in Brewster County to near Redford on the southwest in Presidio County. Overlooking Lajitas and the Rio Grande, just southeast of BBR, is Mesa de Anguila, within the southwest limits of Big Bend National Park (BBNP). Presidio, the nearest major town (population 4,000), is 7 miles west of BBR's west boundary on Highway 170 (the famous River Road).

Despite the park's large size, annual visitation is minuscule. Low visitation is partly a function of the park's remoteness, which is well appreciated, sometimes even celebrated, by those who do visit. The backcountry headquarters, Sauceda, is more than 690 miles from Houston, and 280 miles from El Paso.

BBR remains as inaccessible as it is remote. Outside the river corridor, no paved road accesses the interior, and the main gravel road can be washed out by flash floods in the "monsoon" season. BBR is the "Other" Big Bend, christened the "Other Side of Nowhere," a rugged wilderness outback for the adventuresome. What the park lacks in paved roads is counterbalanced by another kind of accessibility—238 miles of trails for hiking, biking, and horseback riding, and 70 miles of challenging 4WD roads. These trails and roads access the park's greatest asset, its diversity.

Comparison of BBR and BBNP

Many first-time visitors at BBR have previously visited BBNP and know that park well. These visitors come to BBR in search of a new experience, but are uncertain, even a little wary, of what the park has to offer. BBNP is much more developed with good, paved roads and many amenities. In contrast, with no paved roads outside the river corridor, BBR almost advertises remoteness and inaccessibility. Ultimately, much of BBR's allure is that it is relatively undiscovered, wilder, challenging, and slightly intimidating.

BBNP is much larger, both in size and visitation. The national park encompasses 801,163 acres, the largest protected expanse of Chihuahuan Desert in the United States and 118 miles of the Rio Grande corridor. In 2019, park visitation set a record with 463,832 visitors. BBR covers 314,000 acres, with 50,000 visitors. The Chisos Mountains of BBNP are higher, cooler, with more rainfall, and highly diverse flora, including rare orchids and relict species of a wetter climate. Emory Peak (7,832 ft.), BBNP's highest point, and even the Chisos Basin (5,400 ft.), are higher than BBR's highest peak, Oso Mountain (5,135 ft.).

However, BBR is wetter overall, with many more springs and seeps and numerous creeks draining the mountains, which sustain a variety of plant and animal life rivaling that of BBNP. The improbable natural landscape offers a panorama of memorable vistas, especially at sunrise, sunset, and night, when stars light the dark skies. BBR is an International Dark Sky Park, and the stargazing is spectacular.

The Other Side of Nowhere

What BBR has to offer is certainly not limited to its biological resources. The geology and the long history of human habitation, covering 11,000 years, deepen the park's appeal. Remains of historic ranches, cinnabar (mercury) mines, and wax-making factories document a fascinating story of man's efforts to make a living in this portion of El Despoblado ("the deserted, unpopulated place"), a name long applied to the northern CD. Less evident are more than 500 archaeological sites, including camps, caves, rock-shelters, and rock art displays, which provide clues to the park's prehistory.

Geological History

While BBR is inaccessible, its geology could hardly be more accessible. More than 500 million years of geological history are recorded in the landforms. The park's diversity is evident in the contrasting topography, the low river corridor at 2,500 feet in elevation compared to the remote backcountry 2,000 feet higher; steep mountainous terrain alternating with high-walled canyons; and spectacular waterfalls and flowing springs near once fiery volcanic craters.

In the Paleozoic period (520–300 Ma), an ocean basin covered the Big Bend, and marine sedimentary rocks were deposited that are still visible in the Solitario. Folding and faulting during a mountain-building period (300 Ma) uplifted the Ouachita Mountains, the remains of which extend from Arkansas to the Solitario. Deformed and folded rocks in the Solitario's interior walls illustrate this upheaval (Hall 2013).

In the Cretaceous period (115–90 Ma), BBR was again underwater, covered by a shallow sea. Limestone deposited in this environment is exposed in the Contrabando Lowlands and the Solitario rim. A second mountain-building event (ending 50 Ma) constructed the precursor of the Rocky Mountains, stretching from Canada to northern Mexico. In BBR, these compressive forces produced a fold, the Fresno-Terlingua Monocline, marked by an escarpment dividing Terlingua Uplift from the Contrabando Lowlands.

Early volcanic activity (36–35 Ma), a magma intrusion into older Cretaceous and Paleozoic rocks, began to shape the Solitario into a circular igneous dome. After an eruption, the dome collapsed, leaving a basin-like depression (caldera) in the south-central portion of the dome. Resistant Cretaceous rocks were uplifted to form a rim around the collapsed interior.

Volcanic eruptions within the park (32–27 Ma) built the Bofecillos Mountains. Eruptions in the Sierra Rica of Mexico spread volcanic ash and tuff into southern portions of the park. Thick rhyolite lavas, from vents in the central Bofecillos, built volcanic cones and produced the highest mountains, such as cone-shaped Oso Mountain. Thin basalt lavas spread more extensively, forming the Llano plateau (Henry 1998).

Faulting in the western United States (25 Ma) created the Basin and Range Province from Oregon to Mexico. The faulting produced a rift along what is now the Rio Grande. The ancestral Rio Grande, and Rio Conchos in Mexico, spilled over from higher elevations (2 Ma), filled the basins with lakes, washed away sediment, cut into resistant rock, and carved the canyons evident today. The Presidio and Redford basins were breached by the Rio Grande in its struggle to become a through-flowing river. The through-flowing river, and mountain erosion, formed the canyons draining the Bofecillos Mountains (Henry 1998).

Prehistoric Past

The human prehistory of BBR dates to 11,000 years ago, the Paleo-American period (10,000–6500 BC), when nomadic groups began to traverse the area. They hunted large game animals and gathered plant foods to supplement their diet. Temperatures were cooler, rainfall more plentiful, woodlands more extensive, and the desert more constrained.

The Early Archaic period (6500–3000 BC) brought drier conditions and slowly encroaching desert. Smaller game animals were hunted with the atlatl, a dart-throwing handheld tool. Earth ovens were used to cook semisucculent CD plants such as sotol and lechuguilla. The population remained low.

The Middle Archaic period (3000–1000 BC) witnessed some population gains and settlements in more varied topography. Dwellings consisted mostly of open campsites and rock-shelters, usually near sources of water. Plants became a more important source of food, fiber, and construction materials. Rocks were fashioned into grinding implements and chipped stone tools. Agate and chert were prized in toolmaking.

In the Late Archaic period (1000 BC–AD 900), population growth accelerated. Settlements spread to more diverse habitats. New products and tools were fabricated. Small villages were established near the junction of the Rio Grande and Rio Conchos, later known as La Junta de los Rios. The atlatl was replaced by the bow and arrow. Agricultural practices and specialized food processing were slowly adopted.

The Late Prehistoric period (AD 900–1535) saw greater adoption of archery, agriculture, and ceramics, and a cultural transformation, still largely limited to La Junta. More villages were settled along the river, but the hunter-gatherer lifestyle dominated elsewhere. Gradually, trading expanded with other population centers. Agricultural products were traded for pottery, or for animal products brought to La Junta by hunter-gatherers. Reportedly, Cabeza de Vaca traveled through La Junta in 1535. (See Alloway 1995; Sanchez 1999; Ohl and Cloud 2001; Gibbs 2016).

Historic Past

The Protohistoric period (AD 1535–1700) witnessed the arrival of Spanish expeditions, introduction of Spanish culture and religion, and mission establishment. The Antonio de Espejo expedition of 1582–83 identified La Junta residents as "Patarabueyes," and "Jumanos" as hunter-traders who wintered in La Junta but otherwise traveled on hunting-trading expeditions (Kelley 1986). These groups may be descended from hunter-gatherers of the Southern Plains or northwest Chihuahuan Desert.

Apache moved into the Big Bend from the Southern Plains, and by 1650 threatened Spanish control. La Junta asked for missionaries and military assistance, and the Dominguez de Mendoza–Fray Nicolas Lopez expedition of 1683–84 established La Junta missions. Comanche entered the region in the early 1700s. The Spanish made only sporadic efforts to protect La Junta and pacify the Apache and Comanche. Presidio del Norte was established in 1760 near present-day Presidio, and a fort, Presidio de la Junta de los Rios, was constructed. Nevertheless, raids continued. Juan de Ugalde led the last Spanish expedition against the Apache in the Big Bend in 1787, but Comanche raids in northern Mexico soon became an overriding issue. Spanish occupation ended with Mexican independence from Spain in 1821.

In 1839, Dr. Henry Connelly forged the Chihuahua Trail, a trade route from Chihuahua City to La Junta, up Alamito Creek through BBR's northwest panhandle, and on to Missouri. In 1845, US annexation of Texas led to the Mexican-American War. The 1848 Treaty of Guadalupe Hidalgo established the Rio Grande as the US-Mexico boundary, and peace brought new settlers. Ben Leaton opened Leaton Trading Post on the river south of Presidio and was soon trading with all comers—Apaches, Comanches, Mexicans, travelers, soldiers, and merchants.

In 1851–52, Major William Emory headed US Army reconnaissance efforts to survey the boundary between the United States and Mexico. In 1860, Lieutenant William Echols led a camel caravan along Terneros Creek on BBR's northern boundary to test the feasibility of employing camels on the frontier. Chihuahua Trail cattle drives began in the 1860s, and August Santleben started a stage route from San Antonio to Chihuahua City via Presidio.

In the 1870s, railroads entered the Big Bend, spurring ranching and farming activities. Irrigation on the river between Presidio and Redford greatly increased crop yields. By the early 1880s, the Apache were removed from the Big Bend, and a land rush resulted. After the land rush, some large ranches were formed. In the 1910s, the beginning of the Mexican Revolution and the rise of Pancho Villa led to bandit raids on ranches, and the US Army sent troops to Lajitas and elsewhere to protect the border (See Saunders 1976; Burnett 2002; TPWD 2011; Ohl and Cloud 2001; Perry 2008; Alloway 1995).

Park History

The park headquarters, Sauceda, and Howard's Ranch to the southeast, are central to the ranching history of the southern Big Bend. Sauceda was a ranching operation from 1905 until 1988, when the state acquired BBR. George Howard founded a cattle ranch, Howard's (or Chillicothe) Ranch, as co-owner, consolidated the Chillicothe and Saucita properties as Chillicothe-Saucita Ranch, and by 1905 resided in what today is known as Sauceda. Howard constructed a ranch house, water tank, and corral, but in 1909, he sold the property and moved to Marfa (Smith-Savage 1996a).

By 1915, early settler W. W. Bogel had helped his sons, Gus, Gallie, and Graves, secure the land and expand the ranch. By 1923, the ranch exceeded 25,000 acres, but after the onset of the Great Depression, the ranch was sold in 1934. In 1941, the Fowlkes brothers, Edwin H. and J. M. "Mannie," acquired the then 28,000-acre ranch. During the 1940s, the Fowlkes brothers amassed a much larger, mostly sheep and goat ranch, erected most of the stone dams, water tanks, and wire fences remaining today, and put in place an extensive pipeline system transporting well water across the ranch.

The 1950s drought and collapse of wool prices forced the sale of the ranch in 1958, to Midland attorney and oilman Len McCormick. Naming the property Big Bend Ranch, McCormick initiated construction of the bunkhouse, pole barn, and other buildings, but an oil-field accident forced him to sell the ranch to his silent partner, Julian Sprague. Sprague died shortly after, and his wife leased the ranch to Robert O. Anderson's Lincoln Land and Cattle Company. When Anderson purchased the ranch and formed the Diamond A Cattle Company in 1969, he became the largest private landowner in the United States. The ranch prospered, but an early 1980s recession led Anderson to sell half-interest to Houston businessman Walter Mischer, owner of the Lajitas Resort. For a time, the ranch was used as a private hunting preserve.

Efforts of many individuals and conservation groups, spearheaded by General Land Commissioner Bob Armstrong, led the Texas Parks and Wildlife Commission to purchase 215,000-acre BBR in 1988. The property was acquired from the Hondo Oil and Gas Company of Robert Anderson and Walter Mischer at a cost of $8.8 million. First opened to the public as Big Bend Ranch State Natural Area in 1991, BBR was rededicated in 1995 as Big Bend Ranch State Park. The park opened fully to the public in 2009. Subsequent land acquisitions have increased the park's size to 314,000 acres (See TPWD 2011; Saunders 1976; Kohout 2020; Alloway 1995).

Chihuahuan Desert

BBR lies within the CD, arguably the largest warm desert in North America, covering as much as 250,000 square miles. From north to south, the CD stretches from Socorro in south-central New Mexico to San Luis Potosí in Mexico; and from the southeast corner of Arizona to the Pecos River in Texas. The southern two-thirds of the CD lies mostly in the Central Plateau of Mexico. The northern one-third reaches into Trans-Pecos Texas, southern New Mexico, and southeastern Arizona.

In Mexico, the CD is bounded on the west by the high mountainous terrain of the Sierra Madre Occidental, and on the east by mountains of the Sierra Madre Oriental. Since the Sierra Madre Occidental blocks moisture from the Pacific Ocean, and the Sierra Madre Oriental blocks thunderstorms from the Gulf of Mexico, the CD is known as a rain shadow desert, partly created by the barrier mountains.

Rainfall, averaging less than 10 inches annually, is also limited because the CD experiences only one true rainy season, from July to October, with torrential storms and rapid runoff. Because of the interior location and relatively high elevations (3,000–5,000 ft.), desert winters can be cool or even cold, with periodic frosts. Summers, however, are long and hot, with temperatures regularly reaching above 100 degrees during the day and dropping sharply at night.

The CD is a shrub desert, with a largely limestone substrate, notable for its grasslands. Unlike the Sonoran Desert, which is dominated by large columnar cacti and small trees, the CD's principal species are shrubs, especially the ubiquitous creosote bush. Many shrubs, such as mesquites, acacias and mimosas, and allthorn, are armed with spines, thorns, or prickles, which protect them from predators. Semisucculent plants, such as agaves, yuccas, ocotillo, sotol, and candelilla, are adapted to the low rainfall. The widespread lechuguilla, an agave, is endemic, an indicator species of the CD's geographic limits. Small cacti and medium-sized prickly pear are also characteristic of this shrub desert. Grasses are important but diminished from earlier times.

The CD is part of the Basin and Range Province, a physiographic region reaching from southern Oregon to Mexico, with elongated, north–south-trending mountain ranges alternating with flat, arid, warmer valleys. Due to its large size, isolation between barrier mountains, and elevation differences from mountains to valleys, the CD encompasses many varied habitats and ranks as one of the world's most biologically diverse deserts. Many local plant and animal populations have adapted and evolved with minimal outside influence. Up to 3,500 plant species occupy the desert's microhabitats, and 1,000 of these may occur nowhere else (See Hoyt 2002; Ruhlman, Gass, and Middleton 2012; Ohl and Cloud 2001; Sanchez 1999).

Big Bend Ranch State Park Vegetation and Land Use Changes

By David H. Riskind

Over the past two million years, in the geological epoch known as the Pleistocene or "Ice Age," glaciers expanded southward in northern North America several times, and climates in areas south of the glaciers, such as the Big Bend, became cooler and wetter. These "glacial" periods alternated with "interglacial" periods of warmer and drier climate. These climatic cycles caused Big Bend region vegetation and flora to change back and forth between wetter and drier types. Desert species expanded their ranges during warmer, drier periods, and contracted their ranges during cooler, wetter periods. Evidence suggests that characteristic CD species responded to wetter periods in two different ways: lechuguilla, for instance, apparently survived the last glacial period only well to the south in Mexico (Scheinvar et al. 2016), while creosote bush likely survived locally in the warmest, driest areas of the Big Bend (Hunter et al. 2001).

In the past 10,000 years, Big Bend area vegetation and flora have undergone two dramatic changes. The first change, starting about 8,000 years ago, was a natural warming and drying of the climate that led to the expansion of desert species into and within the area; the second, starting in the 1500s but much accelerated from 1900 onward, was human land-use change that furthered expansion of desert species and the introduction of invasive species.

Ten thousand years ago, when the last glacial period was ending, the CD was moister and cooler, with much more widespread piñon-juniper woodland and other, nondesert vegetation (Metcalfe 2006). In the Big Bend, piñon, juniper, and oak were widespread at elevations that now support CD succulent desert scrub. This is based on studies of Late Pleistocene pack rat midden analysis at low CD elevations at Black Gap Wildlife Management Area, Brewster County (Wells 1966, 1977; Van Devender and Spaulding 1979). More recent rat midden analysis from the Livingston Hills of Presidio County (Van Devender, Freeman, and Worthington 1978) drew a similar conclusion. At those moderate elevations, an open, relatively dry piñon-juniper woodland prevailed, compared to today's mid-elevation grass community dominated by grama grasses and sotol.

Since BBR lies between these study sites and within the same elevation range, we can infer a picture of BBR vegetation 10,000 years ago. Because igneous substrates can support more mesic vegetation compared to limestone substrates, we assume that 10,000 years ago igneous substates from 700 meters and above sustained piñon-juniper-oak woodlands and succulents like sotol, agave, and yucca. The Sierra Rica in Chihuahua, a few kilometers south of BBR at slightly higher elevations, now supports such a community on limestone (Riskind 2021). Higher igneous elevations there maintain a denser, taller piñon-oak woodland. Today, the highest BBR elevation in the Cienega Mountains contains a sparse juniper woodland, the remnant of a more widespread, diverse woodland, likely with piñon.

Higher limestone elevations in the Solitario probably sustained oak woodlands at sheltered sites and mid-elevation sotol and grama grasslands with yucca and agave. Low elevations with coarse (gravel, rock, etc.) substrates were likely rich desert grasslands. Plains mid-elevation grasslands probably prevailed on finer upland soils, much like today at well-conserved sites. Perennial waterways would have maintained cottonwood-willow gallery woodlands, with marshland pockets and without invasive exotics such as carrizo and salt cedar. Marshes (ciénagas) would have persisted on uplands along spring-fed watercourses. Periodically flooded, deep alluvial soils likely harbored rich mesquite bosques.

Desertification began to intensify 8,000 years ago (Van Devender, Freeman, and Worthington 1978). As the climate became drier and warmer, woodland stature, density, and diversity decreased, while hardier, drought-adapted species colonized more exposed sites. Until the Spanish Entradas, BBR probably resembled a mosaic of grasslands, with woodlands restricted to the highest elevations, protected canyons, or sites with favorable soil conditions. For example, woodland elements (juniper-oak) persist in the Tascate Hills because the granite-like substrate holds moisture longer.

The arrival of European settlers changed the situation. The Rio Grande/Rio Bravo del Norte riparian corridor from Colorado Canyon upstream, subject to cultivation since the sixteenth century, has been almost completely converted to crops. Extensive cottonwood woodlands, mesquite bosques, wetlands, stands of the native common reed, and bottomland tall grasses have been largely replaced by crops and ruderal and invasive species — salt cedar and carrizo (*Arundo donax*) among the most pernicious. But until the 1900s, it is likely that this was the only park area intensively affected by settlers.

Before the appearance of Anglo ranchers on the Bofecillos Plateau, little large-scale modification of natural vegetation occurred away from the river. Today's iconic CD, dominated by spiny desert shrubs, cacti, ocotillo, and sparse, barren ground, is mostly an artifact of land use after the arrival of Europeans.

Milton Faver, the first Big Bend cattle baron, ran cattle, sheep, and goats in the Cienega and Cibolo Creek drainages, but small rancherias and temporal agriculture had limited habitat effects. In the early 1900s, mining operations in the Fresno-Contrabando Lowlands began to affect natural communities, gradually transforming their character from mostly desert grassland to shrubland. Riparian gallery woodlands and bosques were used for domestic and industrial construction and fuel. Grasslands were consumed by livestock and harvested as fodder for animals used in mining and by the US Cavalry to secure the border.

With the formation of the Howard, Chilicothe, and Saucita ranches on the Bofecillos Plateau (1905), on rich desert plains grasslands of the Llano, grassland habitat declined quickly. Cessation of fires, sheet erosion, and soil loss sped grassland conversion to desert scrub. Gone were grasslands sustaining Hereford stock, moisture-conserving ciénegas, and antelope herds. The shift from a beef cattle to sheep economy further deteriorated habitats. Primitive roads, extended to far-flung reaches, opened heretofore ungrazed habitats.

During World War I, with increased demand for fiber and price supports for wool, intensive sheep ranching operations dominated. Installation of localized livestock water systems with windmills, storage tanks (*pilas*), and water troughs (*bebederos*), enabled greater exploitation of desert plains and desert grasslands. Extension of roads into the backcountry brought shearing pens, sheds, and small camps.

This trend continued into World War II, when another major economic shift occurred. Water systems were extended through pipelines, pumping, pilas, bebederos, and gravity distribution throughout the landscape. Roads and livestock handling facilities were built, along with widely distributed line camps and large wire-fenced pastures. Header dams and dirt *tanques* impounded water and concentrated livestock. Protracted droughts in the 1930s and 1950s took their toll on habitats. Predator control was relentless. Sotol was harvested and converted to cattle feed, using silage choppers with adverse impact on mid-elevation desert grasslands.

In the 1950s through the 1970s, the gradual return to a beef cattle economy compounded a long-term decline in grassland cover and diversity. Riparian and spring habitats, which were not fenced out of pastures, were especially affected by livestock. When the ranching economy shifted to the Texas Parks and Wildlife Department, in 1988, trespass stock became a new problem. The impact of these interlopers in open range country was concentrated in riparian corridors, the principal access to lower desert grazing lands.

One can generalize that the lower the elevation and the deeper the soil, the greater has been the loss or degradation of pre-European habitat. Most intact habitats are at higher elevations with thinner soils. Regardless of elevation, habitats near water usually have been adversely affected. Thus, biodiversity is greatest at higher elevations farthest from water sources. Invasive species are greatest where extensive disturbance and habitat modification has occurred. At BBR, riparian habitats continue to degrade from livestock usage, while mid-elevation desert, desert plains, and succulent grasslands recover, with limestone habitats having the most natural diversity.

In the future, some changes in and deterioration of native vegetation may be partially reversed by conservation-oriented park management. However, some changes seem irreversible, such as the introduction of invasive species. Importantly, global climate change will affect BBR flora and vegetation in ways yet to play out.

Park Plant Communities

Plant composition varies with geology, rainfall and groundwater resources, temperature, and other factors such as soil type, topography, elevation, orientation to the sun, and past land use. Many plant residents have adapted to low rainfall rates, high evaporation rates, and the high temperatures of the northern CD, while other plants have made use of ample water resources from the river, creeks, and springs. Ocotillo, lechuguilla, sotol, and yucca, plants adapted to preserve moisture on barren hills and desert flats, contrast sharply with cottonwoods, willows, and ash, which shade golden columbines, cardinal flowers, even orchids at fern-lined springs. Scattered across much of the park, the springs provide habitat for a broader range of plants than the CD typically supports.

Most park plants fit broadly into the concept of northern CD flora, and the vegetation has been grouped into four habitat categories: desert scrub, desert grassland, juniper roughland, and riparian habitat (Yancey 1997). These plant communities are intergrading, with many species occurring in multiple habitat types and some present almost anywhere. Most notably, the mix of plants within each community varies, with different groups of plants in ascendance.

Desert Scrub

Desert scrub is a shrub-dominant plant community, often accompanied by semisucculent species and a varied mix of herbs and grasses. On hot, dry, desert basins and alluvial flats, creosote can be abundant, at times accompanied by conspicuously black-stemmed tarbush. On rocky bajadas and gentle slopes at slightly higher elevations, plant diversity increases, and frequently, no single species is dominant. Semisucculent, moisture-storing plants, such as ocotillo, leatherstem, and candelilla, can be prominent. Shrubs protected by spiny stems or leaves, including acacia, mimosa, allthorn, agarito, and mesquite, are prevalent.

Typical cacti are the ground-hugging dog cholla, the low desert Christmas cholla, and the taller tree cholla, which store water in their joints; prickly pear, most conspicuously the bulky Engelmann's prickly pear and blind prickly pear, which store water in their pads; and stem cacti, especially strawberry cactus, Texas claret cup cactus, and Texas rainbow cactus, which store water in their stems. Short grasses, especially grama grasses, are part of the mix.

When desert scrub reaches into gravelly arroyo bottoms the plant composition shifts. In this environment, desert olive, netleaf hackberry, Havard's plum, and Apache plume may play a larger role. On exposed limestone slopes, lechuguilla may assume a dominant role, often with Texas false agave and other associated species, including yucca and sotol. Desert flats and low hills with gypseous soils, or gypsum mixed with limestone, attract specialized species. Chihuahuan ringstem can be conspicuous, with gypsum indicator species hairy crinklemat and Texas crinklemat. Threeflower goldenweed and gypsum tansy aster also prefer these substrates.

Over time, mixed desert scrub has become by far the largest community, having gradually encroached on the grassland due to a sequence of major droughts; introduction of goats, sheep, and cattle; some depletion of surface and groundwater resources; and rapid global warming.

Desert Grassland

Consequently, extensive desert grassland communities are limited, often occurring as a transitional plant association on low hills and slopes above desert scrub environs and below juniper roughland at the highest elevations.

The grassland often includes a mix of grama grasses, especially sideoats, chino, black, hairy, and blue grama, and other grasses, such as tobosa; southwestern needlegrass; tanglehead; fluffgrass; threeawn; muhly; bluestem; dropseed; bristlegrass; Arizona cottontop; slim tridens; and green sprangletop.

Numerous larger plants are associated with this community and may appear to dominate the landscape. They often include smooth sotol, yucca (Thompson's yucca, soaptree yucca, and Spanish dagger), beargrass, and prickly pear (especially Engelmann's prickly pear cactus). Other species, such as creosote, western honey mesquite, tarbush, ephedra, tree cholla, ocotillo, catclaw mimosa, and fourwing saltbush, have encroached into the desert grassland in the last 125 years.

Juniper Roughland

Juniper roughland is largely limited to sparse, patchy concentrations on rocky upland terrain, often at elevations above 4,500 feet, such as limestone slopes on the Solitario's north and inner northwest rim, the Blue Range on the Solitario's east rim, the igneous Tascate Hills, and higher, rocky slopes of the Cienega Mountains in the northwest panhandle. The juniper is almost exclusively red berry juniper. Accompanying shrubs may include gray oak, evergreen sumac, western honey mesquite, Roemer's acacia, desert olive, and toothed serviceberry. At some locations, such as Tascate Hills, the roughland might best be described as oak-juniper roughland, with heavy concentrations of gray oak. Semisucculents such as ocotillo, sotol, and beargrass also appear in this circumscribed environment.

Riparian Habitat

Although small in land area, the park's two riparian zones, the Rio Grande, and the network of creeks, springs, and seeps, provide much of the biological diversity. Some species, remnants of a wetter climate, survive solely because of water proximity and could not survive in desert scrub habitat. Water is a prime determinant of plant mix, but other factors are instrumental. Significant differences exist between the riverine plant community on the Rio Grande and the mix of plants associated with creeks and springs at higher elevations, with lower temperatures and more rainfall.

On the river, invasive plants compete with native species. Invasive species such as giant reed, salt cedar, and big saltbush compete with common reed, retama, cheeseweed, tree tobacco, and athel to form a distinct community. Low herbs, including seaside petunia, desert teucrium, and fiddle dock, and some unusual herbs, such as lanceleaf groundcherry and Rio Grande palafoxia, colonize the riverbank.

Creeks and springs support a broader mix of native trees, which provide a sheltering canopy for even more varied herbs and ferns. Typically, a mix of cottonwood, velvet ash, little walnut, western honey mesquite and screwbean mesquite, willow, seepwillow, desert willow, and buttonbush dominates the landscape. Moisture-seeking herbs conspicuous in the understory include golden columbine, cardinal flower, catchfly prairie gentian, and Arizona centaury. Canyon grape is a nearly ubiquitous vine.

The banks of creeks and springs, and rocky outcrops above, provide habitat for ferns. The most common ferns are wavyleaf cloak fern, Standley's cloak fern, the "resurrection fern" flower of stone, and maidenhair fern, which often drapes canyon walls above shaded springs (see Yancey 1997; Powell and Worthington 2018; McKann 1975; Sanchez 1999).

State Park Rules

In exploring BBR, park visitors should remember the golden rule of collecting: Take only memories and photographs. Federal and state laws prohibit collecting plants, animals, and artifacts. Preserve the past for the future by leaving artifacts in place and reporting locations to park staff.

Visitors should observe these specific state park rules on natural and cultural resources:

1. Plant life. It is an offense for any person to willfully mutilate, injure, destroy, pick, cut, remove, or introduce any plant life except by permit issued by the director.

2. Geological features. It is an offense for any person to take, remove, destroy, deface, tamper with, or disturb any rock, earth, soil, gem, mineral, fossil, or other geological deposit except by permit issued by the director.

3. Cultural resources. It is an offense for any person to take, remove, destroy, deface, tamper with, disturb, or otherwise adversely impact any prehistoric or historic resource, including but not limited to buildings, structures, cultural features, rock art, or artifacts, except by written order of the director.

Happy plant hunting!

A TOUR OF BIG BEND RANCH STATE PARK
BY PHYSIOGRAPHIC REGION

In early planning efforts, the Texas Parks and Wildlife Department (TPWD) divided the now 314,000-acre Big Bend Ranch State Park (BBR) into six physiographic regions, and these regions were later used by Christopher Henry in a major study of park geology (Henry 1998). These regions are used in this guide to discuss park features and park locations and as an organizing tool to find and learn about the plants that bring life to this always surprising landscape. Herein, the Bofecillos Mountains have been divided into two distinct regions, Bofecillos Mountains: Highlands, and Bofecillos Mountains: Llano.

Note: In this tour of physiographic regions, the bold numeral following a plant name, for example, soaptree yucca **30**, corresponds to the plant's number in the subsequent Flora section. Plants with a common name followed by the scientific name in parentheses, for example, Thompson's yucca (*Yucca thompsoniana*), are not discussed in the Flora section.

An Important Note on Touring and Hiking the Park

The parkwide and physiographic region maps in this book offer only a "lay-of-the-land" overview of this immense park. One or more additional resources are essential in navigating BBR. These are listed in the following paragraph, and can be obtained through the park's website (https://tpwd.texas.gov/stateparks/big-bend-ranch) and at the park (Barton Warnock Visitor Center, east entrance; Fort Leaton State Historic Site, west entrance; Sauceda Ranger Station, interior).

The Discovery Topographic Map displays the roads, trails, park features, and park locations discussed herein. An even more detailed two-sided, waterproof topographic map, the Exploration Map, is not available online but can be purchased at park headquarters. Visitors must travel through private land, and private inholdings within the park, to reach the interior, and these maps clearly delineate those boundaries. A critical resource for negotiating BBR's difficult jeep routes, Roads to Nowhere, is a guide to the unmaintained 4 X 4 high-clearance roads, with equipment recommendations and safety admonitions. Using the Campsite Guide, campers can choose their campsite from descriptions and photographs of fifty-two campsite locations accessible by vehicle. An online Interactive Map displays BBR campsite, lodging, and permit locations. Numerous other topographic maps, trail maps, and interpretive guides are available.

The "Note on Sources" section of this book draws attention to resources the author used in drafting this work, including park guides, geological and archaeological studies, natural area surveys, and other area studies. Many of these works should prove useful to prospective visitors and can enrich their BBR experience. An example: for those coming to view the park's incredible geological diversity, Chris Henry's *Geology of Big Bend Ranch State Park, Texas* (1998) contains a detailed geology map of BBR and describes the geology of each physiographic region.

For those needing a refresher, the "Glossary of Botanical and Geological Terms" should prove a welcome reference. For those curious about the meaning of Spanish place names or other Spanish terms, the "Glossary of Spanish Terms" has some answers.

The secret to a good tour or hike is preparation. Always consult park rangers in advance about road and trail conditions. Especially during the monsoon season, these conditions can change suddenly. Good tires, good spares, ample fuel (no gas sold at BBR), plenty of water and food, and protection from the desert sun, are minimum requirements. In warm weather months, temperatures typically exceed 100 degrees by late morning, and hikers should stay off trails in the afternoon.

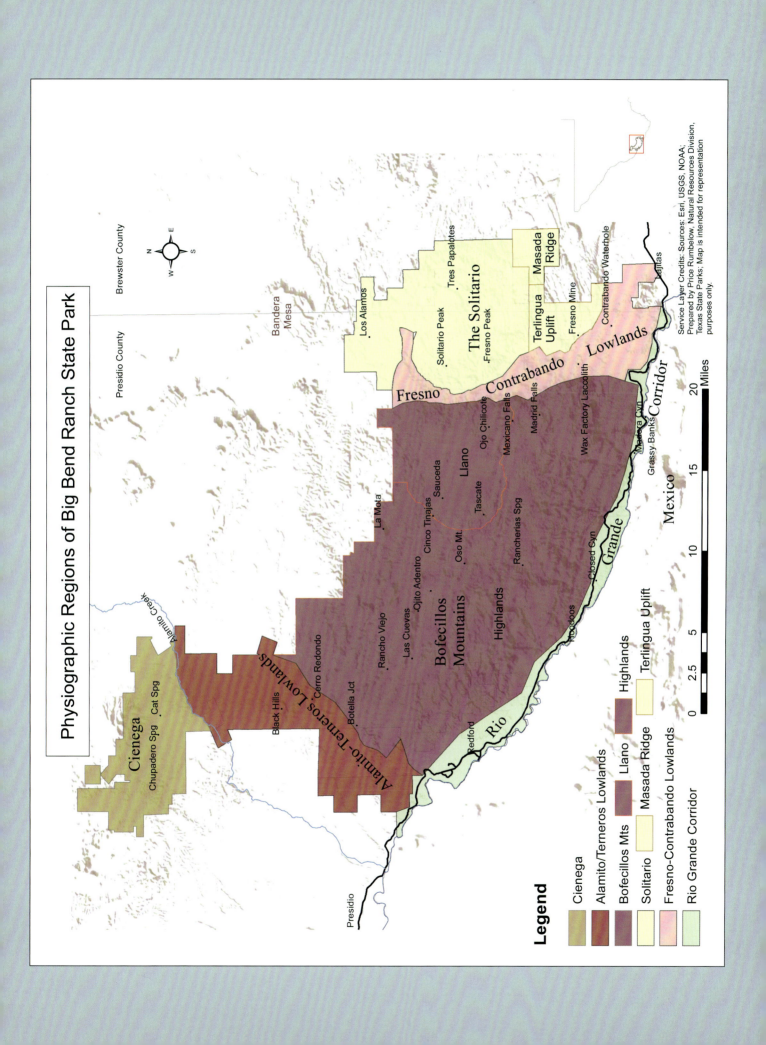

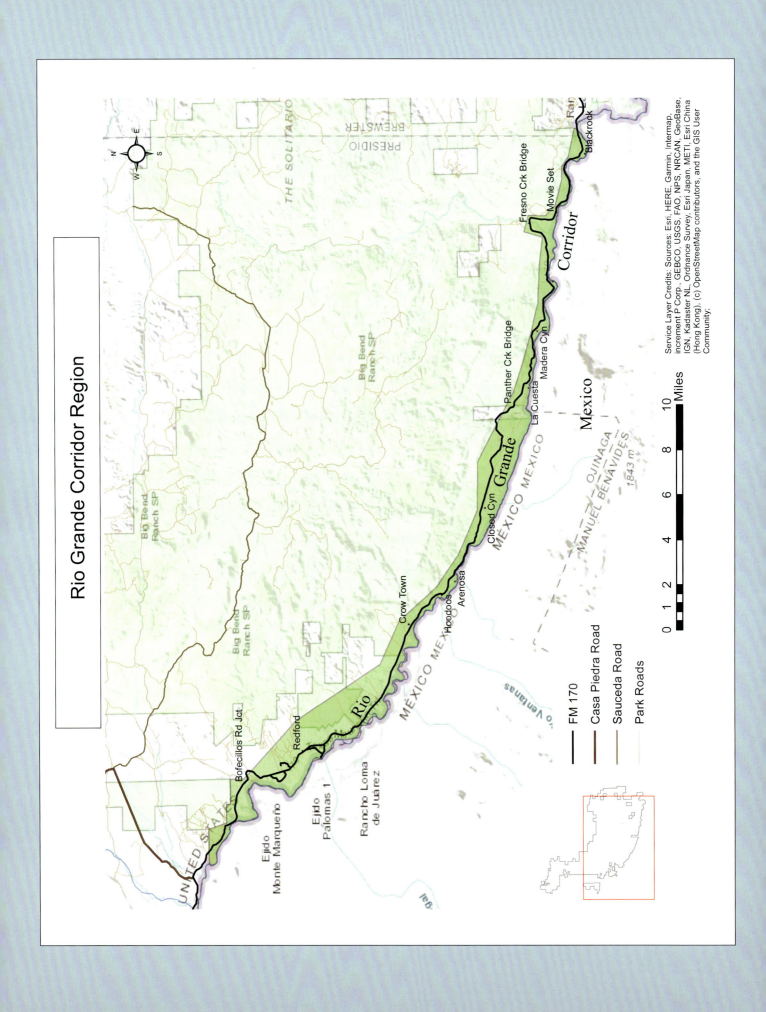

PHYSIOGRAPHIC REGION 1
Rio Grande Corridor

Region 1 centers on the 50-mile river corridor between Lajitas, in southwest Brewster County, and Presidio, in southeast Presidio County. Roughly 25 of these river miles, between Redford and Lajitas, constitute BBR's southern boundary, bordered mostly by mountains on both sides of the US-Mexico border. Corridor elevations range from 2,300 feet at Contrabando Creek to 2,800 feet at Big Hill.

The Rio Grande begins as a snow-fed mountain stream, more than 12,000 feet above sea level, in the Rocky Mountains of southwest Colorado. Traveling 1,900 miles to the Gulf of Mexico, the river runs south into New Mexico, and through Albuquerque and Las Cruces to El Paso. The last 1,200 miles, from El Paso to the Gulf, form the boundary between Texas and Mexico. In the Big Bend, most of the river flow comes from the Rio Conchos in Mexico, which joins the Rio Grande at Ojinaga (across the Rio Grande from Presidio).

Paved and opened to the public in the early 1970s, Highway 170 (FM 170), known as the River Road or El Camino del Rio, parallels the Rio Grande between Lajitas and Presidio and ranks as one of the most spectacular drives in the United States. Steep narrow gorges, Colorado Canyon, and Dark Canyon at the base of Big Hill, have been forged by the river. Eight canyons (Fresno, Monilla, Panther, Rancherías, Tapado, Las Burras, Auras, and Bofecillos) drain the Bofecillos Mountains and feed the Rio Grande. Closed Canyon, a short, narrow slot, knifes through Colorado Mesa to the river. High mountains—the Bofecillos Mountains in Texas (5,100 ft.), and the Sierra Rica (8,000 ft.) and Sierra El Mulato (5,500 ft.) in Mexico—bound the river corridor.

Mesa and basin topography add variety. Lajitas Mesa, partly the product of volcanic activity in BBNP, overlooks Lajitas; Santana Mesa looms over Big Hill; Colorado Mesa forms the north rim of Colorado Canyon; Guale Mesa, flanked by Rancherías and Tapado Canyons, is the quintessential overlook into the river corridor. Near the west park boundary, the corridor widens, and mountain vistas become more distant. The road passes through Redford Basin for 15 miles and then the broad Presidio Basin.

Barton Warnock Visitor Center

One mile east of Lajitas is Barton Warnock Visitor Center, named for Barton Warnock (1911–1998), Sul Ross State University teacher and botanist. Warnock collected 1,600 plant specimens at BBR, and an herbarium of his specimens is housed at the visitor center. The visitor center serves as a park operations and permit office, museum, desert garden, interpretive center, bookstore, and gift shop. Exhibits illustrate Big Bend's human and natural history. A mix of CD plants, including Thompson's yucca (*Yucca thompsoniana*), soaptree yucca **30**, and Spanish dagger **31**, is on display in front of the center and in the garden.

North Lajitas Mesa

Approaching Lajitas from Terlingua, the dominant landscape feature is North Lajitas Mesa (3,672 ft.), across the road from Barton Warnock Visitor Center. Only the mesa's northwest end is part of BBR.

Lajitas, Mesa de Anguila, and view into Mexico from South Lajitas Mesa

Lajitas

Lajitas (little flat rocks) is named for the Boquillas Formation flagstone prominent nearby.

According to Robert T. Hill, who surveyed the Rio Grande in 1899, Spanish explorer Cabeza de Vaca crossed the Rio Grande at or near Lajitas in 1535. A branch of the Comanche Trail crossed at Lajitas.

Lajitas Trading Post opened in 1899, and Lajitas became a substation port of entry in 1900. To counter border incursions by Pancho Villa and his followers, General John "Black Jack" Pershing set up a cavalry post in Lajitas in 1916. The Marfa-Terlingua road and stage route was built, mostly to transport goods to and from Terlingua quicksilver mines.

After World War II, the mines closed, trade through Lajitas declined, and its population dwindled. In the 1970s, Houston developer Walter Mischer rebuilt Lajitas, and Lajitas is now an Old West–style resort community (see Saunders 1976; Alloway 1995; MacLeod 2008).

Lajitas Boat Launch

Just west of Lajitas is the put-in for river trips through Santa Elena Canyon in BBNP and take-out for boaters coming downriver from Madera Canyon, Grassy Banks, and other points in BBR. A goosefoot shrub newly introduced in Texas, big saltbush **97**, 9 feet tall and wide, grows near the boat launch.

Paso Lajitas

Across the river, the Mexican village of Paso Lajitas shares Lajitas's rich history. Before 9/11, a rowboat service ferried people between the communities. I have fond memories of eating Mexican food in Paso Lajitas and being ferried back across the river at dusk in a wobbly rowboat (known as a chalupa, and nearly as fragile), with bats whizzing by overhead. After 9/11, the informal border crossing was closed.

South Lajitas Mesa

Across from the boat launch, South Lajitas Mesa stretches north and west along the Rio Grande to Contrabando Creek, where the mesa forms the east rim of Contrabando Canyon. Much of the mesa lies in BBR, but trail access is through Lajitas Golf Resort and permission is required. The mesa summit offers views west to Contrabando Lowlands, south to Mexican mountains, and east to Mesa de Anguila.

I have hiked the mesa countless times. Notable plant encounters include fleshy tidestromia **8**, spearleaf **21**, tubercled saltbush **95**, Rio Grande phacelia **156**, Coahuila blazingstar **171**, Big Bend ringstem **188**, longcapsule suncup **191**, Boquillas lizardtail **192**, and my favorite cactus, glory of Texas cactus **91**.

River Road East of Stables

Between Lajitas Boat Launch and Lajitas Stables are two secluded spots. The first, 1 mile west of Lajitas, is an open, sandy-gravelly beach. At road marker "Presidio 49," the beach is accessed through a gap in a bluff between FM 170 and the river. The second site, 1.25 miles west of Lajitas, is an old campground, Blackrock, with a scenic overlook of the Rio Grande and a trail to the river's rocky and sandy banks. A faint dirt road runs off FM 170, and a walk up the hill ends at the overlook.

At the first site, trees such as retama **138**, tree tobacco **261**, and salt cedar **269** clog the riverbank. Two uncommon trees/shrubs are in active competition: thicket-forming big saltbush **97**, ensconced near the water; and the handsome tamarisk, athel **268**, cultivated at Presidio and Redford, but self-propagated on this beach. Sand-loving annuals make fleeting appearances, often in response to flooding. They include scaly-crusty, wheelscale saltbush **96**; nettleleaf goosefoot **98**; lanceleaf groundcherry **262**; earlobe mustard (*Dryopetalon auriculatum*), with lobed leaves and four-petal white blooms clustered at the stem tips; little mallow (*Malvella parvifolia*), with large scalloped, deeply lobed leaves obscuring tiny lavender flowers; and prostrate knotweed (*Polygonum aviculare*), with fleshy green stems with reddish joints and greenish flowers with broad white margins.

At Blackrock Campground, Turner's mimosa **137**, a straggly shrub with pink flower heads, shares the bluff with clumped dog cholla **67**, Chisos pricklypoppy **198**, and slimstalk spiderling (*Boerhavia gracillima*). In sand are herbs seldom seen at BBR: Rio Grande palafox **43**, with cylindric pink flower heads; the vine balloonbush **204**, with large bluish blooms; paleflower ipomopsis **217**, with trumpetlike flowers; river dalea (*Dalea glaberrima*), with butterfly-shaped, magenta blooms; and tallow weed (*Plantago hookeriana*), a woolly plantain with slender ascending leaves and cylindric spikes of crowded, four-lobed, membranous blooms.

Lajitas Stables (3.0 miles from Lajitas)

Lajitas Stables furnishes guided tours, horses, and trail rides on the park's backcountry routes suitable for horses. A trail behind the stables, ascending South Lajitas Mesa, offers 360° views.

River Road west of Stables before Movie Set

West of the stables, the road runs southwest and northwest, with a hill in the distance. By the river is a rocky, sandy, open area—an ideal spot to learn about plants that frequent the river corridor. Retama **138**, giant reed **209**, salt cedar **269**, and common reed (*Phragmites australis*) line the river. Across the road yellow rocknettle **168** hangs from cliffsides. Plants spring up on the floodplain, some scarce, such as Mexican gumweed **39**. Water pools after rain offer habitat for giant reed **209** and southern cattail **270**. In 2011, I found locally scarce hooded arrowhead **5** floating in the water. Troublesome invasive weeds like Russian thistle (*Salsola tragus*), a tumbleweed with prickly stems, spine-tipped leaves, and minute greenish flowers, are also a part of this environment.

Contrabando Movie Set (5.0 miles from Lajitas)

For thirty years, an abandoned movie set near the mouth of Contrabando Creek attracted visitors. In 1985, the set was built as a Mexican-style village for the western *Uphill All the Way* and was used in the TV miniseries *Streets of Laredo* (1995). Damaged by flooding in 2008, only one structure remains.

Contrabando Creek

After flowing off Terlingua Uplift and below Lajitas Mesa, Contrabando Creek enters the Rio Grande west of the movie set. With flooding, water backs up to the bridge. As waters recede, cracks in the creek bed make artful designs, and herbs spring up on the banks. On a fruitful day in 2009, I encountered Coahuila twintip **207**, a Mexican annual, on a moist bank below the bridge. In the United States, Coahuila twintip is known only from a few recent discoveries in Brewster, Presidio, and Jeff Davis Counties. Weeks earlier, yerba de tago (*Eclipta prostrata*), an annual, mostly prostrate aster with small white ray and disk florets, was blooming here. This worldwide weed, rare in the CD but occasional along the Rio Grande, is widely used in Ayurvedic medicine.

West Contrabando–Fresno Creek Trailhead

This trailhead provides access to the 25-mile Contrabando Multi-Use Trail, Fresno Canyon hiking, and the Fresno Divide Trail. The Contrabando Lowlands is discussed in "Physiographic Region 6."

Fresno Ranch

Fresno Ranch is soon visible to the southwest, above the Rio Grande and below Mexican cliffs known as the Great Wall of Chihuahua. Formerly a private inholding, now a ranger residence, the 7,000-acre ranch was purchased by BBR in 2008 from the estate of local landscape artist Jeanne Norsworthy.

Fresno Creek (6.5 miles from Lajitas)

Fresno Creek Bridge, west of West Contrabando Trailhead, furnishes a good view of Fresno Creek south to the Rio Grande and north to the Solitario. Fresno Creek is discussed in "Physiographic Region 6."

Grassy Banks Campground (9.6 miles from Lajitas)

Grassy Banks is an open beach-like area along the river, with grassy, sandy banks, sandy, gravelly flats, and boat access. Outfitters use the site for trips downstream to Lajitas or take-outs from upstream access at Madera or Colorado Canyons.

Periodic flooding supports the plants that prosper here. Intriguing sand-loving species include big saltbush **97**, Mexican gumweed **39**, fiddle dock **222**, and many annuals. Annual bastard cabbage (*Rapistrum rugosum*), with wrinkled, lyre-shaped leaves and yellow blooms, lines the riverbank. Plains dozedaisy (*Aphanostephus ramosissimus* var. *humilis*), with lobed oblong leaves, white ray florets, and yellow disk florets; red-tip rabbit-tobacco (*Pseudognaphalium luteoalbum*), a cudweed with tiny flower heads with red-tipped florets and translucent bracts in globular clusters; and silversheath knotweed (*Polygonum argyrocoleon*), apparently scarce in Texas, with slender lance-shaped leaves and crowded spikes of white-margined greenish flowers, occupy sandy flats.

In 2009, a watermelon (*Citrullus lanatus*), with leafy, prostrate, longhaired stems, deeply lobed leaves, dainty yellow flowers, and watermelon fruits, sprang up beside a picnic table.

El Padre al Altar

One-half mile east of Madera Canyon Campground, water, wind, and time have eroded the rocks into whimsical natural designs, or "hoodoos." A favorite of photographers, this jumble of volcanic ash is known as El Padre al Altar because one spire resembles a priest kneeling at an altar (Alloway 1995).

Madera Canyon Campground (11.7 miles from Lajitas)

Madera Canyon Campground is a major put-in, take-out for river rafting and canoeing. Boaters begin at

Colorado Canyon River Access upstream and take out at Madera or put in at Madera and travel a few miles to Grassy Banks, or on to Lajitas. The lower campground provides river access, while the upper campground, on the mesa above, is ideal for sunrise photos, with views west to Big Hill.

Seepwillow **35**, retama **138**, tree tobacco **261**, and salt cedar **269** grow at the river's edge. Thick sand in the campground supports an intriguing plant mix, including Abrams' spurge **118**; longcapsule suncup **191**; doubleclaw **182**; and winged spurge (*Euphorbia serpillifolia*), an ascending annual with winged stems, oblong leaves toothed at tips, flower clusters with white apron-like glandular appendages, and ribbed fruit capsules. A common grass is hairy grama (*Bouteloua hirsuta*), a tufted perennial with basal leaves and comb-like one-sided inflorescences, with the main axis projecting well beyond the spikelets.

Monilla Creek—La Monilla

Monilla is a Spanish name for Mexican buckeye (*Ungnadia speciosa*), which is common in the canyon. Madera (timber) is a misnomer for Monilla Creek, which flows into the Rio Grande west of Madera Canyon Campground. From the bridge, visitors walk up Monilla Creek to La Monilla, the remains of a historic ranchito beside a spring. Behind La Monilla, a steep trail leads to a mesa top above the canyon. Monilla Creek rises at the headwaters of Arroyo Mexicano. Hikes in upper Monilla Canyon begin at Javelin Trailhead on Javelin Road.

Along lower Monilla Creek, yellow-flowered Arizona goldenaster (*Heterotheca arizonica*) grows out of the canyon walls, and Wright's Dutchman's pipe **25** at the base of the walls. Torrey's tievine **103** often drapes over trees and shrubs. Shrubby senna **142** seems out of place here, more characteristic of higher elevations near Sauceda. Pyramid flower **177**, with pinkish blooms and toplike fruits, is plentiful in the drainage. Canyon grape **276** and yellow Indian mallow (*Abutilon malacum*) reside at the spring. Nearby is cluster flaveria (*Flaveria trinervia*), an annual aster with serrate elliptic leaves and tiny yellow blooms in compact clusters of many flower heads, each head with only one ray or disk floret.

Teepees Roadside Park (12.4 miles from Lajitas)

West of Madera Canyon Campground, below Big Hill, is Teepees Roadside Park, with teepee-style shelters and views of Santana Mesa and Mexican mountains. In early spring, Big Bend bluebonnet (*Lupinus havardii*) is prolific, and visitors sometimes find unusual white or pink versions of Big Bend's best-known flower. In later spring, Spanish dagger **31** produces extravagant flower displays.

Big Hill, Santana Mesa, Dark Canyon (13.5 miles from Lajitas)

West of the Teepees, 13.5 miles west of Lajitas, Big Hill (2,836 ft.) is the dominant feature on the River Road, with 15 percent grades. Views from the top may be the best in the Big Bend. On the east skyline, the Chisos Mountains are often visible 75 miles away. West is Colorado Canyon, with the Sierra Rica of Mexico on the horizon. Below, the Rio Grande cuts through Dark Canyon (El Peñasco), a dark-walled gorge.

At Big Hill, Fresno Formation lavas are overlain by tuffs that erupted from the San Carlos and Santana calderas in Mexico. San Carlos Tuff is displayed in the chocolate-colored boulder field at Big Hill and in Dark Canyon below. Santana Mesa (3,925 ft.), named for a Comanche chief, looms above Big Hill. Santana Mesa is the best place to find bighorn sheep, recently reintroduced in the Bofecillos Mountains.

In a good spring, Big Hill is a glorious cactus garden, with Engelmann's prickly pear **89**, blind prickly pear (*Opuntia rufida*), and Big Hill prickly pear **87** most conspicuous. Big Hill prickly pear is a scarce variety, often with yellow, sometimes twisting spines. Scarce Mexican stickleaf **170** also grows atop Big Hill.

Dark Canyon below Big Hill

View west to Colorado Canyon from Big Hill

La Cuesta

La Cuesta, west of Big Hill, consists of a campsite, boat ramp, and sheltered area with a telescope for viewing bighorn sheep on Santana Mesa. Big Hill prickly pear **87** grows here and on Big Hill. Common threesquare (*Schoenoplectus pungens*), a sedge with scaly spikelets and bristlelike blooms, grows by the boat ramp. Retama **138** and common reed (*Phragmites australis*), a grass with flowers in feathery terminal clusters, line the Rio Grande. Cheeseweed **34**, and desert tobacco (*Nicotiana obtusifolia*), with clasping oblong leaves and greenish-white, tubular five-lobed flowers, grow on sandy flats.

Panther Canyon (16.2 miles from Lajitas) and the Mouth of Colorado Canyon

West of Santana Mesa, a bridge crosses Panther Creek. Farther west is a view north up Panther Creek to Panther Canyon, with canyon walls rising 700 feet above the creek (MacLeod 2008). This lower portion of Panther Canyon is a private inholding. The canyon can be hiked from the East Rancherías Trailhead.

From Panther Creek bridge, the creek flows south, entering the Rio Grande just below the mouth of Colorado Canyon. The canyon mouth, and canyon banks, attract an ever-changing mix of herbs. Herbs of interest are California caltrop (*Kallstroemia californica*), a prostrate annual uncommon in the Trans-Pecos, with yellow flowers with early-falling sepals, and tubercled, short-beaked fruits; mimicking matspurge (*Euphorbia simulans*), a vulnerable prostrate annual or perennial with zigzagged stems, lopsided oblong leaves, and flower clusters with unappendaged glands; and Mexican devilweed (*Chloracantha spinosa*), a shrubby perennial aster with erect, nearly leafless, often thorny stems, and flower heads with white rays and yellow disk florets.

Rancherías East/Rancherías Loop Trail (18.5 miles from Lajitas)

West of Panther Canyon, in scrub habitat, is East Rancherías Trailhead. Rancherías Loop Trail, a three-day, two-night, 21-mile loop, begins here or at the west trailhead 2 miles west. The loop passes springs and a seasonal waterfall, winds through rugged Bofecillos canyons, ascends high ridges, and follows old wagon trails. Trail features include ranch ruins, a rock-shelter, Rancherías Dome, and views of 1,000-feet-deep Tapado Canyon, and 8,000-feet-high Sierra Rica.

The ruins of Casa Reza, above a spring, Ojo de León, are a major attraction in Panther Canyon. One century ago, the Reza family maintained an orchard, raised goats, and sold goat cheese and fruits in Redford and Lajitas (Smith-Savage 1996b). The centerpiece of Rancherías Canyon is Rancherías Springs, with a long band of greenery, water pools, and towering cottonwoods. Views from Guale Mesa rival what any US park can offer. East is La Guitarra, a peak above Rancherías Falls; south are the Rio Grande and Sierra Rica; west is Tapado Canyon.

Desert stickleaf (*Mentzelia longiloba* var. *chihuahuensis*), a much-branched perennial with whitish stems, narrow, lobed leaves with sticky barbed hairs, and ten-petal yellow flowers with many protruding stamens, is often plentiful in Acebuches Canyon and near Seep Spring, on the East Rancherías Trail.

In Panther Canyon, watch for Berlandier's wolfberry (*Lycium berlandieri*), a small bee-loving shrub mostly unarmed, with linear leaves, white or lavender tubular blooms, and red berries. Rock phacelia (*Phacelia rupestris*), a woody perennial with branching stems, leaves with lobed leaflets, and bell-shaped white blooms in terminal clusters, often grows under shrubs. Dwarf false pennyroyal (*Hedeoma nana*), an aromatic mint with hairy egg-shaped leaves and spikes of two-lipped lavender-purple blooms, nestles at the base of rocks. Louisiana vetch (*Vicia ludoviciana*), a vinelike trailing legume with curling tendrils, leaves in well-spaced linear segments, pealike lavender flowers, and flattened pods, occupies shoulders along the drainage.

Closed Canyon (19.9 miles from Lajitas)

Closed Canyon is a 350-feet-deep vertical slot, so narrow that you can reach out and nearly touch both walls. Named for the close walls and limited skylight, the canyon cuts through Colorado Mesa, the north rim of Colorado Canyon, to the river. After a 0.7-mile walk toward the Rio Grande, hikers encounter an impassable pour-off. Red-orange Santana Tuff, ash flow welded into rock, forms the canyon walls. The creek bed is gravel and slick rock.

Closed Canyon

Sunlight and soil are limited in the canyon, and vegetation is sparse, but the north wall of Colorado Mesa is promising plant habitat. Gypsum tansy aster **50**, in Texas known only in southeast Presidio County, sometimes grows on the rock wall. Mexican navelwort **56**, with sky-blue flowers; and rock milkwort (*Rhinotropis lindheimeri* var. *parvifolia*), with white rose-tinged blooms, with two large lower wings (petal-like sepals) and three small petals, one keel-like, hide below shrubs. Strawberry hedgehog cactus (*Echinocereus enneacanthus* var. *enneacanthus*), resembling strawberry cactus **73**, with fewer, looser stems and less juicy fruits, grows along Closed Canyon Creek.

Rancherías Canyon (21.4 miles from Lajitas)

A trailhead near Colorado Canyon River Access serves as both the West Rancherías Loop Trailhead and Rancherías Canyon Trailhead. Rancherías Canyon is a 9.6-mile round-trip hike along Rancherías Creek to 80-feet-high Rancherías Falls. With some shady cottonwood and willow canopies and flowing springs, the canyon is wetter than most Bofecillos canyons. Reportedly, the "Battle of Rancherías" took place in the area in 1787, when Juan de Ugalde, Governor of Coahuila, attacked an Apache encampment (Alloway 1995).

The canyon offers habitat for water-loving herbs: catchfly prairie gentian **150**, with chalice-like flowers; knotted rush **159**, with cylindric leaves; Mexican skullcap **165**, with helmet-shaped blooms; yellow rocknettle **168**, in the moist creek bottom; Arizona carlowrightia (*Carlowrightia arizonica*), with orchid-like white blossoms; and James's monkeyflower **200**, in shallow water and mud. Southern cattail **270** may survive in pools of water.

Engelmann's prickly pear **89**, with spines arranged like a bird's foot; blind prickly pear (*Opuntia rufida*), spineless but armed with barbed glochids; and treelike, tree cholla (*Cylindropuntia imbricata* var. *imbricata*), with magenta flowers and knob-like yellow fruits, rest on bluffs above the creek. Desert pincushion cactus **77** is wedged between boulders.

Colorado Canyon River Access (21.7 miles from Lajitas)

After passing through the Presidio and Redford bolsons, valley floodplains to the west, the river enters a confined space, Colorado Canyon, a steep-walled passage of dark volcanic rock. Colorado Canyon River Access, at the canyon head, is the primary put-in for float trips down the 6.4-mile gorge, or day-long trips through Dark Canyon below Big Hill, with take-out at Madera River Access. The site offers fishing, birding, and wildlife watching. The top of Colorado Mesa furnishes canyon and mountain views.

Often, a new botanical prize is discovered at this rest stop. A 9-feet-wide big saltbush **97**, at the river's edge, was the second sighting of this shrub in Texas, washed away three years later. This is the only BBR site where I have encountered western sea purslane **3**, a fig marigold with magenta blooms. Alfalfa (*Medicago sativa*), a roadside escape with trifoliate leaves and clusters of violet-blue, butterfly-shaped blooms, made an appearance in 2009.

Other engrossing riverbank flora: water hyssop **203**, narrowleaf moonpod **185**, buffalobur nightshade **265**, lizardtail gaura **193**; and turkey tangle fogfruit (*Phyla nodiflora*), a prostrate perennial with spatulate leaves toothed above the middle and head-like spikes of white flowers with yellow throats. Big-needle pincushion cactus (*Coryphantha macromeris* var. *macromeris*), with dark green, wrinkled stems and magenta blooms; and numerous annuals, such as spreading spiderling (*Boerhavia triquetra* var. *intermedia*), with much-branched stems, oval leaves, and long-stalked clusters of pinkish-white blooms with yellow throats, occupy the mesa above. One isolated Havard's agave **26**, rare at BBR, grows near the canyon head.

Colorado Canyon from canyon rim

Stream and bluff near Colorado Canyon head

Arenosa (sandy) Campground (23.7 miles from Lajitas)

West of Colorado Canyon, a dirt road leads to Arenosa Group Camping Area. Below bluffs of tuff and conglomerate, the campground is a sandy area with river access and Mexican mountain views. Salt cedar **269** and common reed (*Phragmites australis*) form thickets on the river. Strawberry cactus **73** and diploid purple prickly pear cactus **86** line the gravel road. Of special interest is Rio Grande palafox **43**, growing nearby in thick sand with early spring bloomer, desert Indianwheat **206**, a source of psyllium.

Hoodoos (24.3 miles from Lajitas)

West of Arenosa, the road climbs to a bluff. A parking area overlooks the Rio Grande and a magical complex of pillars—the Hoodoos—at the river's edge. A short loop trail leads down to the river. The Hoodoos, including Balancing Rock and Anvil Rock, are eroded blocks of sedimentary rock, basalt, and ash-flow tuff formed during Basin and Range faulting (Henry 1998). Hoodoo, an African term, was applied to curiously shaped rocks believed to represent evil spirits.

Cane-like grasses, the native common reed (*Phragmites australis*), and the introduced giant reed **209** compete for space with retama **138** beside the river. Distinctive grama grasses, sideoats grama **210** and false grama **211**, grow along the loop trail. Tanglehead **214**, and curlytop knotweed (*Persicaria lapathifolia*), with red stems with swollen joints and nodding spikes of greenish-white to pink blooms form clumps on sandy banks. Cosmopolitan bulrush **110**, with scaly spikelets and florets of barbed bristles, thrives at the river's edge. Buffalobur nightshade **265**, with yellow blooms and bur-like fruits, springs up on gravelly flats above the river. Hartweg's twinevine **18** and shaggy false nightshade (*Chamaesaracha villosa*) cover road shoulders on FM 170.

Tapado Canyon (25.0 miles from Lajitas)

West of the Hoodoos, the deepest, widest canyon draining the Bofecillos Mountains empties into the Rio Grande. The canyon is named Tapado (covered, filled) because it is clogged with boulders. The 8-mile canyon rises near Nopalera Trailhead on Las Burras Road and runs south, west of a Fowlkes era stone dam and Oso Spring, then southwest past five springs, between Guale Mesa and Three Dike Hill, to the river. Guale Trailhead, at the end of Guale Mesa Road, is the canyon's ideal vantage point.

A varied flora is on display, nourished by springs and shade from high canyon walls. At the canyon mouth, cheeseweed **34** fills the drainage. By the river is huisache (*Vachellia farnesiana*), previously unreported at BBR, perhaps this tree's only natural site west of Rio Grande Village in BBNP. Two invasive annual weeds seem quite at home beneath the huisache: London rocket (*Sisymbrium irio*), with huge, pinnately divided leaves, clusters of four-petal yellow blooms, and erect cylindric fruits; and tansy mustard (*Descurainia pinnata*), with much-branched erect stems, deeply segmented leaves, small yellow flowers with four spoon-shaped petals, and upcurving cylindric fruits.

In the canyon's upper reaches, beebrush **271**; Roemer's acacia (*Senegalia roemeriana*); desert hackberry (*Celtis pallida*); shrubby senna **142**, with golden-yellow flowers; and Palmer's bluestar **13**, with tubular white flowers, line the creek bed. Twining over shrubs are climbing wartclub (*Commicarpus scandens*), with funnel-shaped white flowers, and propellerbush **174**, with propellerlike yellow petals. Margined rockdaisy (*Perityle vaseyi*), with yellow flower heads, grows on canyon margins. James's monkeyflower **200** thrives in water. Sticky Mexican clammyweed **99**, with garish blooms, lines the banks. Lyreleaf twistflower (*Streptanthus carinatus* ssp. *carinatus*), with maroon blooms with tiny, twisted petals, is an occasional visitor, and big sandbur (*Cenchrus myosuroides*), a stout grass to 6 feet high, is rare.

Rancho Moreno

West of Tapado Canyon, a crumbling rock house, and defunct windmill, Papalote Sabino, are the only remains of Rancho Moreno. The Moreno family lived here in the early 1900s. Apparently, only one family member survived a fatal attack of dysentery (Alloway 1995).

Crow Town

Three Dike Hill at sunset

Crow Town

West of Rancho Moreno is Crow Town, an old movie set used in the 1995 TV miniseries *The Streets of Laredo*. A curious juxtaposition of Hollywood and geological fantasy, Crow Town sits amid whimsical hoodoos below Three Dike Hill.

Three Dike Hill (26.2 miles from Lajitas)

Three Dike Hill (3,430 ft.) is the subject of geology lessons on the nature of dikes. Capped with dark basaltic rock, the hill is formed from ash-flow tuff and sedimentary deposits that turn gold at sunset. Dark ridges extending down the hill are hardened intrusions (dikes) filling fractures in softer rock. During the US-Mexico Boundary Survey, German immigrant Arthur Schott, surveyor-geologist-plant hunter-artist, prepared illustrations for the 1857 survey report. One illustration was a steel plate of Three Dike Hill.

Cerro de las Burras

Cerro de las Burras (4,345 ft.), at the northern limit of Redford Basin, is capped with basaltic lava and ash-flow tuff (27.1 Ma), with conglomerate, sandstone, and tuff from Fresno Volcano (32 Ma), the oldest Bofecillos Mountains volcano, at the base (MacLeod 2008).

Cerro de las Burras Loop

Climbing above the floodplain, Cerro de las Burras is a 4WD crescent-shaped loop, with access points on FM 170. Along the road, changing views unfold—of Three Dike Hill, Cerro de las Burras, Redford Basin, the farming villages of El Polvo and El Mulato, and Sierra El Mulato in Mexico. Desert willow **53** is plentiful near Crow Town, with funnel-shaped red and pink blooms. Longleaf jointfir **114** grows on flats with Texas claret-cup cactus **71** and Comanche prickly pear (*Opuntia camanchica*), which resembles sweet prickly pear **88** but is shorter with darker spines. Hartweg's twinevine **18** twines on leatherstem **130** and orange caltrop **277**. Goosefoot moonpod **186**, with winged fruits, occupies gravelly mounds.

Las Burras Creek

Las Burras Creek crosses FM 170 west of Cerro de las Burras Loop, outside BBR. Las Burras Canyon is visible to the north, but the land between FM 170 and the canyon is privately owned. Park hikes in the canyon begin above, at Las Burras Trailhead, at the end of Las Burras Road.

Auras Creek

Auras Creek crosses FM 170 east of Polvo. One mile east of the Auras Creek crossing, a gravel road runs north through private land into BBR, to ranch ruins and the mouth of Auras Canyon. This high-walled canyon can be hiked from here, through winding and tightening canyon walls. However, the canyon is usually approached from Sauceda Road, 0.2 miles west of Ojito Adentro. Sprawling tree cholla (*Cylindropuntia imbricata* var. *imbricata*) and sweet prickly pear **88** grow near the canyon mouth.

Redford and El Polvo (33.7 miles from Lajitas)

North of Auras Creek, Redford is a small farming village, with recorded history dating to the sixteenth century. A pueblo, Tapacolmes, was the site of a 1683 Spanish mission. In 1871, after Texas governor Richard Coke offered free land to settlers, El Polvo was founded near the former Tapacolmes. In 1885, the US Army set up Camp Polvo to fend off Apache raids, and troops returned during the Mexican Revolution. In 1911, El Polvo's name was changed to Redford, after a nearby river ford, Vado Colorado (red ford). An adobe-stone church southeast of Redford is the original location of El Polvo (Weiser 2019).

Bofecillos Creek Junction (36.4 miles from Lajitas)

Bofecillos Creek crosses FM 170 and flows into the Rio Grande northwest of Redford, south of the Palo Amarillo Creek crossing and Bofecillos Road. Bofecillos Canyon is discussed in "Physiographic Region 3."

Palo Amarillo Creek Junction (37.5 miles from Lajitas)

Palo Amarillo Creek crosses FM 170 and enters the river west of Bofecillos Creek, east of Bofecillos Road. Palo Amarillo Creek is discussed in "Physiographic Region 3."

Bofecillos Road

Northwest of the Palo Amarillo Creek crossing of FM 170 is the Bofecillos Road junction, 13 miles southeast of Presidio. This gravel road runs north through Alamito Creek–Terneros Creek Lowlands 5.5 miles to a junction with Sauceda Road. The west park entrance, Botella Junction, is 2 miles east; Sauceda is 17 miles east of the park entrance. Bofecillos Road traverses sandy and gravelly flats, creek bottoms, and rolling hills and terraces, with impressive views of surrounding mountains.

Spring cactus displays are dazzling. Purple pads of diploid purple prickly pear **86** and yellow blooms of clumped dog cholla **67** and Texas rainbow cactus **72** color the landscape. Herbs pop up roadside: naked brittlestem (*Psathyrotopsis scaposa*), with pincushion-like yellow flower heads and bumpy leaves resembling a turtle's back; desert evening primrose (*Oenothera primiveris*), with lobed, red-spotted leaves and yellow blooms; and Havard's dayia (*Dayia havardii*), with intricate pink blossoms forming little bouquets.

Terneros/Black Hills Creek Junction (42.2 miles from Lajitas)

The united Black Hills–Terneros Creek intersects FM 170 just east of the Casa Piedra Road junction. Black Hills and Terneros Creeks are discussed in "Physiographic Region 2."

Casa Piedra Road Junction (42.4 miles from Lajitas)

The unpaved county road, Casa Piedra, begins at FM 170, 8 miles east of Presidio. The park headquarters, Sauceda, is 27 miles away. Casa Piedra Road is discussed in "Physiographic Region 2."

Alamito Creek Junction (43.7 miles from Lajitas)

Alamito Creek crosses FM 170 and enters the Rio Grande 1.4 miles west of the Casa Piedra Road junction. See the discussion of Alamito Creek in "Physiographic Region 2" and "Physiographic Region 7."

Fort Leaton State Historical Site (46.5 miles from Lajitas)

In 1848, Ben Leaton opened Fort Leaton as a trading post and private fort on the Chihuahua Trail. Leaton had been a scalp hunter in Chihuahua, where bounties were paid for Apache scalps, but at Fort Leaton he traded with Apaches and Comanches. The fort was used by the US Army in the 1850s and by army patrols until 1880. Texas Parks and Wildlife Department acquired the property in 1967 and opened the site in 1978 (Loden 2012).

Fort Leaton State Historical Site, BBR's west visitor center and one of the largest adobe structures in Texas, has been restored to reflect mid-1800s border life. Visitors obtain backpacking, camping, and river use permits, licenses, and general park information. The site offers guided tours, interpretive exhibits on human and natural history, a gift shop, and outdoor picnic areas.

Desert unicorn plant **181** occupies sandy road shoulders and dunes west of Fort Leaton. This devil's claw has yellow funnel-shaped flowers and beaked fruits that split into two hooked claws. Desert unicorn has been recorded only once in BBR, in Alamito Creek–Terneros Creek Lowlands.

Presidio (49.8 miles from Lajitas)

Presidio is the oldest continuously cultivated area in the United States. When Cabeza de Vaca passed through in 1535, he found Pueblo peoples farming the river floodplain. With a population of 4,067 in 2018, Presidio is a destination for vacationers planning trips on the River Road. The town serves as a Port of Entry for tourists crossing the international bridge to Ojinaga, Mexico, and points south.

PHYSIOGRAPHIC REGION 2
Alamito Creek–Terneros Creek Lowlands

Alamito Creek–Terneros Creek Lowlands stretches from the Rio Grande Corridor north to Alamito Creek in the park's northwest panhandle. The region is a low expanse between the Cienega Mountains to the north, Bofecillos Mountains to the east, and Chinati Mountains to the west, outside BBR. The Lowlands is a product of volcanic eruptions that formed the surrounding mountains and the creeks that drained them. Park elevations range from 3,200 feet along Palo Amarillo Creek to 3,700 feet at Cerro Redondo. Private park inholdings, however, extend to 2,500 feet near Highway 170.

The Lowlands is a hot, dry, gently sloping region of desert foothills, gravel pediments, and elevated terraces, with sparse scrub vegetation. Alamito Creek, with groundwater and some perennial surface water, provides a thin band of greenery and some lush vegetation. In the panhandle, the region is drained by Alamito Creek and its tributary, Alamo Creek, by Black Hills Creek, and Alamo Seco Creek. The Black Hills (3,637 ft.) and smaller Cerro Redondo (3,704 ft.) rise above the surrounding terrain.

The Casa Piedra Road crosses the panhandle, while the Bofecillos Road and another section of Casa Piedra Road cross the southwest Lowlands. Terneros Creek, which runs east to west, draining the northern Bofecillos Mountains, flows west–southwest across Bofecillos Road and the Lowlands' midsection.

The region is blanketed by sedimentary deposits, mostly Perdiz Conglomerate, with volcanic fragments from eruptions in the Chinati and Cienega Mountains. Santa Elena limestone is conspicuous near Alamito Creek just before the creek exits the park. The Black Hills were formed by basaltic lavas and intrusions, concurrent with volcanic eruptions in the Bofecillos Mountains and Basin and Range faulting (Henry 1998).

Casa Piedra Road

From FM 170, an unpaved county road, Casa Piedra, runs north 6 miles, where it intersects Sauceda Road. Sauceda Road travels east 4 miles to the west park entrance, and another 17 miles to Sauceda. Visitors soon realize that BBR is "off the beaten path." From the Sauceda Road junction, Casa Piedra Road continues north, past trailheads for 4WD travel in the northwest panhandle of BBR, exits the park, and eventually connects with US Highway 67 to Marfa.

Casa Piedra Road runs near Alamito Creek, a major route for travelers in prehistoric times, and later part of the Chihuahua Trail. In the 1880s, the road provided a lifeline for farmers and ranchers settling Alamito Creek valley, and during the Mexican Revolution, the road was used by the US Army to protect settlers. Vegetation along the road is typical of the Lowlands. A prominent cactus is diploid purple prickly pear **86**, with striking purple pads. Also conspicuous is devil's claw (*Proboscidea fragrans*), an annual with magenta flowers and clawlike pods. Sacred datura **258** features giant, trumpet-shaped white blooms. Out of place is African rue **184**, a white-flowered exotic perennial extending its range.

Alamito Creek

Alamito Creek rises in the Davis Mountains between Marfa and Fort Davis, runs south through the Marfa Plateau, and eventually enters BBR's northwest pan-

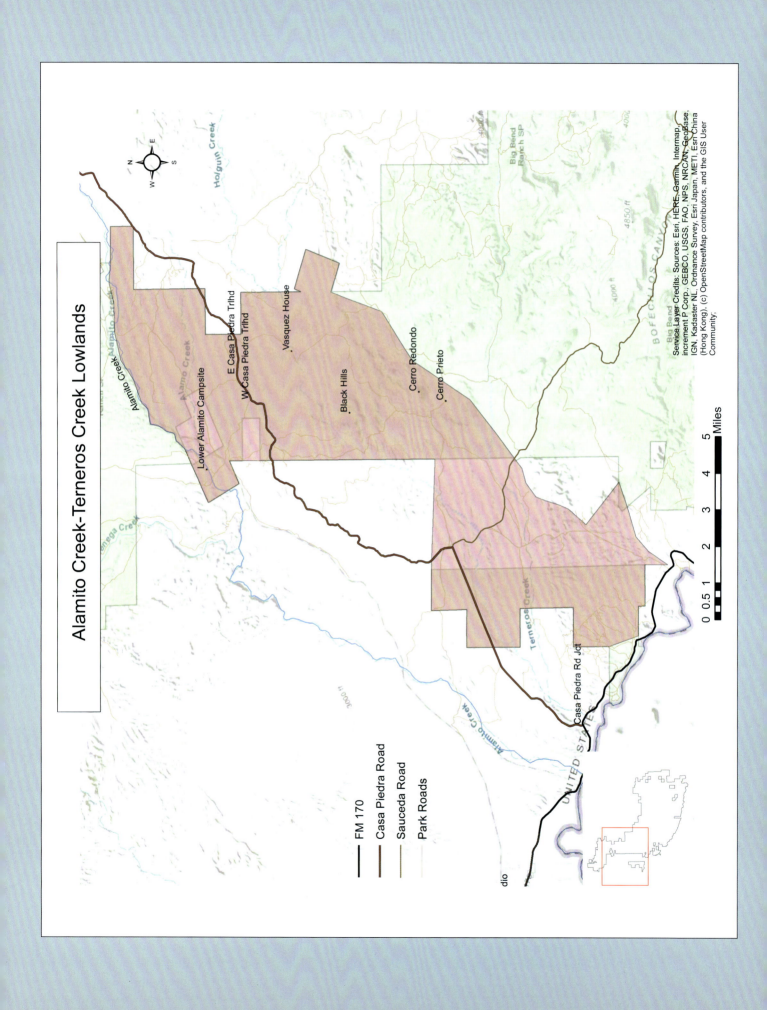

handle. After a 78-mile journey, the creek flows into the Rio Grande 5 miles east of Presidio. The creek was used by indigenous people for farming and for traveling to hunting and trading destinations. Spanish missionaries, soldiers, and explorers came next. In the 1870s, farmers and ranchers began to settle the valley. John Davis built a small creek-side ranch that became Alamito, a farming-ranching community. Cottonwood groves, screwbean mesquite **140**, and willow line the creek bed (Wilde and Platt 2011).

Black Hills Creek

A tributary of Terneros Creek, Black Hills Creek rises outside the park, 7.0 miles northeast of the Black Hills, runs west to Vasquez House in BBR, and southwest 18 miles, on the west flank of the Black Hills, to Terneros Creek. The united drainage travels 1.5 miles to the Rio Grande east of Casa Piedra Road.

Creosote **280**, ocotillo **148**, mariola **44**, leatherstem **130**, and mesquite (*Prosopis*) dominate the sparse landscape. Willow (*Salix*) and catclaw acacia (*Senegalia greggii*) occupy low bluffs along the creek and the mouths of tributary drainages. The creek supports mostly common herbs: bristly nama **155**; Fendler's bladderpod (*Physaria fendleri*), a silverhaired, yellow-flowered mustard; and mesa greggia (*Nerisyrenia camporum*), a white-flowered mustard. A parasitic vine, pretty dodder (*Cuscuta indecora*), drapes on shrubs above the creek. Mesa dropseed (*Sporobolis flexuosus*), a bunchgrass, grows in sand.

Black Hills

The Black Hills (Cerros Prietos), in BBR's northwest panhandle, are located east of Black Hills Creek and southeast of Casa Piedra Road. Points in the Black Hills rise 500 feet above Black Hills Creek, yet the terrain is warmer and drier than much of the park. The 4WD Botella–Black Hills Road runs north from Botella residence 2 miles and then forks. The right fork travels north 8.3 miles past Vasquez House to Casa Piedra Road. The left fork veers west to Cerro Prieto Windmill and across Alamo Seco Creek, then north 7.1 miles, on the west flank of the Black Hills, to Casa Piedra Road.

Shrubs on the rocky slopes include creosote **280**, mariola **44**, and mesquite (*Prosopis* species). Lechuguilla **27**, leatherstem **130**, and ocotillo **148** are prominent semisucculents. Less common are range ratany **161**, a low shrub with large red sepals and tiny petals, and hairy five eyes (*Chamaesaracha sordida*), a glandular-hairy false nightshade with yellow blooms. Strawberry cactus **73** and Warnock's cactus (*Echinomastus warnockii*) grow in the hills, and grama (*Bouteloua*) grasses are common.

Vasquez House

Vasquez House was built by Natividad Vasquez Jr. beginning in 1914, when he settled the property after receiving a land grant. Vasquez constructed a one-room adobe house, well, windmill, livestock pen, and 6 miles of fence. John Humphris acquired the property at auction in 1924 (Smith-Savage 1996b).

Terneros Creek

Terneros Creek rises near the park's northeast corner and flows west outside the park across the Alazan Hills. Reentering the park north of Leyva Dome, the creek continues west through Terneros Ranch into lowlands above Botella, then southwest, joining Black Hills Creek 1.5 miles north of the river. At the river, Terneros Creek has journeyed 26 miles through northern Bofecillos Mountains into desert flats. The creek has an illustrious history as a route to circumvent impassable mountains. The 1747 expedition of Pedro de Rábago y Terán, governor of Coahuila, in search of sites for forts on the Rio Grande, is said to have followed Terneros Creek (Brune 1981). In 1860, Lieutenant William Echols's camel expedition followed Terneros Creek in route from Presidio to Lajitas (Aulbach and Gorski 2000).

West of Botella, diploid purple prickly pear **86** and common herbs ashy-leaf bahia (*Picradeniopsis absinthifolia*), desert marigold (*Baileya multiradiata*), New Mexico dalea (*Dalea neomexicana*), and ever-present dodder (*Cuscuta*), frequent terraces above Terneros Creek. I encountered Abrams's spurge **118**, recently discovered in Texas, in sand and gravel at the Sauceda Road crossing.

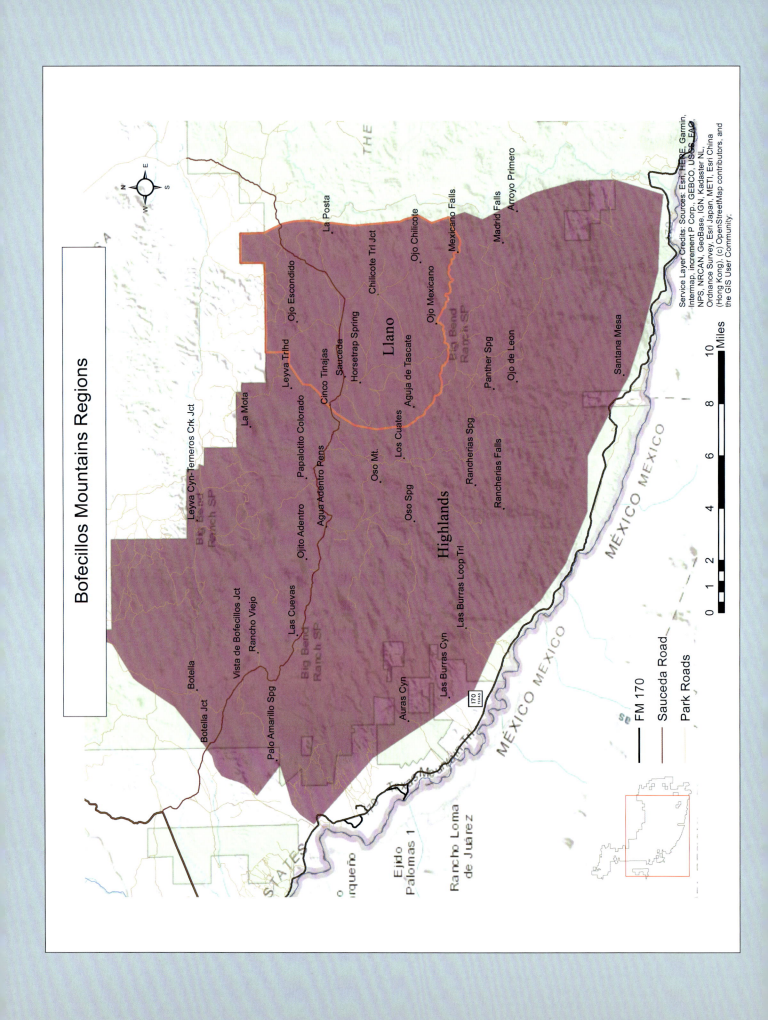

PHYSIOGRAPHIC REGION 3
Bofecillos Mountains: Highlands

The Bofecillos Mountains, bounded by Fresno Canyon on the east, the river to the south, the Alamito Creek–Terneros Creek Lowlands to the west, and the park boundary to the north, form the core of BBR. This mountain complex is the result of volcanic eruptions (32–25 Ma); basin and range faulting, affecting the current path of the Rio Grande (beginning 25 Ma); and formation of a through-flowing river (beginning 2 Ma), which led to incision of river tributaries and continued mountain erosion (Henry 1998).

Elevations range from 2,400 feet west of lower Fresno Canyon to 5,135 feet on Oso Mountain. Other central Bofecillos Mountains are Cerro Elephante (4,938 ft.), off Oso Loop, and Agua Adentro Mountain (4,982 ft.) and Bofecillos Peak (5,002 ft.), visible from Sauceda Road. Panther Mountain (4,922 ft.), on the Highlands-Llano boundary, is seldom explored, in rugged terrain off Rancherías Road. La Mota (5,046 ft.), in the northern Bofecillos, is a park landmark at the end of La Mota Road.

In the last two million years, nine significant drainages were incised. From east to west, Fresno, Monilla, Panther, Rancherías, Tapado, Las Burras, Auras, and Bofecillos Canyons, and Palo Amarillo Creek, drain the central and south Bofecillos Mountains and feed the Rio Grande. Arroyo Mexicano and Arroyo Primero flow into Fresno Canyon. Terneros Creek and its north-flowing tributaries, Yedra and Leyva Creeks, drain the north Bofecillos Mountains. Madrid Falls in Chorro Canyon, and Mexicano Falls in Arroyo Mexicano, are the second- and third-highest waterfalls in Texas (Capote Falls, outside the park in Presidio County, is the highest).

Canyon streams are fed by more than one hundred springs, many flowing all year, and this groundwater resource has played a defining role in shaping the landscape. The springs are oases in the punishing desert, to which flora and fauna gravitate. In shady canyons, microhabitats have formed, centered on the springs. Notable springs include Ojo de León and Panther Spring in Panther Canyon; Rancherías Spring in Rancherías Canyon; Bofecillos Spring, Agua Adentro, and Ojito Adentro in Bofecillos Canyon; Ojo Mexicano in Arroyo Mexicano; and Leyva Escondido in Leyva Canyon.

Abundant man-made resources highlight the Highlands' human past, historic and prehistoric. From the 1890s into the 1980s, ranching dominated economic activity. Scattered remains of ranch residences, livestock facilities, and water storage and distribution facilities illustrate the importance of ranching.

Botella near the park entrance, Rancho Viejo off Sauceda Road, Casa Reza in Panther Canyon, and Madrid House in Arroyo Primero are historic ranch headquarters. Agua Adentro Pens on Sauceda Road, and Javelin Pens on Javelin Road, old livestock holding facilities, are now equestrian campsites. The tall stone corral at Rancho Viejo is a particularly handsome facility.

The Fowlkes era stone dam, off Las Burras Road, and Alamito Dam, off Guale Mesa Road, are works of considerable craftsmanship. Two impressive stone *pilas*, Los Cuates, sit atop a hill off Guale Mesa Road, near La Iglesia. The water distribution system, with a vast pipeline network and windmills pumping water to tanks throughout the Bofecillos, is marveled at by experts (Deal 1976c). Two accessible windmills are the Rancho Viejo windmill, and Papalotito Colorado, a defunct wooden windmill, at Papalote Colorado Campsite. Less obvious are prehistoric remains left by early visitors and settlers. An example is Cueva Larga, a large rock-shelter, with pictographs of human and animal figures, off the Yedra Canyon Trail.

Botella

From Botella Junction, Botella Road runs northeast 2.4 miles to Botella, formerly Botella Camp. The original stone ranch house has not survived, but a restored adobe dwelling is now a park residence. Botella Spring, used to supply water to Botella, is a small spring in hills above the residence. The road continues north as a 4WD road to Terneros Creek, Vasquez House, and Casa Piedra Road.

Sanson Ospital operated Botella Ranch and managed cattle there in the 1890s. In 1917, William T. Davis, a Marfa store owner, purchased the ranch (Smith-Savage 1996b), which was robbed by bandits during the Mexican Revolution (Davis and Davis 1984). The Fowlkes brothers acquired the property in the 1930s, and Edwin Fowlkes and his wife lived for a time at Botella (Smith-Savage 1996b). Ranchers planted shade trees and shrubs, including a few beautiful athel **268**, oleander, and a now dead palm.

Vista de Bofecillos

Vista de Bofecillos is 2.7 miles southeast of the west park entrance. A side road climbs to a high plateau with 360° views of the Bofecillos Mountains, Sierra Rica, and dramatic sunsets. A road trace off the side road leads southwest 3.2 miles to an overlook of Palo Amarillo Spring. Guayacán (*Guaiacum angustifolium*), a shrub with deep purple flowers, and ocotillo **148** grow along the road trace. After monsoon rains, woolly tidestromia **9** blankets the ground. Orange caltrop **277**, Trans-Pecos senna (*Senna pilosior*), bluntscale bahia (*Hymenothrix pedata*), and climbing wartclub (*Commicarpus scandens*) may appear nearby. Annuals pop up, including Yuma spurge **127**, with fringed, petal-like appendages.

Palo Amarillo Creek

Palo Amarillo Creek rises on Agua Adentro Mountain, flows northwest to Rancho Viejo, and spills off the mountain slopes near an old cottonwood. From there, the creek runs west across Sauceda Road and through high canyon walls before entering more open range. The creek then heads southwest, drops into Palo Amarillo Spring outside BBR, and continues to the river.

Palo Amarillo is a Spanish colloquial name for agarito **52**, a shrub common along the creek. In 1835, Mexican Colonel Mecheranero reportedly pursued Comanches, who had raided Presidio del Norte, up the creek. Father Joseph Hoban, Presidio County surveyor, purchased land on the creek in 1880. A 1919 survey shows an adobe house, rock house, peach orchard, and large cottonwoods near Palo Amarillo Spring (Smith-Savage 1996b).

Palo Amarillo Creek is lined with shrubs such as mariola **44**, woolly butterflybush (*Buddleja marrubiifolia*), and trumpetflower (*Tecoma stans*). More common in wetter times, Arizona cottonwood **250** is sparse. A vine with delicate blue flowers, ivyleaf morning-glory **104** grows along the creek. Orange caltrop **277**, with showy orange flowers; and pigweed (*Amaranthus palmeri*), a weedy annual with reddish stems, broadly lance-shaped leaves, and tiny greenish-white blooms in slender compact spikes, can be prolific at the Sauceda Road crossing. Other expected herbs: rough menodora (*Menodora scabra*), sidecluster milkweed **15**, and sacred datura **258**.

Rancho Viejo

The Rancho Viejo turnoff from Sauceda Road, 3.5 miles southeast of the west park entrance, leads to Rancho Viejo Campsite and Old Entrance Road Trailhead. The campsite is located next to a functioning windmill, rock pila, and water troughs. Nolan County obtained the property from the state in 1881. A 6-foot-high stone corral, visible south of the campsite, and two buildings appear on a 1932 map. Reputedly, the corral was built by John Humphris, a Marfa businessman-rancher, in 1930, and used for mules. In 1942, the ranch was sold to the Fowlkes brothers (Smith-Savage 1996b; Nored 2014a).

Old Entrance Road Trail and Rancho Viejo Trail

The 6.9-mile one-way, Old Entrance Road Trail, part of a former road to Sauceda, wraps around the north rim of Palo Amarillo Creek, also known as Cañon de Rancho Viejo. The trail leads east through beautiful backcountry to an old windmill, Papalote Alto, southeast to a junction with a crossover trail to Yedra Trailhead, and on to Sauceda Road east of Agua Adentro Pens. Rancho Viejo Trail runs southeast from the Rancho Viejo turnoff, passes the old corral, and joins Sauceda Road in 1.8 miles.

Spanish dagger 31 and Roemer's acacia (*Senegalia roemeriana*) grow by the ranch and on Old Entrance Road Trail. Havard's plum (*Prunus havardii*), a shrub with small white flowers and wide-spaced petals, is a fixture along the trail. Desert marigold (*Baileya multiradiata*), an aster with long-stalked yellow flowers, and woolly tidestromia 9 bloom prolifically after rains. Rough menodora (*Menodora scabra*), with yellow flowers and paired globelike fruits, grows on rocky flats. Slimstalk spiderling (*Boerhavia gracillima*), with delicate wine-colored flowers on slender stalks, thrives amid boulders.

Las Cuevas

Southeast of Rancho Viejo, Las Cuevas is adjacent to Bofecillos Creek and on bluffs above, on both sides of Sauceda Road. The site is a distinctive exposure of yellow volcanic ash of the Cuevas Amarillas Tuff Formation. Hardened rock flows from Bofecillos volcanos capped the tuff and formed overhangs suited to human habitation. The rock spire atop Las Cuevas was named Sentinel Rock about 1917, by R. S. Dodd, a state surveyor (Nored 2014a).

The waters of Bofecillos Creek and Agua Adentro Spring sustain varied plant life. Vegetation on the yellow tuff and surrounding flats is sparse in comparison. Common trees near the creek are cottonwoods, seepwillow 35, and red berry juniper 109. Mariola 44 and Big Bend silverleaf 253 appear on the surrounding flats, and smooth sotol 28 and ocotillo 148 grow amid the rocks. Canyon grape 276 drapes over shrubs near the spring and creek. Soft twinevine 19 seeks shade between boulders. Comanche prickly pear (*Opuntia camanchica*) occurs at the base of cliffs. Graham's dog cholla (*Corynopuntia grahamii*), with stem joints easily attaching to shoes, grows in gravel around the tuff rock.

Spurges, such as Arizona spurge (*Euphorbia arizonica*), occupy the streambed. Fewflower beggarticks (*Bidens leptocephala*), with tiny three-petal yellow flowers, and horse purslane 4, with pinkish blooms, line the road. Arroyo fameflower 11; crestrib morning glory (*Ipomoea costellata*), a trailing annual with pink, funnel-shaped flowers; and the artful fern, creeping cliff brake (*Pellaea intermedia*), with leathery blue-green leaflets, grow in rock crevices atop the tuff.

Agua Adentro Spring

A short walk south in the usually dry bed of Bofecillos Creek leads to Agua Adentro Spring, on the south side of Sauceda Road. Shaded by cottonwoods and other trees and shrubs draped with grapevines, the spring consists of a low waterfall with shallow spring-fed pools.

Arizona cottonwood 250, seepwillow 35, and netleaf hackberry 93 shade the spring. Beebrush 271 and oreganillo 272 contribute to the understory. Coulter's brickellbush (*Brickellia coulteri*), with toothed leaves and cylindric flower heads, is often resident. Spikerush (*Eleocharis*) and cardinal flower 92 flourish in shallow water below the falls. Canyon grape 276 and Hartweg's twinevine 18 weave through shrubs.

Giant fishhook cactus 81, with lemon-yellow blooms, grows on rocky shoulders above the spring. Woolly tidestromia 9 springs up quickly on flats after rains. Western soapberry 252, with yellowish-white blooms in pyramidal clusters, has established a small grove on the road near the spring. Also roadside is buffalo gourd (*Cucurbita foetidissima*), with huge leaves, large orange flowers, and striped gourds.

Bofecillos Canyon

Bofecillos Canyon is the westernmost canyon draining the Bofecillos Mountains. The 10-mile canyon originates north of Cerro Elephante (4,938 ft.), near the southwest corner of Oso Loop. The canyon follows a circuitous route, northwest to Ojito Adentro, west to Agua Adentro Spring, and southwest to the Rio Grande. Bofecillos Spring is halfway between Agua Adentro Spring and the river. Lacking the steep walls of most Bofecillos canyons, the drainage consists largely of ash-flow tuff, debris flow, and sedimentary deposits. The Cuevas Amarillas Tuff at Las Cuevas also outcrops in Bofecillos Canyon (Sanchez 1999).

Below Ojito Adentro, plant hunters might find siratro **135**, a bean with odd reddish-black flowers; Rio Grande phacelia **156**, with coiled purple flower spikes; and Texas bindweed (*Convolvulus equitans*), a vine with petite white blooms. Near Agua Adentro Spring, Arizona cottonwood **250** lines the creek bed, and the purple grapes of canyon grape **276** are striking. In the creek bed are Yuma spurge **127** and hyssopleaf sandmat (*Euphorbia hyssopifolia*), with long, often blotched leaves. At Bofecillos Spring, screwbean mesquite **140**, with yellow flower spikes and coiled pods, may be the tallest plant around.

Auras Canyon

Auras Creek rises northwest of Nopalera Trailhead on Las Burras Road. The creek travels northwest to foothills near Bofecillos Peak, then southwest, through winding canyon walls to the river. From a car park 0.2 miles west of Ojito Adentro on Sauceda Road, an unmarked route runs southwest 1.8 miles to a junction with Auras Creek and Nopalera Trail. The route follows Nopalera Trail southwest 0.8 miles to Papalote Severo, a defunct windmill, and the creek bed west–southwest into Auras Canyon. The canyon continues southwest 7.5 miles to the river. Cuevas Amarillas Tuff forms the canyon walls.

Cañon de los Bandidos

Off Sauceda Road, just west of Ojito Adentro, is a scenic overlook of Cañon de los Bandidos, with an easily accessible platform for viewing the photogenic small canyon formed by Bandidos Creek. The creek rises off the Oso Loop southwest of Cerro Elephante and travels mostly northwest for 3.0 miles.

Ojito Adentro

East of Las Cuevas is Ojito Adentro, maybe the most popular hike in the park. The green cottonwood canopy of the 1.4-mile round-trip hike is a beckoning attraction from Sauceda Road. A short walk amid desert scrub vegetation leads through a gate to the Bofecillos Canyon drainage. Below ever-present cottonwoods and willows, netleaf hackberry **93**, Roemer's acacia (*Senegalia roemeriana*), Mexican buckeye (*Ungnadia speciosa*), and evergreen sumac (*Rhus virens* var. *virens*) form the understory.

The trail ends in a dense woodland environment at a 35-foot seasonal waterfall surrounded by seeping canyon walls and a garden of maidenhair fern **226**, canyon grape **276**, and northern dewberry **248**. Conspicuous is northern dewberry, only known at this site in the Trans-Pecos, with prickly, reddish stems and large, flimsy white blossoms. Below the pour-off are southern cattail **270**, cardinal flower **92**, and tall Julia's goldenrod (*Solidago juliae*), with yellow flowers in pyramidal clusters. Three species of star-scaled cloak ferns, jimmyfern (*Astrolepis cochisensis*), hybrid cloak fern (*Astrolepis integerrima*), and wavyleaf cloak fern (*Astrolepis sinuata*), flourish here. Bonaire lip fern **229**, with slender erect fronds, and Standley's cloak fern **236**, with pentagonal blades, occupy the slopes.

In desert scrub habitat are giant fishhook cactus **81**; desert pincushion cactus **77**, with small, creamy-white blooms with darker mid-stripes; and desert rosemallow (*Hibiscus coulteri*), with cup-shaped, pale yellow blooms. My most rewarding discovery here was the scarce, erect Purpus's tumblemustard **63**, with large leaves with earlike basal lobes, and white four-petal flowers.

Ojito Adentro, a premier birding site, illustrates how BBR's creeks and springs bring life to the desert.

Cuesta Primo Overlook

From Ojito Adentro, the road east climbs to Cuesta Primo, a bluff on the south side of Sauceda Road. The overlook offers views west to Las Cuevas and the Bofecillos Creek greenbelt. Cuesta Primo is named for Primo Mills, chief of maintenance for Diamond A Ranch (Nored 2014a).

Agua Adentro Pens

Agua Adentro Pens is a major, centrally located campground at the west entrance to Oso Loop. The site, 1.9 miles east of Ojito Adentro, is large and open, with space for RVs and horse trailers. Once a center for sheep shearing and livestock roundups, the campground now is a group site and equestrian staging area. Just east of Agua Adentro Pens, the 6.9-mile Old Entrance Road Trail reaches its east end.

Seeping walls, lush vegetation below seasonal waterfall, Ojito Adentro

Seeping walls, lush vegetation below seasonal waterfall, Ojito Adentro

Oso Loop

Oso Loop is an introduction to BBR's high desert vegetation, in the heart of the Bofecillos Mountains, with 6.1 miles on a rough, crescent-shaped 4WD road, and a 4.0-mile connecting segment on Sauceda Road. From Agua Adentro Pens, the 4WD road travels south to a junction with Las Burras Road, proceeds east to a junction with Guale Mesa Road, and then runs northeast to connect with Sauceda Road. Sauceda Road leads back west to Agua Adentro Pens to complete the loop. Oso Loop passes near Oso Mountain, the park's highest peak, and Cerro Elephante, with superb 360° views. The loop provides access to campsites on Las Burras and Guale Mesa Roads, and hikes in Las Burras, Tapado, and Rancherías Canyons.

Las Burras Road and Las Burras Canyon

The rough Las Burras 4WD Road begins on the southwest side of Oso Loop and travels south, past Nopalera Trailhead, 2.8 miles to Las Burras 1. The campsite sits amid massive boulders, and a short walk leads east to an imposing overlook of high-walled Tapado Canyon. From Las Burras 1, the road travels west 0.9 miles to Las Burras 2, 0.9 miles to Las Burras 3, and then 1.6 miles to Las Burras Trailhead. At the base of a hill, Las Burras 2 offers reminders of ranching days, with livestock pens and a stock tank. With a view west to Las Burras Canyon, Las Burras 3 is an ideal starting point for canyon hiking.

From Las Burras Trailhead, the canyon winds south 1.3 miles to a spring and small drop-off, then west 1.3 miles, around the north foothills of Cerro de las Burras, to the park boundary. At first, the canyon is wide and open, but high canyon walls tighten near the spring. Large spineless cacti, blind prickly pear (*Opuntia rufida*), with pale yellow flowers fading orange, cover the rocky hillsides. After exiting the park, the canyon continues southwest 4.5 miles through private property to the river.

Las Burras Road navigates a resplendent cactus garden, a special treat in early spring. Texas claret-cup **71**, with crimson flowers, can be spotted in the distance. Little known sweet prickly pear **88**, with red-centered yellow blossoms, is a roadside resident. Rare desert night-blooming cereus **90**, with huge trumpet-shaped white blooms, suddenly appeared near the road in 2012, and at two other park locations, then disappeared a few years later. Often growing near the cacti is Pringle's swallow-wort (*Cynanchum pringlei*), a perennial vine with twining stems, oblong leaves, and white urnlike blooms.

Fowlkes Era Stone Dam

A handsome Fowlkes era stone dam sits near the head of Oso Creek, southeast of Nopalera Trailhead on Las Burras Road. The dam was built by master masons employed by Edwin and James "Manny" Fowlkes, who in the 1940s consolidated ranches to form the precursor of BBR (Riskind 2019).

Oso Creek and Oso Spring

Oso Creek rises near Nopalera Trailhead on Las Burras Road, runs southeast to the stone dam, then south to Oso Spring, and through Tapado Canyon, west of Guale Mesa, to the Rio Grande. Oso Spring lies east–southeast of Las Burras 1, but steep drop-offs restrict access from the west. El Oso Ranch, near Oso Spring, was a leasehold property of Mateo Carrasco, an early settler of El Polvo circa 1871. Mateo and his wife Cecilia maintained a small farm and orchard with spring water (Smith-Savage 1996b).

On my late-summer hikes to Oso Spring, desert rosemallow (*Hibiscus coulteri*), with yellow blooms, and paleface rosemallow (*Hibiscus denudatus*), with red-centered light rose flowers, were in full bloom. Here I discovered talayote (*Cynanchum racemosum* var. *unifarium*), a twining vine with greenish-white flowers, not previously reported at BBR, spread across shrubs lining the creek bed.

Tanque Lara

Only 1.6 miles east of Agua Adentro Pens, Sauceda Road crosses an earthen dam that fills Tanque Lara after rains. Built by Diamond A Ranch in the 1960s, the tank attracts wildlife and cattle (Nored 2014a).

Papalote Colorado

Papalote Colorado, east of Tanque Lara, is an ideal campsite for Yedra Canyon hikes. Papalotito Colorado, a defunct wooden windmill, is a short walk north, and a pila, pens, and corrals are nearby.

Guale Mesa Road

From the southeast corner of Oso Loop, Guale Mesa Road travels 1.6 miles, past the Los Cuates Campsite turnoff and Oso Mountain, to a junction with Rancherías 4WD road. From here, Guale Mesa Road runs southwest 0.9 miles to the Guale 1 turnoff, then 2.7 miles across Upper Guale Mesa to the Guale 2 turnoff. A left turnoff leads 0.7 miles to Guale 2; the road continues 0.9 miles to Guale Mesa Trailhead.

Los Cuates ("twins") is named for the campsites' two stone pilas. Nearby are remnants of an old windmill. The site offers views of Oso Mountain to the northwest and La Iglesia to the northeast. La Iglesia, a large dike forming a northward extension of Panther Mountain, is named for its church-like appearance. Below La Iglesia are rock enclosures (*chiqueras*) used to shelter baby goats (Nored 2014b).

Guale 1 is nestled into mountainous terrain at the head of Rancherías Creek. Above the campsite is Alamito Dam, built at the direction of Edwin and Manny Fowlkes during the ranching heyday. Guale 2, at the south end of Upper Guale Mesa, is open, windy, with the best views in the park. Guale 2 looks down onto Lower Guale Mesa, Rancherías Canyon, and La Guitarra, with the Sierra Rica on the southern horizon, above the river. West Rancherías Trail, an old wagon route, crosses Lower Guale Mesa, and a trail running east from the campsite leads to Rancherías Spring.

Guale Trailhead is another spectacular viewpoint. To the west is the rim of Tapado Canyon, with views into a gorge as much as 1,000 feet deep and 2 miles wide. This is the place to ponder sunrises and sunsets, and violent volcanic events. A trail runs east 0.8 miles to join Rancherías Loop Trail. Guale Mesa occupies

Upper and Lower Guale Mesa from Guale trailhead

View west across Tapado Canyon from Guale trailhead

a large expanse between Tapado Canyon on the west, Rancherías Canyon on the east, and Tapado and Oso canyons to the north. Guale is named for Guadalupe Carrasco, who kept sheep on the mesa in the early 1900s (Smith-Savage 1996b).

Rancherías 4WD Loop Road

Rancherías 4WD Loop, an old wagon route, runs left off Guale Mesa Road south of the Los Cuates turnoff. The loop travels south 2.7 miles to Rancherías Spring Trailhead, past Rancherías Campsite. From the trailhead, the rough road proceeds east 4.3 miles to Javelin Pens. Rancherías Campsite, on a rocky bluff with good views, is a convenient base camp for hikes to Rancherías Spring, longer hikes on the 21-mile Rancherías Loop Trail, or travel on Rancherías Loop Road.

Rancherías Spring and Upper Rancherías Canyon

Upper Rancherías Canyon is approached from the west via Guale 2 Campsite or from the east via Rancherías Spring Trailhead on Rancherías Loop. From either direction, the route heads south down Upper Rancherías Canyon. Downstream are Rancherías Spring and a series of smaller springs. At Rancherías Spring, smoke-blackened rock hearths and bedrock mortars provide evidence that others have come before. Hikers can continue down the canyon to Rancherías Falls about 2 miles away. From Guale Mesa or Javelin Road to Rancherías Falls is a 7- to 9-mile round-trip hike, cross-country, following the greenbelt from the spring to the falls.

Arizona cottonwood **250**, with flowers in drooping catkins and cottony seeds, defines the landscape.

Common shrubs include Big Bend silverleaf **253**, with violet flowers; feather dalea (*Dalea formosa*), with butterfly-shaped magenta flowers; and leatherstem **130**, with urnlike white blooms.

A conspicuous perennial is longpetal echeveria **105**, with basal leaves and scapes of nodding orange-red blossoms. Roundleaf wild buckwheat (*Eriogonum rotundifolium*), with basal leaves and tiny pinkish-white blooms, grows in the creek bed. Broom milkwort (*Polygala scoparioides*), with broomlike stems, occupies drier ground. Tanglehead **214**, a grass; flower of stone **254**, a resurrection fern; Christmas mistletoe **274**, a semiparasitic shrub; and canyon grape **276**, a twining vine, add to the diverse flora.

Javelin Pens, Javelin Campsite, Javelin Trailhead

Javelin Pens is situated at the east end of the Rancherías 4WD Loop, and south end of the 2WD, high-clearance Javelin Road from Sauceda. A center for holding and shearing sheep, Javelin Pens also served as the main station pumping water to pastures throughout the southern Bofecillos (Riskind 1993). Evidence of ranching history endures in an abundance of ruins and artifacts. Today, the site is an equestrian staging area and a campsite used by backpackers on the Rancherías Loop Trail.

Just west of Javelin Pens is Javelin Campsite, by an old stock tank. A trail leads northwest to another tank, with views of Pico de las Aguilas (4,882 ft.) and Panther Mountain (4,922 ft.) to the west–northwest. Javelin Trailhead, 0.5 miles south of Javelin Pens, is the staging area for hikes down Monilla Creek into Monilla Canyon, or beyond, to Arroyo Primero; hikes in Panther Canyon on the East Rancherías Trail; or hikes on the Rancherías Loop Road (part of Rancherías Loop Trail) and beyond.

Monilla Canyon

At Javelin Trailhead, a road trace heads southeast and south 1.5 miles, then follows Monilla Creek 2.1 miles to a junction. From here, Monilla Creek flows south through steep-walled Monilla Canyon to the Rio Grande, while the road trace continues east 1.7 miles to Arroyo Primero Trail.

Upper Arroyo Mexicano

Arroyo Mexicano rises north of Panther Mountain and travels east 8.0 miles to Fresno Canyon. An arroyo tributary, a hiking route, flows east across Javelin Road, north of the Madrid Falls Road junction, to join Arroyo Mexicano just south of perennial spring, Ojo Mexicano. The upper arroyo ends at Mexicano Falls, 3.2 miles downstream from Ojo Mexicano. Rawls Formation ash-flow tuffs, erupted millions of years ago, form the canyon walls (Ing 1996b). The drainage is mostly open, with some pour-offs. *Chiqueras* (baby goat pens) attract hikers, but Ojo Mexicano is the key attraction.

The upper arroyo hosts a variety of trees, shrubs, vines, herbs, cacti, and grasses. Ojo Mexicano is the center of this plant world. Texas mulberry **183**, with tasty purple berries, is notable when in fruit. Brownfoot **33**, a perennial with pinkish-white flower clusters, is striking when in bloom. Poison ivy **12** is unremarkable, but in fall the leaves turn red or orange. David's spurge **122**, an erect annual with toothed leaves and terminal flower clusters, grows quite large in this locale. In spring, extravagant displays of strawberry cactus **73**, with lush magenta blooms, appear on terraces above the spring. Later, these plants produce the juiciest cacti fruits in the Trans-Pecos. Needle grama (*Bouteloua aristidoides*), a weak annual grass with wide-spaced spikelets on slender stems (culms), springs up after rains.

Havard's plum (*Prunus havardii*), with white flowers and peach-colored fruits, and tatalencho (*Gymnosperma glutinosum*), with erect stems, sticky linear leaves, and yellow ray and disk flowers, occupy rocky flats in the arroyo. Toothed serviceberry **245**, with rounded leaves and pinkish-white flowers, is plentiful downstream. Scarce at BBR, Texas mountain laurel (*Dermatophyllum secundiflorum*), with fragrant blue-purple blooms, appears above the falls. Notable herbs include Arizona snakecotton **6**, with tiny flowers embedded in cottony spires; sidecluster milkweed **15**, with greenish-yellow blooms providing food for Monarch butterflies; and woolly paintbrush **196**, with colorful orange-red bracts.

Madrid Falls Road and Campsites

After a steep climb up Madrid Hill (Cuesta de los Mexicanos), seven campsites are encountered off Madrid Falls Road. One mile east of Javelin Road is the turnoff for Mexicano 1 and 2. Mexicano 1 rests in an exposed area with hills on three sides, with Pila Mexicano, a steel water tank, on an overlooking hilltop. Mexicano 2 sits at the head of an east-flowing drainage, which leads to Vista del Chisos and across Madrid Falls Road into Chorro Canyon. Vista del Chisos is 1.3 miles southeast of the turnoff to Mexicano 1 and 2, near a water tank and earthen dam. The Chisos Mountains viewed are miles away in BBNP.

Southeast of Vista del Chisos, a short spur leads to Los Hermanos, a secluded campsite on flats within rocky and brushy hills. "Hermanos" refers to the old, stunted mesquite trees near the *bebederos*. Near Los Hermanos turnoff is one of BBR's four extant Havard's agaves **26**. This majestic agave is rare at BBR but common in BBNP and across the river in Mexico.

One-half mile southeast are Primero Trailhead and access roads for La Monilla and Pila de los Muchachos campsites. La Monilla sits in a shallow depression, below a small hill with good views to the east. Pila de los Muchachos rests on a high saddle below an old pila. A jeep road leads to Arroyo Primero. From the saddle are exhilarating views northeast to the Solitario and west into the arroyo.

From Primero Trailhead, Madrid Falls Road continues northeast 1.4 miles to a junction: the left fork leads to Mexicano Falls Trailhead, the right fork to Chorro Vista Trailhead and Chorro Vista Campsite. Chorro Vista is the end point of the 4.9-mile Madrid Falls Road. From the bluff above, views are serene: east to the Solitario, southeast to the Contrabando Lowlands and BBNP landmarks, southwest to Chorro Canyon and the Bofecillos and Sierra Rica Mountains. Many hiking opportunities begin here. Obtain a free Fresno West Rim map at Sauceda before attempting these hikes.

From Chorro Vista Trailhead, one can hike to an overlook of Madrid Falls, or into lower Chorro Canyon and down Arroyo Primero to Madrid House and beyond. Or one can walk the road north 0.6 miles to Mexicano Falls Trailhead, where a 1.0-mile trail leads to an overlook of Mexicano Falls.

Arroyo Primero

Arroyo Primero runs southeast, below La Monilla and Pila de los Muchachos, bottoms out and turns northeast to Ojo Blanco and Chorro Canyon. Arroyo Primero then travels east past Madrid House and two springs before emptying into Fresno Canyon. The trail effectively begins at Pila de los Muchachos, with a steep descent into the arroyo. Madrid House is 5.0 miles away, Fresno Canyon another 1.3 miles. Limestone, absent elsewhere in Arroyo Primero, occurs at the arroyo mouth (McKann 1975).

Many plants in Arroyo Primero grow in or near water. With flowers the color of cardinals' robes, cardinal flower **92** frequents a spring near Madrid House. Other water lovers appear at a spring off Fresno Canyon: a horsetail fern, Ferriss's scouring rush **115**, with infertile spores; golden columbine **238**, with long-tapering yellow flower petals; and western umbrella-sedge (*Fuirena simplex*), with barbed, bristly florets. Farther up the arroyo, giant helleborine **194**, a lovely orchid; Arizona centaury (*Zeltnera arizonica*), with bright pink flowers; and slimjim bean (*Phaseolus filiformis*), a trailing annual with vine-like stems, trifoliate leaves, and pink pealike blooms, form small colonies.

Madrid House

Just off the trail, Madrid House is disintegrating. Early photos show a larger adobe structure, with wooden rafters, doors, and windows (Gerow 2011). In 1880, the land on which the house is situated was patented to Father Joseph Hoban, Presidio County's deputy land surveyor. Reportedly, Father Hoban operated a school for boys nearby. Madrid House was built by Andres and Eusebia Madrid, early El Polvo settlers, in the 1870s. From the mid-1890s to 1916, the Madrid family ran sheep and cattle, raised vegetables, and maintained an orchard with water piped from Chorro Canyon (Brandimarte 2011).

Chorro Canyon and Madrid Falls

Chorro Canyon was long known as Madera (timber) Canyon, an apt name for the moist woodland. In 1973, the Texas General Land Office proposed the name Chorro (spout, place where water issues), which was adopted. That same year, the 100-foot falls, the second-highest in Texas, was named Madrid Falls for early El Polvo settler Andres Madrid (McKann 1975).

The short canyon is lush below Madrid Falls and in the spring-fed pour-off area above, with a cascading stream and pools of water. Upper Madrid Falls drops 100 feet to a large pool surrounded by velvet ash **190**, willow, grapevines **276**, and maidenhair fern **226**. Lower Madrid Falls spills 30 feet to a smaller pool. Madrid Springs, above the falls and below a 25-foot pour-off, is the primary water source. This stretch of Chorro Canyon, the most unspoiled wetland in BBR (McKann 1975), is closed to hikers.

Cottonwood, willow, velvet ash **190**, gray oak **144**, and Mexican blue oak **146** form the canopy, with flowing spring water below. Toothed serviceberry **245** and Havard's plum (*Prunus havardii*) contribute to the understory. Golden columbine **238**, cardinal flower **92**, Julia's goldenrod (*Solidago juliae*), and southern cattail **270** line the creek bed. Ferriss's scouring rush **115**; canyon grape **276** and poison ivy **12**; and prolific, sprawling herbs like sandpaper vervain (*Verbena scabra*), with scabrous herbage, serrated egg-shaped leaves, and slender spikes of tiny lavender blooms, reside on mud and damp banks.

Hardy shrubs, such as desert myrtlecroton (*Bernardia obovata*), with minute, pale green flowers, and feather dalea (*Dalea formosa*) grow along the canyon rim. Varicolor cob cactus **80** nestles in rock crevices. Prickly lip fern **230** occupies slopes above the rim.

With rainfall, low herbs spring up on open flats: ojo de víbora **102**, with pleated, sky-blue flowers; Wright's dalea (*Dalea wrightii*), with butterfly-shaped yellow flowers; common purslane (*Portulaca oleracea*), a prostrate annual with yellow blooms; shaggy portulaca (*Portulaca pilosa*), with dark red blooms; and Stewart's gilia (*Giliastrum stewartii*), with needlelike leaves and pink lavender blossoms.

Lower Arroyo Mexicano and Mexicano Falls

Lower Arroyo Mexicano is lush, varied, and watery, with large pools, seeping arroyo walls, and at times, a shallow stream. The streambed of sand and polished tuff is filled with boulders, huge conglomerates of cemented rock fragments, and exposed, contorted roots of velvet ash and cottonwood. At the lower arroyo head is Mexicano Falls, 80 feet high, the third-highest waterfall in Texas, a pour-off thick with maidenhair fern **226** and moss. The hike from Fresno Canyon to Mexicano Falls is 3.6 miles round trip.

Cottonwood and velvet ash **190** line the arroyo. Agarito **52** and mariola **44** are common in the understory. Toothed serviceberry **245**, gray oak **144**, sandpaper oak **147**, and evergreen sumac (*Rhus virens* var. *virens*) grow on the slopes. Mexican buckeye (*Ungnadia speciosa*), with aromatic pinkish flowers and leathery pods, is a fixture below the falls. One old Texas mountain laurel (*Dermatophyllum segundiflorum*), with fragrant blue-purple flowers, rests on a shelf above the arroyo.

An ever-changing mix of herbs is resident, including brownfoot **33**, Arizona snakecotton **6**, and sidecluster milkweed **15**. Wooton's loco **132**, goosefoot moonpod **186**, and longcapsule suncup **191** appear on arroyo margins. Julia's goldenrod (*Solidago juliae*) and golden columbine **238** are waterfall residents. Slimlobe globeberry (*Ibervillea tenuisecta*), with greenish-yellow blooms, twines over shrubs.

Smith Canyon and Smith Spring

Northwest of the Crawford-Smith House in Fresno Canyon, a drainage leads to Smith Canyon, a narrow, nearly vertical minicanyon with Smith Spring near the top. Water piped from the spring to the Crawford-Smith House supplied a swimming pool and citrus orchard. Now furnishing nonpotable water to hikers, the spring has never run dry (Durst 2015). A trail leads from Mexicano Falls Trailhead down the bluff and across a few hills to an overlook of Smith Spring in 1.4 miles.

Maidenhair fern **226** and golden columbine **238** hang off moist canyon walls at the spring. Southern cattail **270** decorates the spring area. Julia's goldenrod (*Solidago juliae*) stems stand high at the top of the canyon. Desert sumac (*Rhus microphylla*), with tiny creamy-white blooms in early spring, grows lower in the canyon. Wright's Dutchman's pipe **25** is often present at the canyon entrance.

Yedra Canyon

The colloquial name "Yedra" refers to poison ivy **12**, which is plentiful in the canyon. Roughly 1.5 miles east of Agua Adentro Pens, Yedra Road travels north 2.2 miles to Yedra Trailhead. Papalote Colorado Campsite, just east of this turnoff, is often used as a base camp for canyon hikers. Two campsites are located on Yedra Road, Yedra 1, at the base of a north-facing bluff, and Yedra 2, in a more open but rocky area.

From Yedra Trailhead, Yedra Canyon runs north 4.1 miles to Terneros Creek, with a maze of side trails and possible loop hikes along the way. A crossover trail proceeds west from Yedra Trailhead 1.7 miles to join Old Entrance Road Trail. At 0.5 miles on the Yedra Trail, a short side trail runs east to Cueva Larga. At 0.9 miles, a road trace loops west 2.1 miles through mountainous terrain before circling back to Yedra Canyon. Other road traces lead east to Leyva Canyon. Yedra Canyon is also used for longer backpacking routes connecting to Leyva Canyon, Terneros Creek, and the Old Entrance Road Trail.

Just west of Yedra Trailhead is Ojo de Papalote Alto, an occasionally lush spring, with Arizona cottonwood **250**, gray oak **144**, and hybrid oaks forming the canopy. Oreganillo **272**, a small shrub loved by bees, contributes to the understory. Leather flower **240**, with nodding, urn-shaped, purplish blooms, and canyon grape **276** twine across shrubs. Small plants such as desert rosemallow (*Hibiscus coulteri*) and crestrib morning glory (*Ipomoea costellata*) congregate along the creekbed.

The drainage continues northwest, past a side route to Cueva Larga. The willow-like leaves of seepwillow **35** and screw-like pods of screwbean mesquite **140** catch the eye. Threadleaf snakeweed (*Gutierrezia microcephala*), a rounded aster with yellow flower heads; and lacy tansyaster (*Xanthisma spinulosum*), an erect perennial with toothed or lobed bristle-tipped leaves and yellow flower heads with elongate ray florets and many disk florets, brighten the creek bed. Intriguing small herbs pop up, including sticky-haired Organ Mountain blazingstar **169**, with orange blooms, and prostrate mat nama (*Nama torynophyllum*), scarce but sometimes locally common, with pink blooms.

Torrey's tievine **103**, with rose-pink blooms; scarlet morning glory (*Ipomoea cristulata*), with red funnel-shaped blossoms; and siratro **135**, with blackish-red, pea-like flowers, adorn the trees. Grasses are plentiful: a bizarre exotic, spike burgrass (*Tragus berteronianus*), with spikelets with hooked prickles, at disturbed sites; black grama **212**, with abruptly bent, white-haired lower stems, on ledges above the drainage; and bushy bluestem (*Andropogon glomeratus*), with feathery inflorescences, at the springs.

The inner and outer crossover trails connecting Yedra and Leyva canyons often harbor unexpected plants. On an oak-filled bluff above the inner trail is James's wild buckwheat (*Eriogonum jamesii*), scarce at BBR, a shrublet with white flowers aging orange. The crossover from Yedra Trail to Old Entrance Road Trail features a minigorge and spring. Texas snoutbean (*Rhynchosia senna* var. *texana*), with pealike yellow flowers; catnip noseburn (*Tragia ramosa*), with stiff, stinging hairs and fuzzy fruits; and soft twinevine **19** grow in the gorge and spring side.

Cueva Larga

Cueva Larga, one of the largest rock-shelters in the park, looms above the arroyo (Cañon de la Cueva Larga) that intersects Yedra Canyon 0.5 miles north of Yedra Trailhead. Just off Yedra Canyon, the overhang is a popular side trip, with pictographs painted with ground minerals and animal fat (Hoyt 2015). The drawings include human figures, handprints, and animal forms (Hampson 2015). Also present are bedrock mortars for grinding grain and remains of a rock wall likely built by goat herders.

Leyva Creek

Leyva Creek rises northeast of the Papalote Nuevo Campsite turnoff of Sauceda Road, travels west across the road and southwest, crossing La Mota Road south of Ojo Escondido Pens. The creek then meanders west to Leyva Escondido Spring and Leyva Canyon. Leyva Creek, from the La Mota Road crossing to Leyva Escondido Spring, is a rewarding hike with ample plant-hunting opportunities.

Arizona cottonwood **250** and western soapberry **252** line Leyva Creek. Small shrubs include little walnut **158**, desert sumac (*Rhus microphylla*), and shrubby senna **142**. Osage orange (*Maclura pomifera*), with wrinkled, brain-like orange-size fruit, grows at a spring halfway to Leyva Escondido. Curious attractions are the blackish-red flowered vine, siratro **135**, and the whimsical dwarf shrub, Dutchman's breeches (*Thamnosma texana*), with inflated leathery fruits shaped like upside-down pantaloons. Near Leyva Escondido Spring are two intriguing annuals: netted globecherry **260**, with urn-shaped, purple blooms and bladdery fruits; and oakleaf datura **257**, with lobed leaves and thorny fruits.

A rock art site, on a hill by Leyva Escondido Spring, is a 3-mile round-trip hike from Cinco Tinajas. The art includes drawings of humanlike and animal figures, geometric designs, and a horse and rider figure. In 2016, a colony of longpetal echeveria **105** occupied the hill above the pictographs.

Leyva Canyon

Leyva Creek, a north-flowing tributary of Terneros Creek, and Yedra Creek, a tributary of Leyva Creek, drain the northern Bofecillos Mountains. Leyva Canyon can be accessed from Cinco Tinajas or the Leyva Trailhead off La Mota Road. From Cinco Tinajas, Leyva Canyon runs northwest 8.3 miles to Terneros Creek; from Leyva Trailhead, the canyon proceeds west–northwest 7.3 miles to Terneros Creek. Loop trails from Leyva Trailhead and Cinco Tinajas connect the two trailheads. Consult the Discovery Map.

About 1.5 miles from Cinco Tinajas, a box canyon enters Leyva Canyon from the south. This wetland supports a grove of oaks; oak mistletoe (*Phoradendron villosum*), parasitic mostly on gray oak, with hairy yellow green stems and leaves, separate spikes of greenish-yellow male and female blooms, and white or pink berries; and moisture-seeking manybract groundsel **42**, uncommon at BBR. Another wetland is Baños de Leyva, a series of large water pools. The "baths," 2.5 miles from Cinco Tinajas and less than 2.0 miles from Leyva Trailhead, were once a destination for guided horseback trips into the outback.

Leyva Canyon is dotted with springs and cottonwood groves, and the moist, shady environment nurtures intriguing herbs. Wright's stonecrop **106**, with fleshy swollen leaves and pink-tinged white flowers, grows out of rock crevices. Also watch for sinkerleaf purslane **223**, angel trumpets **187**, red cyphomeris (*Cyphomeris gypsophiloides*), and Big Bend bluebonnet (*Lupinus havardii*). Red cyphomeris is a four-o'clock perennial with funnel-shaped reddish blooms and lopsided fruits. Big Bend bluebonnet, the best-known Big Bend wildflower, is an early spring bloomer, with scapes of deep blue flowers.

La Mota Road, Ojo Escondido, La Mota Mountain

Beginning 1.6 miles east of Sauceda, La Mota Road travels 1.8 miles northwest to Ojo Escondido Pens. The 2WD high-clearance road then runs northwest 2.1 miles to the La Mota 1 turnoff, which continues to Leyva Trailhead. From La Mota 1, the road proceeds 1.8 miles to La Mota 2 east of La Mota Mountain.

On open Llano flats, Ojo Escondido Pens is a convenient site for equestrian groups, with pens and water for horses piped from Ojo Escondido Spring. A pila and bebedero are close by. Ojo Escondido Campsite is situated among boulders north of Ojo Escondido Pens. Ojo Escondido is a reliable small spring in the rocks. The Ojo Escondido area lies within "Physiographic Region 4."

La Mota 1 is an open group meeting area and campsite for families and large gatherings. Leyva Trailhead is a short distance southwest. La Mota 2, partly protected by boulders, offers a close view of La Mota (4,922 ft.). La Mota, a park landmark, is visible from long distances in BBR's most remote corners. Seemingly flat-topped, La Mota rises 900 feet above the desert.

Plants off La Mota Road include dayflower (*Commelina erecta* var. *angustifolia*), with two large sky-blue petals and a tiny white petal; Wright's wild buckwheat (*Eriogonum wrightii* var. *wrightii*), a subshrub with small white flowers fading pink; and Texas claret-cup cactus **71**. Uncommon species may crop up. In 2010, scaled cloak fern **233** appeared at Ojo Escondido, outside its limestone habitat. At the road's end, Texas shrub **7** grows on other shrubs. Mexican clammyweed **99** climbs the lower slopes of La Mota.

Upper Terneros Creek

Although little used, Terneros Creek can be rewarding. Hiked from west to east, the creek passes historic Terneros Ranch, riparian woodland with lush greenbelts and cottonwood canopies, several springs, ranch ruins, and connections to Yedra and Leyva canyons for longer backpacking adventures. The hike begins north of Botella, at a junction of Terneros Creek and Botella–Black Hills Road. A road trace along Terneros Creek passes Terneros Ranch in 2.4 miles; a couple of springs and ruins in the next few miles; a junction with Leyva Canyon in 6.6 miles; and a road trace from Leyva Trailhead in 7.8 miles.

Sand spikerush **112**, Abert's wild buckwheat (*Eriogonum abertianum*), and James's monkeyflower **200**, flourish near pools of water, with Big Bend bluebonnet (*Lupinus havardii*), tall wild buckwheat (*Eriogonum tenellum*), scrambled eggs **149** and beebrush **271** along the banks. Abert's wild buckwheat produces spring flowers amid basal leaves, and summer-fall white flowers on erect stems. Tall wild buckwheat forms leafless stems, silvery basal leaves, and tiny white blooms. Heartleaf rockdaisy (*Perityle parryi*), with yellow flower heads, and yellow rocknettle **168** cover the cliffs above.

PHYSIOGRAPHIC REGION 4
Bofecillos Mountains–Llano

The Llano, a 4,300-feet-high plateau, is bounded by the Highlands on the west, park boundary on the north, Fresno Canyon on the east, and Arroyo Mexicano on the south. Elevations range from a low of 3,500 feet on Arroyo Mexicano to a high of 4,882 feet at Pico de las Aguilas in the Tascate Hills. This plateau is the result of volcanic eruptions in the Bofecillos Mountains (27 Ma). The Sauceda Volcano, part of the 25-mile-wide Bofecillos Volcano, produced relatively fluid lavas that spread to form the wide Llano plain.

Mountains in the Highlands are replaced by hills in the Llano. Domelike Cerro Boludo (4,290 ft.), across the road from Cinco Tinajas, and Cerro Chilicote (4,460 ft.), on Cerro Chilicote Loop Trail, are typical. Parts of Tascate Hills, a transition zone between the Highlands and the Llano, reach higher elevations. In addition to Pico de las Aguilas at 4,882 feet, Aguja de Tascate, by Tascate 1, reaches 4,810 feet in elevation.

Other than bordering Fresno Canyon and Arroyo Mexicano, no canyons drain the Llano, and large drainages in the Llano are tributaries of Leyva Canyon in the Highlands. Leyva Creek travels west through the Llano, north of Sauceda, into Leyva Canyon. Fingerlike drainages run north from Llano Loop and Tascate Hills to Sauceda and Los Ojitos, and through artfully sculptured Cinco Tinajas into Leyva Canyon.

Although few springs occur in the Llano, the Llano, with dense soils and former dense grass cover, is the major recharge area for the myriad springs and seeps in the Bofecillos. Los Ojitos is a group of small springs below Los Ojitos Campsite southwest of Sauceda. Horsetrap Spring, on the Horsetrap Spring Loop Trail, once supplied drinking water to Sauceda. Ojo Chilicote, a perennial spring southeast of Puerta Chilicote Trailhead, supplies water to Arroyo Mexicano. Ojo Mexicano, in Arroyo Mexicano, and Ojo Escondido, off La Mota Road, are discussed in "Physiographic Region 3."

Evidence of ranching history is everywhere, with livestock pens at Sauceda and Ojo Escondido, corral ruins at Howard's Ranch, and miles of stone fences. A corral at Sauceda served as a rodeo arena. Ruins of a goat-herding complex remain at Ojo Chilicote. A stone dam survives at Howard's Ranch. A beautiful rock dam goes unnoticed in Tascate Hills. A masonry dam at Los Ojitos traps water after monsoon rains.

Pila de Tascate, at Tascate 1, and Pila de Gato south of Los Ojitos are stone tanks of considerable craftsmanship. Tanque Talique, an earthen tank off Llano Loop, impounds a small lake after rains. Papalote Encino, a windmill at Papalote Encino Campsite, is now a solar pump. Papalote Llano, a metal windmill at Papalote Llano Campsite, is equipped with a new solar-powered water well.

Cinco Tinajas

Cinco Tinajas, 1.2 miles west of Sauceda, is a high-walled canyon named for five bedrock, water-holding depressions in the drainage. The canyon was formed over millions of years by Leyva Canyon tributaries from the Llano plateau and Tascate Hills. One flows north through Tanque Talique and Horsetrap Spring. A more westerly tributary drains Tascate Hills through Los Ojitos and Skeet Canyon.

The dramatic sculpture of Cinco Tinajas is viewed via a 2-mile round-trip trail off Sauceda Road. The trail leads north to a hilltop view into Leyva Canyon. Hikers can follow the trail down to the creek bed and east, upstream, to the slickrock mouth of Cinco Tinajas. The hike can be extended northeast to Leyva Escondido Spring and a rock art site, or northwest deeper into Leyva Canyon. From the hilltop above Leyva Canyon, a side trail runs east to an overlook of Cinco Tinajas and 360° vistas: north to flat-topped La Mota, southeast to the Solitario, southwest to cone-shaped Oso Mountain, and south to dome-like Cerro Boludo above Skeet Canyon. Hikers can return via the trail or the Cinco Tinajas drainage.

Typical shrubs include catclaw mimosa (*Mimosa aculeaticarpa* var. *biuncifera*), with vicious recurved prickles; tarbush (*Flourensia cernua*), with black stems and fall-blooming yellow flower heads; California trixis **48**; black dalea (*Dalea frutescens*), like feather dalea (*Dalea formosa*) but without feathery hairs; and shrubby umbrella thoroughwort **40**.

Ferns are drawn to the water, shade, and protective crevices. Flower of stone **254**, curled into a ball when dry, unfurls with rain. Standley's cloak fern **236** occupies rocky bluffs above the tinajas. Below boulders are Eaton's lip fern (*Myriopteris rufa*), with tufted fronds, matted hairs, and bead-like leaflets (pinnules); Wright's lip fern (*Myriopteris wrightii*), with erect hairless fronds and oblong pinnules; and Alabama lip fern (*Myriopteris alabamensis*), uncommon at BBR, with long hairless, lancelike blades.

Ivy treebine (*Cissus incisa*), with trifoliate leaves, white flowers, and black berries, is a prominent vine. Texas toadflax (*Nuttallanthus texanus*), with spikes of pale blue blooms; redseed plantain (*Plantago rhodosperma*), with large, toothed leaves and parchment-like blooms; and watermelon nightshade **264**, with fearsome prickles, highlight an ever-changing mix of herbs. Mountain mustard (*Hesperidanthus linearifolius*), with four-petal lavender flowers, and Wright's stonecrop **106** grow on rock walls. Trans-Pecos false boneset **36** and bracted bedstraw (*Galium microphyllum*), with tiny, pale yellow blooms and paired round fruits, are creek-side. Rouge plant **202** nestles below shrubs. Orange flameflower (*Talinum aurantiacum*), with orange blooms, red stamens, and yellow anthers, appears on rocky hilltops.

Slickrock drainage from Cinco Tinajas into Leyva Canyon

Skeet Canyon and Cerro Boludo

Arroyo de los Ojitos, locally known as Skeet Canyon, meanders north from Los Ojitos, with low bluffs above the creek bed and a short stretch of higher volcanic walls. The canyon crosses an old skeet shooting range and an old east–west gravel road. The drainage continues north, below Cerro Boludo, crosses Sauceda Road, and flows into Cinco Tinajas. With rain, hills near Cerro Boludo and side drainages into Skeet Canyon offer surprising floral attractions. Cerro Boludo rises 300 feet above the creek bed.

Gray oak **144**, smooth sotol **28**, and Thompson's yucca (*Yucca thompsoniana*) grow atop the hills above Skeet Canyon. Desert olive (*Forestiera angustifolia*), a shrub with greenish-yellow petal-less blooms and blue/black berries; trumpetflower (*Tecoma stans*), with compound leaves, large two-lipped yellow flowers, and long tan pods; and arroyo fameflower **11** occupy low bluffs above the drainage.

Unusual annuals are wingpod purslane **225**; sinkerleaf purslane **223**; New Mexico tickflower (*Desmodium neomexicanum*), with trifoliate leaves and pink pealike blooms; and scarce Coahuila twintip **207**. Perennials, dwarf Indian mallow (*Abutilon parvulum*), with toothed, egg-shaped leaves and yellow orange flowers; and not-to-be-missed plains ironweed **49** grow along the creek bed. The canyon is home to many ferns, such as fairy swords (*Myriopteris lindheimeri*), with lancelike, highly dissected blades and beadlike ultimate segments; hairy bommer **227**; Peruvian spikemoss **255**; and uncommonly, Wright's spikemoss (*Selaginella wrightii*), with mat-forming stems, leaves tipped with hardened bristles, and many spore-bearing cones towering above the mats. Cliff waxwort (*Plagiochasma rupestre*), a liverwort with a flat, branching blue-green body, with purple margins and ringed pores, grows near hairy bommer.

Los Ojitos

Just 0.7 miles southeast of Cinco Tinajas, a turnoff leads to Los Ojitos, a campsite on a bluff above the springs, with easy access and proximity to Sauceda. A stone dam, built by master masons employed by the Fowlkes brothers, traps water after rains. Hikers walk down the road to the spring and continue southwest in the drainage 1.3 miles to a junction with Horsetrap Spring Loop Trail.

Los Ojitos supports cottonwoods and willows; desert sumac (*Rhus microphylla*); splitleaf brickellbush (*Brickellia laciniata*), with cylindric flower heads; Texas mimosa **136**; and heath carlowrightia **1**. Perennials include palmleaf thoroughwort (*Conoclinium dissectum*), with blue flower heads and protruding styles; baby jump-up (*Mecardonia procumbens*), with yellow blooms in a clasping calyx; onionlike crowpoison (*Nothoscordum bivalve*), with six-petal white blooms; and fringed puccoon **55**.

Annuals, redstem stork's bill **152**; bearded flatsedge **111**; longleaf false goldeneye (*Heliomeris longifolia*), an erect sunflower with yellow flower heads; and swollenstalk sneezeweed (*Helenium amphibolum*), with long-stalked flower heads with yellow ray florets and red disk florets, spring up in moist soil. Snapdragon (*Maurandella antirrhiniflora*), with purple, snout-like, two-lipped blooms, and scarlet morning glory (*Ipomoea cristulata*) twine on shrubs. Wright's lip fern (*Myriopteris wrightii*), and jimmyfern (*Astrolepis cochisensis*), with slender fronds and many leaflets, occupy cliffs above the spring.

Horsetrap Spring and Trail

Horsetrap Bike and Hiking Trail navigates an old horse pasture with low hills and rocky outcrops, and views southwest to Tascate Hills, north to La Mota, and southeast to the Solitario. Two trailheads, one just west of Sauceda, one 0.5 miles south on Javelin Road, provide access. From Sauceda Road, the trail runs southwest to a road trace from Los Ojitos, then southeast and northeast to a road trace from Tascate Hills. The trail continues northeast to Horsetrap Spring and Javelin Road Trailhead. A walk on Javelin Road to Sauceda, and west to the Sauceda Trailhead, completes the 4.8-mile loop.

Cottonwoods shade the spring. Beebrush **271**, buttonbush **249**, netleaf hackberry **93**, and Havard's plum (*Prunus havardii*) form the understory. A rare hybrid, lanceleaf cottonwood (*Populus* x *acuminata*), was found years ago. Threadleaf groundsel (*Senecio flaccidus* var. *flaccidus*), with yellow flower heads; and showy menodora (*Menodora longiflora*), with tubular yellow blooms, are small subshrubs. A perennial, linear-leaf four o'clock (*Mirabilis linearis*), with pinkish-white blooms, grows near the trail.

Drainage from Sauceda to Cinco Tinajas

A tributary of Leyva Canyon runs north through Sauceda, by old rock fences, and northwest to Cinco Tinajas. Unknown to most visitors, the drainage is high-walled, photogenic, usually wet, with at least trickles of water and often large pools. From a short spur road just west of Sauceda, hikers can follow this drainage 1.0 mile to Cinco Tinajas and then return on Sauceda Road.

Buttonbush **249**, with pincushion-like flower heads, is plentiful. Infrequent is Chisos Mountain false Indian mallow (*Allowissadula holosericea*), a subshrub with large heart-shaped leaves with felty undersides and orange-yellow blooms. On the bouldery slopes, memorable herbs are Rio Grande phacelia **156**, with purple flowers in coiled clusters; pink baby's-breath **266**, with large leaves and dainty yellow flowers; smooth spiderwort **100**, with three-petal pink blooms in boatlike bracts; and bladder mallow (*Herissantia crispa*), with trailing stems, five-petal white flowers, and inflated fruits.

Sauceda Ranger Station— Bob Armstrong Visitor Center

From 1905 to 1988, Sauceda was owned by ranching families raising cattle, sheep, and goats. On early maps, the property was named Saucita (willow), for nearby, once-flowing Saucita Spring and willows around it. Today, Sauceda includes a ranger station, bunkhouse, historic ranch house, and nature trail. A corral, once used as rodeo grounds, is available for equestrians. Sauceda Ranger Station, a visitor check-in station, also serves as a small store with books, maps, t-shirts, hiking equipment, and exhibits.

Formerly a hunting lodge, the bunkhouse offers lodging for men and women in separate quarters, a large living-dining area, and well-equipped kitchen for visitor use. Until recently, the historic Sauceda ranch house provided more comfortable lodging and an inescapable old ranch atmosphere.

The nature trail introduces the high-desert shrub habitat characteristic of the Llano. The trail crosses a volcanic ridge, with interpretive signs identifying plants. Seepwillow **35**, agarito **52**, desert willow **53**, and catclaw acacia (*Senegalia greggii*), with white flower spikes and catclaw-like prickles, are typical shrubs.

Perennials, Lindheimer's copperleaf (*Acalypha phleoides*) and Wright's Indian mallow (*Abutilon wrightii*), are frequent trail residents. Lindheimer's copperleaf forms flower spikes of male and female flowers, male blooms with red stamens, female blooms with feathery red styles. Wright's Indian mallow produces large velvety leaves and yellow blooms. Wooton's loco **132**, with pink pealike blooms, and tufted cottonflower (*Guilleminea densa*), with white blooms in chaffy bracts, appear after rains.

South Leyva Campground

A short 2WD spur road, 1.2 miles northeast of Sauceda, leads to spacious South Leyva Campground. South Leyva is situated below a volcanic ridge with an old rock wall at the base. In early morning and late afternoon, a ridgetop hike rewards with views southwest to the Bofecillos Mountains and northwest to La Mota. The campground accommodates large groups, with numerous tent sites and picnic tables.

Ocotillo **148** and Thompson's yucca (*Yucca thompsoniana*) are conspicuous on the ridge, while allthorn **160** and skeletonleaf goldeneye (*Sidneya tenuifolia*) grow near the campground. Thompson's yucca is identified by its slender trunk with serrated linear leaves pointing up, white flower spikes, and beaked fruits. Skeletonleaf goldeneye is an aster with linear leaves and long-stalked yellow flower heads.

Papalote Encino Campsite and Encino Loop Trail

Papalote Encino Campsite rests on a spur off Sauceda Road, on a flat, open Llano plain, near a pila and defunct windmill replaced by a solar pump. The campsite is ideal for hiking, biking, and horseback riding.

Encino Nature Trail runs southwest from Papalote Encino 2.9 miles to La Mota Road. The 7.2-mile Encino Loop Trail joins the nature trail with Powerline Trail, which follows the Los Alamos powerline feeder route. From the Nature Trail–La Mota Road junction, the loop continues south on La Mota Road to Powerline

Trail and follows Powerline Trail southeast and abruptly northeast to Sauceda Road. The loop follows Sauceda Road to the Papalote Encino turnoff and takes the spur back to Papalote Encino.

The loop trail exemplifies the Llano, a high plateau formed by basaltic lavas and tuffs of the Rawls Formation (Sanchez 1999). Flats are punctuated by rolling hills, mesas, arroyos, and volcanic outcrops. Leyva Creek and other Leyva Canyon tributaries drain the expanse. Creosote bush **280** and mariola **44** are plentiful. Ocotillo **148**, lechuguilla **27**, and tree cholla (*Cylindropuntia imbricata* var. *imbricata*) are conspicuous. Goosefoot moonpod **186**, and sawtooth spurge (*Euphorbia serrula*), with prostrate stems, serrated oblong leaves, and petal-like appendages longer than the glands, are a treat to find.

Llano Loop

Llano Loop is bounded on the west by Tascate Hills, on the south by Llano Dome, on the east by Fresno Canyon, and on the north by Leyva Creek. The 8.4-mile loop, usually beginning and ending at Sauceda, traverses the Llano, a plain with deep, rich soils overlying thin basalt lavas. The 2WD road runs through open, flat terrain and sparse vegetation. From Sauceda, the route travels on Javelin Road to a junction with West Llano Loop, then east on the loop road to the Puerta Chilicote Trail junction. East Llano Loop runs north to Sauceda Road. The loop is completed by traveling west 2.9 miles to Sauceda.

Overgrazing has depleted the grasslands. Creosote bush **280**, mariola **44**, and acacia dominate, along with soaptree yucca **30**, Thompson's yucca (*Yucca thompsoniana*), smooth sotol **28**, and western honey mesquite **139**. Diploid purple prickly pear **86**, tree cholla (*Cylindropuntia imbricata* var. *imbricata*), desert Christmas cholla **69**, and Texas claret-cup cactus **71** are the principal cacti.

The loop is habitat for three perennials, copper zephyrlily **10**, fragrant heliotrope (*Euploca greggii*), and red dome blanketflower (*Gaillardia pinnatifida*), and two annuals, iron ipomopsis **216** and Rocky Mountain sage (*Salvia reflexa*), seldom encountered elsewhere at BBR. Fragrant heliotrope, with linear leaves and white flowers, colonizes moist gullies. Red dome blanketflower, with a leaf rosette and long-stalked yellow and red flower heads; and Rocky Mountain sage, with dainty blue-white blooms, line the road.

Other perennials are scurfy arrowleaf mallow **176**; desert holly **32**; melon loco **108**; yellow flameflower **267**; Davis Mountains mock vervain (*Glandularia pubera*), with erect stems, compound leaves, and pink blooms; and hog potato (*Hoffmannseggia glauca*), with basal compound leaves and orange-yellow blooms with protruding red stamens. Low annuals, Pope's phacelia **157** and bitterweed (*Hymenoxys odorata*), a toxic, rounded annual with compound leaves and many compact heads of yellow ray and disk flowers, spring up along the loop road. Flatglobe dodder **101** parasitizes plants in seasonal pools. Hooked water clover **180**, a fern with cloverlike leaves, occupies recently inundated flats.

Papalote Llano and Papalote Llano Nuevo Campsites / Llano Dome

Papalote Llano is accessed via Javelin Road 2.8 miles south of Sauceda. The access road leads to Papalote Llano, in an open area with bulky Engelmann's prickly pear **89**. The campsite boasts a windmill, pila, water troughs, solar well, and water faucet. One-half mile east is Papalote Llano Nuevo, below a small bluff. The road continues as a hiking route, with views of Llano Dome. As Bofecillos Mountains eruptions subsided, intrusions like Llano Dome never surfaced, and instead formed dome-shaped uplifts.

The dome supports ferns, such as Wright's cliff brake **237** and small leaf silver fern (*Argyrochosma microphylla*), and cacti, such as Texas claret-cup cactus **71**, in rocky crevices or below boulders. Larger cacti, shrubs such as feather dalea (*Dalea formosa*), and perennials, including shrubby purslane **224**, occupy open areas. Divided into wide-spaced, oval segments, the blue-green blades of small leaf silver fern lack hairs, scales, and farina. Shrubby purslane's showy orange blooms are easy to spot.

Howard's Ranch / Cerro Chilicote Loop / Fresno Rim Trail / Ojo Chilicote

Chilicote is a colloquial name for Lindheimer's senna **141**, a common area perennial. The area is bounded by the Llano on the north and west, Arroyo Mexicano

Localized storm approaching Llano at sunset

on the south, and Fresno Canyon on the east. Puerta Chilicote Trailhead, reached by a 1.5-mile 2WD road from Llano Loop, provides access to trails west of Fresno Canyon. The ruins of Howard's Ranch are visible northeast, by a spring and dam.

From Puerta Chilicote, 3.2-mile Cerro Chilicote Loop is usually followed counterclockwise, southwest around Cerro Chilicote (4,460 ft.), northeast along arroyos to Ojo Chilicote, and northwest back to the trailhead. The loop can be extended south to Arroyo Mexicano. Fresno Rim Trail, a 5.0-mile round-trip hike, runs southeast to Ojo Chilicote, and east to a 700-foot overlook of Fresno Canyon and the Solitario.

Ojo Chilicote, which feeds a tributary of Arroyo Mexicano, is the central plant attraction. The spring once supported extensive sheep and goat operations. Easily recognized shrubs include Arizona cottonwood **250**, Spanish dagger **31**, Apache plume **244**, leatherstem **130**, netleaf hackberry **93**, desert sumac (*Rhus microphylla*), and agarito **52**. Spanish dagger stands out, with a disorganized rosette of bayonet-shaped leaves. Apache plume's feathery fruits, resembling Apache headdresses, are unmistakable.

Blackfoot daisy (*Melampodium leucanthum*), angel trumpets **187**, and woolly paintbrush **196** are common perennials on Fresno Rim Trail. Blackfoot daisy forms white ray florets, and a foot-shaped bract below each ray turns black at maturity. Brownfoot **33** is plentiful on Cerro Chilicote Loop. Tubular slimpod (*Amsonia longiflora*), with long-tubed white blooms, grows at Howard's Ranch. Annuals, common purslane (*Portulaca oleracea*) and sinkerleaf purslane **223**, appear on flats above the arroyos. Tall, dense colonies of deergrass (*Muhlenbergia rigens*), with long narrow, flat leaves, and compact, cylindric flower spikes, fill the arroyo bottoms.

Tascate Hills

Tascate Hills is bounded on the north by Horsetrap Loop Trail, on the east by Javelin Road, on the south by Arroyo Mexicano, and on the southwest by Panther

Tascate Hills north of Tascate 1 near Rock Dam

Mountain. Aguja de Tascate looms above Tascate 1, and Pico de las Aguilas rises near the head of Arroyo Mexicano. Off Javelin Road 3.0 miles south of Sauceda, an access road runs west to Tascate 2 and past a road trace to Tascate 1. Below a bluff covered with ocotillo **148** and yucca, Tascate 2 offers views north to La Mota and east to the Solitario. The road trace travels 2.1 miles north to Horsetrap Bike and Hiking Trail. Cross-country walks west into the hills bring new plant discoveries in an enthralling landscape.

Tascate 1 sits on the east flank of Tascate Hills, beside Pila de Tascate. A rich natural environment fills the hills above. With small east–west drainages cutting across the slopes, the hills sustain prime plant habitat. An added attraction is a rock dam, which traps a large water pool. Gray oak **144** and red berry juniper **109** are notable trees. Thompson's yucca (*Yucca thompsoniana*) and smooth sotol **28** are other landmarks. Midsize shrubs include toothed serviceberry **245**; shrubby senna **142**; Havard's plum (*Prunus havardii*); scarlet bouvardia (*Bouvardia ternifolia*), with tubular scarlet flowers; and silver dalea (*Dalea bicolor*), confined to Tascate Hills in BBR, with butterfly-shaped blooms.

Trans-Pecos croton **117**, a small shrub once believed to be restricted to the Solitario in the United States, is now known to venture outside the Solitario in BBR, and to other sites east of the park. I encountered this dwarf nestled between boulders north of Tascate 1. Common subshrubs are arroyo fameflower **11**; heath carlowrightia **1**; fern acacia (*Acaciella angustissima*), with spineless stems, compound leaves with linear leaflets, and white flower heads; and prickleleaf dogweed (*Thymophylla acerosa*), with clustered needlelike leaves and yellow flower heads with about eight ray florets and up to twenty-five disk florets.

Engaging perennials fill the drainage east of the hills. A morning glory, silver ponyfoot (*Dichondra argentea*), with round silvery leaves and white blooms, and spreading snakeherb (*Dyschoriste linearis* var. *decumbens*), with two-lipped lavender flowers, are uncommon at BBR. Hairy tubetongue (*Justicia pilosella*), with pink two-lipped blooms, Palmer's bluestar **13**, and Wright's Dutchman's pipe **25** are frequent residents. Bigbract verbena (*Verbena bracteata*), with lavender blooms in bristly bracts, grows above the dam, and pink baby's-breath **266** below the dam.

Uncommon park annuals are muleear spurge **124**, not previously reported at BBR; spreading chinchweed (*Pectis prostrata*), with slender bristly leaves and yellow flower heads; and wingpod purslane **225**, in BBR limited to Tascate Hills and Skeet Canyon. Tascate Hills offers exemplary habitat for ferns, including Bonaire lip fern **229**, fairy swords (*Myriopteris lindheimeri*), Wright's cliff brake **237**, and Peruvian spikemoss **255**. Texas claret-cup cactus **71** and varicolor cob cactus **80** are scattered.

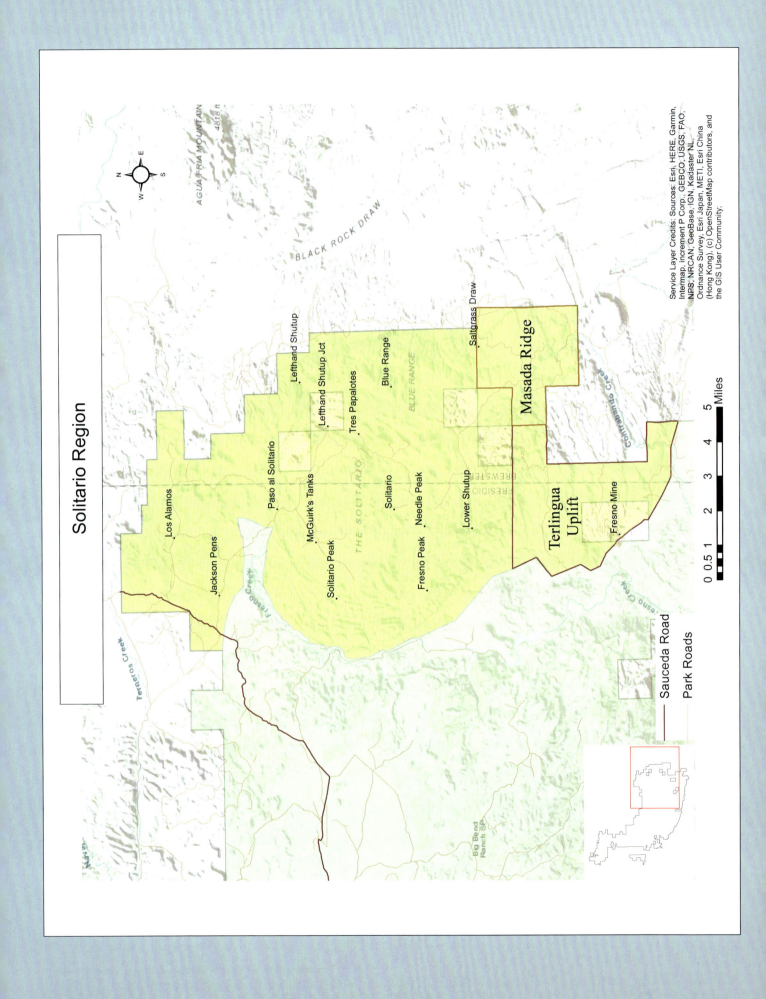

PHYSIOGRAPHIC REGION 5
Solitario

Nearly 10 miles across, the Solitario is a circular igneous dome, collapsed and eroded over millions of years. The dome was formed by an intrusion of magma (36 Ma) into older Cretaceous and Paleozoic rocks. The intrusion produced an eruption that collapsed the dome and created a caldera in the south-central part of the dome. The resistant Cretaceous rocks were uplifted and tilted in concentric circles to form a rim around the interior basin. The Solitario's mountain peaks and canyons pale in comparison to this breathtaking crater, with a steep outer rim (to 5,000 ft.) and flatiron-shaped slopes of mostly Santa Elena limestone (Riskind 2007).

Elevations in the Solitario Region proper (excluding the Terlingua Uplift and Masada Ridge Wilderness subregions), range from 3,500 feet below the Flatirons east of Fresno Canyon to 5,120 feet on Fresno Peak.

Mountains in the Solitario rise hundreds of feet above the basin floor. Cone-shaped Solitario Peak (4,786 ft.) in the northwest interior, and needle-shaped Needle Peak (4,608 ft.) in the south-central basin, are dark rhyolite intrusions. In addition to Fresno Peak

West rim of Solitario. Photo courtesy Texas Parks and Wildlife Department © 2021

View of Flatirons and Fresno Peak from Chorro Vista

above the southwest rim, Eagle Mountain (4,819 ft.) to the southeast, and the little-explored Blue Range (5,001 ft.) to the east, are Cretaceous formations ringing the central basin.

The Righthand Shutup, Los Portales, Lefthand Shutup, and Lower Shutup drain the Solitario. Cowhands strung fences across them to shut in livestock. Above Fresno Canyon, the 2.2-mile Righthand Shutup and 1.8-mile Los Portales drain the Solitario's western interior. Lefthand Shutup Creek runs 3.0 miles northeast from Solitario Road East to empty the north-central–northeast interior. Lower Shutup Creek travels south to Fresno Creek, draining the interior basin.

The shutups are training grounds for aspiring geologists, a glimpse into geological history through a succession of Cretaceous rocks from youngest to oldest as you ascend the drainages. The canyons are storehouses of marine fossils, deposited when the area was beneath the ocean. The canyons lack springs, and water flows mostly during the monsoon season. Springs are replaced by tinajas—bedrock pools that collect rainwater.

In BBR's remote northeast corner, the South Fork of Alamo de Cesario Creek and Panther Canyon are afterthoughts. Yet, the South Fork is one of five major park drainages, which runs east, north of Los Alamos, 2.8 miles to Alamo de Cesario Creek in the Devil's Graveyard badlands north of BBR. Little-known Panther Canyon also runs northeast across the park boundary into Alamo de Cesario Creek, and thence to Terlingua Creek and the Rio Grande. Both canyons drain the Solitario's northern foothills.

Throughout the Solitario are reminders of the human history. Off Sauceda Road, Jackson Pens was used to manage sheep and cattle in Jackson Pasture. McGuirk's Tanks serviced livestock in the central basin. Papalote Ramon pumped water for livestock in the southwest Solitario. In the east-central basin, Tres Papalotes served as a mining camp, and nearby, manganese and uranium mines were prospected. Los Alamos, a prehistoric campsite, became a historic ranch headquarters.

Due to similar geology and vegetation, Terlingua Uplift, Masada Ridge, and Saltgrass Draw are placed in the Solitario physiographic region. Terlingua Uplift is a large rectangular plateau, bounded on the north by the Solitario and on the south by the Contrabando Lowlands. Masada Ridge, part of Terlingua Uplift, is a protected wilderness area, accessed via Terlingua Ranch roads east of the park. Saltgrass Draw, flowing through northern Masada Ridge, drains the Blue Range on the Solitario's east rim.

Solitario Overlook

On Sauceda Road 8.0 miles northeast of Sauceda, this overlook provides a view of the park's signature geological feature. The view is tame compared to the violent geological events that shaped the Solitario. In the distance the Solitario resembles a collection of hills encircling a valley. The view from space, or Google Earth, offers a contrary picture—an otherworldly perspective akin to a giant impact crater.

Solitario Road

From Sauceda Road, 0.8 miles north of Pila Montoya, 4.8-mile Solitario Road runs east 1.0 mile to Jackson Pens, one of five equestrian campsites. The road continues southeast 0.9 miles to a junction with Los Alamos 4WD Loop, which leads northeast to Los Alamos campsite. From this turnoff, Solitario Road proceeds 2.3 miles southeast to Paso al Solitario campsite and a junction with the east end of Los Alamos 4WD Loop. Limestone ridges along this stretch of road offer some of the park's best plant habitat.

From Paso al Solitario, the road continues south 0.6 miles and splits into two branches. The right fork, Solitario Road West, goes southwest 1.2 miles to McGuirk's Tanks and a turnoff to McGuirk's Tanks Campsite. From the turnoff, the road continues southwest 1 mile to Burnt Camp Trailhead. The left fork, Solitario Road East, travels 6.2 miles around the Solitario's eastern interior to Lower Shutup Trailhead. Along the way, the road runs southeast, across Lefthand Shutup Creek at 2.5 miles; then south, to Tres Papalotes at 3.5 miles; and southwest 2.7 miles to Lower Shutup Trailhead.

Jackson Pens

One mile off Sauceda Road, Jackson Pens, now an equestrian campsite, serviced livestock in 2,500-acre Jackson Pasture. State-owned land was acquired by Thomas Rawls (1910–12) and sold to Pearl Andrew

Jackson in 1916. Jackson also leased other land outside the park (Taylor 2011; Cook 2020).

The campsite provides ample parking, a large man-made dirt tank, water trough, and pila atop a hill to the north, on Jackson Pens Trail. The trail, with good views from the pila, runs north 1.6 miles to Los Alamos Road, at an intriguing nonmarine limestone outcrop. The low flats about Jackson Pens are alluvium and alluvial fan deposits, and rocky outcrops nearby are basaltic lava.

Mariola **44** and allthorn **160** are common. Damianita (*Chrysactinia mexicana*), a mounded aromatic shrub with bright yellow flower heads, brightens the scene. Out of place is Havard's dayia (*Dayia havardii*), here growing on sparse flats. This shaggy-haired subshrub is worthy of careful inspection, with odd, pink blooms with red streaks and bluish anthers. Lining the muddy drainage below are lyreleaf parthenium (*Parthenium confertum*), a relative of mariola, with deeply lobed leaves and white flower heads, and golden crownbeard (*Verbesina encelioides*), an aster with yellow flower heads.

Los Alamos 4WD Loop

Just 0.9 miles east of Jackson Pens, Los Alamos Loop runs a mile northeast to Los Alamos Campsite and continues north another mile to Los Alamos. A historic ranch, Los Alamos is now a park residence. On Los Alamos Loop, encounters with gullies, washouts, and overgrown brush should be anticipated (Riskind and Sholly 2009). The campsite is a good base camp for 4WD travelers. In 2015, on mesquite-creosote flats near the Los Alamos turnoff, I encountered Texas milkvine **22**. Widespread in the Trans-Pecos, this vine is rare in BBR, with arrowlike leaves, yellow green flowers, and spindle-shaped pods.

Solitario Northwest Rim

Basaltic lava outcrops continue from Jackson Pens to 1.5 miles northwest of Paso al Solitario, where the lava is abruptly replaced by Santa Elena limestone slopes, rich in plant life. From this point to Paso al Solitario, the Santa Elena Formation gives way to a succession of Cretaceous rocks, from youngest to oldest, first a thin band of Sue Peaks, then Del Carmen, Glen Rose, and Yucca formations (Henry 1998).

In a study of Solitario vegetation (Hardy 2009), Hardy stated that "90% of the Solitario comprises Chihuahuan Desert Scrub or one of its phases." According to Hardy, "thin soil and rapid runoff have given an edge to shrubs and succulent and semisucculent species as opposed to trees, herbs, and grasses." This description fits the slopes northwest of Paso al Solitario, although a variety of herbs and patches of oak and juniper appear, especially along tributaries of the Lefthand Shutup.

Gray oak **144**, red berry juniper **109**, evergreen sumac (*Rhus virens* var. *virens*), desert olive (*Forestiera angustifolia*), and western honey mesquite **139** occupy the slopes and line the drainages. Heath cliffrose **247**, Roemer's acacia (*Senegalia roemeriana*), and allthorn **160** primarily appear on the slopes. Desert willow **53**, beebrush **271**, and catclaw acacia (*Senegalia greggii*) fill the arroyos. Semisucculent species, including Thompson's yucca (*Yucca thompsoniana*), Spanish dagger **31**, smooth sotol **28**, ocotillo **148**, lechuguilla **27**, Texas false agave **65**, beargrass **29**, and candelilla **120** are common.

Many small shrubs are present, including guayule (*Parthenium argentatum*), javelina bush (*Condalia ericoides*), plume tiquilia **59**, spinyleaf zinnia (*Zinnia acerosa*), hyssopleaf asphead **173**, narrowleaf spiderling (*Boerhavia linearifolia*), and Texas crinklemat **58**. Guayule, harvested for rubber during World War I, has silvery leaves and long-stalked, headlike flower clusters. Javelina bush is a squat, thorny shrub with bundled leaves and yellow flowers. Spinyleaf zinnia forms a floral bouquet with white ray and yellow disk florets. Narrowleaf spiderling produces clusters of pink funnel-shaped blooms. Other striking shrubs are white ratany **162**, range ratany **161**, Trans-Pecos croton **117**, Big Bend silverleaf **253**, Rio Grande saddlebush **94**, Emory's mimosa (*Mimosa emoryana*), with pink flower heads and prickly pods, and baccharisleaf penstemon (*Penstemon baccharifolius*), with toothed leaves and tubular scarlet blooms.

A surprising herbal mix includes five perennials: stemmy four-nerved daisy (*Tetraneuris scaposa*), with basal linear leaves and yellow ray and disk flowers on naked scapes; longstalk greenthread (*Thelesperma longipes*), also with mostly basal linear leaves and naked flower stalks, but with only yellow disk florets; needleleaf bluet (*Houstonia acerosa* var. *acerosa*), with clustered spinose leaves and four-lobed white flowers; downy paintbrush **197**; and bristlecup sandmat (*Euphorbia chaetocalyx* var. *chaetocalyx*), often with red stems, and flower clusters with erect glands and notched petal-like appendages. Two annuals, Buckley's centaury **151**, with pink blooms; and Berlandier's flax (*Linum berlandieri* var. *filifolium*), with needlelike leaves and flimsy copper-yellow flowers with dark red centers, brighten the scene.

Plants of Conservation Concern: Texas cone cactus **85** is scattered throughout much of the Solitario and is also present on Masada Ridge. Snow-white silverlace cactus **79** is dispersed across the Solitario, Terlingua Uplift, and Contrabando Lowlands. A single Lloyd's hedgehog cactus (*Echinocereus x roetteri* var. *neomexicanus*), a scarce hybrid of Texas claret-cup **71** and Texas rainbow cactus **72**, was discovered by Jean Hardy Pittman on the far north rim (Hardy 2009). Two subshrubs, Isely's feather duster **133**, a legume, and grasslike Warnock's justicia **2**, an acanthus, might be found anywhere in the Solitario.

Four scarce perennials grow on the northwest rim: Cochise beardtongue **205**, needleleaf gilia **215**, Palmer's cat's eye (*Oreocarya palmeri*), and Thurber's stemsucker **23**. Cochise beardtongue is seldom recorded in the Solitario. In contrast, needleleaf gilia is common. Palmer's cat's eye, with small white flowers in coiled clusters, and Thurber's stemsucker, a stemless, leafless parasite, both vulnerable, were not previously reported at BBR. Although not rare, Bear Mountain milkweed **16** is uncommon, with limited distribution in Texas, and previously unreported in the park, outside its known range.

Paso al Solitario

Solitario Road travels southeast 4.2 miles from Sauceda Road to Paso al Solitario. The campsite, ideal for hiking and biking adventures, rests in an open area with nearby limestone ridges. The challenging Los Alamos 4WD Loop runs north 5.5 miles to Los Alamos. Solitario Road continues south, through alluvium, rim sill rhyolite, and Paleozoic rocks, then splits into west and east branches around the central basin.

Solitario Road West

Just 0.6 miles south of Paso al Solitario, Solitario Road West runs southwest 2.2 miles to Burnt Camp Trailhead, past a junction with McGuirk's Loop (4WD, unmaintained).

McGuirk's Loop and Dos Pilas Spur

Faint, overgrown McGuirk's Loop tracks an old waterline service road east–southeast 2.8 miles to a fork. The north fork runs north 0.8 miles to join Solitario Road East near the Lefthand Shutup crossing. The 0.3-mile east fork, Dos Pilas Spur, climbs to a saddle on a novaculite ridge, with two stone water tanks. The ridge looks down on Tres Papalotes, east to the Blue Range, and west to the Solitario's central basin.

McGuirk's Tanks

From McGuirk's Loop turnoff, the road runs southwest, across Lefthand Shutup Creek and past McGuirk's Tanks, a man-made dirt tank, to a spur road to McGuirk's Tanks Campsite. The tanks are named for H. W. McGuirk, a Big Bend pioneer. McGuirk ranched at Los Alamos (1895–97) and on Contrabando Creek (1897–99) before moving to Lajitas (Smith-Savage 1996a).

The rocky outcrop above the dirt tank is Dagger Flat Sandstone (500 Ma), the oldest rock exposed in the Solitario. This sandstone was deposited in water off what was then the edge of North America (Deal 1976a). The campsite is ideal for visitors hiking the Righthand Shutup and Inner and Outer Loop trails.

Burnt Camp Trailhead

From the campsite turnoff, Solitario Road West continues southwest, through dense brush, to Burnt Camp Trailhead. On the western edge of the Solitario's central basin, the trailhead provides access to the Inner and Outer Loop trails, and the Righthand Shutup.

Inner Loop Trail

From Burnt Camp Trailhead, the Inner Loop follows a jeep trail 3.2 miles southeast to Lower Shutup Trailhead. The loop traverses the southwest edge of the Solitario's inner basin, with interior views. The trail passes between ridges capped with white chert of Caballos Novaculite (350 Ma), underlain by black chert of the Maravillas Formation (450 Ma). The Solitario is the southwesternmost exposure of Caballos Novaculite, part of a mountain range reaching from the Appalachians and beyond (Alloway 1995).

The loop passes an earthen tank and between two hills. At 2.6 miles, a faint jeep road heads north into the central basin. At 3.2 miles, the loop crosses Lower Shutup Creek, ending at Lower Shutup Trailhead. Hikers return via the Inner Loop, or follow the Outer Loop to the Righthand Shutup junction, and back to Burnt Camp Trailhead. Both loops total 9.6 miles. Limestone-loving plants line the Inner Loop Trail and occupy nearby hilltops. Watch for creeping rockvine (*Bonamia repens*), a seldom noticed, trailing morning glory, and two cacti, the diminutive common button cactus **76** and fall-blooming living rock **66**.

Solitario Peak Area

Solitario Peak (4,786 ft.) is an intrusion through Cretaceous rocks (27 Ma) well after Solitario Dome was formed (Henry 1998). Nearly 400 feet above surrounding terrain, the peak is also known as Pico de Fierro (peak of iron), because of the dark red-brown color of the cone-shaped volcanic plug. Highly photogenic at sunrise, the peak emits a fiery glow. On cross-country hikes, visitors can explore the peak and nearby Cretaceous outcrops, the Yucca Formation, and Glen Rose Formation to the west.

Gray oak **144** occupies the peak's lower flanks; strawberry cactus **73** and nipple cactus **83** brighten the base. Wright's snakeroot (*Ageratina wrightii*), a shrub with white disk florets and long styles, and Lemmon's brickellbush (*Brickellia lemmonii* var. *lemmonii*), a perennial with cylindric yellow flower heads, grow in a drainage that wraps around the peak. On limestone flats are two fascinating plants, Warnock's justicia **2**, a grasslike subshrub, and bonsai-like bushy wild buckwheat **221**.

Another area resident is the scarce minishrub Trans-Pecos maidenbush **201**. Trans-Pecos maidenbush is a relict of a wetter climate. A subshrub, plateau rocktrumpet (*Mandevilla macrosiphon*), with long-tubed white flowers, and common button cactus **76**, with tiny pink blooms, grow nearby.

Outer Loop Trail

The Outer Loop travels 6.4 miles from Burnt Camp Trailhead, on the west edge of the Solitario's central basin, to Lower Shutup Trailhead, at the central basin's south end. From the head of the Righthand Shutup to Los Portales, Fresno Peak, and Needle Peak, the loop traverses the east foothills of the Solitario's west rim. The loop first runs southwest to a fork: the right fork proceeds west down the Righthand Shutup to Fresno Canyon; the left fork follows the Outer Loop Trail south to Papalote Ramon, and to the head of Los Portales. The loop then travels southeast to Needle Peak, past the east foothills of Fresno Peak. From Needle Peak, the trail loops back northeast to Lower Shutup Trailhead.

Burnt Camp Trailhead to Righthand Shutup Junction: Slimlobe globeberry (*Ibervillea tenuisecta*), a vine with deeply divided leaves and delicate yellow flowers, sprawls over shrubs in thick vegetation. Ashy sandmat **121**, a prostrate spurge coated with matted hairs, occupies open flats. The purple-tinged, wavy-edged leaves and maroon blooms of wavyleaf twinevine **17** are attractions near the trail.

Righthand Shutup Junction to Papalote Ramon: Ridges at the head of the Righthand Shutup, and along the Outer Loop Trail toward Burnt Camp, are capped with white chert of Caballos Novaculite and black chert of the Maravillas Formation. Texas cone cactus **85**, with magenta flowers, thrives on these rocks. Resident spurges are threadstem sandmat **126**, with a maze of forking maroon branches, and leatherweed **116**, with leathery leaves and hairy clusters of yellow-white male and female blooms.

Papalote Ramon

Papalote Ramon, a defunct windmill, is the only remaining feature of the 1900s Burnt Camp. The state awarded the property to Dallas and Wichita Railroad in 1879. Little is known about the camp and owner-operators until 1944, when the property was sold by Texas & Pacific Railway to G. C. Meriwether and Clay T. Holland, and subsequently purchased by the Fowlkes brothers in 1951 (Smith-Savage 1996b).

Los Portales

Los Portales empties the west interior slopes of the Solitario into Fresno Canyon. The canyon is easily recognized by the high shelter caves cut out of Santa Elena limestone at the canyon mouth. Typically, Los Portales is hiked from below and, despite its botanical and geological interest, is used mostly as a route to the Solitario rim and hikes on Outer Loop Trail. Occasionally, the steep climb is used as part of a counter-clockwise loop route, which goes up Los Portales, north on the Outer Loop, down the Righthand Shutup to Fresno Canyon, and south in the canyon back to the mouth of Los Portales.

Los Portales climbs 1.8 miles from Fresno Canyon to the Solitario rim, where the drainage intersects Outer Loop Trail. The drainage passes through a precipitous slickrock entrance of fossil-filled, white Buda limestone, and canyon walls of uplifted Santa Elena and Del Carmen limestone (Hall and Hunt 2015). Tinajas are common, flowing water less so. Rock scrambling is a requisite and footholds are a premium.

Two small shrubs, narrowleaf fendlerbush **154**, and fascicled bluet (*Arcytophyllum fasciculatum*), with stiff branches and white, four-lobed blooms, are canyon fixtures. A scarce woody subshrub, smallflower milkwort **219**, makes appearances in the canyon. Turner's thistle **37**, juniper globemallow **179**, and shortfruit evening primrose (*Oenothera brachycarpa*), with yellow blooms and long-petioled leaves, are difficult to miss. Rock crevices support prickly lip fern **230**, villous lip fern **231**, and scaled cloak fern **233**.

Fresno Peak

Southeast of Los Portales, 2.7 miles west of Lower Shutup Trailhead, Outer Loop Trail wraps around the northeast foothills of massive Fresno Peak. Composed of Del Carmen limestone, Fresno Peak (5,120 ft.), the second-highest summit in the park, rises 1,400 feet above Fresno Canyon and 500 feet above the central basin. The Flatirons to the west, below Fresno Peak, are composed of younger Santa Elena limestone (Henry 1998). Few have reached the top of Fresno Peak, but a monumental cairn marks the summit.

Fresno Peak is another site of bushy wild buckwheat **221**, a dwarf shrub once considered rare. Another resident shrub is Wright's snakeroot (*Ageratina wrightii*), with white flower heads and long styles. More widespread in the Solitario is the small shrub, narrowleaf fendlerbush **154**. A special pleasure to find and inspect is seldom seen Cochise beardtongue **205**, with artful blue tubular blooms.

Needle Peak

Just north of the trail is Needle Peak (4,608 ft.), 1.2 miles southwest of Lower Shutup Trailhead. The peak is a dark rhyolite intrusion through Tesnus Formation shales (Henry 1996). Nearby is Trans-Pecos ayenia (*Ayenia filiformis*), an erect subshrub with umbrellalike maroon blooms; pale false nightshade (*Chamaesaracha pallida*), a perennial with wavy, lance-shaped leaves and yellow flowers; and gray five eyes (*Chamaesaracha coniodes*), also a false nightshade, with crinkly oblong leaves and yellow blooms.

Righthand Shutup

The 2.2-mile Righthand Shutup drains the Solitario's northwest interior through the west rim. The canyon is open at times, and constricted, with limestone walls to 600 feet high. Tinajas are common, flowing water uncommon. An 80-foot pour-off and drop-offs must be negotiated.

Ridges above the shutup entrance are capped with Caballos Novaculite. At the shutup head, rhyolite rim

Uptilted columnar rock outcrop near Righthand Shutup entrance

sills intrude between Cretaceous rocks. The canyon's descent is a tour through geological time, from the oldest rocks on the Solitario rim, to the youngest at the shutup mouth on Fresno Canyon. Cretaceous rocks encountered, from oldest to youngest, are the Yucca, Glen Rose, Del Carmen, Santa Elena, and Boquillas formations. Marine fossils are visible in the bedrock and cliff walls (Hall and Hunt 2015).

Righthand Shutup Route: Characteristic shrubs are sandpaper oak **147**, toothed serviceberry **245**, desert myrtlecroton (*Bernardia obovata*), fascicled bluet (*Arcytophyllum fasciculatum*), leafy rosemary-mint **163**, damianita (*Chrysactinia mexicana*), and mejorana (*Lantana achyranthifolia*). Mejorana is a low aromatic shrub with hairy leaves and pinkish flower clusters. Notable subshrubs are autumn sage **164**, with two-lipped scarlet blooms; and dwarf ayenia **175**, with maroon petals with slender recurved claws. Perennials, easy to spot in bloom, are velvet bundleflower **134**; bearded dalea (*Dalea pogonathera* var. *pogonathera*), with spikes of purple, butterfly-shaped flowers with feathery calyx lobes; white milkwort (*Polygala alba*), with white blooms with yellow-green centers at the end of long, leafy stems; and hillside vervain (*Verbena hirtella*), with rough incised leaves and slender spikes of hairy lavender flowers.

Solitario rim above the Righthand Shutup: The steep slopes host a wealth of limestone-loving plants, partly protected by how difficult they are to reach. Notable shrubs include downy wolfberry **259**, desert myrtlecroton (*Bernardia obovata*), Big Bend silverleaf **253**, and Trans-Pecos croton **117**. Subshrubs with intricate floral architecture are hyssopleaf asphead **173**; smallflower milkwort **219**; and dense ayenia (*Ayenia microphylla*), with dentate, felty leaves, canopy-like maroon blooms, and tuberculate fruits.

Attractive perennials on the lower slopes include juniper globemallow **179**, spearleaf **21**, and shaggy stenandrium (*Stenandrium barbatum*), with mounds of shaggy leaves and white-streaked pink blooms. Havard's wild buckwheat **220** and downy paintbrush **197** occupy sparse upper slopes. Buckley's centaury **151** and living rock cactus **66** grow in eroded pockets at the crest. Scarce ferns, not previously reported in the park, are Gregg's cloak fern **235** and scaled cloak fern **233**.

Interior slopes north of the Righthand Shutup: The prize of these slopes is mat rock spiraea **246**, which is scattered on peaks in the western states. The shrub is a few inches high, with stubby gnarled branchlets and spikes of white blooms with protruding stamens. Jeff Clark, graduate student at Sul Ross, discovered the shrub in BBR in 1984. I found this dwarf only after months of searching. Other intriguing residents of these remote slopes are hyssopleaf asphead **173** and Bear Mountain milkweed **16**.

Solitario Road East

South of Paso al Solitario, Solitario Road splits into two branches. The left fork, Solitario Road East, runs 2.5 miles southeast to an unmarked crossing of the Lefthand Shutup, and south for another mile to Tres Papalotes. From Tres Papalotes, the 4WD road proceeds southwest 2.7 miles to Lower Shutup Trailhead.

Lefthand Shutup

The shutup is "Lefthand" because it is the leftmost shutup from Solitario Road. Lefthand Shutup Creek rises southwest of Burnt Camp Trailhead on the west side of the central basin and wraps around the north edge of the basin to a point 1 mile north of Tres Papalotes. The creek then veers through the northeast rim for 3.0 miles before exiting the park and emptying into Terlingua Creek.

From Solitario Road East, the shutup drainage at first is an open, gravelly, dry creek bed that winds through a private inholding before entering the high-walled shutup canyon. The inholding is home to Solitario Bar, perched on the drainage edge. Built by an eccentric with a fine sense of humor, the bar is a metal shack with welcoming signs and junk-store-like accoutrements: a makeshift bar, bathtub, and mermaid. A video camera documents visitor reactions online, but the website is no longer in operation.

An imposing rock display dominates the canyon entrance. The rock is rim sill, exposed elsewhere, but here

Tilted columnar jointing near Lefthand Shutup entrance

strongly tilted, with conspicuous columnar jointing. The sill is a thin rhyolite layer intruded into Cretaceous Shutup Conglomerate (Henry 1998).

This shutup is the easiest hike of four canyons draining the Solitario, open in places, with interesting side canyons, but also congested. Flash floods left boulders in the creek bed. Canyon walls tower to 600 feet above the creek (Deal 1976a), exposing Cretaceous rocks deposited when the area was a shallow sea (Corry et al. 1990).

Shrubs fill the creek banks and reach into side canyons. Notable are Texas persimmon **113**, Gregg's ash (*Fraxinus greggii*), sandpaper oak **147**, spiny greasebush **107**, white ratany **162**, leafy rosemary-mint **163**, and Osage orange (*Maclura pomifera*). Gregg's ash possesses leaves with up to seven leaflets, and petal-less flowers. Osage orange is conspicuous at the shutup entrance, with wrinkled, orange-size fruits.

Turner's thistle **37**, Chisos pricklypoppy **198**, and rock flax **167** are eye-catching perennials. Ferns include small leaf silver fern (*Argyrochosma microphylla*), with blue-green fronds and wide-spaced pinnules; villous lip fern **231**, with overlapping scales obscuring the undersides; and Arizona spikemoss (*Selaginella arizonica*), which forms prostrate mats of long creeping stems with upper-side leaves green, under-side leaves tan, and bristle-tips short or broken off. In 1975, on the first major study of Solitario vegetation, Mary Butterwick discovered Fendler's lip fern (*Myriopteris fendleri*) in a shaded side canyon. This fern, with beadlike pinnules, has not been relocated in the park.

Tres Papalotes

One mile south of the Lefthand Shutup crossing, Tres Papalotes rests on the east edge of the Solitario's central basin, west of Blue Range, in a valley eroded from Tesnus Formation shales. Ridges above Tres Papalotes are composed of Caballos Novaculite and green chert of the Maravillas Formation. The former mining camp and pumping station is equipped with a reliable water supply, composting toilets, dilapidated cabin, and defunct windmill. A pumphouse pumps water from a Tesnus Formation reservoir.

Tres Papalotes is an ideal base camp for hikes on the Road to Nowhere, Lower Shutup, Inner and Outer

Colorful Tesnus shale displays are common near Tres Papalotes

Loop trails, or visits to nearby prospect mines. Perennial herbs grow on gravelly flats about Tres Papalotes: the prostrate ashy sandmat **121**; yellow flameflower **267**, on open ground; and a trailing mallow, spearleaf sida (*Rhynchosida physocalyx*), with toothed leaves, yellow flowers, and bladdery fruits. Two attractive perennials often grow south of Tres Papalotes: a silky-haired legume, velvet bundleflower **134**, with small white flower heads and reddish pods; and Hopi tea greenthread **47**, with long-stalked flower heads, usually with disk but no ray florets.

Road to Nowhere

South of Tres Papalotes, the 4WD Road to Nowhere, a narrow prospect road, meanders west to a dead end in the central Solitario. In the first mile, the road climbs a novaculite ridge overlooking Tres Papalotes and reaches a saddle with bedrock mortars and picturesque views. The road then curves along steep slopes 1.4 miles and stops (Riskind and Sholly 2009). A wonderful walk and adventuresome drive, the road encourages attention to small herbs like squareseed spurge **123**, perhaps the most attractive BBR spurge, and spurca spurge (*Euphorbia spurca*). Like Terlingua spurge (*Euphorbia theriaca*), spurca spurge is prostrate and hairless but has stalked glands with petal-like appendages. Skeletonleaf goldeneye (*Sidneya tenuifolia*), a resinous shrub with threadlike leaves and long-stalked single flower heads with yellow ray and disk flowers, is roadside.

Manganese and Uranium Mining Spur Road

Just 1.3 miles south of Tres Papalotes, a 4WD spur leads east to the ruins of undeveloped prospect mines dating to the Cold War, when strategic metals prices were high. The spur soon forks. The left fork runs north 0.5 miles, with tight switchbacks, to a manganese prospect, a horizontal shaft. The right fork travels southeast 0.9 miles to a steep ridgetop, where a uranium prospect shaft is covered with wire mesh (Riskind and Sholly 2009). Prospecting for gold, mercury, and copper also took place in the Solitario.

Lower Shutup

Lower Shutup Trailhead, also the east trailhead for the Inner and Outer Loop Trails, is 2.7 miles south of Tres Papalotes. Lower Shutup Creek rises in the Solitario's central basin, drains the south-central interior through the south rim, and empties into Fresno Creek at Wax Factory Laccolith. One of the state's more spectacular canyons, the shutup is steep, dropping from 4,254 feet at the trailhead to 3,428 feet on exiting the Solitario's outer rim. Ultimately, the creek descends to Fresno Canyon, at an elevation of 2,796 feet. Due to the steep descent, the canyon has more and larger tinajas than the other shutups.

From the trailhead, the canyon runs south 4.5 miles to a cairn, where hikers exit the shutup and follow a trail southwest 0.8 miles to Government Road (see "Physiographic Region 6"). The road then travels northwest 2.7 miles to Rincon 2 off Fresno Canyon. Often, hikers park a car at Rincon 2 and drive a second car to the Lower Shutup Trailhead. The hike through the shutup and back to Rincon 2 is 8.0 miles one way. The shutup is hiked in winter or early spring to avoid flash floods and high temperatures.

At first the shutup is an open sandy-gravelly wash with a few trickles of water and shallow pools, but the walls soon close in. In places, the shutup becomes a tight, snaking slickrock passage, a rather eerie but also awe-inspiring enclosure of sculptured rock. Short pour-offs must be negotiated, and a steep drop-off bypassed. The canyon walls rise as much as 750 feet above the canyon floor (Deal 1976d). Beginning in Tesnus Formation shales, the shutup descends through Cretaceous rocks from oldest to youngest.

In 2012, I encountered a guajillo (*Senegalia berlandieri*), not previously reported at BBR, in the canyon. The shrub, a large legume with prickly branches, compound leaves, creamy-white round flower heads, and velvety pods, has not been relocated. My most surprising plant encounter was date palm **24**, a single small palm surviving in the creek bed, within confining canyon walls.

The rock-crevice-loving subshrub, smallflower milkwort **219** is often sighted in the Lower Shutup.

A characteristic Lower Shutup passage

Growing near the trailhead, threeflower goldenweed (*Xylothamia triantha*) is a small shrub with resinous flower heads, usually with three yellow disk florets. Another aster, shorthorn jefea (*Jefea brevifolia*), a rough-haired low shrub with long-stalked yellow flower heads, is common.

The Lower Shutup is preferred habitat for many perennials, including groovestem bouchea **273**, Trans-Pecos false boneset **36**, Hartweg's evening primrose (*Oenothera hartwegii* ssp. *hartwegii*), rose bladderpod **61**, and shrubby Arizona goldenaster (*Heterotheca arizonica*), with glandular, abruptly tipped leaves and yellow flower heads. Havard's nama (*Nama havardii*), an erect, leafy, much branched annual with prolific, usually pinkish blooms, brightens the creek banks.

Los Alamos Area: Los Alamos

Los Alamos, a historic ranch headquarters, now a park residence, overlooks the South Fork of Alamo de Cesario Creek, which supplies surface water and groundwater (Sanchez 1999). The land was awarded to Galveston, Harrisburg & San Antonio Railway in 1877. Los Alamos is likely the site of Los Alamos de Cesario Ranch, owned and operated by H. W. McGuirk from 1895 to 1897 (Smith-Savage 1996a).

Los Alamos was owned by Clay Holland and Gay Meriwether from 1941 to 1951, when it was purchased by the Fowlkes brothers. Edwin Fowlkes and his wife Frankie moved from Botella to live at Los Alamos (Smith-Savage 1996b). The ranch included a residence, outbuildings, corrals, pond, two water storage tanks, and a spring-fed pipeline network (Sanchez 1999).

Desert sumac (*Rhus microphylla*) is prominent near the Los Alamos turnoff from Sauceda Road, and north to the park boundary. Heath carlowrightia **1**, a low subshrub, grows in the pasture near the residence. Annuals, including the geranium, redstem stork's bill **152**, and the phlox, paleflower ipomopsis **217**, spring up in the pasture after rains. The freshwater limestone outcrop at the north end of Jackson Pens Trail attracts its own collection of plants, including catclaw cactus **82**.

Three plants, scarce in the park, were previously reported near Los Alamos but have not been relocated: escobilla butterflybush (*Buddleja scordioides*), an odd shrub with matted, rust-tipped hairs; and two annuals, whitewhisker fiddleleaf (*Nama undulatum*), with clasping, wavy oblong leaves, and shaggy-haired pearly globe amaranth (*Gomphrena nitida*), with spikes of pinkish-white blooms.

Los Alamos Area: South Fork, Alamo de Cesario Creek

Rising south of Terneros Creek and east of Sauceda Road, the South Fork runs east 2.8 miles, north of and parallel to Los Alamos Road, exits the park, and flows into Alamo de Cesario Creek. A trail from Los Alamos leads north to the South Fork. To the west is a small spring and basaltic rock; to the east, nonmarine deposits, alluvium, and Tascotal Formation sedimentary rock (Henry 1998, geologic map). Little walnut **158**, buttonbush **249**, desert sumac (*Rhus microphylla*), splitleaf brickellbush (*Brickellia laciniata*), Arizona cottonwood **250**, western soapberry **252**, seepwillow **35**, and western honey mesquite **139** line the drainage.

Los Alamos Area: Los Alamos Loop

From Paso al Solitario, the 4WD Los Alamos Loop travels north 3.6 miles to a fork. The west fork runs 1.9 miles to Los Alamos. The east fork traverses eroded terrain 1.4 miles and dead ends at a bluff with views to Devil's Graveyard badlands beyond the park. Panther Canyon flows northeast below the bluff into the badlands. From Los Alamos, the loop runs south 2 miles, past Los Alamos Campsite, to Solitario Road. Paso al Solitario is 2.2 miles southeast on Solitario Road.

Look for Isely's fairy duster **133**, a dwarf shrub with curly pods; eagle-claw cactus **70**, with magenta flowers blooming in unison after rains; pointed sandmat **119**, with spine-tipped leaves; glandleaf dalea (*Dalea lachnostachys*), with woolly flower spikes; and pale false nightshade (*Chamaesaracha pallida*), with yellow, wheel-shaped blooms. Bushy wild buckwheat **221** resides in limestone outcrops.

Terlingua Uplift

Terlingua Uplift is a 100-square-mile rectangular plateau, no more than 4,500 feet high, the uplifted portion of a fold, the Fresno-Terlingua Monocline.

View of Contrabando lowlands and Mexican mountains from Terlingua Uplift rim

View north from rim of Saltgrass Draw

The monocline formed during a mountain-building period 50 Ma, which created a "foldbelt" of Cretaceous mountains from Canada to Mexico (Henry 1998). The monocline extends from the southwest Solitario southeast to Fresno Mine, and east 18 miles into BBNP. A steep escarpment divides the uplift from the Contrabando Lowlands, the downthrust part of the fold. The uplift is overlain by Santa Elena limestone; the Lowlands by more easily eroded Cretaceous rocks (Henry 1998).

There are no established trails on Terlingua Uplift, only a few road traces. Some limited access to, and views of Terlingua Uplift, are provided off Buena Suerte Trail and Contrabandista Spur Trail in the Contrabando Lowlands. See the discussion of these trails in "Physiographic Region 6."

The flora is a mix of Contrabando Lowlands and Solitario species. On the slopes of Terlingua Uplift, heath cliffrose **247**, with showy white flowers, is eye-catching. Roemer's acacia (*Senegalia roemeriana*), woolly butterflybush (*Buddleja marrubiifolia*), and downy wolfberry **259** are common. Colony-forming Texas false agave **65** is abundant. Stinging cevallia (*Cevallia sinuata*), Parry's holdback (*Pomaria melanosticta*), Powell's heliotrope **54**, and Warnock's justicia **2** are intriguing subshrubs.

Noteworthy perennials include creeping rockvine (*Bonamia repens*), Texas crinklemat **58**, glandleaf milkwort (*Polygala macradenia*), purple groundcherry (*Quincula lobata*), Big Bend ringstem **188**, and stinkweed (*Tetraclea coulteri*). Creeping rockvine is a morning glory with trailing stems, silky leaves, and white blooms. The scarce Gregg's cloak fern **235** resides on the uplift's steep slopes. Stem cacti, such as Potts' mammillaria **84**, Texas rainbow **72**, living rock **66**, strawberry cactus **73**, silverlace cactus **79**, and sea urchin cactus (*Coryphantha echinus* var. *echinus*), grow on the slopes. Blind prickly pear (*Opuntia rufida*) and Engelmann's prickly pear **89** are also common.

Masada Ridge / Saltgrass Draw

Masada Ridge Wilderness, accessed from the east via Terlingua Ranch roads, is a high plateau (3,700 ft. to 4,600 ft.) on the park's east boundary, on the southeastern flank of the Solitario. In 2000, this remote property was donated to TPWD by Julius Dieckert, retired Texas A&M University professor, and his wife Marilyne. The Dieckerts named the ridge for the Masada in Israel, an ancient fortress on an isolated plateau by the Dead Sea (Klepper 2007). A faint jeep road can be used for hiking. Saltgrass Draw, which drains the Blue Range, flows northeast through upper Masada Ridge and eventually empties into Terlingua Creek. The draw resembles the Lefthand Shutup to the north. Consult park rangers about access.

On Masada Ridge and in Saltgrass Draw, gray oak **144**, sandpaper oak **147**, red berry juniper **109**, and Texas persimmon **113** are the largest shrubs. Small shrubs include leafy rosemary-mint **163**, spiny greasebush **107**, and fascicled bluet (*Arcytophyllum fasciculatum*). Memorable subshrubs are Powell's heliotrope **54**, autumn sage **164**, and baccharisleaf penstemon (*Penstemon baccharifolius*). Texas cone cactus **85** is plentiful throughout Masada Ridge. In Saltgrass Draw, a perennial, juniper globemallow **179**, grows beneath shrubs; blind prickly pear (*Opuntia rufida*) decorates canyon walls; and slender lip fern (*Myriopteris gracilis*), with lance-shaped blades and beadlike pinnules, emerges from shady crevices.

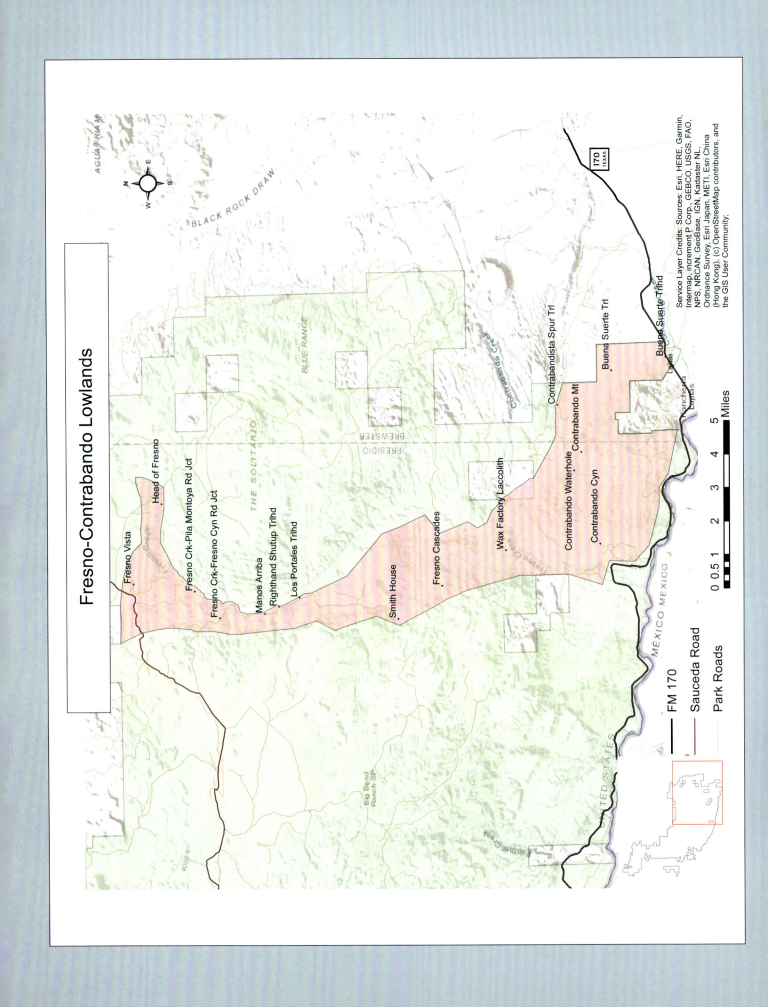

PHYSIOGRAPHIC REGION 6
Fresno Canyon–Contrabando Lowlands

Fresno Canyon–Contrabando Lowlands is dominated by low hills and flats, with easily eroded limestone and clay soils, high temperatures, limited rainfall, and sparse vegetation. Elevations range from 2,400 feet on lower Contrabando Creek and at East Contrabando trailhead, to 4,500 feet at the head of Fresno Canyon. The Contrabando Lowlands subregion is bounded by Fresno Canyon to the west, the southwest corner of the Solitario and Terlingua Uplift to the north, park boundary on the east, and the Rio Grande to the south. Fresno and Contrabando Creeks, and the West Fork of Comanche Creek, drain the region. Contrabando Dome and Wax Factory Laccolith, igneous intrusions into Boquillas limestone, and Lajitas Mesa, a product of eruptions in BBNP, Mexico, and the Bofecillos Mountains, rise above the low expanse.

The area's human history includes ranching, quicksilver mining, small-scale farming, candelilla gathering and processing, smuggling, and military activity. In Fresno Canyon, Crawford-Smith House was an early ranching, wax-making operation, and way station on Marfa-Terlingua Road (TPWD 2009). The remains of Fresno and Whit-Roy quicksilver mines, and miners' quarters at Buena Suerte ghost town, rest on a bluff east of Fresno Canyon. Candelilla wax factory remnants are evident on the Buena Suerte Trail and near Wax Factory Laccolith. Mercury prospect mines are attractions on Contrabando Dome Trail.

The roads themselves are historical reminders. Constructed in the late 1890s, Marfa-Terlingua Road served as a freight and stage route between Marfa and quicksilver mines west of present-day Terlingua. Government Road, built by the US Army about 1916, supported cavalry troops stationed on the river in the turbulent times of Pancho Villa and the first use of army trucks.

Contrabando Lowlands' 25-mile network of multiuse trails provides access to natural resources and glimpses into history.

Before this human history, prehistoric peoples used Fresno Canyon as a passageway through rugged mountainous terrain. Crawford-Smith Ranch was a prehistoric campsite, with human use dating back thousands of years. A rock art site, Manos Arriba, is a popular attraction on Fresno Canyon Road.

Fresno Canyon Area

Fresno Creek rises in the northwest Solitario. A road trace departs Solitario Road below the Los Alamos turnoff and leads south to the head of Fresno Creek near Tanque Blanco. One can follow the creek southwest to a steep pour-off and escape humankind. The creek continues southwest, where it is joined by Pila Montoya Trail and Fresno Canyon Road below iconic La Posta, a knoblike mountain. The creek then runs south 11 miles to Wax Factory Laccolith and a junction with Lower Shutup Creek. At first, the creek is bounded by the Bofecillos Mountains on the west and the Solitario on the east. From Wax Factory Laccolith to the Rio Grande, the creek divides the Bofecillos Mountains from the Contrabando Lowlands.

Fresno Creek below West Contrabando Trailhead

A pleasant short hike begins at Fresno Creek Trailhead and runs south 1.9 miles to the river, with views of Mexican mountains and Fresno Ranch to the south, and Lajitas Mesa to the east. The creek crosses alluvial flats, marine clay deposits, and Chisos Formation outcrops.

Lower Fresno Canyon

A few unusual plants grow along the creek, some at the limits of their range. For example, Ferriss' scouring rush **115** is usually found at higher elevations. Other uncommon species include longcapsule suncup **191** and narrowleaf moonpod **185**, with orangish blooms and winged fruits.

Velvet ash **190**, Texas mimosa **136**, little walnut **158**, catclaw acacia (*Senegalia greggii*), buttonbush **249**, and beebrush **271** are characteristic of this habitat. Jimmyweed (*Isocoma pluriflora*), with narrow leaves and yellow flower heads, is plentiful in sand, silt, and clay. Potts' mammillaria cactus **84** and catclaw cactus **82** grow on the canyon rim.

After rains, a perennial, alkali heliotrope (*Heliotropium curassavicum*), and annuals, such as southern annual saltmarsh aster (*Symphyotrichum divaricatum*), seaside petunia (*Calibrachoa parviflora*), woollyflower spurge (*Euphorbia eriantha*), and desert teucrium **166** spring into bloom. The saltmarsh aster, a much branched annual to 3 feet high with lavender ray florets, grows in the creek bed. On Rio Grande riverbanks, alkali heliotrope develops coiled spikes of funnel-shaped blooms. Prostrate seaside petunia produces plump, fleshy leaves and violet/pink flowers. Woollyflower spurge forms flower heads with straplike appendages over the glands and dangling, ribbed fruits.

Fresno Creek from Trailhead to Wax Factory Laccolith

The hike up Fresno Creek to Wax Factory Laccolith and Buena Suerte Trail is 4.5 miles. The creek runs through alluvium, Pen Formation clay, and Chisos Formation tuff to a low slickrock pour-off, with Mesa de Procton to the west and Contrabando Dome to the east. The creek then travels through rocky terrain, magma intruded by Wax Factory Laccolith, exposed along the creek, sometimes capped by limestone.

At springs, Arizona cottonwood **250** is by far the largest tree. Smaller shrubs growing south of the trailhead, velvet ash **190**, catclaw acacia (*Senegalia greggii*), buttonbush **249**, and beebrush **271**, also occupy this stretch of the canyon. Other shrubs, catclaw mimosa (*Mimosa aculeaticarpa* var. *biuncifera*), downy wolfberry

259, and viscid acacia (*Vachellia vernicosa*) are common. Tree tobacco **261**, with large blue-green leaves and yellow long-tubed blooms, is prevalent.

Perennials include narrowleaf globemallow **178**; Big Bend ringstem **188**; Big Bend croton (*Croton bigbendensis*), with scaly lance-shaped leaves, and male and female blooms on separate plants; Arizona spurge (*Euphorbia arizonica*), with floral cups with tiny red glands and large petal-like appendages; and yellow nut-grass (*Cyperus esculentus* var. *macrostachyus*) and fragrant flatsedge (*Cyperus odoratus*). Yellow nut-grass is a stoloniferous perennial with three-angled stems (culms) to 2 feet high, several flat or V-shaped leaves, loose inflorescences with up to twenty slender linear spikelets, and large, golden yellow floral scales. Fragrant flatsedge has long V-shaped or troughlike leaves and much-branched inflorescences in a bottlebrush arrangement, with up to sixty cylindric spikelets.

The creek supports opportunistic annuals: Trans-Pecos senna (*Senna pilosior*), with bifoliate leaves, yellow flowers, and erect pods; Terlingua milkvetch **131**, with lavender-white blooms; spring-blooming whitlowwort (*Tomostima cuneifolia*), with basal leaves and stalks of four-petal white blooms; Mexican cryptantha (*Cryptantha mexicana*), a borage with long narrow leaves and spring-blooming white flowers; and hairless, prostrate Terlingua spurge (*Euphorbia theriaca*), without petal-like appendages.

Noteworthy vines are old man's beard **239**, with four-sepal white flowers and feathery fruits; and spearleaf **21**, with brownish-green flowers. Blind prickly pear (*Opuntia rufida*), catclaw cactus **82**, Potts' mammillaria cactus **84**, and tree cholla (*Cylindropuntia imbricata* var. *imbricata*) are resident cacti.

Fresno and Whit-Roy Mines

Fresno Mine and Buena Suerte ghost town, private inholdings in BBR, are located at the 7.0-mile point on Buena Suerte Trail. At Fresno and Whit-Roy mines, evidence of mining activity—old buildings, rusted mining equipment, the Buena Suerte Flotation Mill—abounds. Harris Smith, owner of Crawford-Smith House in Fresno Canyon, and Homer Wilson, owner of

Buena Suerte ghost town

Wilson Ranch in BBNP, discovered mercury in 1935 and opened Fresno Mine in 1938. The mine was the largest mercury producer during World War II. Due to mercury price increases, Whit-Roy Mine was opened in 1966, and Anchor Mining operated Fresno and Whit-Roy mines through 1971 (Deal 1976b; Daugherty 1972).

Wax Factory Laccolith

From Whit-Roy Mine, Wax Factory Laccolith looms above Fresno Canyon. Two miles wide and 150 feet thick, the laccolith is a dark volcanic outcrop that intruded into and domed light-colored Boquillas Limestone (Deal 1976b). Ruins of a candelilla wax-making factory are just south in Fresno Canyon.

Government Road Trail

Government Road, built by the US Army about 1916, provided a more reliable truck route than the flood-prone route down Fresno Canyon. The trail begins at the mouth of Lower Shutup Creek, by Wax Factory Laccolith, and runs northwest 5.9 miles to Rincon 2 by Fresno Canyon. The trail follows the Lower Shutup to a junction with a crossover trail running northwest to Fresno Canyon. From here, the Lower Shutup veers northeast, and Government Road Trail continues north, past Chimney Rock, to another junction. A trail from the Lower Shutup joins the road, and the road proceeds northwest to Rincon 2. The route, through Boquillas limestone, offers good views of the Flatirons.

The park's most common beeblossom, scarlet beeblossom (*Oenothera suffrutescens*), is plentiful in the Lower Shutup drainage. This erect perennial has pinkish-white flowers with four paddlelike petals, and short pyramidal fruits. Graham's dog cholla (*Corynopuntia grahamii*), with lemon yellow flowers, is present here and to the west. This cholla resembles clumped dog cholla **67**, but the clublike joints are smaller and easily detached, and new growth occurs near the tips of the previous year's joints.

Fresno Canyon from Wax Factory Laccolith to Pila Montoya Trail

From Wax Factory Laccolith to Pila Montoya Trailhead, vegetation becomes denser and lusher. More springs supply water, in shallow, trickling pools. Arroyo Primero and Arroyo Mexicano bring water from the Bofecillos Mountains, and the Righthand Shutup supplies runoff from the Solitario.

From Wax Factory Laccolith, the trail through Fresno Canyon runs northwest 2.2 miles to the mouth of Arroyo Primero. A spring, harboring a colony of horsetail fern, Ferriss' scouring rush **115**, and the spectacular buttercup, golden columbine **238**, flows at the arroyo mouth. Fresno Cascades, 0.3 miles ahead in Fresno Canyon, is a low slickrock pour-off, much appreciated by Native Americans, who left rock-shelters and bedrock mortars. In 0.2 miles, the trail reaches a junction with Arroyo Primero Trail, which leads west to Madrid House, and up Arroyo Primero to Chorro Canyon and beyond.

From the Arroyo Primero Trail junction, Fresno Canyon runs north 1.9 miles to Crawford-Smith House, where Fresno Canyon Road, coming from the north, ends. Native Americans made camp here, taking advantage of the water resources. In the 1890s, Marfa-Terlingua Road, running through Fresno Canyon, was constructed to transport goods between Terlingua mercury mines and the railroad in Marfa.

From 1914 to 1930, James Crawford and family owned and operated the Crawford-Smith property. Crawford constructed an adobe house, orchard and garden, reservoir, and pipeline transporting water from Fresno Creek and Smith Spring. Road travelers stopped to rest at the ranch. From 1915 to 1920, US Army pack trains used the road to supply military posts on the Rio Grande. Crawford formed Fresno Wax Factory, and in World War I, the wax was used to waterproof tents. Harris Smith bought the ranch in 1930, raised goats, sold mohair, and partnered with Homer Wilson in mercury mining. After their Fresno Mine closed in 1944, Smith sold out and left the Big Bend (Smith-Savage 1996b; TPWD 2009).

From Crawford-Smith House, the road runs north 1.3 miles to Arroyo Mexicano, which joins Fresno Canyon from the west. Massive Rincon Mountain (4,400 ft.) rises at the junction's southwest corner. To the north, on a side road, are Rincon 1 and 2. These campsites are focal points for hikes in the Lower Shutup or Arroyo Mexicano, or exploration of Crawford-Smith ruins. North of the Rincon turnoff, on a bank east of the road, is Fresno Canyon Campsite, suitable for equestrians, with hitching rails and good views.

The road continues north 2.5 miles to Righthand Shutup Trailhead. Along the way, shelter caves on the Solitario rim mark the entrance to 1.8-mile Los Portales shutup, which climbs to the Solitario rim. The Righthand Shutup climbs 2.2 miles to a junction with Outer Loop Trail. North of Righthand Shutup Trailhead is Manos Arriba, a rock art site beside Fresno Canyon Road. The art, depicting red, black, and white hands, is painted on the ceiling of a limestone overhang. A kiosk provides background. Two miles north of Manos Arriba, the road reaches a junction with Pila Montoya Trail.

In this stretch of the canyon, cottonwood groves, higher canyon walls, and shadier tree canopies protect a broader mix of plants. Some shrubs common in the drier lower canyon, such as fourwing saltbush (*Atriplex canescens*), with yellow blooms and four-winged fruiting bracts, continue here. However, many shrubs seldom seen in the lower canyon, such as autumn sage **164**, begin to make appearances.

Many perennials frequent this stretch of Fresno Canyon. Torrey's tievine **103** twines on larger shrubs. Slimleaf bean (*Phaseolus angustissimus*), with trifoliate leaves and pink, pealike blooms, occupies creek shoulders. Limerock brookweed (*Samolus ebracteatus* var. *cuneatus*), with large basal leaves and white flowers on naked stalks, is resident at seeps. Shortfruit evening primrose (*Oenothera brachycarpa*), with a basal rosette of leaves and yellow flowers, prefers stream banks. Captivating annuals include white-flowered Kunth's evening primrose (*Oenothera kunthiana*), bristly nama **155**, and flatspine stickseed (*Lappula occidentalis*), with bluish-white flowers and bizarre fruits with conical white spines.

Fresno Canyon from Pila Montoya Trail Sign to Canyon Head

From Fresno Canyon, Pila Montoya Trail climbs northeast along Fresno Creek, then leaves the creek and climbs north-northwest to Pila Montoya 3. The trail follows Pila Montoya Road to Pila Montoya Trailhead on Sauceda Road. From the 1.2-mile point on Pila Montoya Trail, Fresno Creek runs northeast, below

Boquillas limestone outcrop on Pila Montoya Trail

Fresno Vista Campsite to a steep pour-off. The creek then winds east to Tanque Blanco and the canyon head. Upper Fresno Canyon is usually visited from Solitario Road. Just southeast of the Los Alamos 4WD Loop junction, a road trace leads south from Solitario Road to Tanque Blanco.

Rio Grande saddlebush **94** stands out, with sandpaper-like leaves and white flowers. Beargrass **29** and whitethorn (*Vachellia constricta*), with thorns in pairs at nodes and golden yellow flower heads, grow along the canyon bottom. Lemmon's brickellbush (*Brickellia lemmonii* var. *lemmonii*), with cylindric flower heads with greenish-yellow disk florets, occurs at the pour-off below the canyon head. James's nailwort (*Paronychia jamesii*), with yellow bristle-tipped sepals, grows on rocks near Tanque Blanco.

Pila Montoya Trail Sign to La Posta Campsite and Sauceda Road

From the Pila Montoya Trail sign, Fresno Canyon Road climbs steep Fresno Hill and passes La Posta (4,630 ft.), reaching La Posta Campsite in 2.4 miles and Sauceda Road in 2.9 miles. La Posta, at the eastern limit of the Bofecillos Mountains, is an outpost on the border of the Llano and Fresno Canyon, with exquisite views. La Posta Campsite is a convenient base camp for Fresno Canyon visitors. Tree cholla (*Cylindropuntia imbricata* var. *imbricata*), with spiny joints, magenta blooms, and knob-like yellow fruits, is conspicuous.

Contrabando Lowlands / Multiuse Trail System

Contrabando Lowlands is the hottest, driest region in the park, with relatively scant vegetation. The Lowlands extends to Terlingua Uplift on the north, Fresno Canyon on the west, the Rio Grande on the south, and the park boundary on the east, near FM 170. A 400-feet-high escarpment, along the Fresno-Terlingua Monocline, divides the more eroded, Upper Cretaceous, clay-rich rocks of Contrabando Lowlands from the resistant Lower Cretaceous, Santa Elena limestone on Terlingua Uplift (Henry 1998).

The Spanish words *contrabando* and *contrabandista* have been used since the early 1900s to describe features of Contrabando Lowlands. Goods, especially livestock, wax, and liquor, were smuggled to and from Mexico, on Contrabando Creek, to avoid customs duties. In the early 1900s, Texas ranchers added Mexican livestock to their herds, some brought in unlawfully. In the 1930s, Mexican wax made from candelilla **120** was smuggled across the border, and Mexican smugglers sold liquor to Texas bootleggers. Liquor made from sotol was smuggled in pig bladders (Alloway 1995).

Chihuahuan Desert scrub vegetation is typical of this arid region, and creosote **280** is dominant. The desert is made even sparser by overgrazing. The 25-mile trail complex, with historic waypoints, supports hiking, mountain biking, and horseback riding.

West Contrabando Trail

West Contrabando Trailhead, off Highway 170, provides access to Fresno Canyon hiking, and to the West Contrabando Trail. The 4.7-mile West Contrabando Trail links to the 3.2-mile Fresno Divide Trail, the 8.3-mile Buena Suerte Trail (discussed further on), once a leg on the Marfa-Lajitas stage route, and the 4.5-mile Contrabando Dome spur, which offers insight into Contrabando Dome's geological history.

Thompson's yucca (*Yucca thompsoniana*) and soaptree yucca **30** may be the tallest plants on the trail. Thompson's yucca lacks the filamentous leaf margins of soaptree yucca. Shrubs include downy wolfberry **259**, Schott's acacia **143**, allthorn **160**, and viscid acacia (*Vachellia vernicosa*). Armed with straight, paired white thorns and sticky varnish, viscid acacia forms round yellow flower heads. Gyp daisy **51** and narrowleaf moonpod **185** are smaller subshrubs.

This is a cactus lovers trail, with living rock **66**, glory of Texas cactus **91**, Texas rainbow **72**, Potts' mammillaria **84**, Warnock's cactus (*Echinomastus warnockii*), golf ball cactus (*Mammillaria lasiacantha*), and clumped dog cholla **67**. With a waxy stem and chalky-blue spines, Warnock's cactus produces early-blooming white flowers. Golf ball cactus is golf ball size, with tiny, flexible white spines and early spring white blooms with red mid-stripes.

Perennials, such as spreading sida (*Sida abutilifolia*), pink windmills (*Allionia incarnata*), whitemargin spurge (*Euphorbia albomarginata*), and false broomweed (*Haploësthes greggii* var. *texana*) fill the creek banks. To 2 feet high, false broomweed is woody based,

View of Contrabando lowlands from South Lajitas Mesa

much branched, with slender linear leaves and yellow flower heads sporting five short, rounded ray florets and many disk florets with protruding, branched styles.

Melon loco **108** springs up in the creek after rains. Glandleaf milkwort (*Polygala macradenia*), shaggy stenandrium (*Stenandrium barbatum*), and Mexican tiquilia (*Tiquilia mexicana*) occupy hills above the trail. Locally common perennial sandmat **125** is erect, hairless, with woody stems, glaucous oval leaves, and flower clusters lacking petal-like appendages.

Lower Contrabando Creek fills with annuals after monsoon rains: warty caltrop **278**, a creeping orange-flowered annual; thickleaf drymary (*Drymaria pachyphylla*), a sand-lover with radiating stems, oval leaves, and notched white petals; goathead **281**, with curious fruits with hornlike spines; West Indian nightshade (*Solanum ptycanthum*), with triangular toothed or wavy leaves and white flowers; and bearded flatsedge **111**, with pale green, three-angled stems (culms) and headlike spikes with clustered oblong spikelets. Fluff-grass **213**, with creeping runners and white fuzz and fluff, grows near the creek.

Fresno Divide Trail

Fresno Divide Trail, 6.4 miles round trip, begins at a sign 0.3 miles down West Contrabando Trail. From the sign, Fresno Divide Trail follows an old prospect road north across the desert floor, through igneous rock outcrops, to a saddle overlooking Contrabando Dome. The trail then drops into the depressed dome and joins the Contrabando Dome Trail. The Divide Trail can be part of a loop, by returning via Fresno Creek to the west, or Contrabando Dome and Contrabando trails. Views of Contrabando Dome, Lajitas Mesa, Terlingua Uplift, Fresno Canyon, and Sierra Rica are trail highlights.

In 2016, three plants stood out: Warnock's cactus (*Echinomastus warnockii*) and Big Bend ringstem **188** occupied limestone outcrops, gravelly hills, and clay flats. Bluntscale bahia (*Hymenothrix pedata*), with a naked stem, three-lobed leaves, and yellow flower heads, appeared on desert flats and igneous rocks.

Contrabando Dome

The Contrabando Dome Trail forms a 4.5-mile semicircle off the West and East Contrabando Trails. The up-down-up-down trail runs through desert scrub, low hills and arroyos, baked Boquillas flagstones, sedimentary and igneous rocks, and quicksilver prospect mines developed in the 1930s. Two-mile-wide Contrabando Dome, a mini-Solitario, was formed by a laccolithic intrusion, which pushed Boquillas limestone into a dome that collapsed. Rhyolite intrusions have formed dikes and sills across the dome (Yates and Thompson 1959).

Conspicuous plants are handsome Havard's agave **26** and the artful minitree, bushy wild buckwheat **221**. Once in twenty years, Havard's agave forms a candelabra-like flowering stalk with yellow blooms. Bushy wild buckwheat produces tiny, pinwheel-shaped white blooms with red mid-stripes. Notable perennials are the ubiquitous Havard's wild buckwheat **220** and vulnerable Palmer's cat's eye (*Oreocarya palmeri*), with a woolly leaf rosette and coiled clusters of white flowers with yellow outgrowths in the throat.

A few memorable but petite cacti are Contrabando Dome residents. Watch for golf ball cactus (*Mammillaria lasiacantha*), with tiny overlapping white spines that make its golf-ball-size stems seem smooth to the touch.

Buena Suerte Trail

Across from the Barton Warnock Visitor Center, Buena Suerte Trail travels north-northwest, between the West Fork of Comanche Creek and North Lajitas Mesa, through limestone and clay flats and low hills. Short spurs branch off the trail: Dog Cholla Trail, Rock Quarry Trail, Crystal Trail, and Camino Viejo Trail. At the 4.4-mile point, Contrabandista Spur Road joins from the northeast, and across Contrabando Creek, 0.8-mile Contrabando Waterhole Trail runs southwest along Contrabando Creek. At 5.4 miles, Buena Suerte

Quartzite-covered low hills above Buena Suerte Trail, with Mexican mountains on horizon

Trail reaches a junction with West Contrabando Trail. From this point, the trail continues northwest 2.9 miles to Fresno Mine and past Whit-Roy Mine to Fresno Canyon. Just south of Fresno Mine, a short connecting trail runs southwest, linking Buena Suerte Trail to Contrabando Dome Trail.

Ocotillo **148** and smooth sotol **28**, on desert flats and hilltops, are distinctive perennials. Mexican buckeye (*Ungnadia speciosa*), eye-catching in bloom and in fruit, grows in the creek and arroyos. Big Bend silverleaf **253** springs into bloom after rains. Plentiful along the trail are threeflower goldenweed (*Xylothamia triantha*), Schott's acacia **143**, and Warnock's snakewood **242**. Nearby are longleaf jointfir **114**, and boundary ephedra (*Ephedra aspera*), with green or yellow, nearly naked, jointed stems and two scalelike leaves at nodes.

Notable subshrubs are candelilla **120**, with waxy gray stems; gyp daisy **51**, with violet ray florets and yellow disk florets; and littleleaf moonpod (*Acleisanthes parvifolia*), with brittle white stems and long, greenish-yellow tubular blooms.

Spring cactus displays are a highlight. Texas rainbow **72** is the featured cactus, with voluptuous blooms, usually yellow, but at times rose-pink, orange, salmon, or magenta. Other stem cacti: Potts' mammillaria **84**, with tiny red blooms; eagle-claw cactus **70**, with lush magenta blooms; catclaw cactus **82**, with rust red flowers; giant fishhook cactus **81**, with lemon yellow blooms; and tiny Boke's button (*Epithelantha bokei*), with small pink flowers much larger than the minuscule flowers of common button cactus **76**.

Limestone and gypsum habitats are havens for perennials adapted to these environs. Of interest are Mexican stickleaf **170**, with cuplike or short-cylindric fruit capsules; Chihuahuan ringstem **189**, with long-tubed pink blooms; Havard's wild buckwheat **220**, with yellow, white-woolly blooms on leafless stems; and perennial sandmat **125**. Common perennials are silky-haired New Mexico dalea (*Dalea neomexicana*), with pealike yellow blooms in dense spikes, and Mexican tiquilia (*Tiquilia mexicana*), with gray-haired, bristly leaves and pink, magenta, or purplish blooms. Also present is hairy crinklemat (*Tiquilia hispidissima*), with pink flowers and narrower, greener leaves on short, broad petioles. Fuzzy-haired slimseed spurge **128** is one of many prostrate spurges.

Contrabando Waterhole

Contrabando Waterhole is a historic spring on Contrabando Creek below Contrabando Mountain (3,214 ft.), a flat-topped extension of North Lajitas Mesa. H. W. McGuirk moved his ranch from Los Alamos to the waterhole in 1897 and built a rock house and rock corral (Smith-Savage 1996a). Two years later he left for Lajitas, where he purchased land, farmed cotton, ran a store-saloon, and financed construction of a church and school (Alloway 1995). In the Mexican Revolution, Contrabando Waterhole was as a cavalry substation (Cloud and Mallouf 1996).

From the Buena Suerte Trail, a side trail leads south to the waterhole and low waterfall. Groovestem bouchea **273**, with purple funnel-shaped flowers, leafy California loosestrife **172**, with pink-purple blooms, and tubular slimpod (*Amsonia longiflora*), with blue-white tubular blooms, are plentiful here.

Contrabandista Spur Trail

At 3.2 miles on Buena Suerte Trail, a side road goes right to Rockcrusher Road. Rockcrusher Road runs east to the Contrabandista Spur Trail turnoff, which leads to an unmarked trailhead above Contrabando Canyon. An overgrown road heads down to Contrabando Creek, which can be followed north to a steep pour-off from Terlingua Uplift. The "road" can be hiked up the west bank of Contrabando Canyon and north to Terlingua Uplift, or southwest to the Buena Suerte Trail crossing of Contrabando Creek.

Off the trail are sea urchin cactus (*Coryphantha echinus* var. *echinus*) and Arizona cockroach plant **20**. The cactus, with sulfur-yellow blooms, occupies desert flats above Contrabando Canyon. Long spines projecting at right angles give the cactus a "sea-urchin" look. The poisonous Arizona cockroach plant, with yellow flowers, grows at the pour-off from Terlingua Uplift. A leaf extract is used to kill cockroaches.

Often present nearby are two other perennials, chicken-thief stickleaf (*Mentzelia oligosperma*), with sticky egg-shaped or trilobed leaves with toothed margins and five-petal orange flowers; and Trans-Pecos carlowrightia (*Carlowrightia serpyllifolia*), with small, glandular-hairy leaves and four-lobed pale purple blooms, the upper lobe with a striking, flamelike yellow eye.

Comanche Creek, West Fork

The West Fork of Comanche Creek enters the park northeast of North Lajitas Mesa, continues south on the east side of Buena Suerte Trail, crosses Highway 170 and joins Comanche Creek near the Barton Warnock Visitor Center. The West Fork cuts through Boquillas limestone at elevations from 2,900 to 2,400 feet. A Buena Suerte Trail sign marks the site of a candelilla wax camp on the fork's west bank.

Scarce but locally common, fleshy tidestromia **8** can be prolific. Also scarce is the early-blooming mustard, Texas thelypody **64**, with deeply divided basal leaves and a flower spike with crowded white blooms. Christmas mistletoe **274**, with yellow flowers in segmented spikes and pearly white berries, infects trees on the drainage walls. Clumped dog cholla **67**, with firm, not easily detached joints, occupies desert flats on the West Fork rim.

PHYSIOGRAPHIC REGION 7
Cienega

Cienega, atop BBR's northwest panhandle, is a remote appendage, isolated from the rest of the park, a temptation for the park's more adventuresome visitors. A complex of landforms, Cienega possesses its own special access, a network of some of the most difficult 4WD roads in Texas.

The landforms were shaped by volcanic eruptions across the Cienega Mountains, Chinati Mountains, and Bofecillos Mountains to the southeast. The dominant landmark is the Cienega Mountains, an extrusive rhyolite lava dome 4.0 miles across, rising 1,400 feet above surrounding terrain. The highest point, just north of the park boundary, is 5,223 feet high, higher than the park's tallest peak, Oso Mountain (5,135 ft.). Cienega Region elevations drop to 3,200 feet on lower Alamito Creek, and peak at 4,600 feet in the Cienega Mountains near the northern park boundary.

Alamito and Cienega Creeks flow much of the year. Morita and Tortolo Creeks wind through Cienega's northwest corner, coming from Chinati Mountains foothills to the northwest. A narrow cleft, Cienega Gorge, walls in northern reaches of Cienega Creek. Along the creek, Sierra Blanca Dome is a mini-Solitario, composed of Glen Rose limestone thrusted upward by an igneous intrusion.

Surrounding the mountains is a vast fan of Perdiz Conglomerate, with fragments of Cienega Mountains rhyolite. Fingers of alluvial fan deposits run through the conglomerate south of the Cienegas. Morita Ranch Formation rhyolite and basalt lavas, produced by volcanism in the Chinati Mountains (37–32 Ma), are evident along Cienega Creek above Sierra Blanca Dome, and in Cienega's northwest corner (see Henry 1998; Ing 1996a).

Upper Alamito Creek

Upper Alamito Creek flows into the park's northwest panhandle west of Casa Piedra Road and crosses 4WD Alamo Springs Road below Casa Ramon. Alamito Creek then continues west to the railroad, and southwest, crossing the railroad tracks and the 4WD road from West Casa Piedra Trailhead before exiting the park. The portion of Alamito Creek within BBR is only 7.4 miles of its 78-mile length.

Conspicuous trees along Alamito Creek are Arizona cottonwood **250**, screwbean mesquite **140**, and Goodding's willow **251**, with slender serrated leaves and yellow catkins. Alamito Creek attracts many trees and shrubs, including catclaw acacia (*Senegalia greggii*), Texas persimmon **113**, netleaf hackberry **93**, Warnock's snakewood **242**, Mexican buckeye (*Ungnadia speciosa*), Apache plume **244**, and desert willow **53**. Distinctive herbs are netted globecherry **260**, devilishly attractive oakleaf datura **257**, and mealy goosefoot (*Chenopodium incanum*), with dense clusters (glomerules) of greenish-white blooms.

Cienega Creek and Cienega Gorge

Cienega Creek rises 4.4 miles west of US Highway 67, crosses the road 16 miles north of Shafter, and travels south another 28 miles before flowing into Alamito Creek just west of the park boundary. For nearly 9 miles, Cienega Creek courses through BBR's northwest panhandle, entering near the Cienega park residence, running along the west side of the Cienega Mountains through Cienega Gorge, and continuing south through Sierra Blanca Dome, west of the Cienega Creek 4WD Loop, before exiting the park.

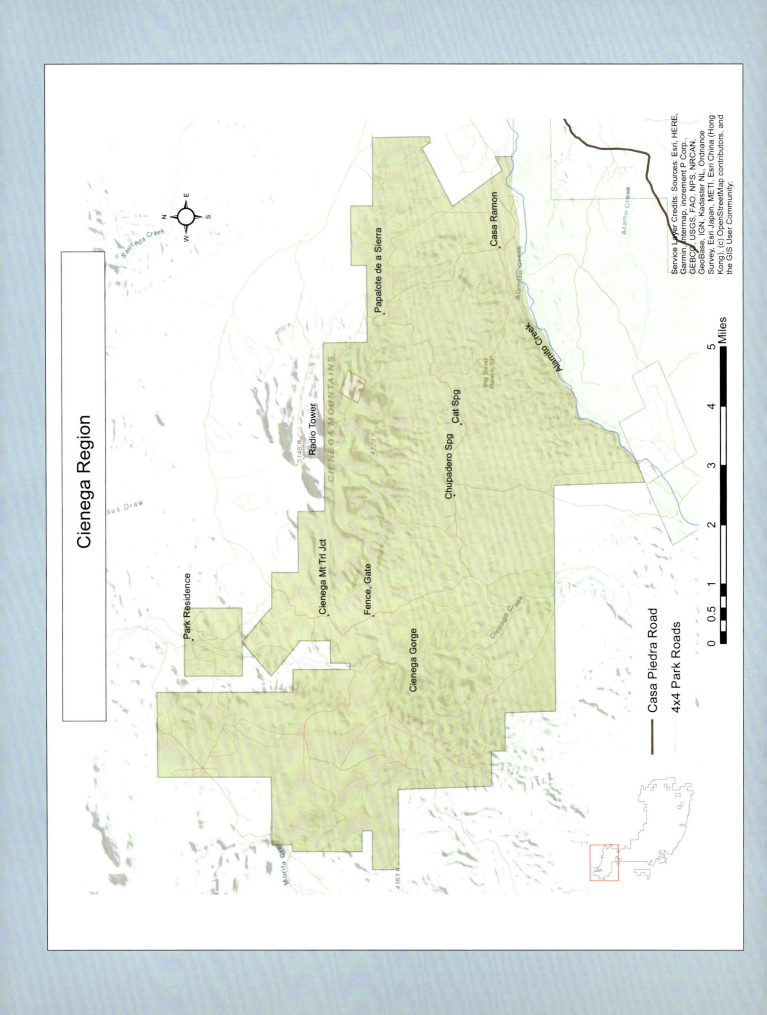

Within BBR, the northern portion of the creek crosses alluvium and terrace deposits, and then Perdiz Conglomerate. The middle portion slices through the narrow confines of Cienega Gorge, which consists largely of rhyolite and basalt lava. The southern portion cuts through Sierra Blanca Dome, exposing Glen Rose and Del Carmen limestones (Henry 1998). A relatively pristine riparian environment with limited human encroachment, the creek supports rare fish and other aquatic life, a wide variety of birds, and some plants unknown elsewhere in the park. The 4WD road, which loops through uplands flanking the creek, provides creek access through small connecting arroyos (Riskind and Scholly 2009).

In 1885 Milton Faver, a pioneering early rancher, established a cattle ranch on Cibolo Creek in the Chinati Mountains. Later, he formed another cattle ranch, La Cienega, on Cienega Creek, on land adjoining the park, and a sheep ranch, La Morita, on Morita Creek. Faver was an archetypal cattle baron and accomplished merchant who sold goods on the Chihuahua Trail and to soldiers at Fort Davis and silver miners at Shafter. In 1885, John Pool Sr. moved to Presidio County, and in 1915, his heirs purchased La Cienega. Reportedly, Pool built Cienega Camp, now the Cienega park residence (Smith-Savage 1996a).

Arizona cottonwood **250**, catclaw mimosa (*Mimosa aculeaticarpa* var. *biuncifera*), velvet ash **190**, screwbean mesquite **140**, and lotebush **243** line Cienega Creek. Bigleaf mistletoe **275**, mostly on cottonwood, and Christmas mistletoe **274** are semiparasitic shrubs infesting older trees. Shrubs mostly on limestone are downy wolfberry **259**, Rio Grande saddlebush **94**, and heath cliffrose **247**.

Semiaquatic perennials, such as often floating water hyssop **203**, with white, bell-shaped blooms; watercress **60**, in shallow water or moist soil, with four-petal white blooms; and water speedwell **208**, with lavender, purple-veined, four-lobed blooms, are fixtures along Cienega Creek. Two terrestrial perennials are blue milkwort **218**, and densely hairy New Mexico silverbush (*Argythamnia serrata*), with elliptic leaves and yellow flower clusters, the male blooms above and female blooms below.

Other perennials are rose bladderpod **61**, Lindheimer's copperleaf (*Acalypha phleoides*), and Arizona centaury (*Zeltnera arizonica*). Longpetal echeveria **105** and candelilla **120** appear on limestone bluffs at Sierra Blanca. Palmer's bluestar **13** and yellow flameflower **267** frequent Cienega Gorge. Knotted rush **159**, and silverleaf nightshade (*Solanum elaeagnifolium*), with blue purple flowers and mature yellow fruits, grow in the creek bed north of the gorge, with water speedwell. Three annuals, oakleaf datura **257**, mat-forming horse purslane **4**, and scrambled eggs **149** are more scattered along the creek.

Cienega Mountains

The Cienega Mountains, also named F Mountain for Milton Faver, dominate the park's panhandle, peaking at 5,223 feet just beyond the park boundary. The mountains are bounded on the east by the South Orient Railway, on the south by Alamito Creek, and on the west by Cienega Creek. The mountains were formed by an extrusive lava dome of Cienega Mountains Rhyolite 4 miles wide. Alluvial fans of Perdiz Conglomerate radiate downward to Alamito and Cienega Creeks (Henry 1998).

Typically, shrubs common in the Cienega Mountains are common at higher elevations throughout BBR, and many shrubs present on upper Alamito Creek also occur in the mountains. They include gray oak **144**, mesquite (*Prosopis*), evergreen sumac (*Rhus virens* var. *virens*), desert sumac (*Rhus microphylla*), desert olive (*Forestiera angustifolia*), guayacán (*Guaiacum angustifolium*), mariola **44**, skeletonleaf goldeneye (*Sidneya tenuifolia*), and the subshrub leatherstem **130**. More scattered shrubs are compact Wright's snakeroot (*Ageratina wrightii*) and spindly Emory's mimosa (*Mimosa emoryana*).

Herbs in many plant families are mountain residents, including the amaranth, Arizona snakecotton **6**, the desertpeony, brownfoot **33**, the birthwort, Wright's Dutchman's pipe **25**, and the milkweed, sidecluster milkweed **15**. Three common perennials are antelope horns **14**, a milkweed with follicles resembling antelope horns; mat-forming oreja de perro **57**, with woolly oval leaves curling like a dog's ear; and hairy spurge **129**, with longhaired, broadly egg-shaped, serrulate leaves.

Cienega 4WD Roads

Cienega is accessed via two trailheads on the Casa Piedra Road. From the west trailhead, a 4WD road leads north to a T intersection, crossing the South Orient Railway and Alamito Creek along the way. From the intersection, a 4WD road runs west to Chupadero Spring and to Cienega Creek 4WD Loop, then north, flanked by Cienega Creek on the west, and Cienega Mountains to the northeast.

Chupadero Ranch, including the spring, was acquired by Ramon Aguirre in 1912 and was added to his home tract, Casa Ramon, but the property soon changed hands. Reportedly, the ranch was named by S. S. "Ted" Harper, who moved there in the early 1930s and built the adobe house, now in ruins. The spring, now depleted, was ringed by cottonwoods and a large velvet ash **190** (Smith-Savage 1996b; Riskind and Sholly 2009).

The 3.7-mile Cienega Creek 4WD Loop travels through uplands east of Cienega Creek and provides creek access. North of the loop, the road dead-ends at a locked gate. The road beyond the gate provides hiking access to the Cienega Mountains on the east. Cienega Gorge is reached cross-country to the west.

From the T intersection near Chupadero Spring, a 4WD road heads east to Cat Spring and roads leading northeast to Papalote de la Sierra and southeast to Casa Ramon. In a rocky, gravelly depression, Cat Spring has limited flow but is attractive for wildlife and wildlife viewing (Riskind and Scholly 2009).

From East Casa Piedra Trailhead, a 4WD road leads northeast through sand to Alamito Creek. Hikes along the creek are rewarding. If the creek is passable, the road can be followed east on the creek's north bank and across railroad tracks to Casa Ramon. Acquired by Ramon Aguirre in 1909 and later used as a line camp, Casa Ramon is a one-room adobe structure, with fences and an outhouse added in the 1980s (Smith-Savage 1996b; Riskind and Sholly 2009).

From Casa Ramon, three difficult 4WD roads lead into the Cienega Mountains. The left road runs west above the railroad tracks, climbs steeply northwest, and joins the road from Chupadero and Cat Springs. That road then climbs northeast to Papalote de la Sierra. The challenging middle road ascends north from Casa Ramon to Papalote de la Sierra, through a boulder-filled arroyo and brush. The right road heads east above the railroad tracks, abruptly climbs north, and loops northwest to Papalote de la Sierra.

Many plants already mentioned in the Cienega region are present off these 4WD roads. Shrubs not discussed previously include Trans-Pecos poreleaf **45**, California trixis **48**, and bush croton (*Croton fruticulosus*), to 3 feet high, with toothed, egg-shaped leaves white-haired below, and tiny male and female flowers in separate clusters. Other subshrubs, such as fern acacia (*Acaciella angustissima*) and desert rosemallow (*Hibiscus coulteri*), are also encountered off these rough jeep roads.

Three plants, not known elsewhere at BBR, were found years ago on these Cienega slopes: Limpia blacksenna (*Seymeria scabra*), a spreading, scabrous perennial, with yellow blooms with reddish markings; New Mexico cliff fern (*Woodsia neomexicana*), with ascending leaves to 1 foot long and pinnae with up to seven pairs of close-spaced, smaller leaflets; and slimlobe rockdaisy (*Perityle dissecta*), an imperiled dwarf shrub with highly dissected leaves and yellow disk florets. I photographed these plants in the Chisos Mountains but have not relocated them at BBR.

A FIELD GUIDE TO THE FLORA

This "Flora" section covers 281 BBR plant species in 90 plant families, 280 photographed by the author. One species, helecho de la Llave **228**, discovered in BBR in 1992 but not so far relocated, was photographed by Carlos Velazco-Macias in Mexico. Notably, 185 other species are briefly mentioned in the "Tour" section but not treated in more detail in the "Flora" section due to space considerations.

The following "Checklist of Plants" lists the 281 species in order by plant family and scientific name, with page numbers for easy reference.

The last complete inventory of BBR plants, *A Floral Inventory of the Big Bend Ranch State Natural Area, Presidio and Brewster Counties, Texas*, was conducted by Richard Worthington in 1995. Other Solitario species were added to the flora by Jean Hardy in her 2009 *Flora and Vegetation of the Solitario Dome, Brewster and Presidio Counties, Texas*.

This work includes 104 species, mentioned in the "Tour" or "Flora" sections, that constitute additions or revisions to the two previous plant lists. They are listed in "Appendix C, Additions or Revisions to the Floral Inventory of BBR."

(*Clockwise*): creeping rockvine (*Bonamia repens*), Havard's plum (*Prunus havardii*), lyreleaf twistflower (*Streptanthus carinatus* ssp. *carinatus*), Boke's button cactus (*Epithelantha bokei*), mimicking matspurge (*Euphorbia simulans*), and littleleaf moonpod (*Acleisanthes parvifolia*). These plants were photographed at Big Bend Ranch State Park, and appear on Texas Parks and Wildlife Department's *Rare, Threatened, and Endangered Species List*, but were not included in the plant descriptions due to space considerations. However, all but Mimicking Matspurge were described in the author's earlier work, *Little Big Bend: Common, Uncommon, and Rare Plants of Big Bend National Park*.

Table 1. Checklist of plants

Family	Scientific Name	Common Name	Page
Acanthaceae—Acanthus	*Carlowrightia linearifolia*	Heath Carlowrightia	
	Justicia warnockii	Warnock's Justicia	
Aizoaceae—Fig Marigold	*Sesuvium verrucosum*	Western Sea Purslane	
	Trianthema portulacastrum	Horse Purslane	
Alismataceae—Water Plantain	*Sagittaria montevidensis ssp. calycina*	Hooded Arrowhead	
Amaranthaceae—Amaranth	*Froelichia arizonica*	Arizona Snakecotton	
	Iresine leptoclada	Texas Shrub	
	Tidestromia carnosa	Fleshy Tidestromia	
	Tidestromia lanuginosa	Woolly Tidestromia	
Amaryllidaceae—Amaryllis	*Habranthus longifolius*	Copper Zephyrlily	
Anacampserotaceae—Arroyo Fameflower	*Talinopsis frutescens*	Arroyo Fameflower	
Anacardiaceae—Sumac	*Toxicodendron radicans*	Poison Ivy	
Apocynaceae—Dogbane	*Amsonia palmeri*	Palmer's Bluestar	
	Asclepias asperula	Antelope Horns	
	Asclepias oenotheroides	Sidecluster Milkweed	
	Asclepias scaposa	Bear Mountain Milkweed	
	Funastrum crispum	Wavyleaf Twinevine	
	Funastrum hartwegii	Hartweg's Twinevine	
	Funastrum torreyi	Soft Twinevine	
	Haplophyton cimicidum var. crooksii	Arizona Cockroach Plant	
	Matelea parvifolia	Spearleaf	
	Matelea producta	Texas Milkvine	
Apodanthaceae—Stemsucker	*Pilostyles thurberi*	Thurber's Stemsucker	
Arecaceae—Palm	*Phoenix dactylifera*	Date Palm	
Aristolochiaceae—Birthwort	*Aristolochia wrightii*	Wright's Dutchman's Pipe	
Asparagaceae—Asparagus	*Agave havardiana*	Havard's Agave	
	Agave lechuguilla	Lechuguilla	
	Dasylirion leiophyllum	Smooth Sotol	
	Nolina erumpens	Beargrass	
	Yucca elata	Soaptree Yucca	
	Yucca torreyi	Spanish Dagger	
Asteraceae—Sunflower	*Acourtia nana*	Desert Holly	
	Acourtia wrightii	Brownfoot	
	Ambrosia monogyra	Cheeseweed	
	Baccharis salicifolia	Seepwillow	

Family	Scientific Name	Common Name	Page
	Brickellia eupatorioides var. chlorolepis	Trans-Pecos False Boneset	
	Cirsium turneri	Turner's Thistle	
	Cirsium undulatum var. undulatum	Wavyleaf Thistle	
	Grindelia oxylepis	Mexican Gumweed	
	Koanophyllon solidaginifolium	Shrubby Umbrella Thoroughwort	
	Machaeranthera tanacetifolia	Tahoka Daisy	
	Packera millelobata	Manybract Groundsel	
	Palafoxia riograndensis	Rio Grande Palafox	
	Parthenium incanum	Mariola	
	Porophyllum scoparium	Trans-Pecos Poreleaf	
	Stephanomeria pauciflora	Wire Lettuce	
	Thelesperma megapotamicum	Hopi Tea Greenthread	
	Trixis californica	California Trixis	
	Vernonia marginata	Plains Ironweed	
	Xanthisma gypsophilum	Gypsum Tansy Aster	
	Xylorhiza wrightii	Gyp Daisy	
Berberidaceae—Barberry	*Berberis trifoliolata*	Agarito	
Bignoniaceae—Catalpa	*Chilopsis linearis*	Desert Willow	
Boraginaceae—Borage	*Euploca powelliorum*	Powell's Heliotrope	
	Lithospermum incisum	Fringed Puccoon	
	Omphalodes alienoides	Mexican Navelwort	
	Tiquilia canescens	Oreja de Perro	
	Tiquilia gossypina	Texas Crinklemat	
	Tiquilia greggii	Plume Tiquilia	
Brassicaceae—Mustard	*Nasturtium officinale*	Watercress	
	Physaria purpurea	Rose Bladderpod	
	Rorippa ramosa	Durango Yellowcress	
	Thelypodiopsis purpusii	Purpus' Tumblemustard	
	Thelypodium texanum	Texas Thelypody	
Bromeliaceae—Bromeliad	*Hechtia texensis*	Texas False Agave	
Cactaceae—Cactus	*Ariocarpus fissuratus*	Living Rock Cactus	
	Corynopuntia aggeria	Clumped Dog Cholla	
	Cylindropuntia kleiniae	Candle Cholla	
	Cylindropuntia leptocaulis	Desert Christmas Cholla	
	Echinocactus horizonthalonius	Eagle-Claw Cactus	
	Echinocereus coccineus var. paucispinus	Texas Claret-Cup Cactus	

Family	Scientific Name	Common Name	Page
	Echinocereus dasyacanthus	Texas Rainbow Cactus	
	Echinocereus stramineus var. stramineus	Strawberry Cactus	
	Echinocereus viridiflorus var. canus	Graybeard Cactus	
	Echinomastus mariposensis	Mariposa Cactus	
	Epithelantha micromeris var. micromeris	Common Button Cactus	
	Escobaria dasyacantha	Desert Pincushion Cactus	
	Escobaria duncanii	Duncan's Pincushion Cactus	
	Escobaria sneedii var. albicolumnaria	Silverlace Cactus	
	Escobaria tuberculosa var. varicolor	Varicolor Cob Cactus	
	Ferocactus hamatacanthus var. hamatacanthus	Giant Fishhook Cactus	
	Glandulicactus uncinatus var. wrightii	Catclaw Cactus	
	Mammillaria meiacantha	Nipple Cactus	
	Mammillaria pottsii	Potts' Mammillaria Cactus	
	Neolloydia conoidea var. conoidea	Texas Cone Cactus	
	Opuntia azurea var. diplopurpurea	Diploid Purple Prickly Pear	
	Opuntia azurea var. discolor	Big Hill Prickly Pear	
	Opuntia dulcis	Sweet Prickly Pear	
	Opuntia engelmannii var. engelmannii	Engelmann's Prickly Pear	
	Peniocereus greggii	Desert Night-blooming Cereus	
	Thelocactus bicolor var. bicolor	Glory of Texas Cactus	
Campanulaceae—Bellflower	*Lobelia cardinalis*	Cardinal Flower	
Cannabaceae—Hemp	*Celtis reticulata*	Netleaf Hackberry	
Celastraceae—Bittersweet	*Mortonia scabrella*	Rio Grande Saddlebush	
Chenopodiaceae—Goosefoot	*Atriplex acanthocarpa*	Tubercled Saltbush	
	Atriplex elegans var. elegans	Wheelscale Saltbush	
	Atriplex lentiformis	Big Saltbush	
	Chenopodiastrum murale	Nettleleaf Goosefoot	
Cleomaceae—Spiderflower	*Polanisia uniglandulosa*	Mexican Clammyweed	
Commelinaceae—Spiderwort	*Tradescantia leiandra*	Smooth Spiderwort	
Convolvulaceae—Morning-Glory	*Cuscuta umbellata*	Flatglobe Dodder	
	Evolvulus alsinoides	Ojo de Vibora	
	Ipomoea cordatotriloba var. torreyana	Torrey's Tievine	
	Ipomoea hederacea	Ivyleaf Morning-Glory	
Crassulaceae—Stonecrop	*Echeveria strictiflora*	Longpetal Echeveria	
	Sedum wrightii	Wright's Stonecrop	

Family	Scientific Name	Common Name	Page
Crossosomataceae—Crossosoma	*Glossopetalon spinescens*	Spiny Greasebush	
Cucurbitaceae—Cucumber	*Apodanthera undulata*	Melon Loco	
Cupressaceae—Cypress	*Juniperus pinchotii*	Red Berry Juniper	
Cyperaceae—Sedge	*Bolboschoenus maritimus ssp. paludosus*	Cosmopolitan Bulrush	
	Cyperus squarrosus	Bearded Flatsedge	
	Eleocharis montevidensis	Sand Spikerush	
Ebenaceae—Persimmon	*Diospyros texana*	Texas Persimmon	
Ephedraceae—Mormon Tea	*Ephedra trifurca*	Longleaf Jointfir	
Equisetaceae—Horsetail	*Equisetum xferrissii*	Ferriss' Scouring Rush	
Euphorbiaceae—Spurge	*Croton pottsii var. pottsii*	Leatherweed	
	Croton sancti-lazari	Trans-Pecos Croton	
	Euphorbia abramsiana	Abrams' Spurge	
	Euphorbia acuta	Pointed Sandmat	
	Euphorbia antisyphilitica	Candelilla	
	Euphorbia cinerascens	Ashy Sandmat	
	Euphorbia davidii	David's Spurge	
	Euphorbia exstipulata	Squareseed Spurge	
	Euphorbia indivisa	Muleear Spurge	
	Euphorbia perennans	Perennial Sandmat	
	Euphorbia revoluta	Threadstem Sandmat	
	Euphorbia setiloba	Yuma Spurge	
	Euphorbia stictospora	Slimseed Spurge	
	Euphorbia villifera	Hairy Spurge	
	Jatropha dioica var. graminea	Leatherstem	
Fabaceae—Bean	*Astragalus terlinguensis*	Terlingua Milkvetch	
	Astragalus wootonii	Wooton's Loco	
	Calliandra iselyi	Isely's Feather Duster	
	Desmanthus velutinus	Velvet Bundleflower	
	Macroptilium atropurpureum	Siratro	
	Mimosa texana	Texas Mimosa	
	Mimosa turneri	Turner's Mimosa	
	Parkinsonia aculeata	Retama	
	Prosopis glandulosa var. torreyana	Western Honey Mesquite	
	Prosopis pubescens	Screwbean Mesquite	
	Senna lindheimeriana	Lindheimer's Senna	
	Senna wislizeni	Shrubby Senna	

Family	Scientific Name	Common Name	Page
	Vachellia schottii	Schott's Acacia	
Fagaceae—Beech	*Quercus grisea*	Gray Oak	
	Quercus hinckleyi	Hinckley's Oak	
	Quercus oblongifolia	Mexican Blue Oak	
	Quercus pungens	Sandpaper Oak	
Fouquieriaceae—Ocotillo	*Fouquieria splendens*	Ocotillo	
Fumariaceae—Fumitory	*Corydalis aurea*	Scrambled Eggs	
Gentianaceae—Gentian	*Eustoma exaltatum*	Catchfly Prairie Gentian	
	Zeltnera calycosa	Buckley's Centaury	
Geraniaceae—Geranium	*Erodium cicutarium*	Redstem Stork's Bill	
	Erodium texanum	Texas Stork's Bill	
Hydrangeaceae—Hydrangea	*Fendlera linearis*	Narrowleaf Fendlerbush	
Hydrophyllaceae—Waterleaf	*Nama hispidum*	Bristly Nama	
	Phacelia infundibuliformis	Rio Grande Phacelia	
	Phacelia popei	Pope's Phacelia	
Juglandaceae—Walnut	*Juglans microcarpa*	Little Walnut	
Juncaceae—Rush	*Juncus nodosus*	Knotted Rush	
Koeberliniaceae—Allthorn	*Koeberlinia spinosa*	Allthorn	
Krameriaceae—Ratany	*Krameria erecta*	Range Ratany	
	Krameria bicolor	White Ratany	
Lamiaceae—Mint	*Poliomintha glabrescens*	Leafy Rosemary-Mint	
	Salvia greggii	Autumn Sage	
	Scutellaria potosina var. tessellata	Mexican Skullcap	
	Teucrium depressum	Desert Teucrium	
Linaceae—Flax	*Linum rupestre*	Rock Flax	
Loasaceae—Stickleaf	*Eucnide bartonioides*	Yellow Rocknettle	
	Mentzelia asperula	Organ Mountain Blazingstar	
	Mentzelia mexicana	Mexican Stickleaf	
	Mentzelia pachyrhiza	Coahuila Blazingstar	
Lythraceae—Loosestrife	*Lythrum californicum*	California Loosestrife	
Malpighiaceae—Barbados Cherry	*Aspicarpa hyssopifolia*	Hyssopleaf Asphead	
	Cottsia gracilis	Propellerbush	
Malvaceae—Mallow	*Ayenia pilosa*	Dwarf Ayenia	
	Malvella sagittifolia	Arrowleaf Mallow	
	Melochia pyramidata	Pyramid Flower	
	Sphaeralcea angustifolia	Narrowleaf Globemallow	

Family	Scientific Name	Common Name	Page
	Sphaeralcea digitata	Juniper Globemallow	
Marsileaceae—Water Clover	*Marsilea vestita*	Hooked Water Clover	
Martyniaceae—Unicorn Plant	*Proboscidea althaeifolia*	Desert Unicorn Plant	
	Proboscidea parviflora	Doubleclaw	
Moraceae—Mulberry	*Morus microphylla*	Texas Mulberry	
Nitrariaceae—Nitre Bush	*Peganum harmala*	African Rue	
Nyctaginaceae—Four-O'clock	*Acleisanthes angustifolia*	Narrowleaf Moonpod	
	Acleisanthes chenopodioides	Goosefoot Moonpod	
	Acleisanthes longiflora	Angel Trumpets	
	Anulocaulis eriosolenus	Big Bend Ringstem	
	Anulocaulis leiosolenus var. lasianthus	Chihuahuan Ringstem	
Oleaceae—Olive	*Fraxinus velutina*	Velvet Ash	
Onagraceae—Evening Primrose	*Eremothera chamaenerioides*	Longcapsule Suncup	
	Oenothera boquillensis	Boquillas Lizardtail	
	Oenothera curtiflora	Lizardtail Gaura	
Orchidaceae—Orchid	*Epipactis gigantea*	Giant Helleborine	
	Hexalectris warnockii	Texas Purplespike	
Orobanchaceae—Broomrape	*Castilleja lanata*	Woolly Paintbrush	
	Castilleja sessiliflora	Downy Paintbrush	
Papaveraceae—Poppy	*Argemone chisosensis*	Chisos Pricklypoppy	
Phrymaceae—Lopseed	*Erythranthe chinatiensis*	Fringed Monkeyflower	
	Erythranthe inamoena	James's Monkeyflower	
Phyllanthaceae—Leaf Flower	*Phyllanthopsis arida*	Trans-Pecos Maidenbush	
Phytolaccaceae—Pokeweed	*Rivina humilis*	Rouge Plant	
Plantaginaceae—Plantain	*Bacopa monnieri*	Water Hyssop	
	Epixiphium wislizeni	Balloonbush	
	Penstemon dasyphyllus	Cochise Beardtongue	
	Plantago ovata	Desert Indianwheat	
	Stemodia coahuilensis	Coahuila Twintip	
	Veronica anagallis-aquatica	Water Speedwell	
Poaceae—Grass	*Arundo donax*	Giant Reed	
	Bouteloua curtipendula	Sideoats Grama	
	Bouteloua erecta	False Grama	
	Bouteloua eriopoda	Black Grama	
	Dasyochloa pulchella	Fluffgrass	
	Heteropogon contortus	Tanglehead	

Family	Scientific Name	Common Name	Page
Polemoniaceae—Phlox	*Giliastrum acerosum*	Needleleaf Gilia	
	Ipomopsis laxiflora	Iron Ipomopsis	
	Ipomopsis longiflora var. neomexicana	Paleflower Ipomopsis	
Polygalaceae—Milkwort	*Hebecarpa barbeyana*	Blue Milkwort	
	Rhinotropis nudata	Smallflower Milkwort	
Polygonaceae—Buckwheat	*Eriogonum havardii*	Havard's Wild Buckwheat	
	Eriogonum suffruticosum	Bushy Wild Buckwheat	
	Rumex pulcher	Fiddle Dock	
Portulacaceae—Purslane	*Portulaca halimoides*	Sinkerleaf Purslane	
	Portulaca suffrutescens	Shrubby Purslane	
	Portulaca umbraticola var. lanceolata	Wingpod Purslane	
Pteridaceae—Maidenhair Fern	*Adiantum capillus-veneris*	Maidenhair Fern	
	Bommeria hispida	Hairy Bommer	
	Llavea cordifolia	Helecho de la Llave	
	Myriopteris aurea	Bonaire Lip Fern	
	Myriopteris scabra	Prickly Lip Fern	
	Myriopteris windhamii	Villous Lip Fern	
	Notholaena aliena	Foreign Cloak Fern	
	Notholaena aschenborniana	Scaled Cloak Fern	
	Notholaena grayi	Gray's Cloak Fern	
	Notholaena greggii	Gregg's Cloak Fern	
	Notholaena standleyi	Standley's Cloak Fern	
	Pellaea wrightiana	Wright's Cliff Brake	
Ranunculaceae—Buttercup	*Aquilegia chrysantha*	Golden Columbine	
	Clematis drummondii	Old Man's Beard	
	Clematis pitcheri var. dictyota	Leather Flower	
Resedaceae—Mignonette	*Oligomeris linifolia*	Desert Spikes	
Rhamnaceae—Buckthorn	*Condalia warnockii var. warnockii*	Warnock's Snakewood	
	Ziziphus obtusifolia	Lotebush	
Rosaceae—Rose	*Fallugia paradoxa*	Apache Plume	
	Malacomeles pringlei	Toothed Serviceberry	
	Petrophytum caespitosum	Mat Rock Spiraea	
	Purshia ericifolia	Heath Cliffrose	
	Rubus flagellaris	Northern Dewberry	
Rubiaceae—Madder	*Cephalanthus occidentalis*	Buttonbush	
Salicaceae—Willow	*Populus fremontii subsp. mesetae*	Arizona Cottonwood	

Family	Scientific Name	Common Name	Page
	Salix gooddingii	Goodding's Willow	
Sapindaceae—Soapberry	*Sapindus saponaria*	Western Soapberry	
Scrophulariaceae—Figwort	*Leucophyllum minus*	Big Bend Silverleaf	
Selaginellaceae—Spikemoss	*Selaginella lepidophylla*	Flower of Stone	
	Selaginella peruviana	Peruvian Spikemoss	
Solanaceae—Nightshade	*Chamaesaracha coronopus*	Greenleaf Five Eyes	
	Datura quercifolia	Oakleaf Datura	
	Datura wrightii	Sacred Datura	
	Lycium puberulum var. puberulum	Downy Wolfberry	
	Margaranthus solanaceus	Netted Globecherry	
	Nicotiana glauca	Tree Tobacco	
	Physalis angulata var. lanceifolia	Lanceleaf Groundcherry	
	Physalis hederifolia	Heartleaf Groundcherry	
	Solanum citrullifolium	Watermelon Nightshade	
	Solanum rostratum	Buffalobur Nightshade	
Talinaceae—Fameflower	*Talinum paniculatum*	Pink Baby's-breath	
	Talinum polygaloides	Yellow Flameflower	
Tamaricaceae—Tamarisk	*Tamarix aphylla*	Athel	
	Tamarix chinensis	Salt Cedar	
Typhaceae—Cattail	*Typha domingensis*	Southern Cattail	
Verbenaceae—Vervain	*Aloysia gratissima*	Beebrush	
	Aloysia wrightii	Oreganillo	
	Bouchea linifolia	Groovestem Bouchea	
Viscaceae—Mistletoe	*Phoradendron leucarpum ssp. tomentosum*	Christmas Mistletoe	
	Phoradendron macrophyllum	Bigleaf Mistletoe	
Vitaceae—Grape	*Vitis arizonica*	Canyon Grape	
Zygophyllaceae—Caltrop	*Kallstroemia grandiflora*	Orange Caltrop	
	Kallstroemia parviflora	Warty Caltrop	
	Kallstroemia perennans	Perennial Caltrop	
	Larrea tridentata	Creosote Bush	
	Tribulus terrestris	Goathead	

1. Heath Carlowrightia (*Carlowrightia linearifolia*)
(Other common names: heath wrightwort, heath carlowright)

Habitat: desert scrub, sandy-gravelly alluvium, rocky drainages, amid boulders in canyons, on mountain slopes.

Geographic range: Texas: Trans-Pecos (excluding Reeves, Terrell Counties). US: Texas, southern New Mexico, southeast Arizona. Mexico: Chihuahua, Coahuila.

Park locations: Los Ojitos, Horsetrap Spring, Tascate Hills, Lefthand Shutup, Los Alamos, Contrabando Creek above Contrabando Waterhole

Heath Carlowrightia is a small, pale green herb or shrub to 2 feet high, with many slender, ascending branches, and a woody crown. Younger stems are hairy, but mature stems are mostly hairless. At times, the branches appear leafless and skeleton-like, but they usually bear narrow, stemless, paired linear leaves 1 inch long.

In summer and fall, with the arrival of monsoon rains, these plants explode with delicate blooms to ½ inch long. Light purple or lavender, four-lobed flowers, with darker purplish veins and a striking yellow eye on the upper lobe, appear in loose, branching clusters. The blooms are accented by two protruding stamens with purple filaments and yellow, arrow-shaped anthers, and an even longer style. Linear bracts and bractlets, and a five-lobed calyx, support each bloom.

The fruit is a stalked, short-beaked capsule, rounded and flattened, somewhat bellows-shaped, which splits explosively to eject four flat black seeds.

Botanical notes: New York botanist John Torrey initially described this species as *Schaueria linearifolia* in 1859, in the botany section of the *Report on the United States and Mexico Boundary*. In 1878, Harvard University botanist Asa Gray moved the species to his new genus *Carlowrightia*, which he named in honor of Charles (also Carlos) Wright. Wright collected the plant in 1849, on the "Culberson-Jeff Davis County border between Chispa Mts & Lobo Siding."

2. Warnock's Justicia (*Justicia warnockii*)
(Other common names: Warnock's water-willow)

Habitat: arid, rocky limestone outcrops in mountains, foothills, canyons.

Geographic range: Texas: Trans-Pecos (south Hudspeth, Presidio, Brewster, Pecos, Terrell Counties); Crockett County. US: Texas. Mexico: Chihuahua, Coahuila.

Park locations: Solitario northwest rim, Solitario Peak, Terlingua Uplift

Of conservation concern: see Appendix A

Less than 1 foot high, this easily overlooked subshrub is grasslike in appearance, with thin, crowded, yellow green stems from a thick, woody base. The branches bear hairless needlelike leaves less than 1/2 inch long.

This acanthus is described as "few-flowered," but at dawn you may find numerous blooms, soon wilted by the sun or otherwise dispatched by the winds. Solitary, stalkless flowers form in leaf axils, supported by two tiny bractlets. Mostly white or suffused with lavender, with reddish purple splotches in the throat, the blossoms are two-lipped, with a distinctly notched upper lip and a broader, spreading, trilobed lower lip.

The clublike fruits are smooth, compressed, stalked capsules with a short beak. Each capsule contains four flat oval seeds.

An early riser who knows where to look may eventually craft a photo that does justice to this elusive and enigmatic plant.

Botanical notes: Linnaeus described *Justicia* in 1753, naming the genus for Scottish horticulturist James Justice (1698–1763). In Scotland, Justice was one of the first gardeners to grow pineapples (1728); he built the first pine stove, pioneered the construction of glass houses, and authored the popular *The Scots' Gardener's Director* in 1754. Sadly, a victim of "Tulip Madness" (in his case, hyacinth madness), Justice went bankrupt and was expelled from the Royal Society. In 1951, University of Texas botanist B. L. Turner named this species for Barton Warnock, Sul Ross State University botanist, who collected the type specimen ten years earlier, "among lechuguilla, abundant on lower limestone slopes, Old Blue, Glass Mts." of Brewster County.

3. Western Sea Purslane (*Sesuvium verrucosum*)
(Other common names: verrucose seapurslane, verdolaga de playa, romerillo, winged sea purslane)

Habitat: often near the Rio Grande; creek beds, saline, alkaline, clay soils.

Geographic range: Texas: Trans-Pecos (El Paso, Hudspeth, Presidio, Brewster, Reeves, Pecos Counties), parts of the Permian Basin, southern Panhandle, lower South Texas Plains. US: southwest US, north to Kansas, Oregon, Louisiana, Arkansas. Elsewhere: northern Mexico; parts of South America.

Park locations: Colorado Canyon River Access

This fig marigold is a grayish green, succulent perennial with fibrous roots and sprawling, hairless stems as much as 3 feet long. The fleshy stems, leaves, and even flowers, are verrucose — dotted with wartlike crystalline protuberances. Often in opposite pairs of unequal size, the leaves are spatulate to oblong and tapered to a broadened base that clasps the stem. To 1 1/2 inch long, the blades are characteristically rolled downward on the margins and backward at the tip.

The solitary magenta flowers, 1/2 inch wide, lack petals and consist of five pointed sepals, with a ring of reddish stamens at the center. Largely stemless, the blooms appear in the axils of upper leaves. Conspicuously, each colorful sepal is beaked by a green dorsal appendage extending well beyond the sepal tip.

The fruit is a cone-shaped capsule 1/5 inch long, with many black, usually smooth and shiny, kidney-shaped seeds.

Botanical notes: Linnaeus named the genus *Sesuvium* in *Systema Naturae* in 1759, but the etymology is uncertain. The species name, from Latin, means verrucose or warty, referring to the herbage. Constantine Samuel Rafinesque, a jack-of-all-trades natural historian and professor of botany at Transylvania University in Kentucky, described this species in his *New Flora and Botany of North America* (1836–38), from a specimen "sent me from Arkansas and the Chacta Country, where it grows near streams on the Yazou and Salt River."

4. Horse Purslane (*Trianthema portulacastrum*)
(Other common names: desert horsepurslane, lowland purslane, verdolaga, verdolaga de cochi, verdolaga de caballo)

Habitat: sand, clay, dunes, depressed or disturbed areas, bottomlands near Rio Grande, arroyos, alkaline, subsaline flats.

Geographic range: Texas: Trans-Pecos (El Paso, Reeves, Jeff Davis, Presidio, Brewster Counties); Loving County; South Texas, sites in East Texas, Gulf Coast. US: southern half of US. Elsewhere: much of Mexico, Central, South America, West Indies, parts of Africa.

Park locations: between Redford and Lajitas, Sauceda Road west of Rancho Viejo, Las Cuevas, Cienega Creek near Cienega residence

After monsoon rains, this succulent annual develops widely branching mats that blanket the ground. The spreading stems are fleshy, nearly hairless, green or turning red or maroon. Spoon-shaped or elliptic leaves form unequal pairs along the stems.

The small, usually solitary flowers lack petals and stalks and consist of magenta-colored, five-lobed calyces that resemble petals, and five to ten white or pale pink stamens. These "reclusive" flowers, partially concealed by sheathing bracts, are only open in the morning. The fruit is a winged cylindrical capsule to 1/5 inch long, with ridged, compressed black seeds. One seed is dispersed in a detachable cap and floats in water.

In Ayurvedic medicine, the herb is used to treat many maladies, including edema, asthma, bronchitis, and stomach problems.

Botanical notes: The genus name, *Trianthema*, is derived from the Greek *treis* (three) and *anthemon* (flower). The species name combines the Latin *Portulaca* (purslane), and the suffix *astrum* (something like), implying similarity. Linnaeus described this genus and species in 1753, noting the plant's habitat as "Jamaica, Curassao." The herb may have been first collected by Sir Hans Sloane on his voyage to Jamaica (1687–89), but the designated lectotype is a drawing in Paul Hermann's *Paradisus batavus*, a description of plants in the Leyden University botanical garden, edited by William Sherard (1698).

5. Hooded Arrowhead (*Sagittaria montevidensis* ssp. *calycina*)
(Other common names: broad-leafed arrowhead, giant arrowhead)

Habitat: muddy flats, stranded pools of water along Rio Grande.

Geographic range: Texas: Trans-Pecos (rare in Presidio County), scattered elsewhere. US: most of central, eastern US; Texas north to South Dakota, west to California; Colorado, Oregon. Mexico: Chihuahua, Coahuila, Tamaulipas, Sinaloa.

Park locations: Rio Grande between stables and movie set; between Alamito Creek and Presidio

I encountered this aquatic herb, with milky sap and rootlike submerged stems, once in a marsh by the river, but never again in frequent revisits. The leaves, to 7 inches long, are broadly arrow-shaped, with large, earlike lobes at the base. These leaves rise above water on thick, fleshy petioles to 20 inches long. Submerged leaves are stalkless, linear in shape.

White three-petal unisexual flowers, to 2 inches wide, usually appear in groups (whorls) of three on elongated flower clusters. The clusters, with male blooms typically in the upper whorls, form on

stalks to 2 feet long or more. Each bloom may contain up to thirty stamens and many pistils, but female blooms form a ring of sterile stamens. Three cupped sepals enclose the fruiting heads, which sit atop clublike, recurving stalks. The globose fruits are achenes (dry, indehiscent, one-seeded), to 1/5 inch long.

Sagittaria tubers were used as food before recorded history and are now used in ponds and aquariums. *Sagittaria longiloba* tubers in Rio Grande lagoons were eaten by Native Americans (Warnock 1970).

Botanical notes: In 1753, Linnaeus named the genus *Sagittaria*, from the Latin *sagitta* (arrow), for the leaf shape. The epithet, *montevidensis*, refers to Montevideo, Uruguay, where the plant was collected by Friedrich Sellow before 1827. German explorer Sellow collected in Brazil and Uruguay (1814–31) before drowning in a river. The subspecies was described by St. Louis botanist George Engelmann as a species, *Sagittaria calycina*, in 1859. The Latin *calycina* (calyx-like) refers to the enclosing calyx. Dr. Joshua Hale, former student of Rafinesque at Transylvania College, collected the subspecies on the Red River near his home in Alexandria, Louisiana.

6. Arizona Snakecotton (*Froelichia arizonica*)
(Other common names: Arizona cotton-top, cotton weed)

Habitat: dry, rocky, gravelly slopes and canyons, arroyos, backcountry roads, limestone and igneous soils.

Geographic range: Texas: Trans-Pecos (El Paso southeast to Brewster County). US: Texas, southwest New Mexico, southern Arizona. Mexico: Chihuahua east to Nuevo León.

Park locations: Acebuches Canyon, Guale Mesa Road, Chorro Canyon, Arroyo Mexicano, Tascate Hills, slope west of Tres Papalotes, Cienega Mountain.

Arizona snakecotton is a stiffly erect perennial to 3 feet high, with one or several stout stems growing from an enlarged woody taproot. Sparingly branched in the upper reaches, the stems are cloaked with long soft, silky hairs.

The paired leaves, to 4 inches long, are narrowly oblong, and white-woolly like the stems, especially underneath. Stemless or short-petioled, the blades are typically massed at the base of the stems, and the sparse upper leaves are much smaller.

Atop long stalks, the flowers are spirally arranged in dense, spikelike cottony clusters. Each tiny bloom, less than 1/5 inch long, consists of five white-woolly, papery tepals (not distinctly petals or sepals), united into a woolly tube, and supported by membranous brown bracts. Within the tepals are five stamens, with united filaments and oblong anthers, and a style with a headlike stigma. Often, the flowers are so cloaked in bright white hairs that only their tips are visible.

The fruit is a winged utricle (bladderlike, one-seeded, indehiscent).

Botanical notes: In 1794, Conrad Moench, director of the Marburg University Botanical Garden in Austria, named this genus for the German botanist and zoologist at Erlangen, Joseph Aloys von Froelich. Von Froelich was an authority on the Gentian plant family. This species was named by John Thornber, dean of the College of Agriculture at the University of Arizona, in 1917, and collected by him and David Griffiths, a cactus authority with the USDA, in 1902, in the Santa Rita Mountains of Arizona.

7. Texas Shrub (*Iresine leptoclada*)
(Other common names: slender-stem bloodleaf)

Habitat: sunny, rocky, brushy slopes, scrub desert; limestone, igneous soils.

Geographic range: Texas: Big Bend (Presidio, Brewster Counties). US: Texas. Mexico: Chihuahua, Coahuila, south to Durango, San Luis Potosí, Zacatecas.

Park locations: Arroyo Mexicano, east of La Mota, Sauceda, McGuirk's Tanks, west of Solitario Peak

With slender, weak, gray-green branches, Texas shrub often grows supported by other shrubs. The stems are thickly coated with long silky hairs. Older branches may become spiny and reddish brown with age.

The alternate short-stemmed leaves are mostly egg-shaped, firm, and thick, silky-haired but hairless with age. The blades turn reddish on the margins and the tips, rarely "blood red" like the leaves of many *Iresine* species.

On elongated, congested flower clusters to 6 inches long, tiny white blooms form in short, few-flowered, silky-haired spikelets. Male and female flowers occur on the same plant. Male blooms are less than 1/10 inch long, female blossoms even smaller. Both are densely hairy. The five-tepal flowers are encased in thin, chaffy, cuplike bracts. Male blooms have five fused stamens with white filaments and protruding golden anthers. Female blooms, lacking stamens, have a short style with threadlike stigma branches.

The flattened egg-shaped fruit (utricle) contains one brown seed.

Botanical notes: The generic name *Iresine*, from the Greek *eiresione* (olive or laurel branch wrapped with wool, carried by singing boys in ancient Greek festivals), alludes to the woolly hairs around the calyx. The epithet *leptoclada*, joining Greek words *lepto-* (thin), and *klados* (branch), refers to the slender branches. Described as *Dicraurus leptocladus* by Joseph Dalton Hooker in 1880, the species was transferred to the genus *Iresine* by James Henrickson and Scott Sundberg in 1986. Charles Wright collected a specimen in 1849, on a military expedition across West Texas.

8. Fleshy Tidestromia (*Tidestromia carnosa*)
(Other common names: fleshy honeysweet)

Habitat: clay flats, dunes, creeks, roads, barren slopes; often gypseous, saline soils.

Geographic range: Texas: Big Bend (Brewster, Presidio Counties near Rio Grande). US: Texas. Mexico: northeast Chihuahua.

Park locations: South Lajitas Mesa, Fresno Creek, West Fork of Comanche Creek

Of conservation concern: see Appendix A

Fleshy Tidestromia is a yellow-green, mostly hairless annual, with prostrate, widely spreading, succulent stems that form extensive mats after summer rains. Although fleshy, the stems can be brittle and membranous when dry. As cooler fall temperatures prevail, the foliage may turn a blushing red. About 1 inch long and nearly as wide, the leaves are often spoon-shaped but variable and at times asymmetrical, with netlike veins.

With late summer and fall rains, flowers appear in clusters of two or three, on stalks at the leaf nodes. Each minute bloom consists of five membranous yellow tepals. Within the tepals are five stamens, with filaments united to form a cup at the base, and oblong yellow anthers with two pollen sacs; a globose ovary; an absent style, and bilobed stigmas. The blooms are supported by three thin, spongy cuplike bracts, which over time redden and thicken. Often, the bracts and flowers bear downy hairs.

The globular, slightly flattened fruits are utricles, less than 1/10 inch long, enclosing a single seed.

Botanical notes: In 1916, Paul Standley, curator of the US National Herbarium, named this genus for Ivar Tidestrom, a Swedish-born American botanist and range plants specialist for the USDA. The Latin epithet, *carnosa*, signifies "fleshy." The species was described by Julian Steyermark in 1932 as a variety (*Cladothrix lanuginosa* var. *carnosa*) and collected by him and John Moore in 1931, near Study Butte in Brewster County. Steyermark later coauthored *Flora of Guatemala* (1946–58) with Paul Standley and authored *Flora of Missouri* (1963). In 1943, Harvard botanist Ivan Johnston elevated Steyermark's variety to species status and placed it in the genus *Tidestromia*.

9. Woolly Tidestromia (*Tidestromia lanuginosa*)
(Other common names: espanta vaqueros, honeymat, woolly honeysweet)

Habitat: gypseous clay, alluvial flats, low hills, desert scrub, grasslands; limestone, igneous soils.

Geographic range: Texas: Trans-Pecos (excluding Pecos, Terrell Counties); much of South Texas; parts of Panhandle, Permian Basin; scattered sites. US: southwest, north to South Dakota, Missouri, Illinois; Louisiana. Mexico: Baja California east to Tamaulipas; Sinaloa to San Luis Potosí. Elsewhere: West Indies.

Park locations: Vista de Bofecillos, Old Abandoned Road Trail, Agua Adentro, lower Arroyo Mexicano, Yedra Canyon, Llano Loop, Road to Nowhere

This rapidly growing, shallow-rooted annual, with widely branching stems, forms extensive patches after rains. The foliage is gray green to ashy white, covered with long, woolly, branching hairs. The stems turn red and hairless with age. Highly variable, the leaves may be oval, broadly spoon-shaped, or rounded, nearly stemless to long-petioled, papery at the base, and occasionally asymmetrical.

Inconspicuous flowers form almost any time, in small headlike clusters (glomerules) in the leaf axils. Each bloom consists of five brownish-yellow tepals, supported by three membranous, woolly bracts. The tepals surround five stamens, with filaments fused to form a basal cup, and projecting, oblong yellow anthers; an ovary; and short style with bilobed stigmas. The fruit is an utricle with one brown seed.

Known in Spanish as *espanta vaqueros*, this herb reportedly spooks horses. In Sonora, the leaves are applied to ant bites and put in a tea to treat measles (Gentry 1998). The Seri of Sonora cook the herbage in water and make a shampoo to cure headaches (Felger and Moser 1974).

Botanical notes: Thomas Nuttall described the species as *Achyranthes lanuginosa* in 1837, after collecting the plant on "sand-beaches of Great Salt river, Arkansas," documented in his *A Journal of Travels into the Arkansas Territory during the Year 1819*. The Latin *lanuginosa* (woolly) refers to hairs on the foliage. In 1916, Paul Standley placed this species in the genus *Tidestromia*.

10. Copper Zephyrlily (*Habranthus longifolius*)
(Other common names: yellow-flowered rain-lily, cebolleta, rain lily, zephyr lily, fairy lily)

Habitat: often on grasslands and alkaline flats in sandy and gravelly soils.

Geographic range: Texas: Trans-Pecos (excluding Reeves, Pecos, Terrell Counties), much of Permian Basin; Garza County. US: Texas, southern New Mexico, Arizona. Mexico: northern Mexico south to Zacatecas, San Luis Potosí.

Park locations: Llano Loop, Righthand Shutup

Plants in this genus are known as rain lilies because they usually bloom quickly after the first rains. This low perennial grows from a dark brown bulb. The plant lacks a stem, but produces a few slender, sheathing, linear leaves that are grasslike and quickly drying and curling at the tips. The smooth blades range from 6 to 12 inches long.

After late spring or summer rains, this lily puts up one or two hollow flowering stalks, as much as 10 inches long, each supporting a single yellow, funnel-shaped flower. Seldom open widely in daytime, the blooms are 1 inch long, typically nodding or suberect, with six tepals of unequal length with pointed tips. Within the tepals are six stamens with orange-yellow anthers, the anthers distinctively not erect but mostly horizontal or downward curving, and a style with a trilobed stigma. The flower is supported and partly enclosed by a tan to pinkish tubular spathe (sheathing bract).

Perhaps 3/5 inch wide and not quite as long, the fruit is a thin-walled, three-chambered capsule, with each chamber enclosing numerous shiny black seeds.

Botanical notes: The genus name, *Habranthus*, from the Greek *habros* (delicate, splendid) and *anthos* (flower), was coined by Reverend William Herbert in 1824 to describe an "elegant little plant" from South America. Herbert, later Dean of Manchester, was a gardener, botanical artist, and bulbous plant expert. In 1837, he authored an extensive work on the Amaryllis plant family, with his own, colored illustrations. The epithet, *longifolia*, from the Latin *longus* (long) and *folium* (leaf), refers to the leaves. In 1880, William Hemsley, keeper of Kew Herbarium in London, described this species as *Zephyranthes longifolia* in a study of new plants from Mexico and Central America. Josiah Gregg collected the plant in 1848, at La Azufrosa in Coahuila, Mexico. Gregg cited a vernacular name of "cebollita," and noted that the herb was a "remedy for cows for maggots in ulcers." In 2010, Raymond Flagg, Gerald Smith, and Alan Meerow moved the plant to the *Habranthus* genus.

11. Arroyo Fameflower (*Talinopsis frutescens*)
(Other common names: arroyo flameflower)

Habitat: rocky slopes, bluffs above desert drainages, on limestone and igneous soils.

Geographic range: Texas: Trans-Pecos (El Paso, Jeff Davis, Presidio, Brewster Counties). US: Texas, New Mexico (Doña Ana County). Mexico: Chihuahua, Coahuila south to Durango, Zacatecas, San Luis Potosí.

Park locations: Las Cuevas, Upper Guale Mesa, Madrid Falls, Cinco Tinajas, Skeet Canyon, Cerro Chilicote Loop, Tascate Hills

Arroyo fameflower is a twiggy subshrub to 2 feet high, with fibrous tuberous roots and slender ascending branches. Weak stemmed, the plant is often supported by other shrubs. Young stems bear tufts of white hairs at enlarged nodes but are otherwise hairless. Older stems are woody, grayish.

Stemless, fleshy round leaves form circular clusters, bundles, or opposite pairs at the stem nodes. The seemingly swollen, often upturned leaves develop hardened red tips, or turn red.

Stemless five-petal flowers cluster at the branch forks. Color varies, and shades change throughout the day. Petals may be pale violet, lavender, white, or white flushed with color. The petals surround five stamens with oblong yellow anthers, a three-chamber ovary, and a short style with three stout white stigma branches.

The fruit is a three-valve capsule (exocarp) to 3/5 inch long, which separates from a six-valve inner part (endocarp). The tiny seeds are hooked or curved, and grainy.

Botanical notes: The genus name joins *Talinum* (the name of a related genus) and the Greek suffix *-opsis* (alike). The species name is the Latin *frutescens* (somewhat shrubby). Harvard botanist Asa Gray described the genus and species in *Plantae Wrightianae* in 1852, from a Charles Wright collection in 1849, on an army expedition across West Texas to New Mexico. Gray cited a plant location in "mountain valleys, seventeen miles east of the Rio Grande, New Mexico, Sept." Most likely, the site was in southern Hudspeth County, Texas. *Plantae Wrightianae* was Gray's account of plants collected largely by Wright on this trip and another in 1851–52 as part of the United States and Mexico Boundary Survey.

12. Poison Ivy (*Toxicodendron radicans*)
(Other common names: eastern poison ivy, poison oak, hiedra mala, sumaque)

Habitat: near springs, other water sources; shaded canyons, arroyos.

Geographic range: Texas: Trans-Pecos (Culberson, Jeff Davis, Presidio, Brewster, Val Verde Counties); much of rest of Texas. US: east half of US, Texas to South Dakota; Arizona. Elsewhere: much of Mexico, western Guatemala; West Indies, Europe, Asia.

Park locations: Ojito Adentro, Arroyo Mexicano, Chorro Canyon, Yedra Canyon, Fresno Canyon

Poison ivy is a shrub to 6 feet high, also a creeping or climbing vine that attaches to trees with dark aerial roots. Reproducing from rhizomes as well as seeds, this robust sumac forms trunks as much as 6 inches wide. Typically, the alternate, long-petioled leaves are divided into three deeply lobed or toothed leaflets. The leaflets, to 8 inches long, may turn red or orange in the fall.

Poison ivy is usually dioecious, with male and female blooms on separate plants. In late spring and summer, yellowish white or greenish white, five-petal blooms, 1/8 inch wide, appear in clusters from the leaf axils. Male blooms have five stamens with white filaments and yellow anthers. Female blooms have smaller petals, a spheric ovary, and a short stout style with a trilobed white stigma. The fruit is a light-colored, berrylike drupe (fleshy, indehiscent, with a stony inner layer), with a bony egg-shaped seed.

Some people are allergic to the oil (urushiol) in the sap, especially after repeated contact, so the plant is commonly regarded as an obnoxious weed. However, the sap has been used as a dye and as an ingredient in indelible ink and varnishes. The founder of the Jamestown settlement, Captain John Smith, published an account of this plant in the early 1600s and named the plant poison ivy (*Wayne's Word: An Online Textbook of Natural History*).

Botanical notes: The genus name, from Greek words *toxikos* (poison) and *dendron* (tree), refers to the plant's toxic oil. The Latin species name is *radicans* (rooting) because the plant reproduces from rootstocks as well as seeds. Philip Miller, superintendent of the Chelsea Physic Garden in London, described the genus in *The Gardeners Dictionary* (1754). Linnaeus named the species *Rhus radicans* in 1753, but in 1891, German botanist Carl Ernst Otto Kuntze moved the plant to the *Toxicodendron* genus.

13. Palmer's Bluestar (*Amsonia palmeri*)
(Other common names: tubular slimpod)

Habitat: sandy, gravelly, rocky arroyos, edges of intermittent creeks and springs, desert canyons, low hills.

Geographic range: Texas: Trans-Pecos (El Paso, Hudspeth, Presidio, Brewster Counties). US: Texas, southwest New Mexico, Arizona. Mexico: Chihuahua.

Park locations: Tapado Canyon, Horse Trap Spring, drainage from Sauceda to Cinco Tinajas, Tascate Hills, Cienega Gorge

This dogbane is an erect, shrubby perennial, approaching 3 feet high, with leafy stems growing from woody roots. The stems are often densely cloaked with soft hairs.

On short petioles, the alternate leaves are notably erect, narrowly lance-shaped, and often shorthaired. To 3 inches long or more, the lower blades are larger, more lancelike, the upper blades smaller and more linear.

Small tubular flowers appear in crowded clusters at the ends of branches. Each funnel-shaped white bloom has a slender greenish or yellowish tube, which is hairy inside, with five abruptly spreading, slightly overlapping lobes at the tip. The white lobes, yellow at the base, are no more than 1/4 inch long, less than half the length of the tube. Not readily visible, the stamens and stigma are included within the floral tube. Also concealed is the pistil, with two ovaries joined by a common style.

The fruits are narrow cylindrical follicles, mostly hairless, as much as 5 inches long. Each follicle contains numerous chestnut brown, corky, cylindric seeds.

Botanical notes: The genus name *Amsonia* was first used in 1757 by John Clayton, county clerk and early plant collector of colonial Virginia. Clayton probably used the name to honor his contemporary John Amson, a physician and mayor of Williamsburg, Virginia. In 1788, British-born American botanist Thomas Walter formally described the genus in his *Flora Caroliniana*, the first flora written by an American, which chronicled his botanical collections in South Carolina. In 1877, Asa Gray named this species for Edward Palmer and described it from specimens raised from seeds collected by Dr. Palmer. British-born Palmer was a prolific botanical, archaeological, and zoological collector throughout the southwest United States and Mexico.

14. Antelope Horns (*Asclepias asperula*)
(Other common names: spider milkweed, green-flowered milkweed, hierba lechosa, talayote, inmortal)

Habitat: rocky, sandy terrain, desert flats, grasslands, mesic mountain slopes; limestone, igneous soils.

Geographic range: Texas: Trans-Pecos, much of Edwards Plateau, north-central, north Texas, Panhandle; scattered in Gulf Coast, Coastal Bend. US: southwest; Idaho, Texas to Nebraska. Mexico: Chihuahua to San Luis Potosí.

Park locations: Lefthand Shutup, north and northwest Solitario rim, upper slopes of Cienega Mountain

Antelope horns is a low perennial to 16 inches high, with often prostrate, radiating stems. Minute hairs render the maroon-tinged stems rough to the touch. Like most milkweeds, the stems ooze a milky sap when damaged or broken. The short-stemmed, pointed leaves are narrowly lance-shaped, often folded upward, and sparsely hairy.

Unusual bowl-shaped flowers form in a crowded terminal cluster to 3 inches wide or more. The flowers have five upward-curving, pale yellow green petals, to 1/2 inch long, accented by five dark purple, tubular hoods. Extending from the stamen column, the hoods turn upward to rounded, bulging tips. The stamen column contains five bright white, fused stamens with knoblike anthers.

This milkweed is named for its erect fruit pods, which taper and curve like antelope horns. On stalks abruptly bent downward, the wrinkled, mostly hairless pods are as much as 5 inches long. A pod stores many tawny egg-shaped seeds, each with a tuft of silky hairs that help to disperse the seed.

Hairs attached to the seeds were used to insulate gloves and vests (*One Hundred Texas Wildflowers* 1993). The herb was used by Native Americans to treat respiratory infections and as an aid in childbirth (Traditional Herbal Blogspot: *Traditional Herbal Medicine Herbs*). Although poisonous to livestock and other animals, the herb is an important host plant of monarch butterfly caterpillars.

Botanical notes: In 1753, Linnaeus named this genus for the Greek god of medicine, Asklepios, due to the medicinal use of some species. The species name, *asperula*, diminutive of the Latin *asper* (rough), refers to slightly rough hairs on the herbage. The species was described in 1844 by French botanist Joseph Decaisne as *Acerates asperula*, from a specimen collected by zoologist Auguste Boniface Ghiesbreght in the Mexican state of Hidalgo. From 1837 to 1840, Ghiesbreght traveled on a Belgian expedition to Cuba and Mexico. In 1954, Robert Woodson placed this species in the genus *Asclepias*.

15. Sidecluster Milkweed (*Asclepias oenotheroides*)
(Other common names: zizotes, hierba de zizotes, longhorn milkweed)

Habitat: rocky desert slopes, flats, brushy arroyos, scrub desert, grasslands, disturbed sites; clay soils at low to mid-elevations.

Geographic range: Texas: Trans-Pecos; much of Edwards Plateau, central and south Texas, Gulf Coast; scattered elsewhere. US: Louisiana west to Arizona; Oklahoma, Colorado. Elsewhere: Mexico to Costa Rica.

Park locations: Fresno Canyon, Palo Amarillo Creek, Arroyo Mexicano, Pila Montoya Trail, Cienega Mountain

Sidecluster milkweed is a 12-inch-high perennial, with one or few sturdy stems growing from a thick, turnip-shaped root. To 6 inches long, the opposite leaves are oblong, wavy on the edges. The blades are covered with short, stiff hairs, especially underneath.

The intricate flowers form on short stalks in rounded clusters. Each bloom has five pale greenish-white or greenish-yellow lobes that are oblong or lance-shaped and abruptly downturned. The five erect hoods are tubular, greenish below, cream-colored, and flaring outward above. Rising above the stamen column, the hoods bear a tiny arching claw at the tip.

The erect fruit pod is a spindle-shaped follicle to 4 inches long, which splits to expose silky-haired oval seeds.

This milkweed, a larval host for monarch butterflies, contains glycosides toxic to many monarch predators. Believing that the milky sap caused skin rashes, the Spanish named the plant *zizotes* (skin sores). However, Native Americans used the sap in a compress to soothe skin irritations.

Botanical notes: The species name *oenotheroides* means "like the genus *Oenothera*." This species was described in 1830 by German botanists Adelbert von Chamisso and Diederich von Schlectendal, with the Royal Botanical Garden in Berlin, and collected by German naturalists and explorers Christian Schiede and Ferdinand Deppe, in the Mexican state of Veracruz in 1828.

16. Bear Mountain Milkweed (*Asclepias scaposa*)
(Other common names: stalked milkweed)

Habitat: dry, open, gravelly hills, flats; on limestone, novaculite.

Geographic range: Texas: Trans-Pecos (Presidio, Brewster, Reeves, Pecos, Terrell Counties). US: Texas; southern New Mexico? Mexico: Coahuila to San Luis Potosí, Zacatecas.

Park locations: Solitario northwest rim, north of Righthand Shutup

This dwarf milkweed, to 8 inches high, was previously unknown in BBR and Presidio County. Known mostly from the Glass Mountains and Marathon Basin of Brewster County, this dwarf has made its way to BBR, where I found it in the Solitario on one lucky day. Bear Mountain milkweed has slender, mostly unbranched stems growing from large turnip-like rootstocks. The thickish stems are covered with soft white hairs.

A few pairs of opposite, petiolate, leaves grow low on the stems. The oblong blades are conspicuously wavy, distinctly veined, and folded upward. As much as 2 1/2 inches long, mature leaves are dark green but turn coppery or maroon on the margins. Leaves on park plants are hairy above and below, especially on the margins.

Flowers form on long hairy scapes, clustered well above the leaves. Each bloom rests on a slender, often sagging stalk. Each flower consists of five petal-like, reflexed lobes, five arching hoods, and a central stamen column. The brown to maroon, oblong lobes are abruptly downturned. The cream-colored hoods are erect, rounded inward, and toothed at the upper edge. A needle-shaped horn protrudes above the hood, arching inward over the cream-colored stamen column.

The fruits are spindle-shaped pods to 2 inches high, with numerous tiny seeds, each tipped with silky hairs.

Botanical notes: The species name is the Latin *scaposa* (scape). Anna Murray Vail, a student of Nathaniel Lord Britton and specialist in the milkweed family, documented the species in 1898 in the *Bulletin of the Torrey Botanical Club*, based on a specimen collected by Charles Wright in 1851 or 1852 in "New Mexico, near Santa Rita." This location is considered dubious—no specimens exist in New Mexico herbaria. Britton was the first director of the New York Botanical Garden, and Vail was the first librarian.

17. Wavyleaf Twinevine (*Funastrum crispum*)
(Other common names: wavyleaf milkweed vine)

Habitat: desert to mesic mountain environs; rocky, gravelly, sandy arroyos, canyons, slopes, open woodlands; limestone, igneous soils.

Geographic range: Texas: Trans-Pecos (excluding El Paso County), much of Edwards Plateau; parts of north-central, north Texas, Permian Basin, Rolling Plains, Panhandle. US: Texas to Arizona, Oklahoma, Colorado. Mexico: south to San Luis Potosí, Aguascalientes.

Park locations: 1 1/2 miles west of Agua Adentro, Righthand Shutup Trail

This is a curious perennial twinevine, strangely scarce in BBR, and a delight to find, with twining or trailing ashy green stems. The color results from short appressed hairs.

The leaves are lance-shaped or linear, to 4 inches long, with remarkably wavy, crinkled margins. Short-petioled, the leathery blades have pronounced white-haired midribs and veins. Leaf color varies, often ashy green with maroon markings.

Small flower clusters form on short stalks. Flower colors have been described as greenish purple, brownish green, and bronze, but park flowers have five maroon lobes, which are white at the base and on the margins. Each bloom has a central stamen column, composed of five stamens with fused filaments, and anthers united in an anther head. The crown consists of a fleshy ring at the stamen column base, and five white bladderlike cavities attached to the anthers.

To 5 inches long, the fruits are narrow spindle-shaped pods, tapered at the tip. A pod contains numerous seeds tipped with white hairs.

Botanical notes: The genus name *Funastrum*, combining the Latin *funis* (rope) and the suffix *-astrum* (somewhat similar), refers to the twining stems. The Latin species name *crispum* (wavy) aptly characterizes the leaves. This twinevine was described by British botanist George Bentham as *Sarcostemma crispum* in 1849, in his *Plantas Hartwegianas*. German explorer Karl Theodor Hartweg, who collected for the Horticultural Society of London (1836–43), found this species in the Mexican state of Aguascalientes in 1837. In 1914, Berlin taxonomist Friedrich Schlecter, curator at the Botanical Museum at Dahlem, placed this species in the *Funastrum* genus.

18. Hartweg's Twinevine (*Funastrum hartwegii*)
(Other common names: climbing or narrowleaf climbing milkweed, fringed twinevine, güirote)

Habitat: sandy, gravelly arroyos, desert flats, brushy canyons; limestone, igneous soils, often near Rio Grande.

Geographic range: Texas: Trans-Pecos (El Paso, Hudspeth, Jeff Davis, Presidio, Brewster Counties). US: southwest US, except Colorado. Mexico: Baja California east to Coahuila, south to Zacatecas, Jalisco.

Park locations: Closed Canyon, Rancherías Canyon, Hoodoos, Tapado Canyon, Cerro de las Burras Loop, Agua Adentro Spring, Lefthand Shutup

This twinevine is a climbing, trailing vine, coiling and ropelike, with slender, branching green stems. The stems are mostly smooth, but white-haired at the nodes. On short petioles, the green or yellow-green blades are linear, tapered to the tip, often with two earlike lobes at the base.

Compact flower clusters form on stalks at the leaf axils. The five-lobed blooms have been described as purple or pink, but park flowers have maroon lobes with dark maroon midribs and white, hairy margins. In the center of each intricate bloom is the stamen column, and around this cap-like mass is the crown. The stamen column consists of five stamens with united filaments and anthers. The crown is composed of a low, fleshy rim at the base of the flower lobes and five waxy bladders attached to the anthers.

The fruit is a narrow spindle-shaped pod to 3 inches long, tapered and grooved. The pods contain numerous seeds, each with a tuft of silky hairs much longer than the seed.

Botanical notes: This species was described by Belgian-born French botanist Joseph Decaisne in 1840 as *Sarcostemma lineare*, in George Bentham's *Plantas Hartwegianas*, a survey of plant collections in Mexico by Karl Theodore Hartweg. Hartweg found this twinevine at León in the Mexican state of Guanajuato. After extensive explorations in Mexico, Hartweg traveled through Guatemala, Ecuador, Peru, and Jamaica before returning to England in 1843. After several name changes, this species was renamed *Funastrum hartwegii* by German taxonomist Friedrich Schlechter in 1914.

19. Soft Twinevine (*Funastrum torreyi*)
(Other common names: Torrey's twinevine)

Habitat: gravelly desert flats, dry, rocky slopes, brushy canyons, more open woodlands.

Geographic range: Texas: Big Bend (Presidio, Brewster Counties). US: Texas. Mexico: Chihuahua to Tamaulipas.

Park locations: Las Cuevas, Arroyo Mexicano, trail from Yedra Trailhead to Old Abandoned Road Trail, Lefthand Shutup, Lower Shutup

Soft twinevine is a climbing perennial vine to 10 feet long, with gray-green, wiry stems, usually densely cloaked with long soft hairs. On petioles to 1 inch long, the lancelike leaves are dark green above, gray green below, with a pointed tip and heart-shaped base. To 2 inches long, the drooping blades are slightly leathery, strongly veined, and hairy, especially on the margins.

Compact flower clusters with wheel-shaped blossoms appear on curving stalks, with each five-lobed bloom on its own short stalk. Flower lobes are bright white, but maroon around the base, with a maroon line down the midrib. The lobes are hairy outside and shaggy-haired on the margins. In the flower center is the cap-like stamen column, with five stamens with fused filaments and pollen-bearing anthers, and the stigma head. Surrounding the stamen column is the crown, composed of a low rim at the base of the flower lobes, and five erect, bladderlike white sacs attached to the anthers.

The fruits are hairy, tapered, spindle-shaped pods to 4 inches long. The pods split open to release slightly flattened seeds, each seed with tiny nipplelike projections and a long silky-haired tail.

Botanical notes: In 1876, Harvard botanist Asa Gray named this species *Philibertia torreyi*, for New York botanist John Torrey, his mentor and collaborator. The type specimen was collected by John Bigelow on the US-Mexico Boundary Survey, in "S. W. Texas," at "rocky hills on the Cibolo, a tributary of the Rio Grande, August, 1852." Bigelow subsequently served as botanist on the Pacific Railroad Surveys and collected extensively in northern California. In 1914, this and other species were moved to the *Funastrum* genus by German taxonomist Friedrich Schlecter.

20. Arizona Cockroach Plant (*Haplophyton cimicidum* var. *crooksii*)
(Other common names: cockroach plant, hierba de la Cucaracha, raiz de la cucaracha, atempatli, actimpatli)

Habitat: rocky slopes, canyons, foothills; sandy, gravelly arroyos, near water; mesquite scrub, desert grassland; igneous, limestone soils.

Geographic range: Texas: Trans-Pecos (El Paso, Hudspeth, Presidio, Brewster Counties). US: Texas, southern New Mexico, southern Arizona. Mexico: Coahuila west to Sonora.

Park locations: Contrabandista Spur Trail

Not previously known in BBR, this dogbane is very sparsely scattered in mountains near the Rio Grande, from the Franklin Mountains in El Paso County to its eastern limits in the Dead Horse Mountains of Brewster County. The plant is a shrubby perennial herb or low subshrub to 2 feet high with slender stems that are green above, often woody below. Erect or sprawling, the stems are moderately branched and coated with fine hairs.

The short-stemmed leaves are lance-shaped, green above, pale green below. Typically, the blades are tapered to the tip, smooth on the edges, but rough, with short, stiff hairs.

From spring to fall, yellow flowers appear singly in the upper leaf axils. Each short-tubed bloom abruptly expands into a flat, five-lobed rim. The oval lobes overlap and twist like a pinwheel. Within the flower's throat are five stamens with four-lobed anthers, and a style with an enlarged headlike stigma. The fruit is a pair of cylindrical, usually upright pods to 4 inches long. The black seeds are linear, with clumps of ashy white hairs at the ends.

Most plant parts are poisonous. A leaf extract is used as an insecticide to kill cockroaches, and the milky sap as a lotion to repel mosquitos (Vines 1960; Powell 1998).

Botanical notes: The genus name, joining the Greek *haploos* (simple) and *phyton* (plant), is explained by Asa Gray in 1886 as "simple plant, alluding to want of calycine glands and disk." Swiss botanist Alphonse Pyramus de Candolle (son of Augustin) described the genus in 1844. In 1942, cactologist Lyman Benson named this variety for Donald Crooks, with the USDA Bureau of Plant Industry. Crooks and Robert Darrow collected the type specimen in 1939 on "Prison Road," in the Santa Catalina Mountains of Arizona. Darrow coauthored with Benson *Trees and Shrubs of the Southwestern Deserts* (1954).

21. Spearleaf (*Matelea parvifolia*)
(Other common names: spearleaf milkvine, anglepod, talayote, littleleaf milkvine)

Habitat: rocky hillsides, above drainages, in rock crevices, below boulders; creosote, desert scrub; igneous, limestone soils, low to mid elevations.

Geographic range: Texas: Trans-Pecos (Jeff Davis, southeast Presidio, south Brewster Counties). US: Texas to California. Mexico: Coahuila, Sonora, Baja California, Baja California Sur.

Park locations: South Lajitas Mesa, one mile west of Lajitas, lower Fresno Canyon rim, Solitario rim above Righthand Shutup

With the smallest flowers and leaves of any Trans-Pecos *Matelea*, spearleaf is challenging to find. The twining, woody-based perennial vine often grows within larger plants, and the flowers are dark, hang downward, and bloom only briefly. Bristly hairs whiten the wiry green stems. The leaves are egg- or lance-shaped, less than 3/4 inch long, with two small lobes at the base. Short-stemmed, the blades bear stiff, bristly hairs like the stems.

The flowers form in stalkless clusters. Each bloom has five triangular lobes, brownish green, or dull shades of green, brown, yellow, or maroon. The bloom is bearded at the base and inside the tube. In the center is a cup-shaped crown, divided into five distinct segments, each with a prominent medial ridge. Within the crown is the stamen column with five fused filaments, a disklike anther head, and an enlarged stigma head.

To 3 1/2 inches long, the fruit is a tapered pod, wider below the middle and grooved lengthwise. The pod contains a small number of silky-tailed seeds.

Botanical notes: In 1775, French botanical explorer Jean Baptiste Aublet coined the genus name *Matelea* for a plant he collected (1762–64) while serving as apothecary botanist in French Guiana. In his seminal work, Aublet described over 208 genera, often using Creole names. *Matelea* was likely one of them. The species name *parvifolia* joins Latin words *parvus* (small) and *folium* (leaf) to emphasize the tiny leaves. The species was described by John Torrey in 1859 as *Gonolobus parvifolius*. Charles Parry collected the lectotype specimen in 1852, on the "sides of hills, cañon of the Rio Grande, below Mt. Carmel." The summit of the Sierra del Carmen in Mexico, above Boquillas Canyon and BBNP, was known as Mount Carmel. In 1941, Robert Woodson moved this species to the genus *Matelea*.

22. Texas Milkvine (*Matelea producta*)
(Other common names: anglepod, trailing hearts)

Habitat: alluvial flats, desert grasslands, mesquite thickets, rocky canyons, mesic mountains.

Geographic range: Texas: Trans-Pecos (excluding Reeves, Pecos, Terrell, Val Verde Counties). US: Texas, New Mexico, Arizona. Mexico: Chihuahua, Durango, Zacatecas.

Park locations: Las Cuevas, Solitario road south of Jackson Pens, Lefthand Shutup

This *Matelea*, common in the Trans-Pecos but scarce in BBR, is a perennial vine twining and climbing from a stout woody base. The plant is hairy: stems, leaf petioles, and flower stalks are covered with long white hairs and short yellow glandular hairs. The vine contains milky sap.

The long dangling, dark green leaves are arrow-shaped, with two earlike lobes at the base and a long-tapered tip. To 3 1/2 inches long on 1 inch petioles, the thick leaves are lined with white veins and at times bristly hairs.

Small flower clusters appear on short stalks at the stem nodes. Reportedly foul-smelling, each pale yellow green bloom forms on its own short, drooping stalk. The flower has five flared, oblong lobes, often densely hairy outside, but hairless within.

Inside the flower tube is a tiny cuplike crown, shallowly divided into five distinct lobes. The crown is attached to the base of the stamen column (five stamens with fused filaments), and surrounds the disklike anther head (fused anthers, joined to a flat stigma head).

The fruit is a smooth spindle-shaped pod tapered at both ends. The pods split to expose large seeds, with long white, silky-haired tails.

Botanical notes: The species name is the Latin *producta* (lengthened), a reference to the long stems. In 1859, John Torrey described this plant as *Gonolobus productus*, from specimens collected by Charles Wright (Jeff Davis County, 1851–52; Hudspeth County, 1852); John Bigelow and Charles Parry (Jeff Davis County); and Josiah Gregg (Cadena, Durango, Mexico, 1847). In 1941, Robert Woodson, curator of the Herbarium at the Missouri Botanical Garden, expanded the genus *Matelea* to include this species as *Matelea producta*.

23. Thurber's Stemsucker (*Pilostyles thurberi*)
(Other common names: Thurber's pilostyles)

Habitat: dependent on host species, open desert scrub, desert washes to mountain woodlands; often rocky limestone environs, low to high elevations.

Geographic range: Texas: Trans-Pecos (excluding Hudspeth, Jeff Davis Counties); parts of Panhandle, Permian Basin, Edwards Plateau, north-central Texas. US: Texas, New Mexico, southwest Arizona, southern California, southern Nevada. Mexico: Nuevo León west to Baja California, south to Durango, San Luis Potosí, Hidalgo, Oaxaca.

Park locations: Solitario northwest rim

I have encountered this tiny stem parasite, which is widely but very sparsely scattered across the southwest United States, only twice in thirty-four years of hiking. Thurber's stemsucker is one of the world's smallest flowering plants, rootless, stemless, leafless, without chlorophyll, incapable of photosynthesis. This intriguing perennial obtains water and nutrients by embedding its vegetative parts in the woody stems of host shrubs (feather dalea [*Dalea formosa*] or black dalea [*Dalea frutescens*] in the Trans-Pecos). The absent leaves are replaced by scalelike bracts.

Erupting through the bark, only the minute flowers, 2 to 3 mm wide, are visible on host stems. In a 1945 issue of *Desert Magazine*, a wise desert denizen, Jerry Laudermilk, likened the blooms to "crumbs of burnt toast." The flowers are unisexual, usually with male and female blooms on separate plants. After drying and falling, the blossoms leave conspicuous crater-like scars in the bark.

Opening as early as January, each bloom consists of four to seven overlapping, scalelike bracts, surrounding four or five bract-like sepals. The blossoms are brown or reddish brown, fleshy when fresh, with a fruity scent. Within the sepals is a central column with an expanded fleshy disk at the tip. In male blooms, nectar-bearing anthers are borne in rows beneath the disk margin. In female blooms, the anthers are replaced by rings of stigmatic hairs.

The fruit is a berrylike capsule, to 1/6 inch long, with many tiny, short-beaked, globose tan seeds.

Botanical notes: French botanist Jean Baptiste Antoine Guillemin, with the Museum of Natural History in Paris, named the genus *Pilostyles* in 1834, from the Greek *pileus* (felt cap) and *stylo* (column), "alluding to the central column of the flower surmounted by a cap in the manner of a little mushroom" [my French translation]. In 1854, in *Plantae Novae Thurberianae*, Asa Gray named the species for George Thurber, botanist on the US-Mexico Boundary Survey, who collected the type specimen in 1850, near the Gila River, probably in Arizona.

24. Date Palm (*Phoenix dactylifera*)
(Other common names: palma (or palmera) datilera, palma común, fénix, támara, dátil)

Habitat: cultivated or escaped near water.

Geographic range: Texas: Big Bend (Brewster, Presidio). US: Texas, Arizona, California, Florida. Elsewhere: Mexico (Baja California); Venezuela, Peru, Chile, West Indies.

Park locations: Lower Shutup

Cultivated in the Trans-Pecos, date palm occurs at Hot Springs and Dugout Wells in BBNP, but how the plant wound up in the Lower Shutup is anybody's guess. Date palm can reach 100 feet high, usually with a single gray trunk. The palm develops upright, arching, featherlike fronds, forming large canopies, and thick skirts of decaying leaves. The collapsing fronds leave prominent canoe-shaped leaf scars on the trunk.

On toothed petioles, the leaves have pinnate blades to 20 feet long, with up to 100+ leaflets in a V-shaped pattern. Each leaflet, to 1 foot long, is linear, stiff, with a sharp-pointed tip.

On thick spikes, three-petal white flowers form in compact, branching clusters. Male flowers with six stamens and female flowers with three pistils appear on separate plants. The fruit is a fleshy, berrylike drupe, with a papery, single-seeded, inner shell. The yellow to dark orange drupes may reach 2 inches long or more.

Human uses of the palm are countless. The seeds are crushed to make cooking oil and coffee, the sap used for syrup and liquor, the fruit for honey and wine, the palm heart in salads. Leaves and fibers are made into mats, roofs, and fences, and trunks into wooden posts, doors, and rafters. In traditional medicine, the juice is used to treat throat and stomach problems, fever, and memory loss, and as a contraceptive, anti-inflammatory, and antioxidant.

Botanical notes: *Phoenix* was the Greek name for date palm used by Theophrastus, possibly a reference to the Phoenicians who may have introduced the palm to the Greeks. The species name joins Greek words dactylos (date) and fero (I bear), meaning date-bearing. Linnaeus described the species in 1753, citing India as the habitat. The lectotype is an illustration by German naturalist Engelbert Kaempfer in *Amoenitatum Exoticarum* of 1712, an account of his travels through Russia, Persia, India, Indonesia, and Japan (1683–93).

25. Wright's Dutchman's Pipe (*Aristolochia wrightii*)
Birthwort or Dutchman's Pipe Family (Aristolochiaceae)

(Other common names: yerba del indio [Tarahumara], Wright pipevine, guaco)

Habitat: cliff bases, rock crevices in canyons, arroyos, rocky hillsides; amid boulders, shrubs, sheltered sites; mostly igneous environs.

Geographic range: Texas: Trans-Pecos (south Culberson, southeast Hudspeth, Jeff Davis, Presidio, Brewster Counties). US: Texas, New Mexico (Luna County). Mexico: Chihuahua, Coahuila, south to Durango.

Park locations: lower Monilla Canyon, Auras Canyon, upper Arroyo Mexicano, Tascate Hills, Pila Montoya Trail, Los Portales

This birthwort is a vinelike, weak perennial, with sprawling stems growing from fleshy, spindle-shaped roots. Commonly, the pale green or maroon stems are cloaked with long yellow, velvety hairs. Thick triangular leaves, with large, earlike basal lobes, are plentiful along the stems. Sometimes 1 1/2 inches long, the blades grow on curving petioles to 1 inch long. Typically, the blades are velvety like the stems, with palmate white veins and white or yellow hairs.

Odd aromatic flowers, reminiscent of upturned pipes, form near the stem tips. Each bloom is a tubular calyx (a flower's usually green outer base), with a long-tapered, throat-like upper lobe and a curving lower tube. Park flowers have a greenish-yellow lower tube, rust-brown upper lobe, and yellow throat. Usually, the yellow throat is marked with rust-brown blotches. The fruit is a silky-haired, globelike capsule to 3/4 inch long. The capsule splits into five valves, releasing many black, three-angled seeds.

This pipevine is host to larvae of the pipevine swallowtail butterfly. The plant contains aristolochic acid, which is harmless to the larvae but toxic to some predators. Consumption of pipevine leaves renders the caterpillars, and later the butterflies, unpalatable to predators.

Botanical notes: The genus name *Aristolochia*, from the Greek *aristos* (best) and *lochia* (delivery), means "best parturition," because one of the first species was used to ease the pain of childbirth. According to the ancient Doctrine of Signatures, resemblance of the curved flower to a fetus properly positioned before birth suggested this plant use. In 1856, in *The Botany of the Voyage of H.M.S. Herald*, the expedition's naturalist Berthold Carl Seemann dedicated this species to "the zealous and indefatigable traveller, Mr. Charles Wright, who collected it in four different localities in New Mexico . . . and between Western Texas and El Paso."

26. Havard's Agave (*Agave havardiana*)
(Other common names: Havard's century plant, Big Bend century plant, Chisos agave)

Habitat: rocky bluffs near Rio Grande, grasslands, wooded mountain slopes; igneous, limestone soils.

Geographic range: Texas: Trans-Pecos (Jeff Davis, Presidio, Brewster, Pecos Counties). US: Texas. Mexico: Chihuahua, Coahuila.

Park locations: west of Rancherías Trailhead, Madrid Falls Road, low hills northeast of Llano Loop, Solitario north rim, Contrabando Dome Trail

This imposing agave is a semisucculent, stemless perennial with rosettes of lance-shaped leaves, with recurving teeth on the edges and a dark spine at the tip. Rosettes are mostly single, but this agave spreads from rhizomes to form clones known as pups. To 2 1/2 feet long, the thick leaves are stiff but fleshy, gray green or blue green, and waxy. Rounded upward, the blades act as gutters to collect water. The leaves may bear bizarre, wavy imprints made by the teeth of leaves pressed together in bud.

Within twenty years, this agave forms a flowering scape to 20 feet high and dies. Climbing up to 18 inches per day, the scape supports the desert's lazy susan: a candelabra-like panicle with lateral branches supporting many blooms. Each tubular flower is greenish yellow, with six tepals, six long stamens with yellow anthers, and a style with a trilobed stigma. The oblong fruit is a chambered, beaked capsule to 2+ inches long. Each chamber stores two rows of black seeds.

Before the Spanish conquest, agave was used for drink, food, and fiber. The sap was made into pulque, an alcoholic beverage, by the Aztecs. Today, mescal and tequila are major Mexican exports. Native Americans roasted the flowering shoots, leaf bases, and pods (Warnock 1970), used the spines as needles and the fibers for cloth, mats, rope, baskets, and sandals.

Botanical notes: In 1753, Linnaeus chose the Greek word *Agave* (noble) for this majestic genus. In 1911, William Trelease, director of the Missouri Botanical Garden, named this species for Valery Havard, who collected it in the Chinati (1880), Guadalupe (1881), and Chisos (1883) Mountains. Charles Wright had collected the plant much earlier, in 1851, in Limpia Canyon of the Davis Mountains.

27. Lechuguilla (*Agave lechuguilla*)
(Other common names: shin dagger, maguey lechuguilla, mescal lechuguilla, tula ixtle)

Habitat: arid grasslands, open rocky (often south-facing) slopes, dry mesas, bajadas, especially limestone soils.

Geographic range: Texas: Trans-Pecos; some adjacent counties in Permian Basin, Edwards Plateau. US: Texas, southern New Mexico. Mexico: northern Mexico south to Mexico City, Hidalgo.

Park locations: Las Burras Road, Chorro Canyon, northeast of Llano Loop, Solitario north rim, Righthand Shutup

Lechuguilla is a stemless perennial with a basal rosette of daggerlike leaves. This semisucculent is also known as shin dagger, from its interaction with hikers' legs. Often spreading from rhizomes by suckering offshoots, lechuguilla can form large, low colonies. Largely within the CD, this agave is an "indicator species," reflecting that desert's geographic limits. The rigid yellowish-green leaves, to 2 feet high, are straight or scythe-like, arching inward. Blades are armed with silvery gray terminal spines and vicious down-curved, prickly teeth on the margins.

Once before dying, in late spring or summer, lechuguilla produces a leafless scape to 16 feet high. The scape bears a long spikelike flower cluster with short lateral branches. Flowers are funnel-shaped, with a short tube and six slender, spreading tepals. Protruding from the flower tube are six stamens with yellow anthers. Each bloom is a stunning mix of colors: light green, pale purple, maroon red, yellow. The dark brown, beaked fruit is an oblong, three-chambered, leathery capsule to 1 inch long. Each chamber contains two rows of many black seeds.

Fibers, ixtle or Tampico fiber, are extracted from the leaves to make rope, twine, mats, and brushes. Native Americans roasted flower stalks for food, but a glycoside in the plant, saponin, is toxic to livestock. Amole (lechuguilla roots containing saponin) has long been used as a soap substitute. Tarahumara, an indigenous tribe in Chihuahua, used saponin in the leaves to poison arrows and put it in water to poison fish (Pennington 1963).

Botanical notes: The species name is the Spanish diminutive of *lechuga* (lettuce), or "little lettuce." John Torrey, professor of chemistry at the College of Physicians and Surgeons in New York, and chief assayer of the US Mint in New York, described this species in 1859. The lectotype was collected by Charles Wright in 1849, on "high prairies of the San Felipe," near Del Rio in Val Verde County, Texas.

28. Smooth Sotol (*Dasylirion leiophyllum*)
(Other common names: desert candle, desert spoon, green sotol, smooth-leaf sotol, sotol brillante)

Habitat: gravelly hills, bajadas, foothills, sotol grasslands, mountains; mid-elevations.

Geographic range: Texas: Trans-Pecos (Jeff Davis, Presidio, Brewster, also Hudspeth, Culberson, Pecos Counties). US: Texas, southeast New Mexico. Mexico: Chihuahua, Coahuila.

Park locations: Las Cuevas, Skeet Canyon, Llano Loop, Tascate Hills, Solitario northwest rim, Buena Suerte Trail

This leaf-succulent shrub grows from a woody, trunk-like stem with a rosette of ribbonlike leaves. The shrub reproduces vegetatively as well as by seeds, with plants sprouting from buds at the leaf base. Sotol grasslands, prominent in the CD, form a vegetation band in foothills below the mountains.

To 1 inch wide, the leaves have downcurved, yellow or red prickles on the edges, fibrous tips, and spoon-shaped bases (hence the name *desert spoon*). The leaves have been described as smooth, flat, grooved, shiny, and twisted, and many are.

From May to August, this sotol develops a scape to 16 feet high, with membranous leaves and spikelike flower clusters with thousands of blooms. Barton Warnock compared these panicles to "giant artist brushes." Male and female blooms appear on separate plants. Each six-lobed bloom is greenish white. Male flowers form six protruding stamens; female blossoms form short styles. The winged fruit is a leathery, three-sided capsule to 1/3 inch long, with a three-angled seed.

Native Americans roasted young flower stalks and sotol hearts in mescal pits. The heart is used by ranchers as an emergency cattle food and is fermented to make the alcoholic beverage, sotol. Leaves are used to make mats, ropes, baskets, sandals, and hats. The stalks, long employed as thatching for roofs, corrals, and pens, are becoming popular as hiking sticks.

Botanical notes: The genus name, from Greek words *dasy-* (dense) and *leirion* (lily), alludes to the flower clusters. The species name joins Greek words *leios* (smooth) and *phyllon* (leaf). The species was described by St. Louis botanist George Engelmann in a 1911 article by William Trelease. US Army surgeon Valery Havard collected the type specimen at "Presidio del Norte" in 1880.

29. Beargrass (*Nolina erumpens*)
(Other common names: mesa sacahuista, basketgrass, foothill beargrass, foothill nolina, palmilla de Chihuahua)

Habitat: grasslands, foothills, exposed hillsides, gravelly, brushy arroyos; limestone, igneous soils.

Geographic range: Texas: Trans-Pecos (southeast Hudspeth, Presidio, Brewster, Pecos, Terrell Counties). US: Texas. Mexico: Coahuila.

Park locations: Black Hills, Solitario northwest rim, Pila Montoya Trail

Like sotol in habit, beargrass is a sprawling, fibrous, grasslike perennial to 8 feet across. From a stout, woody, mostly underground stem, this *Nolina* forms a rosette of clustered, ascending or spilling leaves. Each rosette may consist of several hundred overlapping leaves. To 3 feet long, the yellow-green blades are stiff yet flexible, grooved, serrulate on the margins, and frayed at the tips.

In late spring or early summer, beargrass forms flowering scapes that support pyramidal flower clusters to 3 feet long, with many rigid branches. Formed within the leaves, a cluster contains tiny white, often rose-tinted, flowers. Male and female, six-lobed blooms appear on separate plants. The fruit is a three-chambered capsule to 1/5 inch long, with a dark brown, spherical seed in each chamber.

The Chiricahua and Mescalero roasted young *Nolina* stalks for food (Castetter and Opler 1936). Other Native Americans made mats, sandals, rope, and baskets from the leaves (Maxwell 1967), or brooms and thatched huts (Havard 1885).

Botanical notes: In 1803, French botanist André Michaux named this genus for Abbé Pierre-Charles Nolin (1717–1796), director of the Royal Nurseries of France. While exploring America (1785–96), Michaux collected plants and seeds to send to France, many to Abbé Nolin. The species was described by John Torrey in 1859 as *Dasylirion erumpens*, and collected by Charles Wright, most likely in June 1849 in Uvalde County and June 1852 in Hudspeth County, in "ravines from Rio Grande to Eagle Springs." The species name, from Latin *erumpere* (to break out), may refer to the serrulate leaf margins. In 1879, in a revision of the Lily family, Sereno Watson moved this species to the genus *Nolina*.

30. Soaptree Yucca (*Yucca elata*)
(Other common names: soapweed, amole, palmella, palmilla, palmito, yuca, sota, cortadillo, soyate)

Habitat: hills, mesas, grasslands, gravelly washes, gypseous, silty flats at low elevations.

Geographic range: Texas: Trans-Pecos (excluding Terrell County). US: Texas, southern New Mexico, Arizona, southern Utah, southeastern Nevada. Mexico: Sonora east to Coahuila.

Park locations: Warnock Center, Javelin Road, Llano Loop, Solitario north rim, Contrabando Creek

From stout fibrous roots, soaptree yucca forms a single or sparingly branched, trunk-like stem to 15 feet high or more, with a rosette of ascending leaves and sagging dead leaves below. Some plants propagate suckering shoots and form groups of stems. This yucca is notoriously slow-growing, perhaps 1 inch per year, so a 15-foot plant could be 180 years old. To 3 feet long, the linear leaves are pale green, flexible, often twisting, with a spiny tip and peeling white margins.

From late spring to midsummer, this yucca puts forth flowering scapes as much as 7 feet long, well above the leaves, each scape with a branching flower cluster. A cluster contains many dangling, bell-shaped, creamy-white blooms. The blossoms, to 2 inches long, have three inner and three outer tepals, six stamens with arrow-shaped anthers, a short thick style, and lobed stigmas. Like many yucca species, soaptree yucca does not bloom every year. The fruit is a dehiscent oblong capsule to 3 inches long, with compartments containing winged black seeds.

Flower buds, young stalks, fruits, and seeds were eaten raw, cooked, or ground into meal (Powell 1998). Saponin in the roots and stems was used to make soap, shampoo, and a laxative. The leaves furnished fiber for baskets, mats, brushes, nets, and belts. The Apache used the fibers as dental floss (Little 1980). The seeds are a vital food source for larvae of the yucca moth, *Tegeticula*.

Botanical notes: Linnaeus named this genus *Yucca* in 1753, mistakenly applying the Caribbean name for cassava or *Manihot* to these plants. The species name, the Latin *elata* (exalted), refers to the tall trunks. This species was described by St. Louis botanist George Engelmann in the *Botanical Gazette* of 1882 and collected by US Army surgeon-botanist Joseph Rothrock in 1874, near Camp Grant in Graham County, Arizona. Rothrock collected on Lieutenant George Wheeler's surveys west of the 100th meridian.

31. Spanish Dagger (*Yucca torreyi*)
(Other common names: Torrey's yucca, old shag, Spanish bayonet, palma)

Habitat: desert scrub, grasslands, dry, brushy, or rocky slopes in foothills; low to mid-elevations.

Geographic range: Texas: Trans-Pecos, much of Edwards Plateau; parts of Permian Basin, South Texas Plains; Garza County. US: Texas, southern New Mexico. Mexico: Chihuahua, Coahuila.

Park locations: Warnock Center, Teepees Roadside Park, Old Abandoned Road Trail, Ojo Chilicote, Solitario northwest rim

Spanish dagger has a shaggy and disheveled, awkward habit, hence the name "old shag." The older trunklike stems are covered with withered leaves, and rosettes of green leaves are often asymmetrical, with blades diverging haphazardly. Seldom more than 7 feet high, Spanish dagger forms one to several stems that develop obvious trunks with age, and at times a few branches. The daggerlike leaves, to 4 feet long, are thick, rough, fibrous, and peeling on the edges. Leaf color is gray-, light-, yellow-, or blue-green.

Often in the spring of alternate years, this yucca forms short stalks with large flower clusters. The dense clusters with many branchlets are held partly above or within the leaf rosettes. Each creamy-white, bell-shaped flower is waxy, fragrant, often flushed with pale green or pink. The dangling blooms have six tepals, six stamens, one pistil, and distinct stigmas. The indehiscent fruits are spongy, short-tipped cylindrical capsules, 4 inches long, with many black seeds.

The fruits were eaten raw, cooked, or ground into meal. Yucca roots and stems were made into soap and a laxative (Powell 1998), leaves were used as fiber, and trunks as construction material. Insects lay eggs in the fruits (Warnock 1970), flowers are pollinated by the yucca moth (*Tegeticula* genus), and seeds are food for moth larvae (Powell 1998).

Botanical notes: In 1859, John Torrey described this species as *Yucca baccata* var. *macrocarpa*, but in 1908, John Shafer, museum custodian at the New York Botanical Garden, treated the yucca as a distinct species, naming it in honor of Torrey. Torrey, the foremost American botanist of his time, promulgated in America British botanist John Lindley's natural system of plant classification, which replaced Linnaeus's sexual system. By 1831, Torrey had cataloged North American genera according to the new system.

ASPARAGACEAE — ASPARAGUS FAMILY

32. Desert Holly (*Acourtia nana*)
(Other common names: dwarf desert peony, dwarf desert holly)

Habitat: creosote scrub, grasslands, rocky foothills, woodlands; sandy, clayey, loamy flats, gravelly washes, disturbed areas; various soils.

Geographic range: Texas: Trans-Pecos (excluding Reeves, Val Verde Counties), much of Permian Basin, western Edwards Plateau. US: Texas to Arizona. Mexico: Sonora east to Nuevo León, Durango south to San Luis Potosí.

Park locations: Llano Loop, Solitario Road west of Los Alamos turnoff, south of Tres Papalotes

Desert holly is a low perennial less than 1 foot high, which grows from a woody crown and may spread by rhizomes. The slender stems, with a distinctive basal cluster of rust brown, woolly hairs, form expansive patches blanketing the ground. The most striking plant feature is the holly-like leaves, since the herb is not often seen in bloom. Stiff, yet wavy, the stemless, clasping blades are light green but soon dried and tan. To 2 inches long, the alternate leaves have netlike veins and spiny projecting teeth along the margins.

A single sweetly fragrant flower head appears at the stem tip. The head is supported by a bell-shaped collection of leaflike bracts (phyllaries) in overlapping series. Mostly green with purplish tips, the phyllaries bear silky glandular hairs along the margins. A flower head contains up to twenty-four two-lipped florets, lavender pink or white, each to 2/3 inch long.

The fruits are oblong cypselae (dry, one-seeded, like achenes, but surrounded by an adherent calyx). The cypselae, 1/3 inch long, are grooved, ribbed, and covered with glandular hairs. Hairlike bristles at the cypselae tips produce feathery, globelike seed heads (see photo) and encourage wind dispersal.

Botanical notes: In 1830, Scottish botanist David Don, librarian at the Linnaean Society of London, "dedicated this . . . genus to Mrs A'Court, of Heytesbury House, Wilts, whose botanical taste and knowledge have long merited for her this compliment." Apparently, Mrs. A'Court was Mary Gibbs A'Court, wife of Lieutenant General Charles Ashe à Court, the MP for Heytesbury. In 1849, in *Plantae Fendlerianae*, Asa Gray named the species *Perezia nana*, from a collection by Josiah Gregg in 1847, "on a high and dry valley near Chihuahua." The Latin species name, *nana* (dwarf), refers to the low-growing habit.

33. Brownfoot (*Acourtia wrightii*)
(Other common names: pink perezia, Wright's desert peony, fluffroot, Wright's acourtia)

Habitat: sand, clay, grassy flats to wooded mountain slopes, near springs, along desert arroyos; limestone and igneous soils.

Geographic range: Texas: Trans-Pecos (excluding Terrell County), parts of Permian Basin, western Edwards Plateau, South Texas Plains. US: Texas to Arizona, Utah, Nevada. Mexico: Sonora east to Nuevo León, south to Zacatecas, San Luis Potosí.

Park locations: Arroyo Mexicano, Ojo Chilicote, near Pila Montoya Trailhead, Cienega Mountain.

Brownfoot is a stout, upright, much branched perennial to 4 feet high, growing from a large, bulbous crown. Like other *Acourtia* species, the woody crown and basal stems are covered with tufts of rust brown hairs. The oblong leaves, as much as 5 inches long, are tough, yet thin and papery. Often clasping, the stemless blades are coarsely serrate with wavy margins and obvious veins.

Flower heads form in rounded clusters at the branch tips. Glandular-hairy bracts support the flower head in two or three overlapping series. Each head contains up to a dozen honey-scented, two-lipped florets, pink, rose-red, or white. Each floret has a two-lobed upper lip and a toothed or notched oblong lower lip. The fruits, to 1/4 inch long, are slender spindle-shaped cypselae, which are ribbed and usually glandular-hairy. Buff bristles, at the fruit tips, form brushlike seed heads (see photo).

Native Americans made an astringent from the root (Warnock 1977). Kayenta Navajo used the plant as a gynecological aid for difficult labor and postpartum medicine (Moerman 2009).

Botanical notes: This species was described by Asa Gray in 1852, in *Plantae Wrightianae*, as *Perezia wrightii*, and named for Charles Wright, who in 1849 collected the plant "On the Rio Seco and westward; also on the Rio Grande, Texas; June." The Rio Seco is Seco Creek, a tributary of Hondo Creek. The specimen date would place Wright's party on the Leona River at Camp Leona (later Fort Inge), a frontier outpost south of Uvalde. In 1973, University of Maryland botanist James Reveal, and Robert King, with the US National Herbarium, moved this species into the genus Acourtia.

34. Cheeseweed (*Ambrosia monogyra*)
(Other common names: singlewhorl burrobrush, burrobush, needle-leaf burrobush, jecota, romerillo)

Habitat: sandy banks, mouths of desert canyons; drainages from desert to foothills, alluvial fans.

Geographic range: Texas: Trans-Pecos (excluding El Paso, Pecos Counties); Uvalde, Travis Counties. US: Texas to southern California, Nevada. Mexico: Baja California east to Chihuahua, south to Sinaloa.

Park locations: Fresno Canyon, Tapado Canyon, La Cuesta, South Fork of Alamo de Cesario Creek

Cheeseweed is a fast-growing, densely branched shrub, with leaves that emit a cheese-like odor when crushed. Seldom more than 5 feet long, the slender stems are hairless but resinous. Mostly upright, the buff or tan branches tend to arch and bow toward the ground. Persistent in deep sand and flooding, this shrub forms extensive thickets along water courses.

To 3 inches long, the alternate green leaves are stemless, threadlike, and sometimes lobed. The blades are involute and grooved above, and resinous below.

In summer and fall, small white flower heads form in leafy spikelike clusters with male and female heads. To 1/6 inch long, the male heads, with up to twelve tubular florets, are supported by shiny overlapping bracts. Female flower heads contain one floret and an ovary, with papery white bracts forming a saucerlike base.

The female flower head becomes a "bur," a fruit-bearing flying saucer. The spindle-shaped burs, 1/5 inch long, with wings around the middle, disperse the black fruits.

Native Americans made arrow shafts from the branches (Warnock 1977).

Botanical notes: The import of the Greek genus name *Ambrosia* (food of the gods) is unclear. The epithet *monogyra* (one turn or ring), refers to the fruit's single whorl of wings. The species was described as *Hymenoclea monogyra* by John Torrey and Asa Gray in 1849 but was moved to the *Ambrosia* genus by John Strother and Bruce Baldwin in 2002. The type specimens were collected by Lieutenant William Emory in Arizona, in the "valley of the Gila," and by Dr. Josiah Gregg "at Ojito [probably in New Mexico]."

35. Seepwillow (*Baccharis salicifolia*)
(Other common names: water wally, mule's fat, jara, jarilla, willowleaf baccharis, hierba del carbonero)

Habitat: sandy banks of Rio Grande, creeks, springs, floodplains, arroyos, disturbed sites.

Geographic range: Texas: Trans-Pecos (excluding Reeves, Pecos Counties), parts of South Texas Plains, west Edwards Plateau. Elsewhere: Mexico to Central, South America.

Park locations: Madera Canyon Campground, Las Cuevas, Agua Adentro Spring, Yedra Canyon, Sauceda, South Fork of Alamo de Cesario Creek

Seepwillow is a large thicket-forming shrub that grows along watercourses and is a reliable indicator of available groundwater. The ascending green stems, often grooved and resinous, grow in tight, leafy clusters from extensive roots. Willowlike, the arching leaves are elliptic, to 4 inches long, mostly stemless, finely toothed and glandular.

Typically flowering in summer and fall, seepwillow is not a willow at all, but a sunflower. The flower heads appear in many-branched, rounded clusters at the branch tips. Supported by bracts, the heads lack ray florets and consist of tiny white disk florets. The bracts are thin, dry, papery, often purple-tinged, and ciliate on the margins.

The plants are unisexual, with male or female flowers. Female flower heads contain up to 150 threadlike florets, each five-toothed. Male flower heads, with up to forty-eight florets, are slender below but funnel-shaped and five-lobed above. Mature female heads produce fine white bristles, resembling shaving brushes, which are more impressive than the blooms. The fruits, five-ribbed oblong cypselae to 1/16 inch long, bear plumelike pappus bristles at the tips.

Native Americans used the stems as arrow shafts and as matting for mud roofs. Stems were chewed to cure toothaches (Warnock 1974), and leaves were made into eyewash (Vines 1960).

Botanical notes: *Baccharis* (a plant named for Bacchus, Greek god of wine) was chosen for unknown reasons to name this genus. The epithet, *salicifolia*, implies "*Salix*-like leaves." Hipolito Ruiz López and Jose Antonio Pavón found this plant on their 1798 Peru/Chile expedition and named it *Molina salicifolia*. In 1807, Christiaan Persoon, in his *Synopsis Plantarum*, gave the plant its current name.

36. Trans-Pecos False Boneset (*Brickellia eupatorioides* var. *chlorolepis*)
(Other common names: false boneset)

Habitat: flats, grasslands, rocky canyons, mountain woodlands, near water; limestone, igneous soils.

Geographic range: Texas: Trans-Pecos (excluding Reeves County); part of Permian Basin, western Edwards Plateau. US: Texas west to Arizona, northern Colorado, Utah. Mexico: Sonora east to Tamaulipas, south to Michoacán.

Park locations: Cinco Tinajas, Righthand Shutup, Lower Shutup, near Los Alamos, Cienega Creek

This perennial sunflower is seldom more than 30" high, with clustered stems growing from a deep taproot. Erect or sprawling, the green, leafy stems are often branched above, and sometimes thickly coated with fine, short hairs. The stemless green leaves are linear, usually less than 2 inches long. The blades are finely hairy above, gland dotted below, and frequently folded along the margins.

In late summer and fall, flower heads form in small clusters, on leafy stalks at the branch tips. Each flower head is supported by slender, overlapping green bracts. The bracts are membranous, ciliate, often purple-tinged. Outer bracts are gland dotted. Each flower head contains up to thirty five-lobed, tubular florets. Less than 1/3 inch long, the florets are varicolored (pale yellow, creamy white, pinkish lavender, maroon red).

The ten-ribbed, one-seeded fruits are oblong brown cypselae less than 1/5 inch long. At the cypsela tip is the pappus (modified calyx), consisting of featherlike white bristles.

Botanical notes: In 1824, in *A Sketch of the Botany of South Carolina and Georgia*, South Carolina botanist Stephen Elliott named this genus for his friend John Brickell, Irish-born Georgia physician. Brickell, a plant collector in Georgia and South Carolina, was also a plantation owner, state legislator, and natural history professor. The species name connotes "like plants of the genus *Eupatorium*." The variety name, from the Greek *chloro* (green) and *lepis* (scale), refers to the bracts (phyllaries). Elmer Wooton and Paul Standley described this plant as *Kuhnia chlorolepis* in 1913, from a collection by Wooton's student, Orrick Metcalfe, at Mangas Springs, New Mexico, in 1903. In 1989, University of Texas botanist B. L. Turner gave this plant its current name.

37. Turner's Thistle (*Cirsium turneri*)
(Other common names: cliff thistle)

Habitat: crevices of vertical limestone cliffs, ceilings of overhangs, usually in partial shade.

Geographic range: Texas: Trans-Pecos (Presidio County east to Val Verde County). US: Texas. Mexico: adjacent Chihuahua and Coahuila.

Park locations: Los Portales, Lefthand Shutup, Terlingua Uplift

Of conservation concern: see Appendix A

This is the scarcest thistle in the Trans-Pecos, ranging from the Chinati Mountains in southern Presidio County east to Amistad National Recreation Area near Del Rio. Turner's thistle is robust, sprawling, and cliff-loving, with perennial roots and annual stems. Slender stems, to 18 inches long, grow from a woody crown. The pale green, grooved branches are cloaked with conspicuous, cobweb-like hairs.

As much as 1 foot long, the vicious-looking oblong leaves are yellow green and crowded on the stems. The blades are pinnately divided into triangular lobes, spiny at the tips and along the margins, and clasping at the base. While the leaves are mostly hairless, the mid-stem is often troughlike, filled with matted hairs.

Mostly in summer, flower heads appear singly at the stem tips, or in short terminal clusters. Spiny and spine-tipped bracts, in overlapping series, support the flower heads. The slender bracts are pinkish, reddish, or purplish in the upper half, ciliate on the margins, with some long hairs. A flower head contains as many as eighty tiny florets, each 1 inch long or more, with five slender lobes. Floret color has been described as rose purple, magenta, red, and reddish purple.

The ribbed fruits are buff oblong cypselae, less than 1/4 inch long. The cypselae bear a pappus of unequal, silky-white bristles at the tip.

Botanical notes: The genus name is borrowed from the Greek *kirsion* (thistle), a term used by Dioscorides. *Kirsion* is from *kirsos* (swollen vein), because some thistles were believed to cure swollen veins. Sul Ross State University botanist Barton Warnock discovered the plant in 1948, in Doubtful Canyon of the Del Norte Mountains southeast of Alpine, and described it in 1960 in *The Southwestern Naturalist*. Warnock named the thistle for his former student, Billie Turner, professor emeritus at the University of Texas at Austin. Turner authored *The Legumes of Texas* in 1959 and was principal author of *Atlas of the Vascular Plants of Texas* in 2003.

38. Wavyleaf Thistle (*Cirsium undulatum* var. *undulatum*)
(Other common names: pasture thistle, gray thistle, nodding thistle, plumed thistle)

Habitat: shrub desert, grassland, canyons, woodlands, roadsides, disturbed sites; limestone, igneous soils.

Geographic range: Texas: Trans-Pecos (excluding El Paso, Pecos, Terrell, Val Verde Counties); much of Edwards Plateau, north-central, north Texas; parts of Panhandle. US: west half, most of Midwest; Pennsylvania, Georgia. Elsewhere: Canada, Mexico (Sonora east to Coahuila, south to Durango).

Park locations: Skeet Canyon, near Sauceda, Lower Shutup, Pila Montoya Trail

The most common thistle in the park and in the Trans-Pecos, wavyleaf thistle is a gray green perennial growing from stout, creeping roots. Less than 3 feet high, the sturdy, leafy stems are single or sparsely branched above and coated with soft hairs.

Gray green like the stems, the alternate oblong leaves may reach to 8 inches or more from a clasping base. The crowded blades are rigid, yet wavy or even twisted, with yellow-spined, triangular lobes on the margins, and a prominent white-haired midrib. Both leaf surfaces may be blanketed with entangled, cottony hairs.

From spring to fall, up to ten or more flower heads form in terminal clusters, on stalks to 10 inches high. Each bell-shaped flower head contains many tiny disk florets in shades of lavender, pink, purple, and white. Green bracts, in as many as twelve overlapping series, support the flower head. Each bract bears a white glandular midrib and a yellow spine tip. The slender florets are tubular, with five unequal lobes.

The fruit is an oblong, slightly compressed, brown cypsela, 1/4 inch long, tipped with a pappus of feathery bristles much longer than the fruit.

Botanical notes: The species name, the Latin *undulatum* (wavy), refers to the leaf margins. Thomas Nuttall, English botanist, plant hunter, and Philadelphia resident, described this species as *Carduus undulatus* in 1818, in *The Genera of North American Plants*. According to Nuttall, the thistle was found "on the calcareous islands of Lake Huron, and on the plains of Upper Louisiana." In 1826, Curt Sprengel, German physician-botanist at the University of Halle, placed the species in the genus *Cirsium*.

39. Mexican Gumweed (*Grindelia oxylepis*)
(Other common names: none known)

Habitat: sandy soils near Rio Grande, moist sites at low elevations; desert flats, open fields, roadsides.

Geographic range: Texas: Trans-Pecos (Presidio County). US: Texas, southern New Mexico. Mexico: Chihuahua east to Coahuila, south to Durango, Zacatecas, San Luis Potosí.

Park locations: River Road west of stables before movie set, Grassy Banks

Of conservation concern in Texas: see Appendix B

Although more widespread in northern Mexico, this aster scarcely reaches into the United States at a few sites along the Rio Grande in Presidio County, Texas, and Doña Ana County, New Mexico. Mexican gumweed is an annual to 2 feet high, with a single or sparsely branched stem from sturdy taproots. The pale erect branches and other plant parts are hairless. Instead, the plant is protected by glandular protrusions on the flower heads and the leaves. The glands secrete a sticky resin (hence the name *gumweed*) that wards off predators.

Seldom more than 1 inch long, the thick oblong leaves have resinous, rounded teeth on the edges and a base that clasps the stem.

In summer, yellow flower heads with ray and disk florets appear singly or in clusters at the branch tips. Supported by overlapping lance-shaped bracts, each flower head contains up to thirty strap-like ray florets and many tubular disk florets. The triangular, often recurved, gland-dotted bracts produce gummy white resin. The oblong fruits are four-sided cypselae, 1/10 inch long, with a pappus of two or three bristles.

Native Americans used the gummy resin of some *Grindelia* species as glue, and to treat skin rashes. Dried leaves and flowers were used in a tea to treat bronchitis (Kleiman, Earle, and Wolff 1966).

Botanical notes: In 1807, German botanist Carl Ludwig von Willdenow named this genus for David Hieronymus Grindel, who founded the first Russian pharmaceutical journal. The epithet *oxylepis* combines Greek words *oxus* (sharp) and *lepis* (scale), possibly referring to the inner bracts enclosing the flower. In 1899, Edward Lee Greene described the species in his journal *Pittonia*, from a collection by Cyrus Pringle in Chihuahua in 1886. An Episcopal priest, Greene became the first botany professor at the University of California, Berkeley, and was later at Catholic University and the Smithsonian.

40. Shrubby Umbrella Thoroughwort (*Koanophyllon solidaginifolium*)
(Other common names: shrubby thoroughwort)

Habitat: igneous and limestone slopes, canyons, in partial shade, near water sources.

Geographic range: Texas: Trans-Pecos (Jeff Davis, Presidio, Brewster Counties). US: Texas to Arizona. Mexico: Chihuahua to Coahuila, south to Zacatecas.

Park locations: Las Burras Road, Cinco Tinajas, Skeet Canyon, Tascate Hills

This thoroughwort is a small shrub or subshrub to 3 feet high, with a strong goatlike aroma. The several clustered stems are stiffly erect and branched above. Tan or green, the stems are often coated with curved or bent hairs.

To 3 1/2 inches long, the leaves are broadly lance-shaped, tapered, sometimes with small, rounded teeth on the margins. The yellow green, short-petioled blades are paired, hairless, often shiny, with netlike veins.

In late summer or fall, small flower heads with only disk florets form in compact clusters at the stem tips. A head usually contains three to five white disk florets, to 1/8 inch long, often purple-tinged. Protruding from enclosing bracts, each narrow floret is tubular, with five glandular lobes. The slender bracts around the florets are lance-shaped, sharp-tipped, and membranous. Only 1/10 inch long, the fruits are five-ribbed cypselae with a pappus of thirty to thirty-five white, barbed bristles.

Botanical notes: The origin of *Koanophyllon* is unclear, possibly from Greek words *choano-* (funnel) and *phyllon* (leaf), for the funnel-shaped leaves. The name may also relate to the leaves as a source of dye. Manoel Arruda da Cámara named the genus in 1816, in pamphlets translated by Henry Koster in *Travels in Brazil*. The type species was a dye plant known as anil de Pernambuco. Brazilian botanist Arruda conducted expeditions in northeast Brazil. Koster, Portuguese-born British explorer, made extensive trips to Brazil. The species was described by Asa Gray in 1852 as *Eupatorium solidaginifolium*, from a specimen collected by Charles Wright in 1849, probably near Van Horn, Texas. The species name *solidaginifolium* combines the genus name *Solidago* and the Latin *folium* (leaf), implying "with *Solidago*-like leaves."

41. Tahoka Daisy (*Machaeranthera tanacetifolia*)
(Other common names: tansy or tansyleaf aster)

Habitat: gravelly, sandy soils, gypseous outcrops, desert flats to grasslands, rocky, wooded slopes; disturbed sites.

Geographic range: Texas: Trans-Pecos, most of Permian Basin, much of Panhandle; sites in north-central, north Texas. US: much of west half of US; Illinois, New York. Elsewhere: Alberta, Canada; Mexico (Chihuahua east to Nuevo León, south to Durango, Zacatecas, San Luis Potosí).

Park locations: mouth of Monilla Canyon, Grassy Banks

Seeds of this attractive aster, found growing near Tahoka, Texas, in 1898, later came to be sold commercially under the name Tahoka daisy. This herb is a leafy annual or biennial to 2 feet high, with erect or spreading, much branched stems growing from a taproot. The stems can be densely hairy and dotted with sticky stalked glands. The crowded oblong leaves, to 4 inches long, are dissected into narrow lobes, each tipped with a small hard spine. The alternate, stemless blades are hairy and usually glandular.

From spring to fall, single flower heads form on stalks at the branch tips. Each head contains up to forty lavender or purple ray florets and many yellow disk florets. The slender ray floret is straplike, while the five-lobed disk floret is tubular. Linear to lance-shaped, membranous bracts support the flower head.

To 1/6 inch long, the oblong tan fruit (cypsela) is ribbed, hairy, with a pappus of up to eighty barbed bristles.

Botanical notes: The epithet *tanacetifolia* joins the genus name *Tanacetum* and the Latin *folia* (leaves), alluding to "*Tanacetum*-leaves." This species was described by Karl Kunth in 1818 as *Aster tanacetifolius*, from collections by German explorer Alexander von Humboldt and French botanist Aimé Bonpland in Mexico about 1803. German taxonomist Kunth described 4,500+ plants collected by Humboldt-Bonpland on their five-year journey. In 1832, University of Breslau botanist Christian Gottfried Nees placed this aster in his new genus, *Machaeranthera*.

42. Manybract Groundsel (*Packera millelobata*)
(Other common names: Uinta ragwort, Uinta groundsel)

Habitat: cool, shady canyons, intermittent creeks, rocky slopes, woodlands, mid-high elevations.

Geographic range: Texas: Trans-Pecos (Hudspeth, Culberson, Jeff Davis, Presidio, Brewster Counties). US: Texas. Mexico: Coahuila, central Chihuahua.

Park locations: Agua Adentro Spring, oak-filled box canyon west of Cinco Tinajas

Manybract groundsel is a mostly hairless perennial to 20 inches high, which grows from persistent, often woody bases or occasionally creeping rhizomes. Up to a half dozen clustered stems arise from the roots.

The lacy fernlike leaves are a prominent feature of this sunflower. Short-stemmed, alternate on the branches, the lance-shaped leaves are up to 4 inches long and deeply dissected into twelve-plus segment pairs. The lateral lobes are usually toothed on the margins and larger than the terminal lobe.

From March to September, flower heads appear singly or in large clusters on hairless stalks. The yellow heads are flower collections, with female ray florets and minuscule bisexual florets crowded into a central disk. A head contains up to thirteen ray florets, each 1/3 inch long, and fifty-five tubular disk florets, each with a gradually expanded five-lobed border. A bell-shaped group of yellow-tipped green bracts supports each flower head.

The fruit is a five- to ten-ribbed cylindrical cypsela, tipped with a pappus of about sixty barbed bristles.

Botanical notes: Áskell and Doris Löve, authorities on alpine and arctic flora, described *Packera* in 1975, naming the genus for Canadian botanist John Packer, curator of the University of Alberta Vascular Plant Herbarium. The epithet *millelobata* joins the Latin *mille* (thousand) and *lobatus* (lobed), a reference to the leaves. In 1900, Per Axel Rydberg, Herbarium curator at the New York Botanical Garden, described this species as *Senecio millelobatus*, from a collection by Charles Wright in 1851 or 1852, in Limpia Canyon of Jeff Davis County. In 1981, University of Colorado natural historian William Alfred Weber and Áskell Löve moved this and other North American *Senecio* to the *Packera* genus.

43. Rio Grande Palafox (*Palafoxia riograndensis*)
(Other common names: Rio Grande palafoxia)

Habitat: dunes near Rio Grande, sandy or silty soils.

Geographic range: Texas: Big Bend (southern Presidio, southern Brewster Counties). US: Texas. Mexico: Chihuahua, Coahuila.

Park locations: near the Hoodoos, River Road near old Blackrock Campground, off Bofecillos Road

Rio Grande palafox enters the United States only in the southern Big Bend and does not venture far from the river. Few notice the arrival of this stout taprooted annual, with thick ascending stems to 2 feet high or more. Moderately branched, the stems are a strikingly bright, greenish yellow above, and a bright, grayish white near the base. Lower branches at times are cracking or peeling and strongly ridged or grooved. Upper stems may be coated with long white hairs.

The mostly alternate, short-petioled leaves are up to 3 inches long, narrowly lance-shaped, and hairy.

In late summer or fall, cylindric flower heads, with only disk florets, cluster on stalks at the branch tips. The stalks may be dotted with sticky red glands. A flower head contains up to twenty-five disk florets, each to 1/3 inch long. Each floret has a slender tube and a funnel-like throat with five recurving pink lobes. Branching styles protrude from the floret's throat. Each flower head is supported by overlapping, often densely hairy bracts. The thickish bracts are green with red tips.

Shaped like upturned pyramids, the black fruits are four-angled cypselae to 1/2 inch long, with stiff, rough hairs. Each cypsela is tipped with a pappus of lance-shaped scales.

Botanical notes: Mariano Lagasca y Segura, director of the Real Jardín Botánico in Madrid, named this genus in 1816, for José de Palafox y Melzi, captain general of Aragon, who led the defense of Zaragoza against invading French armies. Southern Methodist University botanist Victor Cory (at the time range botanist with the Agricultural Experiment Station in Sonora, Texas) discovered this plant in Presidio County in 1938, "in a flat wash near the Rio Grande." In 1946, Cory named the species *riograndensis* for obvious reasons.

ASTERACEAE — SUNFLOWER FAMILY

44. Mariola (*Parthenium incanum*)
(Other common names: New Mexico rubber plant, crowded rayweed, hierba ceniza, copalillo)

Habitat: scrub, grasslands, woodlands, arid slopes; limestone, igneous soils.

Geographic range: Texas: Trans-Pecos, parts of Permian Basin; sites in Edwards Plateau, South Texas Plains, southern Panhandle. US: Texas to Arizona, north to Utah, Nevada. Mexico: Sonora east to Nuevo León, south to Michoacán, Mexico City, Hidalgo.

Park locations: Black Hills, Palo Amarillo Creek, Las Cuevas, lower Arroyo Mexicano, Skeet Canyon, Encino Loop Trail, Llano Loop, Jackson Pens, Cienega Mountains

Mariola is a compact, extensively branched, aromatic shrub to 3 feet high, heat and drought tolerant. The stems are stiffly erect, gray with long crinkled hairs. Short-stemmed, the oblong leaves are often lyrate, with a large terminal lobe and small lateral lobes. The curling, ashy blades are gland dotted, and coated with matted hairs.

Usually in summer or fall, flat-topped flower heads appear in clusters on short branching stalks. Each head contains five ray florets and up to thirty-plus disk florets. The short ray florets are whitish, notched at the tip. The small disk florets are dull yellow to white, funnel-shaped, and five-lobed. In two overlapping series, bracts form a hemispheric base around the flower head.

The fruit is a black, inversely egg-shaped cypsela, 1/12 inch long, with two or three awl-shaped scales resembling pappus bristles.

Botanical notes: The genus name may be taken from the Greek *parthenion* (feverfew), which was prescribed by Dioscorides to treat female ailments. Or the name may originate from the Greek *parthenos* (virgin), because the ray florets are virgin-white, or because only female ray florets are fertile. The Latin epithet *incanum* (gray) refers to the hairs on stems and leaves. This species was described by German botanist Karl Kunth in 1820, from a plant collected by Baron von Humboldt and Aimé Bonpland in Mexico City about 1803 and cultivated at the Royal Botanic Garden. From 1799 to 1804, Humboldt and Bonpland explored parts of South America, Cuba, and Mexico, with a brief excursion to the United States.

45. Trans-Pecos Poreleaf (*Porophyllum scoparium*)
(Other common names: poreleaf, shrubby or broom poreleaf, hierba del venado, jarilla, romerillo, pomerillo)

Habitat: desert scrub, sandy, gravelly arroyos, foothills, roadsides, low-mid elevations; limestone, igneous soils.

Geographic range: Texas: Trans-Pecos (excluding Reeves, Jeff Davis, Pecos Counties); Crockett County. US: Texas, southern New Mexico. Mexico: Chihuahua to Nuevo León.

Park locations: Monilla Canyon, lower Panther Canyon, Arroyo Mexicano, Cienega Mountain

This poreleaf is an aromatic shrub or subshrub to 3 feet high, with many, pale green, at times glaucous, stems. Slender but erect, hairless, with sparse thread-like leaves, the stems give the plant a naked, broomlike aspect. The leaves, to 1 1/2 inches long, are stemless, fleshy, and dotted with tiny, translucent oil glands.

Yellow flower heads, without ray florets but with up to eighty disk florets, appear singly on short stalks at the branch tips, or in loosely arranged, terminal clusters. Each minuscule floret consists of a short tube, a long throat, and five short, triangular lobes. Up to ten thick, oblong yellow green bracts support the flower heads. Like the leaves, the bracts bear oil glands, or in the words of Asa Gray in 1852, "one or two purplish glandular lines or spots."

The fruit capsules are narrow, at times hairy cypselae, tipped with up to fifty or more tawny, sometimes barbed, bristles.

This plant is used in Mexican traditional medicine to treat fever, rheumatism, and gastrointestinal problems (Vines 1960).

Botanical notes: The genus name *Porophyllum*, from the Greek *poros* (pore) and *phylum* (leaf), refers to the gland dotted leaves. The Latin species name *scoparium* (broomlike) pertains to the stems. This species was described by Asa Gray in 1852, in *Plantae Wrightianae*, and collected by Charles Wright in 1849, on "rocky bluffs of a creek 3 miles beyond the crossing of Devil's River" in Val Verde County.

ASTERACEAE — SUNFLOWER FAMILY

46. Wire Lettuce (*Stephanomeria pauciflora*)
(Other common names: desert straw, brownplume or few-flowered wirelettuce, prairie skeletonplant, pionilla [little peonia])

Habitat: desert flats, gravelly arroyos, grasslands, rocky, wooded slopes; igneous, limestone, gypsum soils.

Geographic range: Texas: Trans-Pecos (excluding Reeves, Pecos, Terrell Counties); much of Permian Basin, Panhandle; west edge of north Texas. US: southwest; Oklahoma, Kansas, Wyoming. Mexico: Baja California east to Chihuahua.

Park locations: South Lajitas Mesa Trail

Wire lettuce is a rounded perennial to 2 feet high, with wiry, densely branched, mostly bare stems and sparse foliage. Growing from a woody crown, the slender branches are smooth, hairless but notably glaucous, gray green or blue green. A bed of tan dead stems often forms below the new growth.

To 2 1/2 inches long, the hairless, mostly basal leaves are linear, pointed at the tips, sharply lobed on the margins, with white midribs. Blades on the upper branches are much smaller, scalelike. All leaves are ephemeral, dead or dying when the plant blooms.

Mostly in late spring and summer, short-stalked or stalkless flower heads form singly on the branches. A head contains five or six pink, white, or lavender ray florets 2/5 inch long, five-lobed, with conspicuous purple styles and stigmas. Flower heads are supported by four to six inner bracts and short outer bractlets. The tan fruit is a five-angled, columnar cypsela, 1/5 inch long, with a pappus of fifteen to twenty bristles.

Botanical notes: Thomas Nuttall named the genus in 1841, joining the Greek *stephanos* (crown) and *meris* (part), probably a reference to the plumelike bristles atop the fruit. The species name, from the Latin *pauci* (few) and *floris* (flower), means few-flowered. New York botanist John Torrey described the species in 1827 as *Prenanthes pauciflora*, from a specimen collected by Edwin James, "at the base of the Rocky Mountains," during Major Stephen Long's 1819–20 expedition to the Rocky Mountains. University of Wyoming botanist Aven Nelson placed this species in the *Stephanomeria* genus in 1909, in the *New Manual of Botany of the Central Rocky Mountains* (coauthored with John Coulter).

47. Hopi Tea Greenthread (*Thelesperma megapotamicum*)
(Other common names: rayless greenthread, Navajo tea, cota, Rio Grande greenthread, Zuni tea)

Habitat: desert lowlands, sandy banks, arroyos, grasslands, mountain slopes, various soils.

Geographic range: Texas: Trans-Pecos; most of Panhandle, Permian Basin, west parts of Edwards Plateau, north Texas. US: Texas to Arizona, north to Utah, Wyoming, South Dakota; Arkansas. Elsewhere: Mexico (Chihuahua to Tamaulipas, south to Zacatecas); parts of South America.

Park locations: near mouth of Panther Canyon, Arroyo Primero, near Sauceda, near Tres Papalotes

This greenthread is a slender perennial to 30 inches high, delicate and spindly, with one or few upright stems from stout roots. Sparsely branched, the smooth, hairless stems are green or blue-green and glaucous. To 4 inches long, the short-stalked or stemless leaves are at times dissected into linear segments. The blades are smaller, sparser on the upper branches.

From late spring to early fall, thick urn-shaped flower heads appear on leafless stalks to 10 inches long or more. Each head contains up to 100+ yellow to reddish-brown, five-lobed disk florets. The flower heads are supported by small, egg-shaped outer bractlets with curving tips, and larger, broadly lance-shaped inner bracts. The bracts are gray green or blue green, but often yellow, red, purple, or black at the tips.

To 1/3 inch long, the fruits are dark red-brown, oblong cypselae, usually covered with nipplelike projections. The cypsela is tipped with a pappus of two barbed bristles.

For centuries, Native Americans in the southwest US have made a tea from this species and used the flowers to make a reddish-brown dye (Kirkpatrick 1992; Whiting 1939; Fewkes 1896).

Botanical notes: The genus name, from the Greek *thele* (nipple) and *sperma* (seed), refers to the nipplelike bumps on the fruits. The species name, from the Greek *mega* (large), *potamos* (river), and suffix *-icum* (relating to), implies "by a large river." German botanist Curt Sprengel, professor at Halle, described this species as *Bidens megapotamica* in 1826, from a collection by Brazilian explorer Friedrich Sellow (circa 1823), with the notation "Rio Grande." Assuming the notation referred to a river, Sprengel coined the name *megapotamica*, but Sellow likely meant the Brazilian state of Rio Grande do Sul.

48. California Trixis (*Trixis californica*)
(Other common names: California or American threefold, American trixis, plumilla, cachano, hierba de aire)

Habitat: desert flats, rocky slopes, canyons, low-mid elevations; limestone, igneous soils.

Geographic range: Texas: Trans-Pecos (excluding Reeves, Pecos, Terrell, Val Verde Counties). US: Texas west to California. Mexico: Baja California east to Nuevo León, south to Zacatecas, San Luis Potosí.

Park locations: Monilla Canyon, Cinco Tinajas, Needle Peak, Contrabandista Spur Trail, Cienega Mountain

California trixis is a compact, densely branched shrub to 3 feet high, with somewhat glandular and hairy herbage. Woody at the base, the pale green, leafy stems are typically erect. Often growing amid other plants, this tough little shrub likes sun, tolerates heat, and prefers southern exposures.

To 4 inches long, the stemless or short-stemmed leaves are lance-shaped, with smooth or incised, inrolled margins, and a broad white midvein below. The alternate blades are mostly erect.

Especially in spring and after rains, yellow flower heads appear singly or clustered near the branch tips. A head lacks ray florets but contains up to twenty-five two-lipped disk florets. Each floret has a large three-lobed outer lip resembling a ray flower, a small inner lip with two recurving lobes, and protruding anthers. Up to ten linear green bracts and fewer outer, basal bractlets support each head.

The brown fruits are five-ribbed, cylindric or spindle-shaped capsules to 2/5 inch long, coated with mucilaginous hairs, and tipped with a pappus of sixty to eighty barbed bristles.

Botanical notes: The Greek *Trixis* (threefold) alludes to the floret's tri-cleft outer lip. In 1756, Irish-born Jamaican physician-botanist Patrick Browne, in *Civil and Natural History of Jamaica*, described this and 103 other new genera. Browne's masterwork contained fifty copperplate illustrations by the foremost botanical artist of the day, Georg Dionysius Ehret. The species was described by California botanist-illustrator Albert Kellogg in 1862, from a plant collected on Cedros Island, off the coast of Baja California, by John Veatch, curator of conchology at the California Academy of Sciences.

49. Plains Ironweed (*Vernonia marginata*)
(Other common names: James' ironweed, narrow-leaf ironweed, ironweed)

Habitat: canyons, intermittent creeks, springs, moist, low-lying areas, roadside ditches; igneous, limestone soils.

Geographic range: Texas: Trans-Pecos (excluding El Paso, Reeves, Terrell, Val Verde Counties); most of Panhandle into Edwards Plateau; part of western north Texas. US: Texas, New Mexico, north to Kansas, Colorado. Mexico: Coahuila.

Park locations: Cinco Tinajas, Skeet Canyon

Rather uncommon in BBR, plains ironweed is an erect perennial to 2 feet high or more, suitable for a flower garden, with intricate flowers highly attractive to butterflies, hummingbirds, and plant photographers. Largely unbranched, the mostly hairless stems grow from spreading rhizomes and form colonies in periodically inundated areas. To 6 inches long, the stemless leaves are narrowly lance-shaped, pointed at the tips, smooth or sparsely dentate, and sometimes revolute on the margins. The blades are mostly hairless, but strongly pitted with glands underneath.

In summer, compact flower clusters, with few flower heads, appear on short stalks near the branch tips. Each flower head consists of up to twenty-five or more purple or reddish-purple disk florets. The tubular florets have funnel-shaped throats with five straplike lobes and conspicuously protruding styles. Supporting the flower head is a bell-shaped group of bracts in graduated series. The bracts are purplish, sharp-pointed, and ciliate, with long cobwebby hairs on the margins.

The yellow-tan fruits are ten-ribbed cypselae 1/6 inch long, often dotted with yellow oil globules. The pappus at the tip consists of two series of coarse bristles, purple or tan.

Botanical notes: In 1791, German botanist Johann Schreber, director of the Erlangen Museum of Natural History, named this genus for British botanist-entomologist William Vernon. Vernon traveled to Maryland in 1698 and collected plants (especially mosses), fossils, shells, and insects for the Royal Society in London. The Latin epithet *marginata* (margined) refers to the denticulate leaves. In 1827, John Torrey described the plant as a variety, *Vernonia altissima* var. *marginata*, from a collection by Edwin James in 1820 on Long's Expedition to the Rocky Mountains. In 1832, in his *Atlantic Journal and Friend of Knowledge*, Constantine Samuel Rafinesque gave this ironweed species status.

50. Gypsum Tansy Aster (*Xanthisma gypsophilum*)
(Other common names: gypsum sleepy daisy, gyp sleep-daisy)

Habitat: amid rocks, on rock walls, sandy environs, canyon mouths; igneous, limestone, gypseous soils.

Geographic range: Texas: Big Bend (southeast Presidio County). US: Texas, New Mexico. Mexico: Chihuahua east to Coahuila, south to Durango, San Luis Potosí.

Park locations: Panther Canyon, Closed Canyon, Rancherías Creek south of Highway 170

Of conservation concern in Texas: see Appendix B

Only known in Texas from the southeast corner of Presidio County, this aster is a shrubby, taprooted perennial or subshrub to 2 feet high, with a few erect to sprawling stems. Sparingly branched above, the rather stout stems are often gland dotted and rough with coarse, stiff hairs.

The leaves, to 2 1/2 inches long, may be oblong and stemless, almost clasping the stem, or spatulate, with a long-tapered base. The blades bear stiff hairs, sometimes stalked glands, and prominent bristle-tipped teeth on the margins.

From spring to fall, showy, long-stalked flower heads are borne at the stem tips. Each head consists of up to twenty-eight white ray florets, often flushed with pink or purple underneath, and up to 120 yellow disk florets. The heads are supported by a goblet-shaped collection of overlapping green bracts. The bracts are linear to narrowly lance-shaped, densely glandular and bristle-tipped, with membranous white margins.

To 1/10 inch long, the fruit capsules are oblong to egg-shaped cypselae, eight- to fourteen-ribbed, with white to tan hairs. The fruit is tipped with a pappus of up to 100 tawny-white bristles.

Botanical notes: In 1836, Swiss taxonomist Augustin de Candolle formed the genus *Xanthisma* when he described an aster with yellow ray florets (*Xanthisma texanum*), which was collected by naturalist Jean Louis Berlandier in South Texas. The genus name, from Greek words *xanthos* (yellow) and the suffix *-ismos* (condition), refers to the yellow florets. The species name, *gypsophilum*, from the Greek *gypsos* (gypsum) and *philos* (loving), alludes to the aster's preference for gyp habitats. Billie Turner, professor emeritus at the University of Texas at Austin, described the species as *Machaeranthera gypsophila* in 1973, from his collection in 1970 in Coahuila, south of Cuatro Ciénegas, on "blowing dunes of gypsum." In 2003, botanists David Morgan and Ronald Hartman moved this and many other *Machaeranthera* species to the genus *Xanthisma*.

ASTERACEAE — SUNFLOWER FAMILY

51. Gyp Daisy (*Xylorhiza wrightii*)
(Other common names: Big Bend woodyaster, gypsum daisy, Wright's woody-aster, Wright's desert daisy, Terlingua or Big Bend aster)

Habitat: arid, sparse desert hills, flats, rocky slopes, on clay, gypsum, shale; limestone soils.

Geographic range: Texas: Trans-Pecos (near Rio Grande in Hudspeth, Culberson, Jeff Davis, Presidio, Brewster Counties). US: Texas. Mexico: northern Chihuahua, northern Coahuila.

Park locations: West Contrabando Trail, Buena Suerte Trail

Of conservation concern: see Appendix A

A borderlands endemic with limited range, gyp daisy is a long-lived perennial or subshrub to 20 inches high. Rising from a stout taproot and a spreading woody base, the thick purplish stems are extensively branched below, with resinous glands and sometimes, hairs.

The alternate leaves are variable — spoon-shaped, spatulate, oblong, abruptly spine-tipped, stemless or tapered at the base, and glandular. The blade margins may be lined with hairs and spinelike teeth. To 3 inches long, the leaves are smaller and scattered on upper branches.

Mostly from March to May, showy flower heads 2 inches wide appear singly on naked stalks. A head contains up to thirty-two ray florets, notched at the tip, and 130+ disk florets. Ray color is variable — blue, lavender, violet, purple, or white. Each yellow tubular disk floret has a funnel-like throat with five triangular lobes. The flower head is supported by a bell-shaped group of many leaflike green bracts in overlapping series. The lance-shaped bracts are often hairy and glandular, with spreading or curling tips, and fringed margins.

The fruits, to 1/3 inch long, are oblong cypselae, slightly compressed, silky-haired, and tipped with a pappus of up to ninety tawny-white bristles.

Botanical notes: In 1840, while working at the Academy of Natural Sciences in Philadelphia, Thomas Nuttall named this genus, joining Greek words, *xylon* (wood), and *rhiza* (root), a reference to the large woody taproots. In 1853, in *Plantae Wrightianae*, Asa Gray named the species *Aster wrightii*, for Charles Wright. Wright discovered the plant in 1852, on "stony hills on the Rio Grande 60 to 70 miles below El Paso, near the canon through which passes the road to San Antonio." In 1896, Catholic University botanist Edward Lee Greene moved this species to the *Xylorhiza* genus.

52. Agarito (*Berberis trifoliolata*)
(Other common names: agarita, algerita, agrito, agritos, agrillo, currant-of-Texas, wild currant)

Habitat: desert flats, arroyos, canyons, rocky slopes, grasslands, woodlands, varied soils.

Geographic range: Texas: Trans-Pecos, Edwards Plateau; parts of South Texas Plains, Coastal Bend, Panhandle; scattered sites. US: Texas, southern New Mexico, Arizona. Mexico: Chihuahua east to Tamaulipas, south to Durango, San Luis Potosí.

Park locations: Black Hills, Palo Amarillo Creek, Arroyo Mexicano, Sauceda, Ojo Chilicote

Agarito is a sturdy evergreen shrub to 10 feet high, with long spreading branches and many short lateral shoots. The hairless stems are gray to reddish-brown, with peeling bark and yellow wood underneath.

This thicket-forming barberry is a maze of jagged points that fends off predators and yet provides cover for animals. The blue-green leaves are divided into three daggerlike leaflets, with toothed lobes and a spiny tip. Stiff and leathery, the leaflets bear a waxy coat that slows water loss.

In spring, this early bloomer is filled with small clusters of yellow flowers, on short, terminal shoots or in the axils of upper branches. Each sweet-scented, cuplike flower has six petals, six yellow sepals, and six sensitive stamens around a single pistil. The stamens, when touched by insects, propel pollen toward the pistil and onto the insects, which the insects take with them to the pistils of nearby plants.

Edible but tart red fruits form from April to June. The pea-size berries, with one or few seeds, are juicy, aromatic, a traditional ingredient in jam, wine, and pie.

Native Americans made a yellow dye from the stems (Warnock 1974). In frontier days, a toothache remedy was derived from the roots and roasted seeds provided a coffee substitute (Vines 1960).

Botanical notes: The genus name *Berberis* is from the old Latin *barbaris*, an Arabic term for North Africa and the barberry fruit. The species name, *trifoliolata*, refers to the tripart leaves. This species was described in 1841 by Stefano Moricand, one of a group of Swiss botanists who sent Jean Louis Berlandier to collect plants on the US-Mexico Boundary Survey. Berlandier collected agarito near the Frio River, between Laredo and San Antonio, in 1828.

53. Desert Willow (*Chilopsis linearis*)
(Other common names: desert or willowleaf catalpa, flowering or bow or false willow, flor de mimbre, jano)

Habitat: desert washes, sandy arroyos, creeks, springs, low elevations, near Rio Grande.

Geographic range: Texas: Trans-Pecos, Hill Country, spotty elsewhere. US: southwest. Mexico: Baja California to Tamaulipas, south to Zacatecas, San Luis Potosí.

Park locations: Cerro de las Burras Loop, Panther Canyon, Sauceda, Solitario northwest rim, upper Alamito Creek

Desert willow is a rapidly growing, long-lived, deciduous shrub or tree to 30 feet high, with dark brown or gray bark. The hairless, often resinous stems are ascending or occasionally spreading, wand-like, or drooping.

About 5 inches long, the slender leaves are willowlike, linear or narrowly lance-shaped. The stemless, pale green blades are often covered with a waxy, sometimes sticky, coating.

From April to September, often after rains, fragrant, broadly funnel-shaped flowers appear in elongated clusters at the branch tips. The ornamental blooms, with decidedly ruffled margins, are two-lipped, with a pinkish white, bilobed upper lip, and a multicolored, trilobed lower lip. The lower lip has striking maroon-red splotches and lines, and two hairy yellow ridges leading into the throat. The lines attract large bees and hummingbirds to nectar. The dangling brown fruits are slender, two-chamber round capsules to 10 inches long, with many tan, oval seeds.

The wood was used for firewood and fence posts, and Native Americans wove baskets from the twigs. A flower decoction is used in Mexico to treat coughs (Vines 1960).

Botanical notes: In 1823, David Don, secretary of the Linnaean Society, coined the genus name, combining Greek words *cheilos* (lip) and the suffix *-opsis* (resembling), emphasizing the two-lipped flowers. The Latin species name *linearis* (linear) refers to the leaves. Spanish taxonomist Antonio José Cavanilles, director of the Real Jardín Botánico in Madrid, described and illustrated this species as *Bignonia linearis* in 1794. The type specimen is Cavanilles's drawing of a plant cultivated in the Madrid gardens from Mexican seeds. In *Sweet's Hortus Britannicus* (1827), a catalog of plants cultivated in British gardens, British horticulturist Robert Sweet moved this species to the *Chilopsis* genus.

54. Powell's Heliotrope (*Euploca powelliorum*)
(Other common names: none known)

Habitat: rocky desert hills, mesas, with lechuguilla, grama grass; low-mid elevations, limestone soils.

Geographic range: Texas: Big Bend (Brewster, Presidio Counties). US: Texas. Mexico: Chihuahua east to Coahuila, south to San Luis Potosí.

Park locations: Horsetrap Spring area, Lefthand Shutup, Terlingua Uplift, Masada Ridge Road

This heliotrope is an upright, gray green shrublet to 18 inches high, with slender, rough stems and numerous ascending branchlets with sharp white hairs.

To 1 inch long, the leaves are linear or narrowly lance-shaped, pointed at the tip, often rolled under on the edges. The blades are thickly coated with stiff, sharp, close-pressed hairs.

Small greenish-yellow flowers appear in long, narrow, sparse, loosely arranged clusters to 5 inches long. Each five-lobed, tubular blossom is only 1/7 inch long. The fruit, 1/8 inch wide, has four distinct lobes that split at maturity into four hairy, rounded nutlets, each with a single seed.

Before 2007, these plants were named Torrey's heliotrope (*Heliotropium torreyi*, now *Euploca torreyi*), which grows east of the Dead Horse Mountains in BBNP. Powell's heliotrope has smaller blooms, longer but sparser flower clusters, larger but sparser leaves, and more uniformly hairy leaves and fruits. The species do not intergrade.

Botanical notes: In 1836, Thomas Nuttall coined the genus *Euploca* from a Greek word meaning "to plait," alluding to the "peculiar character of the corolla," with a "plaited" flower tube and "connivent" and "plaited" anthers. However, in 1857, Asa Gray placed *Euploca* species in the genus *Heliotropium*. In 2007, University of Texas plant taxonomist Billie Turner named this species for his "academic son," A. Michael Powell, emeritus professor of biology and director of the Sul Ross State University Herbarium, and his wife Shirley, author of studies on Texas flora. Now, based on new molecular studies, many *Heliotropium* species are being transferred to *Euploca*. Oregon State University botanists Christian Feuillet and Richard Halse moved this species to *Euploca* in 2017.

55. Fringed Puccoon (*Lithospermum incisum*)
(Other common names: cutleaf or narrowleaf puccoon, narrowleaf gromwell, narrowleaf stoneseed)

Habitat: rocky terrain, grasslands, woodlands, open fields, disturbed sites, in sand or gravel; igneous, limestone soils.

Geographic range: Texas: Trans-Pecos, most of state. US: midwest and west (excluding Washington, Oregon, Idaho); Tennessee, Florida. Elsewhere: Canada, Mexico (Chihuahua, Coahuila).

Park locations: Terneros Creek, junction of Guale Mesa Road and Rancherías 4WD Road, Los Ojitos

This puccoon is a perennial to 1 foot high, which grows from a sturdy taproot and a woody crown. At first, the plant forms a basal leaf cluster, then several hairy, erect, bunched stems. By late spring or summer, the basal leaves have died, but the stems have become quite leafy.

The long, slender, basal leaves are pointed at the tips, and fringed with stiff hairs. The shorter stem leaves, to 2 inches long, are narrowly oblong, pointed at the tips, and stemless at the base. All leaves are coated with close-pressed hairs.

In early spring, yellow to gold, trumpet-shaped flowers form in leafy clusters at the branch tips. Each bloom has a long slender tube, which abruptly flares into five spreading lobes with conspicuously ruffled margins. Although striking, these flowers are infertile. The heavy lifting is done in late spring and summer by tiny, self-fertilizing blossoms that never open. Seldom noticed, these blossoms are hidden among the floral bracts. The fruit consists of four or fewer hard egg-shaped nutlets, to 1/4 inch long, each enclosing a single seed.

Native Americans prepared a purple dye from the roots (Warnock 1977).

Botanical notes: Greek physician Pedanius Dioscorides, in his famous first-century AD herbal, *De Materia Medica*, named this genus, combining the Greek *lithos* (stone) and *sperma* (seed), to characterize the hard nutlets. The Latin epithet *incisum* (incised) refers to the fringed petals. In 1814, the species was described by Frederick Pursh in *Flora Americae Septentrionalis* as *Batschia longiflora*, from a collection by Thomas Nuttall on the Missouri River. British botanist Nuttall spent thirty-three years in the United States and engaged in extensive expeditions west of the Mississippi. In 1818, German botanist Johann Georg Christian Lehmann, director of the Hamburg Botanical Garden, named this species *Lithospermum incisum*.

56. Mexican Navelwort (*Omphalodes alienoides*)
(Other common names: Mexican navelseed)

Habitat: arid mountains, canyons, rocky slopes, partial shade of rock crevices and walls, mostly on limestone.

Geographic range: Texas: Trans-Pecos (south Presidio, Brewster, Terrell, Val Verde Counties). US: Texas. Mexico: Coahuila.

Park locations: lower Fresno Canyon, Closed Canyon, Rancherías Creek south of Highway 170

An artful annual to 1 feet high, Mexican navelwort grows from a thin taproot to form delightful forget-me-not flowers and bizarre, sea-anemone-like fruits. The slender stems are weakly erect or sprawling, sparsely branched, and sparingly hairy.

To 1 1/2 inches long, the leaves form low on the branches, on petioles often longer than the blades. The alternate leaves are broadly egg-shaped or rounded, with abrupt tips, heart-shaped bases, and coarse hairs.

In early spring, narrow clusters of wide-spaced flowers form at the ends of branches. Each short-tubed bloom abruptly expands into five petal-like lobes. The lobes are sky blue, but white, constricted and pleatlike near the base, and slightly wavy or curling on the edges. At the mouth of the floral tube, elevated above the lobes, are five conspicuous, swollen, and knoblike yellow tubercles.

The fruit consists of four nutlets, each comblike on the margins, with erect "teeth." The nutlets have two distinct forms: one mostly hairless, with shorter, clublike, blunt-tipped teeth; the other hairy, with longer, acute teeth. A single fruit may contain nutlets of one or both types.

Botanical notes: In 1754, in *The Gardeners Dictionary*, Philip Miller coined the genus name *Omphalodes* from the Greek *omphalos* (navel) and *eidos* (resemblance), meaning "navel-like," for the fruit's hollow depressions. This navelwort was formerly classified as *Omphalodes aliena*, a plant discovered by British-born freelance collector Edward Palmer at Monterrey, Mexico, in 1880. *O. aliena* was described by Asa Gray in 1882 in British botanist William Hemsley's *Biologia Centrali-Americani*.

In 2013, Guy Nesom placed plants from Texas and parts of Coahuila, Mexico, in a new species, *Omphalodes alienoides*, based on differences in fruit structure and hairs on stems and petioles. Nesom joined *aliena* and the suffix *-oides* (similar to) to form the new name, meaning "like the species *aliena*."

57. Oreja de Perro (*Tiquilia canescens*)
(Other common names: woody crinklemat, dog's ear, shrubby, woody, or gray tiquilia, ratear, hierba de la virgen, shrubby coldenia)

Habitat: desert flats, creosote scrub, sandy banks, wooded sites; limestone, gypsum, clay, caliche.

Geographic range: Texas: Trans-Pecos, parts of South Texas Plains, Edwards Plateau, Permian Basin. US: southwest (excluding Colorado). Mexico: Baja California east to Tamaulipas, south to Hidalgo.

Park locations: Grassy Banks, West and East Contrabando Trails, Terlingua Uplift, Cienega Mountains

Oreja de perro (dog's ear) is a perennial subshrub with a woody taproot and woody lower stems. This crinklemat forms extensive mats or low mounds seldom more than a few inches high. The branching stems are covered with woolly hairs.

On stout petioles, the thick oval leaves are 1/3 inch long, crowded on short shoots. The gray-green blades have a dense cover of soft hairs below, close-pressed hairs above, and bristles, mostly on the margins. The leaves curl like floppy dog ears.

From spring to fall, pale pink or white flowers form in the leaf axils or cluster on short, leafy branchlets. Each bloom is funnel-shaped, with five rounded lobes and, often, a yellow throat. The morning bloomers are fragile, falling before the day ends. To 1/10 inch long, the fruits are spheric, with vertical grooves. The fruits split to form four brown one-seeded nutlets with tiny, nipple-like projections.

In Mexico, a stem and leaf extract is used to wash wounds, and prevent sleepiness (Richardson 1977).

Botanical notes: In 1799, Spanish botanists Hipólito Ruiz López and José Antonio Pavon described *Lithospermum dichotomum*, a plant with the native vernacular name of Tiquil-Tiquil, in *Flora Peruviana et Chilensis* — a description of flora from their 1777–88 expedition. In 1805, Christiaan Persoon renamed the plant *Tiquilia dichotoma*, creating the genus name from the vernacular name. In 1845, Swiss botanist Alphonse Pyramus de Candolle described this species as *Coldenia canescens*, from a specimen collected by Jean Louis Berlandier in the Mexican state of Tamaulipas. The Latin epithet *canescens* (gray) refers to the gray hairs on leaves and stems. In 1976, botanist Alfred Richardson, at the University of Texas at Brownsville, moved all New World species of *Coldenia* to the genus *Tiquilia*.

58. Texas Crinklemat (*Tiquilia gossypina*)
(Other common names: Texas tiquilia)

Habitat: sandy desert, alluvial flats, limestone hills; gypseous, calcareous soils.

Geographic range: Texas: Trans-Pecos (Presidio, Brewster, also El Paso, Hudspeth, Reeves, Pecos Counties). US: Texas, New Mexico (Doña Ana County). Mexico: Chihuahua, Coahuila, northern Durango.

Park locations: Solitario northwest rim, Terlingua Uplift, hills north of Buena Suerte Trail

This crinklemat is a low perennial or subshrub growing from a woody taproot, and like oreja de perro, forming mats several feet across. Especially the young branches are covered with stiff hairs.

Most leaves form in tight bundles on short shoots off the branches. To 1/2 inch long, the leaves are narrowly oblong, with short thick, bristly petioles as wide as the blades. Upper leaf surfaces bear prickly bristles and usually, a layer of hairs. Lower surfaces may be nearly hidden by the rolled leaf margins.

From spring to fall, mostly after rains, pink or magenta flowers, to 1/2 inch long, appear singly in the leaf axils, or in small clusters on short, leafy branchlets. Each bloom is supported by shorthaired calyces, often with some long bristles on the edges. The funnel-shaped flower has five spreading, rounded lobes at the tip and frequently, a yellow throat. Distinctively, this species has winged veins within the floral tube, below the base of the five stamens, and the buds may be dotted with stalked glands. Like other Trans-Pecos *Tiquilia* species, the fragile blossoms open in the morning and fall by evening.

The four-lobed fruit separates to form four one-seeded nutlets. Each minuscule blackish nutlet is egg-shaped and covered with tiny cyst-like protuberances.

Botanical notes: In 1913, Elmer Wooton and Paul Standley described this species as *Eddya gossypina*, noting that the plant "has a white, cottony appearance." The species name joins the Latin *gossypium* (cotton plant) and the suffix *-ina* (likeness), implying "covered with cottony hairs." Wooton collected the species in 1894 on Tortugas Mountain in New Mexico. In 1976, Alfred Richardson moved this and other plants into the genus *Tiquilia*.

59. Plume Tiquilia (*Tiquilia greggii*)
(Other common names: plumed crinklemat, Gregg's tiquilia, plume coldenia, hierba del cenizo)

Habitat: desert scrub, gravelly or rocky slopes and canyons, low to mid-elevations; limestone or gypseous outcrops.

Geographic range: Texas: Trans-Pecos (excluding Pecos County), mostly along Rio Grande; Crane County. US: Texas, southern New Mexico. Mexico: eastern Chihuahua to Nuevo León, south to north Zacatecas, north San Luis Potosí.

Park locations: lower Monilla Canyon, Closed Canyon, Solitario northwest rim

Plume tiquilia is a stiff, erect shrub, seldom more than 2 feet high, extensively branched yet compact, with short, leafy branchlets. The herbage is strikingly gray, densely cloaked with hairs and at least some stiff bristles. On short narrow petioles, the many small, oval leaves are up to 1/3 inch long, with margins often rolled under. The blades bear a coat of short hairs and bristles with enlarged mineralized bases.

From spring to fall, especially after the first monsoon rain, pink-magenta flowers appear in rounded, headlike clusters at the branch tips. Each small bloom, no more than 1/3 inch long, is funnel-shaped, with five rounded lobes. The smoky appearance of the flower clusters results from the threadlike, maroon calyx lobes at the base of each bloom, which are covered with feathery, smoky-white bristles.

Unique in this species, three of the four fruit nutlets are aborted, and the calyx, enclosing the remaining nutlet, falls when mature. The feathery calyx acts as an instrument for wind dispersal of the seed.

In Mexico, this shrub has been used in the treatment of gonorrhea (Richardson 1977).

Botanical notes: In 1855, in the *Pacific Railroad Surveys* on explorations "to ascertain the most practicable and economical route for a railroad from the Mississippi River to the Pacific Ocean," John Torrey named this species *Ptilocalyx greggii*, in honor of Josiah Gregg, merchant trader and natural historian of the Santa Fe and Chihuahua Trails. Gregg collected the plant in 1847, at the site of a major battle of the Mexican-American War, Buena Vista, in the Mexican state of Coahuila. During the war, Gregg played new roles as guide and interpreter for the US Army, and as a war correspondent.

60. Watercress (*Nasturtium officinale*)
(Other common names: true or common watercress, nasturtium, berro, berro de agua)

Habitat: streambeds, intermittent creeks, shallow slow-moving water, seeps, ditches.

Geographic range: Texas: Trans-Pecos (excluding Hudspeth, Reeves Counties); much of Edwards Plateau, central Texas; scattered in Panhandle, North Texas. US: throughout. Elsewhere: Canada, Mexico to Panama; Europe, Asia, North Africa.

Park locations: Arroyo Mexicano, Madrid House, Cienega Gorge

Watercress is a semiaquatic perennial with hollow, creeping stems to 3 feet long or more. Usually floating in shallow water or prostrate on wet ground, this mostly hairless herb is rhizomatous, frequently rooting at the nodes and forming extensive mats. Often, the fleshy stems are maroon, and notably grooved. The succulent leaves are alternate, with up to nine pairs of opposite leaflets and a larger terminal leaflet. The stemless leaflets are egg-shaped, oblong, or round.

Blooming year round, small white blossoms form compact clusters at the branch tips, often floating just above the water. A bloom consists of four erect green sepals, four rounded petals, four long and two short stamens with purple filaments and yellow anthers, and a short style.

The fruits are slender, short-beaked, cylindric seedpods to 3/4 inch long, on stalks as long as the pods. The pods contain minuscule red brown seeds with honeycomb-like surfaces.

The peppery leaves, high in vitamin C, were cultivated for centuries in Europe. In a ninth-century edict, Charlemagne listed watercress as a culinary-medical herb to be grown in Imperial gardens (Katzer 2012). Pliny the Elder (23–79 CE) cited medicinal uses, one an antidote to scorpion's venom.

Botanical notes: The genus name, from the Latin *nasus* (nose) and *tortus* (twisted), or "twisted nose," alludes to the pungent taste and odor. The species name, *officinale*, was often applied to plants with assumed medicinal properties. This species was described by Linnaeus in 1753 as *Sisymbrium nasturtium-aquaticum*. In 1812, William Townsend Aiton renamed the plant *Nasturtium officinale*, in an enlarged edition of his father's *Hortus Kewensis*, a catalog of plants at Kew Gardens in London.

61. Rose Bladderpod (*Physaria purpurea*)
(Other common names: purple or white or western white bladderpod)

Habitat: sandy, gravelly arroyos, rocky drainages, fissures in rock walls, limestone or volcanic slopes, usually partial shade.

Geographic range: Texas: Trans-Pecos (excluding Reeves, Pecos, Terrell Counties). US: Texas, southern New Mexico, southern Arizona. Mexico: Sonora east to Coahuila.

Park locations: Black Hills, Monilla Canyon, Madrid Falls Trail, oak-filled box canyon west of Cinco Tinajas, Los Ojitos, Lower Shutup, Cienega Creek

Growing from a thick woody crown, this perennial develops erect or sprawling, unbranched stems to 28 inches long, with mostly basal leaves. Typically, stems and leaves are thickly covered by stellate hairs with radiating branches.

To 6 inches long, the basal leaves are often elliptic or oblong, at times cleft, lobed, or fiddle-shaped, and mostly smooth on the margins. The leaves rest on petioles at times much longer than the blades. Stem leaves are smaller, stemless or short-stemmed.

From spring to fall, flowers form in clusters at the ends of long, nearly naked stems. Each bloom has four white, frequently pink-tinged, petals with purple or maroon streaks at the clawlike base.

The fruiting stalks, to 1 inch long, are ascending to downward-curving, not distinctly S-shaped as in some species. The fruits are hairless spheric seedpods to 1/3 inch long, inflated or appearing so, tipped with a pin-like persistent style. At first green, the pods turn ruddy brown or purplish with age.

Botanical notes: The genus name, from the Greek *physa* (bladder), refers to the inflated fruits. The Latin species name *purpurea* (purple) may relate to the flowers or the fruits. In 1853 in *Plantae Wrightianae*, Asa Gray named this species *Vesicaria purpurea*, from a collection by Charles Wright in 1852, on the US-Mexico Boundary Survey, on "stony hills near El Paso." Gray, professor of Natural History at Harvard, was a preeminent taxonomist, tireless recruiter of botanical explorers, and prolific author of botanical works, culminating with *Synoptical Flora of North America* (1878–86). In 2002, Steve O'Kane and Ihsan Al-Shehbaz transferred this and many other species formerly in the *Vesicaria* and *Lesquerella* genera to the genus *Physaria*.

BRASSICACEAE — MUSTARD FAMILY

62. Durango Yellowcress (*Rorippa ramosa*)
(Other common names: branched or canyon watercress)

Habitat: moist alluvial soils, floodplains, creeks, sandy arroyos, damp disturbed habitats.

Geographic range: Texas: Big Bend (Brewster, Presidio Counties). US: Texas. Mexico: Coahuila, Chihuahua, Durango.

Park locations: Along the Rio Grande

Of conservation concern: see Appendix A

After reading about the decades-old discovery of Durango yellowcress near Santa Elena Canyon, I searched the River Road in BBNP every spring unsuccessfully, only to stumble across this elusive mustard years later in BBR. Prostrate, profusely branched, this perennial grows from horizontal roots and forms circular mats to 3 feet across. All plant parts are sparsely hairy, with odd, bladderlike hairs.

Oblong and pinnately lobed, or merely wavy, the leaves may reach 2 inches long or more. The stemless gray-green leaves are alternate, thickish, with oblong lobes. Typically, the blades are nearly hairless above, with bladderlike hairs on the veins and midribs underneath.

From March to May, four-petal flowers bloom in short clusters. Each flower rests on a short, often slightly curved stalk. The pale-yellow petals soon fade to white on the margins. Within the petals are six stamens, with short filaments and oblong anthers, and a persistent style. Below the petals are four early falling, oblong sepals.

The fruit is a bivalved, dehiscent pod to 2/5 inch long, oblong to lance-shaped, often hooked at the tip. The two valves split to expose many plump, rough-surfaced seeds.

Botanical notes: In 1760, *Tyrolean naturalist Giovanni Scopoli named this genus Rorippa,* from an Old Saxon term, *rorippen,* used by German herbalist Euricius Cordus (1486–1535) for similar plants. The Latin epithet *ramosa* (branched) refers to the stems. Reed Rollins, Gray Herbarium director at Harvard University, described the mustard in 1961, from a 1959 collection by Donovan Correll and Ivan Johnston in Durango, Mexico. The herb was collected earlier by Josiah Gregg (1847), Edward Palmer (1880), and Eula Whitehouse (1948, below Santa Elena Canyon in BBNP). Herbarium curator at Southern Methodist University, Whitehouse authored and illustrated *Texas Wildflowers in Natural Colors* (1936).

63. Purpus' Tumblemustard (*Thelypodiopsis purpusii*)
(Other common names: unknown)

Habitat: shade of rocks, shrubs, on arid rocky slopes; mid-elevations, limestone, igneous soils.

Geographic range: Texas: Trans-Pecos (Hudspeth, Presidio Counties). US: Texas, New Mexico, Arizona (Grand Canyon). Mexico: Coahuila.

Park locations: Ojito Adentro, Smith Canyon, Fresno Canyon head

Purpus' tumblemustard is an erect, hairless annual to 2 feet high, with one or several stems unbranched or only branched in the upper reaches. The stems are leafy, and often, stems and leaves are glaucous. Known from a few collections in Hudspeth and Presidio Counties, this mustard is rare in Texas, but apparently secure globally.

Withering before the plants mature, the basal leaves may approach 5 inches in length. Basal and lower stem blades are oblong, lobed, or only toothed and wavy, with an obvious midrib. Middle and upper stem leaves are smaller, stemless, with earlike basal lobes that often clasp the stem.

From February to May, few-flowered clusters of small flowers terminate the branches. On a spreading or ascending stalk, each bloom has four white or lavender-tinged, spatulate petals. Within the petals are six stamens, four long and two shorter, with egg-shaped yellow anthers, and a greenish yellow or maroon style with a knoblike greenish yellow stigma. Below the petals are four erect oblong sepals, greenish yellow to maroon.

To 2 1/2 inches long, the fruit is a slender cylindric pod, spreading to nearly erect, straight or slightly curving, with oblong seeds.

Botanical notes: This species was described as *Thelypodium purpusii* by California botanist Townshend Brandegee in 1906, from a 1905 collection by freelance collector Carl Purpus at Sierra de Pata Galana and Parras in Coahuila. The generic name, from Greek words *thely* (female) and *podion* (little foot), refers to the pistil's short stalk. In 1976, mustard expert Reed Rollins moved the plant to the related *Thelypodiopsis* genus. *Thelypodiopsis* combines *Thelypodium* with the Greek suffix *-opsis* (resembling), meaning "like plants of the *Thelypodium* genus."

64. Texas Thelypody (*Thelypodium texanum*)
(Other common names: Texas Thelypodium)

Habitat: drainages, low bluffs near the Rio Grande; gypseous, clay flats, dunes, bentonite mounds.

Geographic range: Texas: Trans-Pecos (Culberson, south Presidio, south Brewster Counties). US: Texas, New Mexico. Mexico: expected.

Park locations: South Lajitas Mesa, Fresno Canyon, Terlingua Uplift, West Fork of Comanche Creek

Of conservation concern: see Appendix A

This mustard is a scarce — but sometimes locally common — annual, seldom more than 20 inches high, often with a single erect, stout stem and a rosette of basal leaves. Stems and leaves are sometimes glaucous, with a waxy, whitish coating.

As much as 6 inches long, the leaves in rosettes and on the lower stems are alternate, petioled, broadly oblong. The blades are pinnately divided into oblong lobes, which are thick, fleshy, and often bluntly toothed or wavy on the margins. Upper stem leaves are smaller, with slender linear lobes.

From January to April, this early bloomer produces a dense, unbranched, elongated flower cluster at the stem tips. White four-petal blossoms, often flushed with pale lavender or purple colors, appear on spreading stalks to 1/2 inch long. Each striking bloom has slender spatula-shaped petals with a clawlike base and protruding stamens with downward curving purple anthers.

To 2 inches long, the fruits are slender cylindric seedpods, spreading or upward curving, and beaked, with many plump oblong seeds.

Botanical notes: The genus name *Thelypodium*, joining the Greek *thely* (female) and *podion* (little foot), refers to the short stalk of the pistil. Victor Cory discovered this plant in 1936 "on the narrow flood-plain of Terlingua Creek, at about 18 miles on an airline north of Terlingua, Texas." In 1937, in *Some New Plants from Texas* in the journal *Rhodora*, Cory described the species as *Sisymbrium texanum*. Cory collected thousands of plants, and named many of them, while working at the Agricultural Experiment Station in Sonora, Texas, at Texas A&M University, and at Southern Methodist University herbarium. In 1946 Stanford botanist and mustard authority Reed Rollins moved this species to the genus *Thelypodium*.

65. Texas False Agave (*Hechtia texensis*)
(Other common names: false agave, hechtia, Big Bend hechtia, guapilla, guapilla de Texas, bromelia)

Habitat: arid, open, rocky or gravelly limestone bluffs, bajadas, often with lechuguilla, low to mid elevations.

Geographic range: Texas: Big Bend (south Presidio and south Brewster Counties). US: Texas. Mexico: Chihuahua east to Nuevo León; Zacatecas.

Park locations: Bofecillos Mountains, Solitario northwest rim, Terlingua Uplift

Texas false agave is a perennial or subshrub with thick, somewhat fleshy roots but without obvious stems. This pineapple produces rosettes of crowded leaves and may form extensive colonies vegetatively by clonal growth from rhizomes and development of "pups."

To 16 inches long, the leaves are spine-tipped and daggerlike, or recurving and sicklelike, with sharp curved, red-blotched teeth. The blades are usually pale-yellow green, and scurfy on the backsides, with a coat of gray-white scales. However, especially in fall, the leaves may turn a stunning dark red.

In spring, fragrant blooms appear in many-flowered, branched clusters on stalks to 6 inches high. Male and female flowers are borne on separate plants. Each bloom has three scaly, rose-colored sepals, three white petals, and six stamens with yellow anthers.

The fruits are berrylike capsules to 1/2 inch long, scaly but smooth with age. The three-chamber, dehiscent capsules contain numerous oblong seeds.

Botanical notes: In 1835, in the German garden magazine *Allgemeine Gartenzeitung*, Johann Klotzsch, curator of the Berlin Herbarium, named the genus *Hechtia* for Julius Gottfried Conrad Hecht (1771–1837), counselor to the King of Prussia. Klotzsch described a plant cultivated in the Berlin Botanical Garden from material collected in Mexico by German botanist-explorers

Christian Schiede and Ferdinand Deppe. Sereno Watson, curator of the Gray Herbarium at Harvard, described *H. texensis* in 1885, from a collection by US Army surgeon Valery Havard "in the Great Bend of the Rio Grande," in 1883. Havard, while stationed at San Antonio, and Fort Duncan near Eagle Pass, traveled with exploring expeditions in the upper Rio Grande valley of West Texas.

66. Living Rock Cactus (*Ariocarpus fissuratus* var. *fissuratus*)
(Other common names: false peyote, chautle, star cactus, dry whiskey, sunami, hikuli sunami, peyote cimarron [Tarahumara], tsuwiri [Huichol])

Habitat: barren hills, desert scrub, rocky, gravelly limestone, clay or gyp, low-mid elevations.

Geographic range: Texas: Trans-Pecos (excluding El Paso, Culberson, Jeff Davis, Reeves Counties); Maverick County. US: Texas. Mexico: northern parts of Chihuahua, Coahuila, Durango.

Park locations: Solitario Inner Loop, hills above Righthand Shutup, Terlingua Uplift, West Contrabando Trail

Often overlooked, this cactus forms flat, gray-green, spineless stems flush with the ground, blending into the surroundings. To 5 inches across, each stem grows from a turniplike taproot that serves as a water storage tank. The stem surface is covered with triangular, overlapping tubercles in a star-shaped pattern. Each tubercle is fissured, warty, and tough-skinned. Grooves in tubercles (areoles) are filled with matted hairs.

Unlike other Big Bend cacti, the flowers of this cactus bloom in fall, near the white-woolly stem tip. The inner tepals (petal-like lobes) are pink, rose, or magenta, and the outer tepals are brown to green. Each bloom is broadly funnel-shaped, with golden anthers and a multilobed, protruding white stigma. To 1 inch long, the globose or clublike fruits are white to pale green, with shiny black seeds.

Tarahumara in Chihuahua ate living rock in religious ceremonies, although the plant lacks mescaline (Schultes et al. 1998). Tarahumara long-distance runners drank the juice or chewed the stems to increase stamina (Pennington 1963). In the borderlands, the cactus was used as a maize beer supplement, often with highly intoxicating, unpredictable consequences (Havard 1896).

Botanical notes: In 1838, Brussels taxonomist Michel Scheidweiler named the genus *Ariocarpus*, combining the Greek *aria* (the genus *Aria*) and *karpos* (fruit), for the similar pear-like fruits. The plant was collected by Belgian botanist Henri Galeotti, who explored Mexico from 1835 to 1840. The Latin species name, *fissuratus* (fissured), refers to the tubercles. St. Louis botanist George Engelmann described the species as *Mammillaria fissurata* in 1856, from a collection by Arthur Schott in 1852, near "Rio Bravo del Norte," during the US-Mexico Boundary Survey. Schott was a surveyor, geologist, collector, and artist: his landscapes and portraits of Native Americans appeared in Major William Emory's survey report.

67. Clumped Dog Cholla (*Corynopuntia aggeria*)
(Other common names: Big Bend prickly pear, club cholla, mound-forming opuntia)

Habitat: creosote scrub, desert, alluvial flats, sandy, gravelly low hills; limestone, gyp, igneous substrates near river.

Geographic range: Texas: Big Bend (south Brewster, southeast Presidio Counties; probably El Paso County). US: Texas. Mexico: Coahuila; likely northeast Chihuahua.

Park locations: 1 1/4 miles west of Lajitas, Bofecillos Road, Sauceda Road east of Bofecillos Road turnoff, Old Government Road, Contrabando Creek, West Contrabando Trail

Growing from thick tuberlike roots, this cholla forms patches or low mounds of clublike, warty joints to 3 feet or more across. Not easily detached, the joints bear clumps of yellowish white glochids, three or four large central spines, often chalky-white and flattened, and two to four small thin radial spines, mostly grayish white.

In late March and April, at noon or later, yellow flowers to 2 inches wide appear for a single day. Each bloom has three or four circular layers of tepals, the inner bright yellow, the outer yellow but often green at the base and pink at the tips. The filaments are red to pale orange with yellow anthers, and the style, with five or more stigma lobes, is pale green to yellow.

To 2 inches long, the pale-yellow fruits are clublike and spineless, but dotted with tufts of woolly glochids. The fleshy fruits contain brown to yellow, disklike seeds.

Botanical notes: In 1935, Danish Count Frederik Knuth, in a work coauthored with Curt Backeberg, *Kaktus-ABC*, segregated prickly pear species with club-shaped stems from the *Opuntia* genus into his new *Corynopuntia* genus. Knuth combined the Greek *koryne* (club) and *Opuntia*, implying "clublike *Opuntia*." In 1989, Barbara Ralston and Richard Hilsenbeck named this cholla *Opuntia aggeria*, from the Latin *agger* (mound), noting that the name was "chosen to describe the clumped or aggregated growth habit of this mound-forming species." Margery Anthony, ecologist at Chico State College in California, had previously collected the species in 1948, on Tornillo Flats in BBNP, and described it in 1956 in the *American Midland Naturalist* as a hybrid, *Opuntia grahamii x schottii*.

68. Candle Cholla (*Cylindropuntia kleiniae*)
(Other common names: Klein's cholla or pencil cholla or tasajillo, cholla tasajillo macho)

Habitat: sandy, silty alluvial flats, floodplain, desert scrub, rocky bluffs; limestone hills by river to rocky mountain slopes.

Geographic range: Texas: Trans-Pecos Texas (excluding Reeves, Pecos, Terrell, Val Verde Counties); Eastland, Lampasas Counties. US: Texas, south-central New Mexico, Oklahoma. Mexico: Chihuahua to Nuevo León, south to Durango.

Park locations: west of Colorado Canyon River Access, near Cienega Creek

Candle cholla is a slender shrub to 6 feet high, erect but scraggly, with loosely branched, cylindric stems. The ultimate stem segments (joints) are at most 1/2 inch thick, thicker than desert Christmas cholla (*Cylindropuntia leptocaulis*) joints, thinner than tree cholla (*Cylindropuntia imbricata* var. *imbricata*) joints. This cholla is likely a hybrid of the other species.

Long low tubercles, needlelike spines, and small glochids are borne on the joints. On plants near the Rio Grande, a single spine grows from wide-spaced, white-felty areoles, each spine with a loose, papery sheath. Typically, the spines are downward-pointing, gray or gray and reddish-brown with yellow tips.

In late spring, stunning blossoms form with subtle color variations. The flowers have a bronze or pinkish-bronze aspect but are distinctly multicolored. Each tepal may display shades of pink, cream, pale greenish yellow, maroonish red and white. Each bloom has five inner and five outer rounded tepals, pale green filaments with yellow anthers, and a protruding pink style with a buff multilobed stigma. To 1 1/3 inches long, the spineless ripe fruit is red or orangish red, inversely egg-shaped, with some glochids.

Botanical notes: In 1930, Frederik Knuth, Count of Knuthenborg, in *Den nye kaktusbog* (The new cactus book), moved prickly pear cacti with cylindric stems into his new *Cylindropuntia* genus. The name joined the Latin *cylindrus* (cylinder) and *Opuntia*, implying "cylindric *Opuntia*." In 1828, Swiss botanist Augustin Pyramus de Candolle named the species *Opuntia kleiniae*, apparently because of the resemblance of the cholla's "big-finger-sized" stems to those of an aster, *Cacalia kleinia*. Candle cholla was collected by Irish botanical explorer and student of de Candolle, Thomas Coulter, at the time a physician with the Real del Monte Mining Company in Mexico. After extensive collecting in Mexico, in 1828 Coulter sent fifty-seven species of living cacti to de Candolle, 47 of which de Candolle described as new species (Coville 1895).

69. Desert Christmas Cholla (*Cylindropuntia leptocaulis*)
(Other common names: tasajillo, pencil or Christmas cholla, desert Christmas cactus, garambullo, agujilla)

Habitat: bottomlands, washes, thickets; sandy, gravelly, clayey flats; also bajadas, grasslands, wooded terrain.

Geographic range: Texas: Trans-Pecos, west 2/3 of state. US: Texas to Arizona, Oklahoma. Mexico: Sonora east to Tamaulipas, south to Durango, Zacatecas.

Park locations: Colorado Canyon River Access, Llano Loop

Desert Christmas cholla is a thicket-forming shrub to 3 feet high, erect or sprawling, and often densely branched. A menace to ranchers and hikers, the plant reproduces vegetatively, as easily detached joints take root and form clones.

The slender stem joints are green or yellowish green, with highly reflective spines and felty tufts of glochids, but otherwise smooth. Needlelike or slightly flattened, the spines are often gray near the base, yellow or reddish-brown above, and wrapped in a papery sheath. Typically, a single spine grows from an areole, and many short lateral joints are spineless.

In late spring to early fall, pale greenish-yellow flowers form toward the ends of longer joints. Opening widely in late afternoon, each bloom is somewhat wheel-like, with thin but lustrous spoon-shaped tepals, which are abruptly pointed at the tips. The blossoms have pale greenish-yellow filaments and yellow anthers and a yellowish style with a bulky multilobed stigma.

This cholla is named for its bright red, persistent fruits that decorate the plant like Christmas ornaments. Pear-shaped but grape-size, the spineless fruit is fleshy, with a smooth surface marked with cinnamon-colored, brushlike glochids. The cheery fruits are a vital food for birds, deer, and small mammals.

Botanical notes: The species name joins the Greek *lepto* (thin) and the Latin *caulis* (stem), for the pencil-size stems. In 1828, Swiss botanist Augustin Pyramus de Candolle described this and other species collected by Irish botanist Thomas Coulter in Mexico. From 1824 to 1834, Coulter explored Mexico, California, and Arizona and sent specimens to de Candolle in Geneva and to Trinity College in Dublin.

70. Eagle-Claw Cactus (*Echinocactus horizonthalonius* var. *horizonthalonius*)
(Other common names: blue barrel cactus, devil's or Turk's head, biznaga meloncillo, manca caballo, manca mula)

Habitat: desert scrub, dry, rocky slopes, typically limestone, also igneous, clay, gypsum soils.

Geographic range: Texas: Trans-Pecos; Kinney County. US: Texas, south-central New Mexico, south-central Arizona. Mexico: Chihuahua east to Nuevo León, south to San Luis Potosí.

Park locations: Upper Guale Mesa, Paso al Solitario 4WD Road to Los Alamos, Buena Suerte Trail

The bulky stems of this cactus may grow low to the ground and domelike, or to 1 foot high or more and barrel shaped. The stems are gray green to blue gray, glaucous, with eight thick, broadly rounded, smooth ribs. The ribs are arranged vertically or in a spiral curving toward the stem tip.

Reminiscent of eagle claws, projecting or down-curved, cross-ribbed spines line the ribs. Up to three central spines and seven smaller radial spines, gray, brown, pink, red, or black, grow from each areole.

From April to September, usually after rains, lush and lustrous funnel-shaped blooms, 3 inches across, form on new growth at the woolly stem tip. Opening at midday, the tepals are magenta to rose-pink but darker and more saturated near the base. Often, inner tepals are fringed and outer tepals abruptly bristle-tipped or tapered to a spinelike tip. The stamens are yellow, and the style and multilobed stigma are pink.

To 1 inch long, the red or pink fruits are spheric, nestled in woolly hairs at the stem tip. Spineless but somewhat scaly, the fleshy fruits contain gray to black seeds.

Botanical notes: The genus name *Echinocactus* combines the Greek *echinos* (hedgehog) and *Cactus* (an old genus name), from the Greek *kaktos* (spiny plant). This species was described by French botanical author Charles Lemaire in 1839 and collected by Belgian botanist Henri Galeotti in central Mexico. Lemaire named the species *horizonthalonius* because the areoles are "horizontaliter dispositas," or horizontally arranged (my Latin translation). Professor of classical literature at the University of Paris, Lemaire became interested in botany and cactology and later edited French horticultural periodicals.

71. Texas Claret-Cup Cactus (*Echinocereus coccineus* var. *paucispinus*)
(Other common names: claret cup or scarlet hedgehog cactus, organo-pequeño de espinas escarlatas)

Habitat: desert scrub, grassland, rocky slopes, oak-juniper roughland; igneous, limestone soils.

Geographic range: Texas: Trans-Pecos (excluding El Paso, Hudspeth Counties); Edwards Plateau. US: Texas, southeast New Mexico. Mexico: northeast Coahuila.

Park locations: Cerro de las Burras Loop, Las Burras Road, La Mota Road, Llano Loop, Llano Dome, Tascate Hills

Of conservation concern: see Appendix A

Texas claret-cup is a profusely branching cactus, with stout cylindrical stems forming mounds to 3 feet across. The shiny green stems, to 16 inches high, have seven or eight bulky, often wrinkled ribs and clusters of round spines that do not hide the stem. The areoles bear zero to two central spines, and four to eight smaller radial spines, mostly grayish white, and frequently dark-tipped.

Appearing in March and April, funnel-shaped scarlet flowers, open day and night for a few special days, light up the landscape. The prolific blooms have several whorls of spoon-shaped tepals, each rather thick and waxy, with a rounded tip and cream-colored to pale yellow base. Within the tepals are white stamens with pink or maroon anthers, and a protruding style with a multilobed, greenish yellow stigma. Male flowers with sterile stigmas, and female flowers with sterile stamens, appear on separate plants.

About 1 inch long, the red fruits are globose to cylindric, edible and juicy, with some small spine clusters. The fruits bear tiny black seeds with pimple-like projections.

Botanical notes: The genus name combines the Greek *echinos* (hedgehog, by implication spiny) and *Cereus* (a columnar cactus genus), from the Latin *cereus* (wax candle), because the stems resemble spiny candles. This species, *coccineus* (scarlet), is named for the flower color. The varietal name joins the Latin *pauci-* (few) and *spinus* (spine), since this variety has fewer spines per areole than other varieties. This claret-cup was described by George Engelmann in 1856 as a species (*Cereus paucispinus*), but in 1989, David Ferguson, in the *Cactus and Succulent Journal*, gave the plant its current botanical name.

72. Texas Rainbow Cactus (*Echinocereus dasyacanthus*)
(Other common names: spiny hedgehog cactus, golden rainbow hedgehog, yellow or yellow-flowered pitaya, alicoche arcoiris de Texas)

Habitat: desert scrub, grasslands, rocky ridges and canyons, primarily limestone substrates below 4,500 feet.

Geographic range: Texas: Trans-Pecos (excluding Val Verde County); Mitchell County. US: Texas to southern New Mexico. Mexico: northeast Chihuahua, north Coahuila.

Park locations: Bofecillos Road, Solitario northwest rim, Terlingua Uplift, West Contrabando and Buena Suerte Trails

Texas rainbow cactus has robust stems, more cylindric and taller with age, to 1 foot high or more. Usually, the stems are single or sparsely branched, but older plants may form large clumps of ten plus stems. Each stem is segmented into fifteen or more wavy ribs, hidden by dense, overlapping short spines.

Up to twelve projecting central spines and twenty-five radial spines are borne in each areole. Spine colors are variable: grayish white, tan, pink, or reddish brown. Some plants exhibit rainbow-like banding: new spines grow horizontally around the stem in contrasting colors each year.

In spring, this cactus forms resplendent funnel-shaped flowers to 4 inches wide. Emerging from one-year-old growth on the stem sides, near the stem tip, the sweet-scented blooms typically are yellow with greenish throats. Rarely, blooms appear in shades of pink, rose, salmon, and magenta, possibly from hybridization with Texas claret-cup cactus (Powell and Worthington 2018). The waxy tepals are elongate, with pointed tips and green bases. Within the tepals are many stamens with yellow anthers, and a style with a green, multilobed stigma.

To 2 inches long, the globose green fruits, purple when ripe, are juicy but acidic, with spine-filled areoles and dark seeds.

Botanical notes: In 1848, in Friedrich Adolf Wislizenus's *Memoir of a Tour to Northern Mexico*, George Engelmann named this species *dasyacanthus*, from the Greek *dasy* (thick) and *akantha* (spine), for the dense spines. Wislizenus collected the cactus in "mountains about El Paso." German-born St. Louis physician trained by Engelmann, Wislizenus joined a trading party from St. Louis to Santa Fe and Chihuahua, where he was imprisoned at the onset of the Mexican-American War. Freed by Colonel Alexander Doniphan's invading American troops, Wislizenus traveled with them across Mexico to New Orleans.

73. Strawberry Cactus (*Echinocereus stramineus* var. *stramineus*)
(Other common names: strawberry pitaya, strawpile hedgehog, spinemound, spiny strawberry or porcupine hedgehog cactus, agosteña, alicoche sanjuanero)

Habitat: scrub, rocky hills; limestone, novaculite, igneous substrates.

Geographic range: Texas: Trans-Pecos (excluding Reeves, Jeff Davis, Val Verde Counties); Upton, Crockett Counties. US: Texas, southern New Mexico. Mexico: Chihuahua east to Nuevo León.

Park locations: Closed Canyon, Arenosa Campground, Black Hills, upper Arroyo Mexicano, Terlingua Uplift, Solitario Peak

Strawberry cactus forms domelike mounds to 3 feet wide, with up to one-hundred-plus stems. The erect stems are narrowly egg-shaped, often tapered at the tip, with ten to sixteen sinuous ribs.

The ribs are partly hidden below reflective, straw-colored spines, often pink-tinged, which turn white and fragile with age. Each areole bears two to four projecting central spines to 3 inches long, and up to fourteen smaller radial spines. In morning or evening light of the desert landscape, the spines become a glistening, glowing mass.

From late March to May, sometimes into summer, dozens of huge, funnel-shaped satiny flowers to 5 inches across, open before noon and blanket the cactus. Each tepal is magenta to rose-pink, darker near the base and pale pink near the often-toothed or frayed tips. Many stamens, with red filaments and yellow anthers, and a long style, with a prominent multilobed green stigma, add color to the bloom.

This cactus is named for its delectable fruits, 2 inches long, with the scent and flavor of strawberries. The ripe oval fruits are pinkish brown, with a few easily removed spines, juicy pinkish pulp, and tiny black seeds. These delicacies are a tradition, eaten as dessert with sugar and cream, or made into preserves.

Botanical notes: In 1856, George Engelmann named this species *stramineus* (made of straw) for the straw-colored spines. The plant was collected by Charles Wright in June 1851, during the US-Mexico Boundary Survey, on Escondido Creek, a tributary of the Pecos River, in Pecos County, Texas.

74. Graybeard Cactus (*Echinocereus viridiflorus* var. *canus*)
(Other common names: none known)

Habitat: novaculite slopes, quartz- and silica-covered exposures, igneous outcrops; elevations from 4,400 to 4,800 feet (Powell and Weedin, 2004).

Geographic range: Texas: south Presidio County.

Park locations: Solitario

Of conservation concern: see Appendix A

Many of my early ventures in the BBR backcountry were focused on finding populations of this rare, enigmatic cactus variety, which is endemic to the Solitario and isolated from other close relatives. Graybeard cactus stems are less than 6 inches high, egg-shaped to cylindric, with fourteen to sixteen ribs concealed by a thick cloak of white spines. Each areole bears up to fifteen central spines and forty-eight radial spines.

The spreading central spines, to 1 inch long, are thin, flexuous, white but at times reddish near the tips. The tiny radial spines are extremely thin, no more than 1/3 inch long, usually close-pressed against the stem. Curiously, the seedlings bear long, soft, hairlike spines that blanket juvenile plants like a beard.

In March and April, funnel-shaped, lemon-scented flowers appear on the stem sides. The yellow-green tepals, some with prominent reddish-brown mid-stripes, are spatulate, with an abrupt short-spined or triangular long-spined tip. Within the tepals are stamens with yellow anthers, and a yellow green style with a multilobed green stigma.

Egg-shaped to oblong, the spiny green fruits bear tiny, pear-shaped black seeds with pimple-like protuberances.

Botanical notes: The species name joins the Latin *viridis* (green) and *flora* (flower), meaning "green-flowered," and the varietal name is the Latin *canus* (grayish white), the spine color. This species was collected by Adolf Wislizenus in 1846, near Wolf Creek on the Santa Fe Road in New Mexico and described by George Engelmann in *Memoir of a Tour to Northern Mexico*, Wislizenus's account of his travels. Variety *canus* was described by A. Michael Powell and James Weedin in 2004 in *Cacti of the Trans-Pecos*. In 1984, while working on his master's thesis, James Jeff Clark, a Sul Ross State University student, discovered this variety on a novaculite slope of the Solitario.

75. Mariposa Cactus (*Echinomastus mariposensis*)
(Other common names: Lloyd's mariposa or Lloyd's fishhook cactus, golfball cactus, biznaga bola de mariposa)

Habitat: desert scrub, low sparse hills, rocky limestone outcrops; low-mid elevations.

Geographic range: Texas: Big Bend (Brewster, Presidio Counties). US: Texas. Mexico: Coahuila.

Park locations: Withheld — Threatened species

Of conservation concern: listed as Threatened species in 1979; see Appendix A

This rare cactus, not previously recorded at BBR, is now primarily known from the Dead Horse Mountains of BBNP and the Black Gap Wildlife Management Area. However, the plant was discovered near Terlingua, and occurs in scattered limestone outcrops across southern Brewster County. Mariposa cactus has an erect, unbranched, spherical stem to 4 inches high, golf-ball to tennis-ball size, usually with twenty-one ribs. The ribs, blue-green and glaucous, are concealed below a dense layer of spines.

Commonly, each areole holds four long central spines and twenty-six to thirty-two shorter, thinner radial spines. The radials are ashy-white, pressed against the stem and overlapping. The three upper central spines, to 3/4 inch long, may be tan or pink but often chalky-blue, and notably upswept. The upper centrals form a distinctive bluish tuft of longer, darker, ascending spines at the stem apex. In contrast, the shorter, lower central spine angles sharply downward.

In February and March, this early bloomer displays modest white flowers 1 inch wide, on recent growth near the stem tip. The tepals are white, with prominent mid-stripes in pale shades of pink, rose pink, tan, brown, or reddish brown. Within the flower's throat are many cream-colored or yellow stamens, and a pale green style with an elevated, multilobed yellow green stigma.

Growing within spine clusters at the stem tip, the fruits are green or yellow green, globose to oblong, 2/5 inch long. Each fruit contains tiny black seeds dotted with pimply bumps.

Botanical notes: In *Cactaceae* (volume 3, 1922), Nathaniel Britton and Joseph Rose named this genus *Echinomastus*, joining the Greek *echinos* (hedgehog, spiny) and *mastos* (breast), to emphasize the plant's spiny tubercles. In 1945, J. Pinckney Hester, cactus-succulent enthusiast from Fredonia, Arizona, discovered this species at Mariposa Mines, a cinnabar mining company near Terlingua, and gave it the company name. Mariposa is the Spanish word for butterfly. Hester, a collector of southwest US plants, authored articles on new species in the *Cactus and Succulent Journal* and *Desert Plant Life*.

76. Common Button Cactus (*Epithelantha micromeris*)
(Other common names: button or pingpong-ball cactus, mulatto, tapone, biznaga-blanca chilona, híkuli mulato [Tarahumara])

Habitat: dry rocky slopes, low bluffs, sparse desert flats, grasslands; limestone, caliche, volcanic soils.

Geographic range: Texas: Trans-Pecos (excluding Jeff Davis, Reeves Counties), east to Medina, Bandera Counties; also, Upton, Howard Counties. US: Texas, southern New Mexico, southeast Arizona. Mexico: Coahuila.

Park locations: Solitario Peak area, Inner Loop, slopes near Righthand Shutup, near Cienega Creek

This dwarf is often compared to a button, mushroom, ping-pong ball, golf ball, and a fuzzy ball, since the tiny, tight-pressed spines are about as rough as fuzz. Usually, the stems are single, to 2 inches high, globose to oblong, and slightly sunken at the tip. Lacking defined ribs, the stems have numerous small tubercles, obscured by the spines.

Up to forty overlapping spines grow in each areole. The longest spines, to 1/2 inch long, form a dense upswept tuft at the stem tip. Older spines are shorter (the tips break off), close-pressed, and brown basally, creating a prominent brown spot at the center of each spine cluster.

From February to April, tiny flowers emerge from new growth at the stem tip, partly hidden within the spines and woolly hairs. Open about noon for only a few hours, the funnel-shaped blooms are the smallest of any cactus species in the United States or the Chihuahuan Desert. The tepals are pale pink or peach, lighter near the tips and sometimes jagged on the margins. Within the tepals are up to sixteen stamens with yellow anthers and a style with a multilobed white stigma. The bright red fruits — known as *chilitos* — resemble little chili peppers. To 3/4 inch long, the fruit is spineless, cylindric, with a few shiny black seeds.

Boke's button cactus (*Epithelantha bokei*) also occurs in the park. Compared to common button cactus, Boke's button has whiter spines, more spines per areole, and flowers three times as wide.

Botanical notes: The genus name *Epithelantha* joins Greek words meaning "flower upon nipple," because the blooms form at the tubercle tips. The epithet *micromeris*, from the Greek *mikros* (small) and *meris* (part), emphasizes the small plant parts. This cactus was described by George Engelmann in 1856 as *Mammillaria micromeris*, from collections by Charles Wright in 1849, 1851, and 1852, "from El Paso eastward to the San Pedro [Devil's] River."

77. Desert Pincushion Cactus (*Escobaria dasyacantha*)
(Other common names: Big Bend or Big Bend foxtail cactus, dense mammillaria, dense cory or stout spine cactus, biznaga-escobar de espinas gruesas)

Habitat: desert flats, creosote, mesquite scrub, grassland, bajadas, rocky hills, woodlands; igneous, limestone soils.

Geographic range: Texas: Trans-Pecos (El Paso, Hudspeth, Jeff Davis, Presidio, Brewster, Pecos Counties). US: Texas. Mexico: expected.

Park locations: Big Hill, Rancherías Canyon, Ojito Adentro, Los Ojitos, Cienega Creek

Of conservation concern: see Appendix A

This scarce, reclusive cactus occurs at some park sites, mostly near the Rio Grande. Usually, desert pincushion has a single unbranched stem, spheric or short-cylindric, to 4 inches long, but older plants may be few-branched, with corncob-like, basal tubercles. The stems are covered with small, protruding green tubercles. Four to nine central spines, to 2/3 inch long, and up to thirty-one radial spines half as long, are borne in each areole, nearly hiding the stem. The spines are white to brown, but central spines may be dark tipped.

From March to May, small funnel-shaped flowers, to 3/4 inch across, form on new growth near the stem tips. The flowers resemble blooms of Duncan's pincushion cactus (*Escobaria duncanii*). The tepals are creamy white on the margins and spine-tipped, with dull mid-stripes in shades of pink, peach, or brown. Outer tepals are lighter in color and fringed. Within the tepals are many stamens with yellow anthers and a greenish style with a multilobed green or yellow-green stigma.

The spineless fruits are red when ripe, club-like to cylindric, to 1 inch long, with obvious floral remains. The fruits contain tiny black, pitted seeds.

Botanical notes: In 1923, Nathaniel Lord Britton and Joseph Nelson Rose named the genus *Escobaria* for Mexican brothers and agricultural engineers Numa and Romulo Escobar Zerman, who founded an agricultural school in Ciudad Juárez in 1906. The species name combines the Greek *dasys* (thick) and *akantha* (thorn) to emphasize the plant's dense spines. George Engelmann described this species as *Mammillaria dasyacantha* in 1856, from collections by Charles Wright near "El Paso and eastward" in 1851 and 1852. Britton and Rose moved the plant to their new genus *Escobaria* in 1923.

78. Duncan's Pincushion Cactus (*Escobaria duncanii*)
(Other common names: Duncan's snowball or Duncan's foxtail or Duncan's cory cactus, beehive cactus)

Habitat: driest, hottest parts of CD, near Rio Grande; desert scrub, rocky ledges, bluffs, crevices, limestone fissures.

Geographic range: Texas: Big Bend (south Brewster, southeast Presidio Counties). US: Texas, New Mexico (Sierra County; reported in Doña Ana County). Mexico: expected in adjacent Mexico.

Park locations: Solitario, Terlingua Uplift

Of conservation concern: see Appendix A

Duncan's pincushion is a rarity, previously unknown at BBR, and not conveniently located on a trail or hiker's route. The cactus is difficult to spot in the field because its bristly spines blend with the surrounding limestone habitat, and its favored abode is a tight rock crevice.

Mostly solitary, rarely branching, the green stems are globelike, or short oblong, to 2 inches high. Growing from a thick, fleshy taproot, the stems are covered with small, protruding tubercles, grooved on the top. The tubercles are invisible beneath dense spines, up to seventy-five per areole. The thin spines are flexible and fragile, often twisted or bent from overcrowding. All spines are snowy white, some with darker tips.

Nearly identical to desert pincushion flowers but blooming even earlier, the small funnel-shaped blossoms open in February and March. The blooms form on new growth near the stem tip. Each tepal is white or cream-colored with a spiny tip and prominent midrib in shades of pink, peach, brown, or tan. The outer tepals are lighter in color with fringed margins. Within the tepals are numerous stamens with yellow anthers and a style with a multilobed green or yellow green stigma.

Cylindric to clublike, the red fruits, 1/2 inch long, contain shiny black, pitted seeds.

Botanical notes: In 1941, in *Desert Plant Life*, Big Bend explorer J. Pinckney Hester named this cactus *Escobesseya duncanii*, for Frank Duncan, a prospector-rockhound with mineral rights near Terlingua and a professional photographer with a studio in Marfa. Duncan sold postcards, portraits of ranchers and their ranches, and panoramic photographs of the Big Bend landscape. For added income, Duncan and his daughter sold rocks in *Desert Magazine* and in *Harry Oliver's Desert Rat Scrap Book*. German horticulturist Curt Backeberg moved the species to the *Escobaria* genus in 1961. Backeberg operated an export-import business and made extensive collecting trips to Mexico and Central and South America.

79. Silverlace Cactus (*Escobaria sneedii* var. *albicolumnaria*)
(Other common names: snowcone nipple cactus, white column, silverlace cob or white-spine cob cactus, white column or column foxtail cactus)

Habitat: gravelly alluvial flats, desert scrub, arid, rocky limestone hillsides.

Geographic range: Texas: Big Bend (south Brewster, southeast Presidio, southeast Pecos Counties). US: Texas. Mexico: Chihuahua, Coahuila.

Park locations: Solitario, Terlingua Uplift, Contrabando Lowlands

Of conservation concern: see Appendix A

Although limited in range, this CD endemic is probably the most plentiful variety in a large complex of rare cacti. Typically, silverlace cactus produces a single cylindric, stiffly erect stem to 10 inches tall. Older plants may form clusters of numerous stems. Protruding tubercles are arranged in a spiral around the stems, concealed below long, bristling spines.

As many as fifty-six spines are borne in each areole, at least twenty-five snowy white radial spines, and ten central spines, snowy white or chalky pink, to 1 inch long or more. The long, radiating central spines, often darker near the tips — pink, red-brown, or black — give this cactus a distinct bristly aspect.

From March to June, in early afternoon, small funnel-shaped flowers open, seldom widely, at the stem tips. The slender tepals are cream-colored, but darker at the mid-regions and the base—in shades of pink, magenta, lavender, apricot, salmon, or tan. All tepals are sharp-pointed, and outer tepals are usually fringed. Commonly, park plants have pink to red filaments with golden anthers and a pale pink style with a yellowish multilobed stigma.

The green or red fruits are cylindric or clublike, to 3/4 inch long, with minuscule pitted, reddish-brown seeds.

Botanical notes: In 1923, Nathaniel Lord Britton, director of the New York Botanical Garden, and Joseph Nelson Rose, curator with the Smithsonian Institution, named the species *sneedii* for J. R. Sneed. Sneed collected the plant in the Franklin Mountains near El Paso in 1921. The varietal name combines the Latin *albus* (white), *columna* (column), and -*aria* (pertaining to), because the white-spine-covered stems resemble white columns. J. Pinckney Hester described this cactus as a species (*Escobaria albicolumnaria*) in 1941, in *Desert Plant Life*. In 2018, Sul Ross State University botanist A. Michael Powell, Richard Worthington, and Shirley Powell decided that Hester's species is better treated as a variety of *Escobaria sneedii* in their new flora, *Flowering Plants of Trans-Pecos Texas and Adjacent Areas*.

80. Varicolor Cob Cactus (*Escobaria tuberculosa* var. *varicolor*)
(Other common names: mountain cob cactus, varicolor cory cactus)

Habitat: igneous mountain slopes, also novaculite; rocky terrain in scrub, grasslands, woodlands.

Geographic range: Texas: Trans-Pecos (Brewster, Presidio, Jeff Davis, Pecos Counties). US: Texas. Mexico: Chihuahua, Coahuila.

Park locations: Chorro Canyon rim, Los Ojitos, upper Cienega Gorge

Varicolor cob stems are solitary, almost never branched, as much as 6 inches tall. At first more globular, this cactus elongates with age, often becoming broadly cylindric, with a blunt, not tapered tip.

While the more common cob cactus is found mostly in limestone environs, varicolor cob is largely restricted to igneous or novaculite habitats. Compared to the common cob, varicolor cob has thicker stems, more widely spaced tubercles, and fewer spines. With up to twenty-four radial spines compared to thirty-six for the common cob, the spines do not fully hide the green stems.

On late spring and summer afternoons, delicate white or pale pink flowers appear at the stem tip. The slender, elongate inner tepals have sharp-pointed tips and often darker shades at the middle and the base. Outer tepals are more creamy-white and frayed on the edges. In the flower's throat are many stamens with cream-colored filaments and yellow anthers, and a style with a multilobed white stigma.

The fruits are red at maturity, as much as 1 inch long, broadly oblong to cylindric, with obvious floral remnants at the base. The pitted seeds are reddish-brown.

Botanical notes: The Latin epithet *tuberculosa* (tuberculate) is an apt name for the stems of this cactus. The varietal name *varicolor* (varicolored) likely refers to the spines. George Engelmann described the species in 1856 as *Mammillaria tuberculosa*, from a collection by John Bigelow in 1852, in Chihuahua south of El Paso. In 1932, in a Berlin cactus periodical, German cactus enthusiast Ernst Tiegel described this variety as a species, *Coryphantha varicolor*. In 1988, Kenneth Heil and Steven Brack reclassified the plant as a variety of *Escobaria tuberculosa* in the *Cactus and Succulent Journal*.

81. Giant Fishhook Cactus (*Ferocactus hamatacanthus* var. *hamatacanthus*)
(Other common names: Turk's head, Texas barrel or Mexican fruit cactus, whiskered barrel, biznaga de limilla or de tuna, biznaga barril costillona)

Habitat: desert scrub, alluvial flats near the Rio Grande, steep bluffs, oak roughland, rocky mountainsides, volcanic and calcareous soils.

Geographic range: Texas: Trans-Pecos (excluding Hudspeth, Culberson, Terrell Counties), lower South Texas Plains, Coastal Bend. US: Texas, southern New Mexico. Mexico: Chihuahua, Coahuila south to Zacatecas, San Luis Potosí.

Park locations: Agua Adentro Spring, Ojito Adentro, Solitario north rim, inner Solitario, slopes above Righthand Shutup, Buena Suerte Trail

Although a large, bulky stem succulent with vicious-looking hooked spines, giant fishhook is not gigantic when compared to others in this genus. To 20 inches high and 10 inches wide, this cactus is one of the smaller barrel cacti in the CD and America's desert southwest. Typically, the dark green stems are solitary, spherical to barrel-shaped with age, erect or notably leaning, and segmented into thirteen to seventeen rounded, wrinkled, or wavy ribs.

Commonly, ten to fourteen radial spines and four longer central spines are borne in each areole, the longest more than 4 inches long, usually flattened, flexible, twisted, with a prominent hooked tip. The spines are varicolored, at times multicolored — pinkish, reddish, tan, pale yellow — and grayish with age. Occasionally, larger plants to 3 feet tall, or multistemmed clusters to 6 feet wide, are encountered. On high canyon slopes, the expansive clusters create a memorable landscape attraction.

In the afternoon, lemon-yellow flowers bloom in a ring around the stem tip, on the season's new growth. The large blooms are remarkably luminous and fragrant. The tepals are spatulate, with a narrow base and rounded, often jagged tip with a spinelike point. Within the tepals are many stamens with yellow anthers and a yellow style and protruding yellow stigma with numerous fingerlike lobes. Although scaly, the oval to oblong fruits are edible and juicy, to 2 inches long, green but maroon or brown with age. The lustrous black seeds are egg-shaped and covered with tiny depressions.

Botanical notes: In 1922, Nathaniel Britton and Joseph Rose created the genus *Ferocactus*, stating that "the generic name is from *ferus*, wild, fierce, and *cactus*, referring to the very spiny character of the plants." In 1846, in the German garden magazine *Allgemeine Gartenzeitung*, Hanover botanist Friedrich Mühlenpfordt named this species *Echinocactus hamatacanthus*, joining the Latin *hamatus* (hooked) and Greek *akantha* (thorn), to describe the hooked central spines. Britton and Rose moved this species to the *Ferocactus* genus in 1922.

82. Catclaw Cactus (*Glandulicactus uncinatus* var. *wrightii*)
(Other common names: Chihuahuan fishhook or Wright's fishhook cactus, Turk's head, eagle-claw cactus, brown flowered hedgehog)

Habitat: gravelly desert flats, desert scrub, grasslands, foothills; limestone, igneous soils.

Geographic range: Texas: Trans-Pecos; Crockett, Starr, Victoria Counties. US: Texas, south-central New Mexico. Mexico: Sonora east to Nuevo León, south to Durango, Zacatecas.

Park locations: inner Solitario west of Tres Papalotes, lower Fresno Creek, Contrabando Creek

Mostly solitary, catclaw stems are seldom more than 6 inches high, green or blue green, with a chalky, waxy bloom. The stems are shortly columnar, with nine to thirteen deeply grooved and wrinkled ribs.

Customarily, up to four central spines (one hooked) and up to ten radial spines (three hooked) are borne in each areole. The pale-yellow, hooked central spine may reach 4 inches high, well above the stem. All the spines, in shades of tan, gray, pink, red, and yellow, seem to form a cage around the stem but do not hide the bulky tubercles. This catclaw often grows within grasses and is easy to overlook because the ascending hooked spines blend with the blades of grass.

From March to May, curious dark blooms, short and narrowly funnel-shaped, are tightly clustered around the stem tip. The oblong tepals are brick red or maroon, or lighter orange red with darker midribs. Outer tepals may have white or transparent margins and slightly fringed or irregularly toothed tips. Within the tepals are stamens with yellow anthers and a red style with a yellow to orange-yellow, multilobed stigma. Unusual for a cactus, the flowers are long-lasting — open for a week or longer.

Within spines at the stem apex, the red fruits are egg-shaped to globose, to 1 inch long. Fleshy at maturity, the fruits are spineless but scaly, with many black seeds.

Botanical notes: German horticulturist Curt Backeberg, prolific collector and author on cacti, created the genus *Glandulicactus* in 1938, from the Latin *glandula* (gland) and *Cactus* (an old genus name), apparently emphasizing the prominent areolar glands of these plants. Belgian botanist Henri Galeotti named the species *Echinocactus uncinatus* in 1848, using the Latin word *uncinatus* (hooked), for the hooked spines. Galeotti, who sold rare plants at his nursery outside Brussels, introduced this fishhook from Mexico. In 1856, George Engelmann named the variety for Charles Wright, who collected the type specimen in 1852 on "stony hills near Frontera," about 8 miles north of El Paso.

83. Nipple Cactus (*Mammillaria meiacantha*)
(Other common names: little nipple cactus, pancake pincushion, little chilis, biznaga de pocas espinas, chilitos, biznaga de chilitos)

Habitat: desert scrub, grasslands, woodlands, xeric to mesic environs, low-high elevations.

Geographic range: Texas: Trans-Pecos (El Paso, Culberson, Jeff Davis, Presidio, Brewster, Terrell Counties). US: Texas, New Mexico. Mexico: Coahuila, perhaps parts of Chihuahua and Zacatecas.

Park locations: Upper Guale Mesa, one mile northwest of Sauceda, near Solitario Peak

Seeking protection amid rocks and beneath other plants, this cactus squeezes into the tightest spaces and goes unnoticed until blooming. Nipple cactus has a single round, dark green stem to 6 inches across, nearly flush with the ground, or more dome-like, 2 inches high. The stems are covered with projecting nipple-like tubercles, each with a small cluster of radiating spines at the tip and milky sap within. Usually, zero or one central spine and five to seven radial spines are borne at the tubercle tips and do not obscure the stem. The spines are white, gray, tan, pink or red tinged, often with darker tips.

In mid- to late spring, delicate flowers form a circle around the stem apex, below the newest plant growth. The flower tepals are creamy white, with darker midribs in shades of pink or greenish brown, and sharply pointed tips. Within the tepals are many stamens with cream-colored or pale yellow anthers, and a green multilobed stigma, elevated above the stamens.

To 1 1/4 inches long, the fruits are rose-pink to red, broadly club-shaped, with reddish-brown, pitted seeds.

Botanical notes: British gardener-botanist Adrian Haworth, a succulent plant specialist, created the *Mammillaria* genus in 1812, from the Latin *mamilla* (nipple) and the suffix *-aria* (like), to describe the nipplelike tubercles. George Engelmann named the species in 1856, using the Greek *meion* (less) and *akantha* (thorn) to distinguish this species from the similar *M. heyderi*, partially "by the fewer and stouter spines." The lectotype was collected by John Bigelow in 1853, on "cedar plains near the Llano Estacado to the Pecos." Bigelow served as botanist on Lieutenant A. W. Whipple's Pacific Survey Expedition, in search of a railroad route from the Mississippi River to the Pacific Ocean.

84. Potts' Mammillaria Cactus (*Mammillaria pottsii*)
(Other common names: Potts' nipple or foxtail cactus, rat-tail nipple or rat-tail pincushion cactus, biznaga de Potts)

Habitat: calcareous, sedimentary, gravelly desert flats, creosote and lechuguilla scrub; arid, rocky, limestone bluffs, mostly below 4000 feet.

Geographic range: Texas: Big Bend (Presidio, Brewster Counties). US: Texas. Mexico: Chihuahua east to Nuevo León, south to Durango, Zacatecas.

Park locations: Solitario north rim, Terlingua Uplift, Fresno Creek, West Contrabando, Contrabando Dome, and Buena Suerte Trails

This nipple cactus has narrow cylindrical stems to 6 inches high, single or clustered, branching below ground from spreading roots. The stems are stiffly erect or stiffly leaning, but older, longer stems may grow nearly parallel to the ground. The small, tough, blue-green tubercles are organized in spirals over the stem, and the gaps are filled with woolly hairs. A dense cover of overlapping spines hides the stems.

Up to forty-nine thin, stiff radial spines and twelve stout central spines are borne in each areole. The radial spines are white, ashy-white, gray, or tan. The longer central spines, especially upswept centrals near the stem tip, are more colorful, darker near the tips, in shades of red, tan, chalky blue, or grayish black.

As early as February, tiny bell-shaped flowers appear on growth of the previous year, in a ring around but below the stem tip. Seldom fully open, the tepals are rust red to maroon, often with darker midribs and lighter margins. Within the tepals are stamens with pale yellow anthers and a red style with a multilobed orange-yellow stigma. To 3/4 inch long, mature fruits are red, club-like or cylindric, and spineless, with tiny brown, pitted seeds.

Botanical notes: Frederick Scheer, German-born London merchant, described this species in 1850 from a plant cultivated in the garden of Prince Joseph Salm-Reifferscheidt-Dyck, near Düsseldorf. Like Salm-Dyck, Scheer was an avid cactus cultivator-importer with a large living collection. Scheer named this cactus for John Potts, lessee of the Chihuahua mint, and/or his brother Frederick, a mining engineer in the Sierra Madre. From 1842 to 1850, John Potts sent plants to Scheer, some collected by Frederick.

85. Texas Cone Cactus (*Neolloydia conoidea* var. *conoidea*)
(Other common names: Chihuahuan beehive, cone cactus, biznaga cónica)

Habitat: desert scrub, dry rocky hills, on limestone.

Geographic range: Texas: Trans-Pecos (excluding Hudspeth, Jeff Davis, Reeves Counties); Edwards County. US: Texas. Mexico: Coahuila east to Tamaulipas, Durango southeast to San Luis Potosí, Querétaro.

Park locations: Solitario northwest rim, Outer Loop, Masada Ridge

Texas cone cactus is single or often branched to form clusters of stems. To 5 inches high, the erect, dull green stems are spheric to cylindric with age. Lacking distinct ribs, the stems are covered with large, close-spaced, conic tubercles arranged in spiraling rows. The deeply grooved tubercles, filled with woolly hairs, are not completely hidden by dense spines.

Typically, two to four widely projecting central spines, and up to seventeen close-pressed radial spines are borne in each areole, all with enlarged, bulbous bases. The radials are white but frequently dark-tipped; the centrals, to 1 inch long, are dark brown or black.

Primarily in late spring and early summer, intense magenta or rose-pink flowers, 2 inches wide, form on new growth near the stem apex. Opening in the morning, the blooms may be nearly closed by midafternoon. The tepals have an abrupt, bristlelike tip and often a distinctly darker, red base. Outer tepals may be paler, especially on the margins. Within the flower's throat are stamens with golden anthers and a style elevated well above the stamens, with a multilobed creamy-white stigma.

In late fall, the spineless, rounded green fruit, to 1/3 inch long, becomes dry and tan, with a curious red ring at the top, a prominent floral remnant. The pear-shaped seeds are black, with nipplelike projections.

Botanical notes: In 1922, Nathaniel Britton and Joseph Rose named this genus for American-Canadian botanist Francis Lloyd. Lloyd, at McGill University in Montreal for many years, is remembered for an early work on *Guayule* (1911), the rubber plant, and a seminal study of *Carnivorous Plants* (1942). In 1828, Swiss taxonomist Augustin de Candolle named the species *Mammillaria conoidea*, from the Greek *konos* (cone) and the suffix *-oidea* (resembling), likely a reference to the slightly conic stems: he described the plant as "simplex, ovata, conica" (simple, ovate, conic).

86. Diploid Purple Prickly Pear Cactus (*Opuntia azurea* var. *diplopurpurea*)
(Other common names: purple prickly pear, coyotillo)

Habitat: Rio Grande to mountain slopes, scrub, grassland; igneous and limestone substrates.

Geographic range: Texas: Trans-Pecos (Hudspeth, Jeff Davis, Presidio, Brewster Counties). US: Texas. Mexico: Chihuahua.

Park locations: near Closed Canyon, Arenosa Campground, Bofecillos Road, Casa Piedra Road, Terneros Creek, Llano Loop

Diploid purple prickly pear is seldom more than 2 feet high, but often 3 feet or more across, with a low, spreading habit. Compared to Big Bend purplish prickly pear (*Opuntia azurea* var. *parva*), which is prevalent in BBNP, this cactus has fewer but larger and rounder pads. The pads, to 6 inches long and wide, are purple or reddish-purple much of the year, especially during cold or drought stress. Usually two to six central spines, to 4 inches long or more, are borne in each areole, mostly on the upper pads or upper edges. Radial spines are lacking, but ruddy-brown glochids may fill the areoles. The spines are reddish black, reddish brown, or nearly black, lighter near the tips.

Opulent yellow blossoms with deep red centers, many 3 inches wide, bloom mostly in April and into May. Dozens of flowers may bloom simultaneously on a single plant. The tepals, especially the outer ones, are spine-tipped and lighter in color on the edges. Yellow stamens, and a style with a cream-colored to pale green, multilobed stigma, fill the blossoms' red throats. At peak bloom, the purple and yellow colors dominate the desert landscape for miles.

To 1 inch long, the inversely egg-shaped, mature fruit is fleshy, purplish red, and spineless, with a flat rim. The fruit dries and shrinks with age. The tan seeds are compressed and unevenly shaped.

Botanical notes: Philip Miller, superintendent of Chelsea Physic Garden, named the genus *Opuntia* in *The Gardeners Dictionary* in 1754, and later gave the reason: "This plant is called *Opuntia*, because Theophrastus writes, that it grows about *Opuntium*." Theophrastus's plant, not a cactus, was confused with newly introduced prickly pears from America. In 1909, Joseph Rose, curator at the US National Herbarium, named *O. azurea* (blue, azure) for its blue-green pads. A. Michael Powell, professor emeritus of biology at Sul Ross State University, and James Weedin, with the Chihuahuan Desert Research Institute, described variety *diplopurpurea* in 2004 as a "diploid taxon with purple stems." The type specimen was collected by Weedin in 1997, on Sul Ross Hill in Alpine, Texas.

87. Big Hill Prickly Pear (*Opuntia azurea* var. *discolor*)
(Other common names: none)

Habitat: mixed desert scrub; rocky, silty, gravelly soils; at base of and on igneous desert slopes.

Geographic range: Texas: Presidio County, at BBR, on Big Hill above Rio Grande. US: Texas. Mexico: adjacent Chihuahua.

Park locations: Big Hill, La Cuesta at west base of Big Hill

Big Hill prickly pear is an unusual variety of *Opuntia azurea*, lacking purplish stems, in the United States found only at the base of and on igneous slopes of Santana Mesa, at Big Hill on the Rio Grande. Upright or spreading, the plants may reach 3 feet high and wide or more. The light green or slightly bluish-green prickly pear pads are broad, inversely egg-shaped to nearly round. Only the areoles on the pads may turn purplish. While varied, the spines are often long, yellow, gold, or white, disparately protruding, and at times, notably curling or twisting. One to several spines per areole are most prevalent on the upper portion of the pads, especially on the margins.

Yellow blossoms with red centers, like variety *diplopurpurea* blooms, form from March to mid-May. Within each flower's red throat are many stamens with brilliant yellow anthers and a style with a cream-colored to light green, bulky multilobed stigma. On Big Hill, varieties *discolor* and *diplopurpurea*, occur without apparent hybridization (Powell and Weedin 2004).

Light red to pinkish-red fruits, like variety *diplopurpurea* fruits in size and shape, appear in June and remain into August or later. To 1 inch long, the mature fruit is spineless or nearly so, with a flat, slightly depressed apex. At first fleshy, the fruit soon shrivels with age.

Botanical notes: *O. azurea*, known as "nopalito" in Mexico, was collected by American-Canadian botanist Francis Lloyd in Zacatecas in 1908 and described by Joseph Rose in 1909. In 2004, James Weedin named variety *discolor* (of different colors), for the "different spine colors and lack of purple pigment in the stems and spines (Powell and Weedin 2004)." A. Michael Powell and Shirley Powell collected the holotype of variety *discolor* on Big Hill in 1993.

88. Sweet Prickly Pear (*Opuntia dulcis*)
(Other common names: sweet opuntia)

Habitat: near Rio Grande below 3,500 feet; rocky, sandy soils, desert and alluvial flats, bajadas, grasslands, low mountains.

Geographic range: Texas: Trans-Pecos (excluding Reeves, Pecos, Val Verde Counties). US: Texas, New Mexico. Mexico: probably Chihuahua, Coahuila.

Park locations: Auras Canyon, west of Vista de Bofecillos, west of Las Burras Campsite 2

To 6 feet high, upright or spreading, sweet prickly pear is intermediate between the slightly smaller Comanche prickly pear (*Opuntia camanchica*) and the slightly larger Engelmann's prickly pear (*Opuntia engelmannii*). To 7 inches long, the pads are inversely egg-shaped or rounded, with two central and at times two radial spines in each areole.

The uppermost central spine, projecting out or down, is dark at the base and white or gray above. The much shorter, lower central spine is chalky white and extended down. Radial spines, less than 1/8 inch long, are whitish and angled down. Reddish-brown glochids are concentrated in areoles on the pad's rim.

Appearing in April and May, the large flowers are yellow (aging orange), with red centers. The tepals are spatulate, with a short spinelike tip. Within the tepals are stamens with yellow anthers and a style with a green multilobed stigma.

The red or maroon fruits are inversely egg-shaped, to 2 inches long, with a depressed white circle at the rim. Smooth-skinned and spineless, the fruits are sweet and juicy, with tan disklike seeds.

Botanical notes: The Latin epithet *dulcis* (sweet) refers to the fruits. In 1856, George Engelmann described this species as "a doubtful plant, of which we have not material enough. It has been found near the middle course of the Rio Grande, near Presidio del Norte, &c. It resembles *O. engelmanni* and may be a form of it; but it is lower, more spreading, with a similar but very sweet fruit, and small, regular seeds." The type specimen labeled "El Paso?" was collected by Charles Wright in 1852. Engelmann described six-hundred-plus new species, which form the nucleus of the Missouri Botanical Garden.

89. Engelmann's Prickly Pear (*Opuntia engelmannii* var. *engelmannii*)
(Other common names: cactus apple, purple-fruited prickly pear, nopal, abrojo, vela de coyote, joconostle)

Habitat: desert, sandy flats, canyon bottoms, rocky cliffs, grasslands, woodlands.

Geographic range: Texas: Trans-Pecos. US: southwest (excluding Colorado). Mexico: Sonora east to Coahuila, south to Durango.

Park locations: Big Hill, Rancherías Canyon, Llano Dome, Terlingua Uplift

Engelmann's prickly pear is one of the largest, most widespread prickly pears in the Southwest. The massive cactus grows erect, sometimes exceeding human height, or sprawls, with trunkless pads many feet across. Pale green, less often gray green or blue green, the pads are thick, bulky, to 14 inches long and 12 inches wide, shaped like a big teardrop, or merely round.

The spine clusters are widely spaced across the pad. Commonly, three or four downward-angled, white central spines, to 1 3/4 inches long, and one or two much shorter white radial spines, are borne in each areole. Distinctively, the central spines are often organized in a bird's-foot pattern.

Flowers, to 3 inches wide, appear from April to July. The blooms, lasting a single day, change color from yellow to orange or peach by late afternoon. The tepals are spatulate, darker near the base, lighter on the edges, and sharp-pointed. Within the tepals are stamens with yellow anthers, and a style with a green, multilobed stigma.

The plump fruits, to 3 inches long, are barrel-shaped, maroon to purplish, with a depression on top. Largely spineless, the fruits ("tunas") are dotted with wool- or glochid-filled areoles. Juicy and sweet tasting, but seedy, the tunas bear many tan disklike seeds.

In Mexico and the southwest United States, the tunas are harvested commercially to make syrups and jellies (Kurz 1979) and sold in grocery stores (Warnock 1977) along with young prickly pear pads (nopalitos).

Botanical notes: In 1850, Prince Joseph Salm-Reifferscheidt-Dyck named this cactus for George Engelmann, in *Plantae Lindheimerianae*, mostly an account of Texas plants collected by Ferdinand Lindheimer and described by George Engelmann and Asa Gray. Engelmann noted that this cactus was collected by "Dr. Wislizenus." Adolf Wizlizenus traveled the Santa Fe and Chihuahua Trails and found this species in 1846 "north of Chihuahua, common as high up as El Paso."

90. Desert Night-Blooming Cereus (*Peniocereus greggii*)
(Other common names: nightblooming cereus, queen of the night, reina de la noche, deerhorn or sweet potato cactus, huevos de venado, saramatraca)

Habitat: sandy desert flats, desert scrub, alluvial plains, arroyos, bajadas; igneous, limestone soils.

Geographic range: Texas: Trans-Pecos. US: Texas, southwest New Mexico, southeast Arizona. Mexico: Chihuahua to Coahuila, south to Durango, Zacatecas.

Park locations: scarce in mid-elevation desert grasslands

Of conservation concern: see Appendix A

This elusive cactus grows within woody "nurse" shrubs and blends into them for protection, support, and shade. To 2 feet high or more, the stems are only 1/4 inch wide at the base, so the cactus leans on other plants for support and is buffeted against them by desert winds.

Single or few-branched, the stems are gray green, lead gray, or purple tinged, with four to six ribs and minute velvety hairs. Curiously, the weak stems have bulky taproots, some weighing up to 125 pounds. Growing from areoles filled with cobwebby hairs, the blackish spines are also small and thin and gray with age. The spines are borne in spiderlike clusters on the rib crests, with one or two spreading central spines, and six to nine close-pressed radial spines.

In April and May, aromatic, trumpetlike flowers open at dusk and close at sunrise the following day. Pollinated by hawk moths, the white blooms, to 4 inches across, last a single night. The spreading tepals are

slender, sharp-tipped, the inner tepals often pink-tinged, the outer flushed with maroon. Within the tepals are many stamens with white filaments and yellow anthers and a white style with a multilobed white stigma.

To 3 inches long, the mature fruits are red to orange-red, spindle-shaped or globose and beaked. Although juicy and edible, the fruits bear some short spines and black, egg-shaped seeds covered with tuberlike swellings.

Botanical notes: The genus name, from the Greek *penios* (thread) and *Cereus* (the former genus), refers to the slender stems. In 1848, George Engelmann, in Adolph Wislizenus's *Memoir of a Tour to Northern Mexico*, named this cactus *Cereus greggii*, for Josiah Gregg, who collected the species "near Cadena, south of Chihuahua," in 1847. Merchant-trader and historian of the Santa Fe and Chihuahua Trails, Gregg served as a US Army guide during the Mexican-American War. In 1909, Nathaniel Britton and Joseph Rose moved the species into their new genus *Peniocereus*.

91. Glory of Texas Cactus (*Thelocactus bicolor* var. *bicolor*)
(Other common names: Texas pride, straw spine or bicolor cactus, biznaga pezón bicolor, gloria de Texas)

Habitat: silty desert flats, desert scrub, alluvial fans, gravelly bluffs, low rocky hills, sedimentary and igneous soils.

Geographic range: Texas: Big Bend (south Presidio, south Brewster Counties); Starr County. US: Texas. Mexico: Chihuahua east to Tamaulipas, south to Zacatecas, San Luis Potosí.

Park locations: South Lajitas Mesa, 1 1/4 miles west of Lajitas, Solitario northwest rim, West Contrabando and Buena Suerte Trails

The green stems, to 7 inches tall, are egg-shaped to cylindric, bulky but not deeply seated in the ground, solitary or much branched, with many-stemmed clumps several feet across. In mature plants, tubercles covering the stem surface merge into eight to thirteen low, rounded ribs, each organized in vertical or spiraling rows.

Up to four central spines, eighteen shorter radial spines, and two flat, upper bladelike spines, are borne in each areole. Bicolored or tricolored, the central and radial spines are pink or red, with yellow to tan tips, and gray or white bases. The two bladelike spines, to 3 inches long, are gray or white, often curving, curling, or twisting. The spines reflect almost as much light as the glossy flowers.

From spring to fall, satiny flowers 3 inches across form on new growth at the stem tip. The lustrous funnel-shaped blooms are magenta, fuchsia, pink, or rose-pink, with a deep scarlet throat. The tepals are spatulate, at times jagged at the tips. Outer tepals may be whitish. In the flower's throat are stamens with golden anthers and a white style with a multilobed yellow, orange, or red stigma.

Spineless but scaly, with persistent floral remnants, the fruits are green to brownish red, globose or egg-shaped, to 3/4 inch long. The many black seeds are covered with raised, netlike projections.

Botanical notes: In 1922, Nathaniel Britton and Joseph Rose moved this *Echinocactus* species into their new *Thelocactus* genus. Borrowed from the Greek *thele* (nipple) and *Cactus* (an old genus name), *Thelocactus* refers to the nipplelike tubercles. Belgian botanist Henri Galeotti named the species *bicolor* in 1848, in *Illustrations and Descriptions of Blooming Cacti*, by Ludwig Pfeiffer and Friedrich Otto. The epithet may refer to the bicolor spines, or in the opinion of Pfeiffer and Otto, to the bicolor flowers. After introducing this and many other cacti to Europe from Mexico, Galeotti's import-nursery business failed, and in 1853, he became director of the Brussels Botanical Garden.

92. Cardinal Flower (*Lobelia cardinalis*)
(Other common names: scarlet or red lobelia, lobelia, lobelia escarlata)

Habitat: near water sources in partial shade; springs, creeks, ponds, seeps, moist depressions.

Geographic range: Texas: Trans-Pecos (Culberson, Jeff Davis, Presidio, Brewster, Val Verde Counties), much of Edwards Plateau, East Texas; small parts of Panhandle; scattered elsewhere. US: most of continental US (excluding South and North Dakota west to Oregon, Washington). Elsewhere: eastern Canada, Mexico, Central America.

Park locations: Agua Adentro, Ojito Adentro, Arroyo Primero, Chorro Canyon, Arroyo Mexicano

Cardinal flower is a perennial with erect, leafy, unbranched stems to 6 feet high. The plants grow from short rootstocks and spread by offshoots. Often growing in clusters, the densely short-hairy stems are grooved, and contain milky sap.

The alternate green leaves are linear to narrowly lance-shaped, to 5" long. Mostly stemless, the blades may be finely or irregularly toothed along the margins.

From spring to fall, brilliant scarlet flowers appear in narrow, leafy clusters as much as 20 inches long on the upper third of a flowering stem. The intricate blooms are tubular, and two-lipped at the tip, the erect upper lip with two slender linear lobes, and the spreading lower lip with three broader, narrowly oblong lobes. The five stamens, with red filaments and white-haired anther heads, are united into a long, arching tube that extends above the upper lip and around the style.

The fruit is a two-chamber capsule, 1/3 inch long, which splits into halves, exposing many tiny, wrinkled brown seeds.

Botanical notes: In 1753, Linnaeus named the genus *Lobelia* for Matthias de l'Obel, Belgian-born physician-botanist to James I of England. Charles Plumier, in his *Nova Plantarum Americanarum Genera* of 1703, first coined the term *Lobelia* to describe a plant from the West Indies. Plumier was an admirer of l'Obel, who in 1570 authored a landmark early effort to classify plants by natural affinities rather than medicinal uses. The epithet, *cardinalis*, refers both to the color (the red of a cardinal's robe) and shape of the flower (like a bishop's cap). Linnaeus described this species, as well as genus, in 1753, in *Species Plantarum*.

93. Netleaf Hackberry (*Celtis reticulata*)
(Other common names: palo blanco, western or canyon hackberry, netleaf sugar hackberry, acibuche, cumbro, palo mulato)

Habitat: by springs, creeks, arroyos, tanks, ponds; moist canyons; limestone, basalt, other igneous exposures.

Geographic range: Texas: Trans-Pecos (excluding Reeves County); much of Edwards Plateau, Panhandle, central, north Texas; sites in south, northeast Texas. US: most of west half; Louisiana. Mexico: Baja California east to Coahuila.

Park locations: Agua Adentro, Ojito Adentro, Horsetrap Spring and Trail, Ojo Chilicote, Lower Shutup, Upper Alamito Creek, Cienega Mountains

This hackberry is a deciduous small tree or large shrub, to 25 feet tall, which grows from an extensive root system and forms a trunk to 12 inches wide. Often, the gray bark is covered with odd corky knobs. Frequently, these trees are intricately branched, with many short branchlets. Younger branches may be dark maroon, and hairy.

The alternate leaves are egg-shaped, to 2 inches long, with a pointed or blunt tip and a rounded to heart-shaped base. On hairy yellow petioles, the blades are hairy, strikingly net-veined, with a rough upper surface, a yellow-green lower surface, and smooth margins.

In spring, tiny greenish blooms appear with the leaves, in small clusters on new branchlets. The flowers are unisexual, but male and female blooms occur on the same plant, the male low on new growth, the female above. Each blossom has a cupped five-lobed calyx, five stamens, and a conspicuous style, split into two curving, bristle-haired lobes. The fruit is a round berry to 2/5 inch across, reddish, and beaked, with a single thick-walled nutlet. Ripening in fall, the fruit is sweet and edible, with thin, dry pulp.

Botanical notes: In 1753, Linnaeus named this genus *Celtis*, a term used by Pliny the Elder for a lotus tree in northern Africa. In Greek mythology, lotus flowers and fruits were the narcotic, sleep-inducing food of the Lotus-eaters on an island off northern Africa. Somehow, early botanists — Caspar Bauhin, Joachim Camerarius, Mathias de l'Obel, and later Linnaeus — saw a resemblance between the lotus and the hackberry. The epithet *reticulata* (netlike) refers to the leaves. This species was described by John Torrey in 1824 from a collection by his student, Edwin James, in 1820, on Major Stephen Long's Rocky Mountains expedition. James later settled in Iowa and ran a station on the Underground Railroad.

94. Rio Grande Saddlebush (*Mortonia scabrella*)
(Other common names: rough mortonia, sandpaper bush, Trans-Pecos saddlebush, cucharita, tickbush)

Habitat: arid, rocky, limestone hills, slopes, canyons, desert scrub to oak-juniper communities.

Geographic range: Texas: Trans-Pecos (excluding Jeff Davis, Reeves, Terrell, Val Verde Counties). US: Texas, southwest New Mexico, Arizona. Mexico: Chihuahua, Sonora.

Park locations: Solitario northwest rim, Pila Montoya Trail, Lefthand Shutup, near Sierra Blanca Dome

Rio Grande saddlebush is a rigidly erect shrub to 6 feet tall, with many slender, whitish, leafy branches. The often spine-tipped stems are stiff, brittle, with smooth gray bark and crowded yellow green leaves.

Short petioled, the small leaves are alternate, oval, to 2/5 inch long. The blades are thick, leathery, with edges thickened and curling in the shape of a saddle. Typically ascending and pressed against the stem, the leaves usually have smooth margins and rounded or blunt tips. The surfaces of these curious leaves are extremely rough, scratchy like sandpaper, from a thick coat of short, stiff hairs.

From spring to early fall, many small white flowers only 1/4 inch wide appear in clusters near the branch tips. Each bloom has five rounded petals, with ciliate margins and a short, clawed base. Within the petals are five stamens with white filaments and yellow anthers and a short style with a tiny multilobed stigma.

The fruit is a tan oblong capsule 1/6" inch long, with a beak formed by the persistent style. Dry, hard, indehiscent, the capsule contains a single tan, oblong seed.

Botanical notes: In 1852, in *Plantae Wrightianae*, Asa Gray dedicated this genus "to the memory of that most eminent American naturalist, the late Dr. Samuel Morton, author of the *Crania Americana*, &c., and President of the Academy of Natural Sciences, Philadelphia." The species name *scabrella*, from the Latin *scabra* (rough) and the diminutive *-ella* (slightly), refers to the leaf texture. In 1853, in a continuation of *Plantae Wrightianae*, Asa Gray described the species from a collection by Charles Wright in 1851, on mountainsides near the San Pedro River, in the Mexican state of Sonora.

95. Tubercled Saltbush (*Atriplex acanthocarpa* ssp. *acanthocarpa*)
(Other common names: armed or spiny-fruited saltbush, burscale, huaha, saladillo, quelite)

Habitat: alkaline, gypseous clay flats, mounds, low hills, low elevations in hottest desert.

Geographic range: Texas: Trans-Pecos (El Paso, Hudspeth, Jeff Davis, Presidio, Brewster Counties). US: Texas, southwest New Mexico, southeast Arizona. Mexico: Chihuahua, Coahuila.

Park locations: South Lajitas Mesa, Madrid House, Rockcrusher Road, West Fork of Comanche Creek

This goosefoot is an evergreen shrub or subshrub to 3 feet tall and broad, with erect branches. Growing from a woody root, tubercled saltbush is also woody at the base, with gray bark on the lower branches. The stems are greenish yellow or whitish (farinose, with a mealy powder).

The short-stemmed leaves, to 1 1/2 inches long, are egg- or diamond-shaped or oblong. Commonly, the blades have wavy, curling, and irregularly toothed margins. Alternate on upper stems, opposite below, the leaves are mostly hairless like the stems, but densely farinose and, consequently, silvery to blue gray.

Blooming in summer and fall, the obscure flowers are unisexual, with male and female flowers on separate plants. The male blooms form compact clusters less than 1/5 inch across, in spikes at the branch tips. Each male bloom has a three- to five-lobed membranous calyx and three to five stamens. The female blooms consist of a single pistil, supported by two bracts that expand to enclose the fruit.

The fruits are utricles (thin-walled, one-seeded, bladderlike), enclosed by expanded fruiting bracts with many protruding tubercles. The odd tubercles are irregularly shaped, flattened, conical, or cylindric, and bluntly toothed or lobed.

Botanical notes: The genus name, *Atriplex*, used by Linnaeus in 1753, is the ancient Latin name (from the Greek *atraphaxes*) for mountain spinach. The epithet *acanthocarpa*, from the Greek *akantha* (thorn) and *karpos* (fruit), emphasizes the spiny fruit. New York botanist John Torrey described the species as *Obione acanthocarpa* in 1859, citing specimens from John Bigelow, Charles Parry, George Thurber, and Charles Wright, but the earliest collection was by Josiah Gregg in 1848, near Saltillo in Coahuila, Mexico. After practicing medicine in Saltillo in 1847 and 1848, Gregg joined the California Gold Rush.

96. Wheelscale Saltbush (*Atriplex elegans* var. *elegans*)
(Other common names: whitescale saltbush, wheelscale or mecca orach, wheelscale, chamizo cenizo)

Habitat: full sun, alkaline or moderately saline, dry or damp soils, desert and alkali flats, floodplains, disturbed places.

Geographic range: Texas: Trans-Pecos (El Paso, south Hudspeth, Jeff Davis, Presidio, Brewster, Reeves Counties); Mitchell County. US: southwest (excluding Colorado). Mexico: Sonora, Chihuahua.

Park locations: 1 mile west of Lajitas, Cienega Creek, Cienega park residence

This curious annual is a sight to behold, at times completely caked with flaky, branlike scales. Woody at the base, to 18 inches high, the ascending, branching stems are greenish yellow or whitish, and densely scaly.

The many stemless or short-stalked, alternate leaves are oblong to lance-shaped, usually no more than 1 inch long. Often wavy and toothed on the margins, the blades are silvery-white underneath with crusty scales, but greener and darker above.

Appearing from spring to early fall, the tiny greenish blooms are unisexual, with male and female flowers on the same plant. On a branchlet, the lower flower clusters bear only female blossoms, the upper male and female or only male blooms. The bractless male flowers have a three- to five-lobed calyx and three to five stamens. Female blossoms have one pistil eventually enclosed by two bracts.

The disklike fruits are compressed bracts to 1/6 inch long, with spikelike teeth on the margins. Pale greenish yellow, scaly, and farinose, the flat bract faces enclose one tiny brown seed.

Early Americans boiled the leaves in water and mixed them with corn meal to make a pudding (Warnock 1977), or used them as a food flavoring.

Botanical notes: The species name is the Latin *elegans* (elegant). Alfred Moquin-Tandon described the species as *Obione elegans* in 1849, from a collection between 1824 and 1831 by Irish botanical explorer Thomas Coulter in Sonora, Mexico. Moquin-Tandon was chair of medical natural history at the Faculty of Medicine of Paris. In 1852, German botanist David Nathaniel Friedrich Dietrich moved this and other *Obione* species to the *Atriplex* genus. Dietrich, University of Gena Herbarium curator and botanical illustrator, authored *Flora Universalis* (1831–54), with plant descriptions and 4,760 hand-colored copperplate engravings.

97. Big Saltbush (*Atriplex lentiformis* ssp. *lentiformis*)
(Other common names: quailbush, quailbrush, big saltbrush, lens scale, lens scale saltbush, white thistle)

Habitat: Rio Grande, with tamarisk, seepwillow, retama, mesquite, sandy or silty soils.

Geographic range: Texas: Presidio County. US: Texas, California, Arizona, Utah, Nevada, Hawaii. Mexico: Baja California, Sonora.

Park locations: Lajitas Boat Launch, 1 to 1 1/4 miles west of Lajitas, Grassy Banks, head of Colorado Canyon

This invasive shrub is new to Texas, known at a half-dozen sites on the Rio Grande in Presidio County. Big saltbush is big, to 10 feet high and wide. The stems are erect, with many spreading, slender branches and scaly branchlets.

The alternate, petioled leaves are up to 2 inches long, triangular, oblong, or egg-shaped. The thickish blades have a blunt or rounded tip, wavy margins, and fine scales that coat the leaves gray, white, or blue.

Blooming from spring to fall, the flowers are unisexual, with male and female flowers on separate plants (dioecious), or at times on the same plants (monoecious). This saltbush adapts: dioecious shrubs may become monoecious and monoecious shrubs dioecious; less often, male shrubs turn female or female shrubs turn male. The yellow male flowers form in slender clusters to 20 inches long. Clusters of female blooms are shorter and denser. The bractless male blooms have a three- to five-lobed calyx and three to five stamens; female blooms consist of a single pistil with ultimately enclosing bracts.

The fruits are utricles, enclosed in a pair of flattened, round fruiting bracts less than 1/5 inch long, with thick, wrinkled margins. The single brown seed is also compressed.

Botanical notes: The species name, *lentiformis*, from Latin words meaning "lens-shaped," alludes to the compressed fruits. John Torrey described the species in 1853 as *Obione lentiformis*, in Captain Lorenzo Sitgreaves's *Report of an Expedition down the Zuni and Colorado Rivers.* The lectotype was collected on the Colorado River in 1851 (*probably in* Mojave County, Arizona), *by* Samuel Woodhouse, naturalist on the Sitgreaves Expedition. In 1874, Sereno Watson placed this saltbush in the *Atriplex* genus.

98. Nettleleaf Goosefoot (*Chenopodiastrum murale*)
(Other common names: sowbane, Australian spinach, salt-green, quelite de perro, quelite de puerco, quinoa negra)

Habitat: sandy, moist, open, disturbed environs; fields, pastures, yards, roadsides, waste sites.

Geographic range: Texas: Trans-Pecos (excluding El Paso, Hudspeth, Reeves Counties); parts of South Texas Plains, west Edwards Plateau, Coastal Bend; scattered sites. US: most of US. Elsewhere: native to Europe, Asia, North Africa.

Park locations: 1 mile west of Lajitas, head of Colorado Canyon, mouth of Tapado Canyon

This goosefoot is a stout but fleshy annual to 3 feet high, with reclining to erect stems branched at the base. The thick stems are yellow green and at times maroon tinged. Young stems may be strikingly "mealy," dotted with salt-excreting, bladderlike hairs. Apparently, the bladdery hairs shrivel when dry, and older plants become hairless.

Alternate, slightly fleshy, the leaves are up to 3 inches long, triangular, diamond- or egg-shaped, or lancelike. On long petioles, the blades are often tapered to a sharp, abrupt tip, wedge-shaped at the base, and toothed, lobed, or curling on the margins. Dark green above, paler below, the blades may be coated with scaly hairs like the stems.

Mostly blooming in spring in headlike clusters, the tiny flowers are green and mealy white. Each bloom lacks petals and consists of a five-lobed calyx. Within the lobes are five protruding stamens and two short, threadlike stigmas. Later, the lobes enclose developing fruit. The fruit is an achene, egg-shaped but compressed, with a membranous outer wall adherent to the disklike black seed.

Botanical notes: In 2012, botanists Suzy Fuentes Bazan, Pertti Uotila, and Thomas Borsch formed the genus *Chenopodiastrum* and placed this species in it. The genus name joins *Chenopodium* (the former genus) with the Latin suffix *-astrum* (similar): the herbage, unlike *Chenopodium*, is hairless at maturity. Linnaeus created the genus *Chenopodium* in 1753 from the Greek *chen* (goose) and *pous* (foot), alluding to the leaf shape, and the species *murale* from the Latin *muralis* (growing on walls). The first collection is unknown, but Josiah Gregg obtained a specimen in 1847, in Saltillo, Coahuila, and cited a vernacular name, quelite de perro. According to Gregg, the plant was used to treat "eruptions on the head."

99. Mexican Clammyweed (*Polanisia uniglandulosa*)
(Other common names: one-gland clammyweed, hierba del coyote, ortiga)

Habitat: river corridor, moist depressions, creek beds, canyons, desert flats, disturbed sites; grasslands, woodlands, rocky slopes.

Geographic range: Texas: Trans-Pecos (excluding Reeves, Pecos, Terrell Counties). US: Texas, New Mexico. Mexico: northern Mexico to Oaxaca.

Park locations: Arenosa Campground, Tapado Canyon, Las Cuevas, Arroyo Mexicano, La Mota

This clammyweed is a robust perennial to 2 feet high or more, with upright stems. With leafy stems coated with sticky, gland-tipped hairs, the plant is malodorous, glandular-hairy, and clammy. The alternate, long-petioled leaves are divided into three elliptic leaflets, to 1 1/2 inches long, with a common base. Leaflets are blunt or abruptly pointed at the tip, and smooth on the margins.

Blooming from late spring to fall, the flowers appear in congested, leafy clusters, to 8 inches long, at the branch tips. Each bloom has four purple sepals, four erect white petals in two pairs of unequal length, and many long projecting stamens.

The spatula-shaped petals taper to a purple, clawlike base. The purple stamens are much longer than the petals, and the purple style, with a purple stigma, approaches the stamens in length. Prominent at the petals' base is an orange, nectar-producing gland.

The mostly ascending, narrow fruit capsules are up to 4 inches long, slightly inflated, and glandular-hairy. The capsules split at the tip to expose numerous round or short oblong, reddish brown seeds.

Botanical notes: In 1819, Constantine Rafinesque moved *Cleome dodecandra* into his new genus *Polanisia*, joining the Greek *polus* (many) and *anisos* (unequal). The consensus is that *Polanisia* refers to the many stamens of unequal length, but according to Rafinesque, "The etymology . . . derives from *many irregularities*." The plant differed in many ways from plants of the genus *Cleome*. Antonio Cavanilles, director of the Madrid Botanical Garden, described this species as a *Cleome* in 1797, using the Latin epithet *uniglandulosa* (uniglandular) to refer to the yellow-orange gland at the base of the ovary. The plant was grown in the Madrid garden from Mexican seeds. Augustin Pyramus de Candolle moved this species to the genus *Polanisia* in 1824.

100. Smooth Spiderwort (*Tradescantia leiandra*)
(Other common names: canyon spiderwort)

Habitat: rocky, moist sites in partial shade, ledges, crevices of canyon walls, springs, intermittent creeks.

Geographic range: Texas: Trans-Pecos (Jeff Davis, Presidio, Brewster Counties). US: Texas. Mexico: Coahuila, Nuevo León.

Park locations: Cinco Tinajas, drainage from Sauceda to Cinco Tinajas

Smooth spiderwort is a scarce perennial growing from thick, fibrous roots, not forming rhizomes. To 20 inches long, the smooth, erect stems are single or sparsely branched. The narrow lance-shaped leaves, to 6 inches long, are fleshy, mostly hairless, with pointed tips and often, ciliate margins.

In summer and fall, pale rose or pink, three-petal flowers appear in clusters at the branch tips. The clusters are supported by united floral bracts, paired and boatlike. The rounded petals are narrowed at the base, and united into a short tube below. Six stamens have pale purple filaments and yellow anthers.

The oval fruit is a hairless, three-lobed, stalked capsule to 1/4 inch long. Each lobe splits to expose two slightly wrinkled seeds.

Native Americans ate these plants raw in salads or boiled with butter and vinegar (Warnock 1977).

Botanical notes: In 1753, Linnaeus named genus *Tradescantia* for John Tradescant the elder, the younger, or both. Apparently, the name was first used by herbalist John Parkinson in 1629, to dedicate a spiderwort to John Tradescant the elder, "who first received it of a friend, that brought it out of Virginia." The Tradescants, gardeners of King Charles I, collected flora, fauna,

and natural history objects worldwide. Their house of curiosities was known as Tradescant's Ark. John Torrey described *T. leiandra* in 1859, joining the Greek *leios* (smooth) and *andros* (male, stamen), alluding to the hairless stamens. During the US-Mexico Boundary Survey, John Bigelow collected the herb on "mountains and moist, rocky places, Puerto de Paysano," Sonora, Mexico.

101. Flatglobe Dodder (*Cuscuta umbellata*)
(Other common names: umbrella or Santa Fe dodder, zacatlascal, zacatlaxcale [Nahuatl])

Habitat: varied, often overflow or disturbed areas; parasitic on small herbs such as spurge, milkwort, purslane, pigweed, goosefoot.

Geographic range: Texas: Trans-Pecos (excluding Jeff Davis, Reeves, Pecos, Terrell Counties); part of Permian Basin; near Rio Grande, Laredo to Brownsville. US: Louisiana west to Arizona, Texas north to Kansas, Colorado. Elsewhere: Mexico, Panama, parts of Caribbean, South America.

Park locations: west park entrance, 1 1/2 miles west of Agua Adentro, Arroyo Mexicano, Llano Loop

This dodder is a parasitic annual vine with stringy orange-yellow stems and a few scalelike leaves. Lacking enough chlorophyll to make nutrients from sunlight, the twining, trailing stems form tangled networks over host plants, penetrating them with rootlike suckers (haustoria).

Small bell-shaped flowers with five spreading lancelike lobes cluster along the stems. From summer to fall, white or nearly colorless blooms appear on stalks as long as or longer than the flowers. Each bloom has two styles of unequal length and stamens with white filaments and oblong yellow anthers.

The fruit is a small translucent capsule, 1/10 inch long, spheric but compressed, with conspicuous floral remnants. The capsule splits to release four rounded or broadly elliptic, edible seeds.

Botanical notes: The genus name, *Cuscuta*, is the ancient Latin name for dodder, presumably from the Arabic, *kushkut*, meaning "a tangled wisp of hair." Dodder, the common name, refers to the weak or "doddering" stems. The specific epithet *umbellata* (umbellate) refers to the inflorescence shape, in which flower stalks of similar length radiate from a central point. This species was described by Karl Kunth in 1818, from a collection of Prussian explorer Baron Alexander von Humboldt and his French colleague and botanist Aimé Bonpland, in Mexico, between Querétaro and Salamanca, about 1803.

102. Ojo de Vibora (*Evolvulus alsinoides*)
(Other common names: slender dwarf morning glory, slender evolvulus, dwarf morning glory, blue eyes, ojitos azules, oreja de ratón, fulgencia)

Habitat: sandy washes, caliche bluffs, open, sunny grasslands, rocky slopes, mesic canyons; limestone and igneous substrates.

Geographic range: Texas: Trans-Pecos (excluding Culberson, Reeves, Pecos Counties); most of south Texas; sites in central Texas. US: Texas to Arizona, Alabama, Florida, Puerto Rico, Virgin Islands. Elsewhere: Mexico to South America, West Indies, Old World.

Park locations: Agua Adentro, upper Guale Mesa, upper Arroyo Mexicano area, Madrid Falls Trail, Chorro Canyon

Ojo de vibora is a low perennial with weak, silky-haired stems to 20 inches long. From a small, woody base, the wiry stems branch with a trailing or creeping habit but do not twine or root at the nodes. The variable leaves are stemless or short-petioled, oblong to lance-shaped, with long soft hairs above and below.

From spring to fall, delicate pale to deep blue flowers, in few-flowered clusters, grow from the leaf axils. The clusters form on hairy, threadlike stalks. Each five-lobed bloom, 1/4 inch across, is flat and pleated, the sky-blue color accented by a white circle at the throat. Within the flower are five white stamens and two styles, each with two slender branches. The fruit is a rounded, four-valved, dehiscent capsule, 1/10 inch long, with four or fewer brown oval seeds.

This species is used extensively in traditional medicine, in Africa, India, the Philippines, and other parts of East Asia. The plant extract is used as a "brain tonic" to treat Alzheimer's and dementia (Singh 2008).

Botanical notes: The genus name *Evolvulus* is derived from the Latin *evolvere* (to unroll or untwist), because the plants do not twine or climb like some morning-glory species. The specific epithet *alsinoides* suggests "like plants of the genus *Alsine*." Linnaeus described this species as *Convolvulus alsinoides* in 1753 but moved it to his new *Evolvulus* genus in 1762. As a source, Linnaeus cited Dutch botanist Johannes Burman's *Thesaurus Zeylanicus* of 1737, a flora on the plants of Ceylon (Sri Lanka). The young Linnaeus, while traveling through Holland, had been hired by Burman to help him complete this work.

103. Torrey's Tievine (*Ipomoea cordatotriloba* var. *torreyana*)
(Other common names: purple bindweed, cotton morning glory)

Habitat: open fields to desertic mountains; draped on trees, shrubs, near creeks, springs, roads, disturbed sites.

Geographic range: Texas: Trans-Pecos (Presidio, Brewster, Val Verde Counties); much of Edwards Plateau, south Texas; parts of central to north Texas. US: Texas. Mexico: Chihuahua, Coahuila, Tamaulipas.

Park locations: Monilla Canyon, Yedra Canyon, Fresno Canyon

Seldom found west of the Pecos River, this bindweed is a trailing, twining, climbing perennial, to 3 feet long or more, with all plant parts hairless or nearly so. The many leaves may be heart-shaped and scarcely lobed, or trilobed, with earlike lobes at the base. Up to 3 inches long, the blades are often wavy or wrinkled on the margins.

In summer and fall, purple-rose to rose-pink, funnel-shaped flowers, with a dark magenta throat, bloom in clusters of one to ten. The clusters are borne on long stalks from the leaf axils. To 2 inches across, each bloom forms on its own short stalk that may be noticeably rough and warty. The flower is supported below by papery sepals. In the flower's throat is a slender style with a rounded white stigma and stamens of unequal length with pollen-bearing anthers.

The spheric to egg-shaped fruits, 1/3 inch long, are two-valved dehiscent capsules with four brown seeds.

Botanical notes: Some argue that the genus name *Ipomoea*, from the Greek *ips* (worm, bindweed) and *homoios* (like), refers to the plants' wormlike twining habit. However, Linnaeus, in *Genera Plantarum* of 1737, indicates that he chose *Ipomoea* for the plants' resemblance to bindweed (genus *Convolvulus*). The epithet *cordatotriloba*, from Latin words *cordatus* (heart-shaped) and *trilobatus* (trilobed), refers to the leaves. The varietal name, *torreyana*, honors New York botanist John Torrey. Asa Gray described this variety as *Ipomoea trifida* var. *torreyana* in 1878, citing Charles Wright, John Bigelow, Ferdinand Lindheimer, and Arthur Schott as collectors in West and South Texas. Wright collected the lectotype in 1849, on Zacate Creek (now Mud Creek) in Kinney County, near the Rio Grande. In 1988, Daniel Austin treated the plant as a variety of *I. cordatotriloba*.

104. Ivyleaf Morning Glory (*Ipomoea hederacea*)
(Other common names: ivy-leaved, blue, woolly, Mexican, or entire leaf morning glory, trompillo morado, flor de verano, manto de la virgen)

Habitat: desert, riparian, xeric mountain environs, creek banks, arroyos, thickets, cultivated and abandoned fields, roadside ditches, waste places.

Geographic range: Texas: Trans-Pecos (El Paso, Presidio Counties); Crockett County; scattered throughout much of Texas. US: east half; Texas west to Arizona, north to North Dakota. Elsewhere: Canada (Ontario); Mexico (Sonora east to Tamaulipas; parts of southern Mexico); parts of Central, South America, West Indies.

Park locations: Palo Amarillo Creek

Native to tropical America, this morning glory is a twining, climbing annual vine, which grows from a taproot and forms a single stem to 10 feet long. On long petioles, the alternate leaves, to 5 inches long, are round to egg-shaped in outline, often trilobed, and heart-shaped at the base.

After summer monsoon rains, funnel-shaped, white-throated blue flowers form on one- to few-flowered stalks. The abruptly flaring blooms, to 2 inches wide, are supported by slender, attenuate green sepals with recurving tips. Within the flower's throat are five white stamens and a white style with a bilobed stigma. The plant is largely self-pollinating.

This vine is strikingly hairy: stems and petioles are covered with long and short hairs, some bristly and pustulate; the leaves bear short, stiff hairs; flower stalks are adorned with reflexed hairs; and sepals bear long bristly hairs with enlarged bases.

About 1/3 inch long, the spherical brown fruit is a three-valve dehiscent capsule with four to six seeds. Half the length of the fruits, the brown or black seeds are pyramid-shaped and often densely hairy.

Botanical notes: The species name *hederacea*, from *Hedera* (the genus name of ivy), implies "relating to ivy," for the trilobed, ivylike leaves. Austrian botanist Nicolaus von Jacquin described this species in *Collectanae ad botanicam . . .*, a five-volume work published from 1786 to 1796. From 1755 to 1759, Jacquin traveled to the West Indies, Venezuela, and Colombia on behalf of Emperor Francis I, collecting plants for the Schoenbrunn Palace gardens. Jacquin later documented and illustrated these collections in lavish publications, including the *Collectanae*.

105. Longpetal Echeveria (*Echeveria strictiflora*)
(Other common names: desert savior, siempreviva, live forever, Big Bend echeveria)

Habitat: rocky ledges, crevices, amid boulders, rockslides, talus slopes; volcanic, novaculite, limestone exposures, mid-high elevations.

Geographic range: Texas: Trans-Pecos (Jeff Davis, Presidio, Brewster Counties). US: Texas. Mexico: Chihuahua east to Nuevo León.

Park locations: Oso Creek above Oso Spring, Rancherías Spring, Arroyo Mexicano, Madrid Falls Trail, hill near Leyva Escondido, Sierra Blanca Dome

Longpetal echeveria is a perennial, to 1 1/2 inches high in bloom, which grows from a short crown and forms a stout basal leaf rosette. The crowded spatulate leaves, to more than 4 inches long, are notably thick and fleshy, also smooth, hairless, and stemless. The blade color is often pale green, or gray green and glaucous. Often, the leaves develop a reddish tinge on the margins, especially at the tips.

This succulent produces as many as four flowering scapes to 16 inches high, well above the leaf rosette. Atop the reddish scapes are long, one-sided, usually nodding flower clusters with many crowded blooms. The scapes may also bear some lance-shaped leaves, to 2 inches long, much smaller than the basal leaves.

Blooming after summer rains and into fall, the flowers are orange-red, yellowish inside, with five lancelike lobes united in the shape of a five-angled pyramid. The nearly closed, short-stalked blooms are no more than 2/3 inch long. Within each bloom are ten stamens with oblong anthers, and green styles. The flowers are supported by pinkish-red, oblong sepals of varying lengths. The fruits are erect oblong pods with a persistent style, and oval seeds.

Botanical notes: In 1828, Swiss taxonomist Augustin Pyramus de Candolle, in his seventeen-volume *Prodromus* (an effort to document all seed plants), named this genus for Mexican artist Atanasio Echeverria y Godoy. Echeverria was an accomplished botanical illustrator who traveled with the Sessé and Mociño natural history expedition (1787–1803) to Mexico, Guatemala, Cuba, and Puerto Rico. In 1852, in *Plantae Wrightianae*, Asa Gray named this species *strictiflora*, using the Latin *stricti-* (straight, close) and *flos* (flower), a reference to "a very strict and close secund raceme or spike." Charles Wright collected the plant in 1849 on "mountains west of the pass of the Limpia," in the Davis Mountains of Jeff Davis County.

106. Wright's Stonecrop (*Sedum wrightii*)
(Other common names: siempreviva de Wright)

Habitat: canyons, shaded springs, amid boulders, rock crevices, moist mountain bluffs; igneous, limestone soils.

Geographic range: Texas: Trans-Pecos (excluding Hudspeth, Reeves, Terrell Counties); Uvalde County. US: Texas, New Mexico. Mexico: Chihuahua east to Nuevo León, south to Zacatecas, San Luis Potosí.

Park locations: Leyva Canyon, Cinco Tinajas

This stonecrop is a succulent perennial to 9 inches long or more, with a basal leaf rosette and widely branching, leafy stems. The stems may be erect, sprawling, straggling, or dangling from rock walls.

The early falling, crowded basal leaves are no than 1/2 inch long. Stem leaves are smaller. All leaves are stemless, hairless, and rounded or blunt at the tip. The fleshy, swollen blades are varied, often oblong.

From late summer to fall, white flowers, often pink-tinged, form on long flowering stalks, in leafy clusters of many blooms. Each short-stalked bloom, to 3/5 inch across, has five spreading oblong petals. The bloom is supported by oblong green sepals of varied lengths. Within the flower are ten stamens with white filaments and red anthers. The blossoms exude a sharp, musky scent.

The fruits are erect, tan pods, 1/5 inch long, with a needlelike, persistent style, and numerous small seeds.

Young stems and leaves of stonecrop species are edible, used in salads and as potherbs (Warnock 1977). Stonecrop plants are actively cultivated as ornamentals, and popular in rock gardens.

Botanical notes: In the first century AD, Lucius Junius Moderatus Columella, in a massive work on Mediterranean agriculture, *De Re Rustica*, published the name *Sedum*, describing an "on the ground" or "sitting plant, which grows on roofs." Apparently, the name is from the Latin *sedere* (to sit), for the plant's habit, rather than the Latin *sedare* (to calm, heal), as some botanists have believed. In 1852, Harvard botanist Asa Gray named this species for Charles Wright, who collected it in 1849, on "hills of Devils River [near Del Rio], in crevices of rocks, and summit of mountains near El Paso." Wright traveled on foot with a military expedition from San Antonio to El Paso and collected plants along the way.

107. Spiny Greasebush (*Glossopetalon spinescens* var. *spinescens*)
(Other common names: spiny or Nevada greasewood, thorny greasebush)

Habitat: rocky canyon walls, rims, steep bluffs of arid draws, sotol-yucca environs, on limestone.

Geographic range: Texas: Trans-Pecos (El Paso, Hudspeth, Culberson, Presidio, Brewster Counties). US: Texas to Arizona; Oklahoma. Mexico: Chihuahua.

Park locations: Lefthand Shutup, Saltgrass Draw

Spiny greasebush is a compact, rounded shrub, seldom more than 3 feet high, with many green stems and stiff, thorn-tipped branchlets. The many short branchlets are hairless, often angled or arching, nearly naked because the leaves are early falling. When leafless, the green stems replace the leaves and serve as instruments of photosynthesis.

If present, the light green, alternate leaves are elliptic or spatulate, to 2/5 inch long, and pointed at the tips. The short-stemmed blades are mostly hairless but often glaucous, with a powdery coating.

In spring, odd, visually stunning white flowers, mostly five-petaled, proliferate in the leaf axils. The petals are strikingly slender, spatulate to linear, frequently curling or twisting. Within each bloom are six to eight stamens with white filaments and yellow anthers and a single style with a headlike stigma. The blooms are supported by bracts and usually five sepals, often of varying lengths. The hairless sepals are green to brownish purple, rounded or blunt at the tips and transparent on the margins.

The leathery fruit is a green to brown, egg-shaped pod to 1/5 inch long, usually beaked, with conspicuous lengthwise ribs or streaks. Each pod contains one or two shiny oval seeds.

Botanical notes: The genus name *Glossopetalon* joins Greek words *glosso* (tongue) and *petalon* (petal), alluding to the narrow "tonguelike" petals of these plants. The epithet *spinescens* is Latin for "spiny." Harvard botanist Asa Gray described the genus and species in 1853, in *Plantae Wrightianae*. In 1852, Charles Wright obtained a specimen "on mountain sides near Frontera [a few miles north of El Paso]."

108. Melón Loco (*Apodanthera undulata*)
(Other common names: melón loco, melón de coyote, coyote or loco melon, wild cucurbit, melón or calabaza hedionda, calabaza loca)

Habitat: sandy dunes, clay mounds, alluvial flats, roadsides, grasslands; limestone, igneous soils.

Geographic range: Texas: Trans-Pecos (excluding Reeves, Pecos, Terrell Counties). US: Texas to Arizona. Mexico: Chihuahua east to Coahuila, south to Jalisco, Guanajuato.

Park locations: Arroyo Mexicano, Llano Loop, Sauceda Road to Fresno Canyon turnoff, Encino area, lower Contrabando Creek

Melón loco is a densely hairy, coarse perennial vine, with prostrate stems to 10 feet long and short paired tendrils. The whitish stems grow from a bulky root. Dark green above, and gray, with short, stiff hairs below, the kidney-shaped leaves are shallowly lobed and sometimes sparingly toothed. Long-petioled, the blades are up to 6 inches wide, with white ruffled margins.

From late spring to early fall, yellow funnel-shaped flowers form on long stalks from the leaf axils. Male and female blooms appear on the same plant, the male blooms in long clusters from lower axils, and the larger female blooms single on upper branches. Each flower has five spreading, spatulate lobes, three stamens with stalkless anthers, and a style with a trilobed stigma. Only larger plants produce female blooms (Delesalle 1989).

The green gourdlike fruit is a hard-shelled berry (pepo). To 4 inches long, the hairless fruit is round to oblong, with raised, lengthwise ribs, and light brown seeds.

Reportedly, the fruit has an unpleasant taste and odor, but Native Americans ate the roots and made a mush from the seeds and pioneers used crushed roots to wash clothes (Warnock 1974). In Guanajuato, Mexico, the fruit's pulp is used to treat urinary problems, and in Zacatecas, street vendors sell the roasted seeds (Lira and Caballero 2002).

Botanical notes: The genus name *Apodanthera*, from the Greek *a* (without), *podos* (foot), and *anthera* (anther), refers to the stalkless anthers. The Latin epithet *undulata* (wavy) references the wavy leaves. This species was described by Asa Gray in 1853 and collected by Charles Wright in 1852, in valleys between Eagle Spring, a stage stop in southeast Hudspeth County, and Limpia Creek in Jeff Davis County.

109. Red Berry Juniper (*Juniperus pinchotii*)
(Other common names: Pinchot's or Texas juniper, Christmas berry juniper, táscate, enebro de fruto rojo)

Habitat: sandy flats, rocky hills and canyons, broken, eroded terrain, depleted grasslands; gravelly limestone, gypseous soils.

Geographic range: Texas: Trans-Pecos, much of west Edwards Plateau, Panhandle, north-central Texas. US: Texas, western Oklahoma, southeast New Mexico. Mexico: Coahuila.

Park locations: Las Cuevas, Tascate Hills, Solitario northwest rim, Lower Shutup, Masada Ridge

This juniper is an evergreen shrub or small tree to 25 feet high, with a branching trunk and shallow, spreading roots. The ascending branches form dense shrubs with many slender branchlets (whips). The stems are covered with smooth gray or brown bark, which in time flakes and peels.

Most of the yellow-green leaves are scalelike. In tight groups of two or three, the blades are egg-shaped to triangular, often with glands that secrete a white exudate. Known as whip leaves, blades on terminal branchlets are longer, to 1/4 inch long, and distinctively, not glaucous on the upper surface.

The reddish-brown or copper-colored "juniper berries" are fleshy cones resembling berries. This juniper is dioecious, with male and female cones on separate plants. The pollen cones are oblong, to 1/4 inch long, and consist of scales (sporophylls) bearing pollen sacs. The pollen is shed in the fall, as the female cones open for pollination. The female cones, to 2/5 inch long, are spheric, juicy, and sweet, with one chestnut-brown, oval seed.

The wood is used for fence posts and firewood. The Comanche consumed dried and crushed roots to ease menstrual problems (Jones 1968). Distilled fruits were used to make gin (Warnock 1970).

Botanical notes: *Juniperus* is the classical Latin name for juniper. In 1905, George Sudworth named this species for forest conservation pioneer Gifford Pinchot, the first head of the US Forest Service. Sudworth was chief dendrologist of the Forest Service and author in 1897 of a classic study of American trees. *George Clothier, a Forest Service employee and later professor of forestry at Washington State University, collected this tree in Palo Duro Canyon in the Texas Panhandle in 1905.*

110. Cosmopolitan Bulrush (*Bolboschoenus maritimus* subsp. *paludosus*)
(Other common names: saltmarsh, alkali, prairie, or seaside bulrush, bayonet grass, saltmarsh club-rush)

Habitat: Rio Grande, sandy banks, standing water, springs, depressions, ditches, alkaline, saline sites.

Geographic range: Texas: Trans-Pecos (excluding Culberson, Jeff Davis Counties); scattered in Edwards Plateau, Permian Basin, Panhandle, north-central Texas. US: west half; much of Midwest, northeast seaboard. Elsewhere: Canada; Mexico, South America, West Indies.

Park locations: Hoodoos, Chupadero Spring

This bulrush is an aquatic perennial growing from an extensive system of horizontal roots. The erect stems (culms), three-angled and flat on the sides, may approach 5 feet in length. The stems are often clumped at the roots, with an enlarged, bulblike base.

Typically, the smooth culms bear folded basal leaves and at least a few thin, flattish stem leaves, which are keeled below, pointed at the tip, and minutely toothed on the margins and the midrib. The leaves are partly covered by membranous tubular sheaths.

Flower clusters form near the stem tips, with numerous stalkless spikelets supported by a few partly sheathed bracts. Each narrowly egg-shaped spikelet consists of floral scales, each with a tiny floret. The scales are brown to colorless, membranous, with a two-branched, bristlelike appendage. A spikelet may bear twenty-five-plus florets, each with barbed bristles, three protruding stamens with yellow anthers, and a two-branched, threadlike style.

The brown fruits are light, floatable achenes (dry, one-seeded, indehiscent) to 1/6 inch long, egg-shaped but compressed or three-angled, blunt-tipped, and beaked.

Botanical notes: The genus name, from the Greek *bolbos* (bulb), refers to the bulbous plant bases, and *Schoenus* (bulrush) is an old genus. The Latin epithet *maritimus* (maritime) and subspecies name *paludosus* (marshy) relate to the habitat. The species, described by Linnaeus in 1753 as *Scirpus maritimus*, was moved by Austrian botanist Eduard Palla to his newly formed *Bolboschoenus* genus in 1905. This subspecies, given species status as *Scirpus paludosus* by Wyoming botanist Aven Nelson in 1899, was most recently treated by Japanese sedge authority Tetsuo Koyama as a subspecies of *B. maritimus* in 1980. Nelson collected the first specimen in 1896, on Salt Creek, near Newcastle, Wyoming.

111. Bearded Flatsedge (*Cyperus squarrosus*)
(Other common names: awned flatsedge, bearded nutgrass, awned cyperus, awned umbrella-sedge, tulillo, apoyamate [from Nahuatl])

Habitat: Rio Grande floodplain, alluvial flats, creek banks, ponds, seeps, muddy ditches.

Geographic range: Texas: Trans-Pecos (El Paso, Jeff Davis, Presidio, Brewster, Terrell, Val Verde Counties); parts of Edwards Plateau, South Texas Plains, Coastal Prairies, north and east Texas. US: contiguous forty-eight states. Elsewhere: Canada to Argentina; West Indies; Australia, Africa, Eurasia.

Park locations: Contrabando Creek, Cinco Tinajas, Agua Adentro, near Sauceda, Arroyo Mexicano

This sedge is a small clumped annual, with narrow, light green stems (culms) to 6 inches long. Notably three-angled, smooth and hairless, the culms grow from shallow, fibrous roots and seldom bear more than three leaves, always near the base. The alternate leaves are short, to 1/2 inch long, and slender, flat, or V-shaped. At times forming large stands, this summer annual may exude a distinct curry-powder aroma.

The inflorescence is a headlike spike to 1 1/2 inches long, seldom with more than twenty spikelets. A spike is supported by one to four slender, light green, leafy bracts, some the length of the culms. The oblong spikelets, green/tan/reddish brown, are up to 3/4 inch long, and compressed. Each spikelet bears as many as eight florets, each enclosed in a floral scale. The scales are oblong or lancelike, prominently ribbed, with an outward-curving, bristlelike tip. The floret consists of a single stamen with an oblong anther, and a linear, three-branched style.

The fruits are minute, three-angled achenes, tan to black, egg-shaped to triangular, and punctate.

Native Americans ate the roots as food (Castetter 1935; Swank 1932).

Botanical notes: The Latin species name *squarrosus* (rough, scaly) refers to the bristlelike appendages of the floral scales. Linnaeus described this species in 1756 in his *Centuria II. Plantarum* ... The type specimen was collected in India by Johann Gerhard Koenig, Linnaeus's pupil, who in 1778 became the first naturalist of the British East India Company. Credited with introducing Linnaean botany to India, Koenig collected many economic and medicinal plants.

112. Sand Spikerush (*Eleocharis montevidensis*)
(Other common names: sand spikesedge, tule, giant hairgrass, hairgrass, Montevideo or slender or Dombey's spikerush, slender creeping spikerush)

Habitat: sandy, moist, muddy, wet soils of floodplains, creek beds, low-lying, seasonally flooded areas, sites near springs, tanks, seeps.

Geographic range: Texas: Trans-Pecos (excluding El Paso, Hudspeth, Reeves Counties); Ward, Winkler Counties; most of Edwards Plateau, central Texas, Gulf Coast; much of the rest of Texas. US: South Carolina to Florida, Florida west to California; Oklahoma, Kansas, Colorado. Elsewhere: Mexico; Guatemala, Honduras; South America.

Park locations: Panther Canyon, Rancherías Canyon, Arroyo Mexicano, upper Terneros Creek, Skeet Canyon, Cienega Creek

This spikesedge is an erect terrestrial or semiaquatic perennial, with long slender, widely spreading rhizomes, and green, nearly leafless culms to 20 inches high. The culms are round, angled, or rectangular, firm but spongy. Reduced to persistent basal sheaths, the culm leaves are purplish brown to reddish brown below, papery above, usually with a slender, spiny tip.

The inflorescence is an erect, bractless spikelet at the culm tips, with up to 100 florets. Ovoid or oblong, the spikelets are up to 1/2 inch long, with blunt or pointed tips. The bisexual florets consist of five or six tan bristles, three stamens with pale yellow anthers, and a three-branched style. Each floret is covered by overlapping oblong scales (bracts), which are rounded and wrinkled. The membranous scales are orange brown or maroonish brown, often paler at the midrib, and translucent on the edges.

The dark brown fruits are achenes, less than 1/25 inch long, egg-shaped to pyramidal, weakly three-sided and compressed, with a conic tubercle.

Botanical notes: The genus name is formed from the Greek *heleios* (marsh-growing) and *charis* (grace). The species name, *montevidensis*, refers to Montevideo, Uruguay. University of Berlin professor Karl Kunth described this species in 1837, from a specimen collected by German naturalist-botanical explorer Friedrich Sellow in Uruguay. A student of Carl Willdenow at the Berlin Botanical Garden, Sellow participated in expeditions in Brazil and Uruguay from 1814 until his death by drowning in 1831.

113. Texas Persimmon (*Diospyros texana*)
(Other common names: Mexican or black persimmon, chapote, chapote prieto, chapote manzano, chapote negro, zapote prieto, tzapotl (Nahuatl))

Habitat: stream beds, gravelly arroyos, open woodlands, rocky slopes and ravines; shallow limestone and occasionally, igneous soils.

Geographic range: Texas: Trans-Pecos (Presidio, Brewster, Pecos, Terrell, Val Verde Counties), east through parts of central, south Texas, Gulf Coast. US: Texas. Mexico: northeast Chihuahua east to Tamaulipas.

Park locations: Lefthand Shutup, Lower Shutup, Masada Ridge, Saltgrass Draw, upper Alamito Creek

This persimmon is a semi-evergreen shrub or tree to 16 feet high, with hard dark wood. The dense, intricate branches are covered with smooth, gray or reddish gray, often flaking or peeling, bark. Mostly stemless, the alternate leathery leaves are as much as 2 inches long, spatulate to oblong, rounded or blunt at the tip, and curled under on the margins. The leaf undersides, especially, may be coated with soft, velvety hairs.

In spring, small clusters of creamy-white urnlike flowers, each with five recurving lobes, dangle from the leaf axils. About 1/2 inch long, the sweetly fragrant, silky-haired male and female flowers appear on separate plants. A little smaller than the female, the male blooms have sixteen stamens with short filaments and rather long yellow anthers. The female blooms have up to eight sterile stamens that produce no pollen, and a four-branched style.

The fruit is a fuzzy round berry 4/5 inch wide, black, sweet, and edible when ripe, with three to eight seeds.

The wood is used to make tool handles and engraving blocks (Vines 1960). In Mexico, the fruit juice is used to dye animal hides, and the berries are widely used in Mexico and Texas to make jelly, wine, pudding, and custard (Standley 1920–26; Powell 1998; Hartung 2016).

Botanical notes: In 1753, Linnaeus named the genus *Diosporos*, from the Greek *Dios* (divine) and *pyros* (wheat), implying "food of the gods." German botanist Adolph Scheele described this species in 1849, from a specimen collected by Ferdinand Lindheimer at New Braunfels, Texas, in 1846. The plant was collected even earlier, by Jean Louis Berlandier, in Mexico in 1836.

114. Longleaf Jointfir (*Ephedra trifurca*)
(Other common names: longleaf ephedra, Mexican tea, longleaf Mormon tea, longleaf teabush, cañutillo, popotillo, cañatilla, tepopote, hierba de la coyuntura)

Habitat: low to mid-elevations, sandy, gravelly flats, arroyos, gypseous dunes, creosote scrub, grasslands, rocky slopes.

Geographic range: Texas: Trans-Pecos; Loving, Ward Counties. US: Texas to California. Mexico: Baja California east to Coahuila.

Park locations: Cerro de las Burras Loop, Black Hills area, Cienega Creek area

Longleaf jointfir is a densely branched but largely leafless shrub, to 5 feet high, with green or yellow-green stems that provide photosynthesis in place of leaves, and cones that replace fruits in reproduction. The round stems are stiff, notably jointed, sharp-tipped, hairless and smooth, but finely grooved. Both stems and branches turn ashy gray in age, and the bark cracks and fissures.

The leaves are scalelike or sheathlike, relatively long and persistent. To 3/5 inch long, the stemless lance-shaped blades, with bristlelike tips, form in whorls of three per node. At maturity, the leaves remaining are ashy-white and frayed.

This species is dioecious, with male pollen cones and female seed cones on separate plants. Short-stalked, inversely egg-shaped cones form at the nodes. The male cones, to 2/5 inch long, consist of reddish-brown, membranous bracts in overlapping whorls of three and protruding white fused stamens with yellow anthers. The seed cones are larger, to 3/5 inch long, with fewer whorls of bracts. The bracts are papery, translucent, and reddish brown at the base. The bracts envelop one smooth, tan seed.

Fossil evidence of ephedra dates to the Early Cretaceous period, 145–100 Ma (Wang and Zheng 2010). Ephedra has been found at Neanderthal burial sites more than 60,000 years old (Lietava 1992). CD species are sold in Mexican markets to treat venereal disease and kidney problems (Henrickson and Johnston unpublished).

Botanical notes: The genus name *Ephedra*, from the Greek *ep* (upon) and *hedra* (seat) implies "sitting upon." Pliny used the term for horsetail (*Equisetum*) fern with jointed stems. Describing *Ephedra* in 1753, Linnaeus saw a resemblance between "jointfirs" and Pliny's horsetails. The Latin epithet *trifurca* (three-forked) refers to the leaves. In 1848, the species was named provisionally by John Torrey. The type locality was "the region between the Del Norte [Rio Grande] and the Gila" in New Mexico. Sereno Watson validly published the name in 1871.

115. Ferriss' Scouring Rush (*Equisetum xferrissii*)
(Other common names: Ferriss' horsetail, intermediate horsetail, intermediate scouring rush)

Habitat: creeks, springs, ponds, moist canyons, standing water, mud, wet depressions, marshy fields; sand, silt, clay.

Geographic range: Texas: Trans-Pecos (El Paso, Culberson, Jeff Davis, Presidio Counties). US: contiguous forty-eight (excluding Mississippi to South Carolina, Florida). Elsewhere: Canada, Mexico (Baja California, Chihuahua, Coahuila, Durango).

Park locations: Arroyo Primero, Chorro Canyon, Fresno Canyon

Ferriss's scouring rush is a sterile hybrid of smooth scouring rush (*E. laevigatum*) and scouring rush (*E. hyemale*) and is intermediate between the two. The fern commonly spreads through vegetative rather than sexual reproduction but occurs outside the known range of smooth scouring rush.

Like the parents, this rush is terrestrial, colony-forming, with erect, jointed, hollow, mostly unbranched, photosynthetic green stems, and spreading and creeping rhizomes. Like smooth scouring rush, the stems are lined with fine vertical ridges, and bear scalelike leaves fused into sheaths at the stem nodes.

However, this hybrid's stems are larger, more persistent, less smooth. The leaf sheaths seem stouter, and their teeth are often lighter in color and more persistent. The cones tend to dry and shrink without releasing spores. The spores are white, not green, and usually misshapen, not spheric.

The sheaths, especially those lower on the stems, may bear two bands, a narrow black band below the teeth (like smooth scouring rush) and a broader black band near the sheath base (like scouring rush), with white between the bands.

Botanical notes: The genus name *Equisetum*, from the Latin *equus* (horse) and *seta* (bristle), implying "horsetail," may refer to the coarse roots or bristly branches of some species. The species name, from *x* (hybrid) and *ferrissii* honors naturalist James Henry Ferriss, an avid collector of ferns and land shells throughout the United States. Ferriss, newspaper owner-editor in Joliet, Illinois, formed a collection of living US ferns for the park in Joliet and authored articles on shell collecting for the journal *Nautilus*.

EQUISETACEAE — HORSETAIL, SCOURING RUSH FAMILY

116. Leatherweed (*Croton pottsii* var. *pottsii*)
(Other common names: leatherweed or leatherleaf or leather croton, encinilla, Potts' leatherweed)

Habitat: sandy flats, desert scrub, grasslands, rocky mountain slopes; limestone, igneous, novaculite strata.

Geographic range: Texas: Trans-Pecos; parts of Permian Basin, southern Panhandle, Edwards Plateau, South Texas Plains. US: Texas to Arizona. Mexico: Sonora east to Coahuila, south to Durango, or farther south.

Park locations: Monilla Canyon, Arroyo Mexicano, Outer Loop Trail, Cienega Mountain

Leatherweed is a perennial to 2 feet high, with erect, leafy stems growing from woody roots. Scarcely branched below, the herb may form branches above that extend beyond the flowers. The branches are gray, coated with radiating, star-shaped (stellate) hairs. To 2 inches long, the gray-green leaves are oval, pointed at the tip, and smooth on the edges, with stellate hairs. The thick leathery leaves alternate along the stems or form bundles below the flowers.

From spring to fall, flower clusters form at the branch tips. The densely hairy blooms are unisexual, with male and female blooms on the same plant, commonly in the same clusters, the male above, the female below. At times, clusters contain flowers of only one sex. A male bloom, to 1/4 inch across, consists of five lance-shaped or triangular sepals; five small brownish-yellow petals; a disk with five orange or yellow-orange, bulblike glands opposite the sepals; and up to fifteen protruding stamens. A female bloom, lacking petals, has five oblong sepals and a pistil with three brown, two-branched styles.

The fruit is an oblong, trilobed capsule, 1/5 inch long. The tick-like gray or brownish seeds are oblong, with a yellow outgrowth (caruncle) at the seed's point of attachment to the fruit.

Botanical notes: In 1753, Linnaeus took the genus name from *kroton* (tick), a Greek name for the castor oil plant (*Ricinus communis*), for the seed's resemblance to a tick. In 1853, Johann Klotzsch, curator of the Berlin Royal Herbarium, named the species *Lasiogyne pottsii*, for the discoverer, John Potts, manager of the Chihuahua Mint in Mexico. Potts sent specimens to Berlin and to Kew Botanic Gardens in London. Swiss botanist Johannes Müller Argoviensis moved the plant to the *Croton* genus in 1866.

117. Trans-Pecos Croton (*Croton sancti-lazari*)
(Other common names: none)

Habitat: rocky outcrops, bluffs, xeric shrubland; limestone, novaculite, igneous substrates.

Geographic range: Texas: Big Bend (southern parts of Presidio and Brewster Counties). US: Texas. Mexico: Chihuahua, Coahuila, sites in Durango, Zacatecas.

Park locations: Upper Guale Mesa, Tascate Hills, Solitario northwest rim, northwest of Solitario Peak, Righthand Shutup, hills north of Righthand Shutup

This croton is a compact shrub to 2 feet high, with ascending stems from a thick woody crown and leafy branches. Stems and leaves are grayish white, coated with stellate hairs. The alternate leaves are oval, to 2 inches long, with a pointed or blunt tip and short thick petioles.

Trans-Pecos croton is dioecious, with male and female flowers appearing from spring to fall on separate plants. Several male blooms form in short flower clusters at the stem tips. Typically, each male bloom consists of a yellow or white, bell-shaped calyx with five triangular sepals; five white, nearly transparent petals, inversely lance-shaped, alternating with the sepals; five minuscule orange glands opposite the sepals; and normally eleven stamens. The male bloom is only 1/6 inch long.

Usually, two female blooms appear in minute flower clusters. Each petal-less female bloom consists of five sepals; five tiny, purplish-brown glands opposite the sepals; and a spheric pistil with three purplish-brown styles, each with two diverging branches. Both male and female flowers are densely hairy.

The globose fruits are yellowish-green or whitish capsules to 1/4 inch long, with dense stellate hairs. The gray oval seeds have a white bladderlike appendage at the seed's point of attachment to the fruit.

Botanical notes: French-Italian botanist Léon Croizat described this species in 1945, from a specimen collected by Cornelius Muller and Frederick Wynd in 1936, in Coahuila, Mexico. Croizat named the species *sancti-lazari* for the plant's location: "south of Castaños, rocky slopes of El Puerto de San Lazaro." Croizat is best known for developing panbiogeography, an effort to explain the evolution and geographic distribution of species over time.

118. Abrams' Spurge (*Euphorbia abramsiana*)
(Other common names: Abrams' sandmat or broomspurge or prostrate spurge)

Habitat: sandy, alluvial flats, loam, flooded sites.

Geographic range: Texas: Trans-Pecos (south Presidio, south Brewster, Reeves Counties). US: Texas to California. Mexico: Chihuahua, Sonora.

Park locations: 1 mile west of Lajitas, Madera Canyon, Terneros Creek at Sauceda Road, Contrabando Canyon

This sandmat was collected for many years in the Trans-Pecos but misidentified. I often photographed the plant in BBNP and BBR, unaware of its true origins. In 2015, Nathan Taylor, Sul Ross State University student, recognized the species as Abrams' spurge, a California native, and corrected this case of mistaken identity (Taylor and Terry 2015). This spurge is a prostrate annual that branches, often profusely, from a central root. The short main stem and longer branches bear short stiff hairs, unlike similar species. Nearly stemless, the paired leaves are oval to oblong, to 2/5 inch long, sparsely shorthaired to hairless. Frequently, the leaves are marked with reddish blotches and pale lines along the veins. The blades are rounded at the tip and lopsided, with sides of unequal lengths.

The flower clusters (cyathia) are solitary at the nodes but often closely grouped on short lateral branchlets. The cyathium is a cup-shaped ring of bracts (involucre) with a female flower (pistil) in the center surrounded by three to five male blooms (stamens). Four fleshy nectary glands, round or oblong, pink or yellow, rest on the involucre rim, each with a white petal-like appendage usually wider than the gland. The fruit is a three-angled, three-chamber, round or oblong capsule 1/14 inch long, with one white seed per chamber. Transverse ridges on the seed, and a protuberance at one end, are distinctive keys.

Botanical notes: According to Pliny the Elder in his *Natural History*, Juba II, King of Mauretania, named a succulent medicinal plant from Mt. Atlas in northern Africa for his corpulent physician Euphorbus. The name, from the Greek *eu* (well) and *phorbe* (food) implied "well fed." In 1753, Linnaeus named the genus *Euphorbia* for the large physician. In 1934, California botanist Louis Wheeler named the species for Stanford University botanist Leroy R. Abrams, who collected the plant at Heber in Imperial County, California, in 1904. Abrams co-authored *An Illustrated Flora of the Pacific States (1923–60)*.

119. Pointed Sandmat (*Euphorbia acuta*)
(Other common names: pointed or sharpleaf spurge, pointed broomspurge)

Habitat: desert scrub, dry rocky hills, broken limestone and clay mounds, sandy flats.

Geographic range: Texas: Trans-Pecos (mostly Brewster, Pecos, Terrell, Val Verde, also southeast Presidio, northeast Hudspeth Counties); Permian Basin, Edwards Plateau. US: Texas, southern New Mexico. Mexico: Chihuahua, Coahuila.

Park locations: peak east of Pico 4522, Paso al Solitario 4WD Road, below Terlingua Uplift

Many spurges are called sandmats because they are mostly prostrate, frequently grow in sand, and form extensive mats. This sandmat is atypical, often upright and occurring on a variety of soils. Pointed sandmat is a stout perennial with milky sap and ascending stems to 1 foot long or more, densely coated with long white hairs. The stems arise from a woody crown and widely spreading, woody rhizomes.

To 3/4 inch long, mostly stemless, the paired leaves are lance-shaped to egg-shaped, with a sharp, spinelike tip and noticeably thick margins. The stiff leaves bear a thick cover of close-pressed, matted hairs.

Male and female flowers appear together on the same plant in small clusters (cyathia) that resemble a single bisexual flower. Each cluster consists of a solitary female pistil, with three branching styles, in the center, surrounded by twenty to twenty-five male blooms consisting of a single stamen. The blooms appear in an urn-shaped, white-haired, calyx-like involucre with four yellow glands on the rim. The glands bear white, petal-like appendages that are erect and shallowly notched at the tip. The fruit is a trilobed, whitehaired capsule, 1/8 inch long, with three smooth, four-angled, white seeds.

Botanical notes: The species name, from the Latin *acuta* (acute), refers to the sharply pointed leaf tips. St. Louis botanist George Engelmann described this species in 1859, in Major Emory's *Report on the United States and Mexican Boundary*, from collections by Charles Wright in 1849, on "prairies west of the Sabinal" in Uvalde County, and in 1852, in the "Big Bend of the Devil's River," in Val Verde County.

120. Candelilla (*Euphorbia antisyphilitica*)
(Other common names: wax plant)

Habitat: desert scrub, barren desert flats, rocky cliffs; mostly limestone, low to mid-elevations.

Geographic range: Texas: Trans-Pecos (Hudspeth, Presidio, Brewster, Terrell, Val Verde Counties); southeast to Webb, Starr Counties. US: Texas, southern New Mexico. Mexico: scattered, east Chihuahua to Nuevo León, south to Querétaro, Hidalgo.

Park locations: limestone slopes southeast of Righthand Shutup, Cienega Creek area, scattered elsewhere in Solitario, Contrabando Creek near Contrabando Waterhole

Candelilla is a low succulent perennial or subshrub to 2 feet high, with many tightly clustered, stiffly upright, waxy stems growing from spreading rhizomes. The stems are cylindric, usually leafless and unbranched, gray or gray green from wax exuded from the pores. These odd stems have been described as pencil-like, broomlike, and rush-like, and in 1859, St. Louis botanist George Engelmann likened them to horsetail fern (*Equisetum*) and *Ephedra*.

Mostly in spring and summer, flower clusters (cyathia) appear singly or in small groups near the stem tips. Each cyathium is a collection of fifty to seventy male blooms (stamens) encircling one protruding female bloom (pistil) with three two-branched styles. The flower collection rests in a bell-shaped, whitehaired floral cup with five red glands around the rim, four glands oblong, and one smaller and rounded. Each gland bears a white, seldom pink, triangular or lance-shaped, petal-like appendage, which is strongly notched at the tip. The bright appendage attracts pollinating insects.

Often dangling from a long, curving stalk, the fruit is an oblong green capsule, 1/6 inch long. Each capsule is tri-lobed, with three whitish seeds, each with a yellow, bladderlike appendage (caruncle) at one end.

Candelilla wax is used to make many commercial products, including candles, chewing gum, soap; floor, auto, and shoe wax; phonograph records, carbon paper, and paints (Powell 1998; McDougall and Sperry 1951). Wax factory remains are scattered throughout the park.

Botanical notes: Because of the shape of the stems and their use in candle making, this spurge has long been known by the Spanish name candelilla (little candle). In 1832, Munich botanist Joseph Zuccarini described the species from a specimen collected by Bavarian explorer and mining engineer Baron Wilhelm Friedrich Karwinsky. Karwinsky found the plant at Tolimán in the Mexican state of Jalisco, between 1827 and 1832. Zuccarini named the species for its traditional use in Mexico to treat syphilis.

121. Ashy Sandmat (*Euphorbia cinerascens*)
(Other common names: fuzzyfruit or ashy spurge, ashy broomspurge)

Habitat: desert scrub, rocky, gravelly limestone hills, arroyos, igneous mountain drainages, roadsides.

Geographic range: Texas: Trans-Pecos (excluding El Paso, Culberson, Reeves Counties); west Edwards Plateau, South Texas Plains; Ward County. US: Texas. Mexico: Chihuahua east to Tamaulipas, south to Durango, Zacatecas, San Luis Potosí.

Park locations: Solitario northwest rim, Outer Loop Trail, Tres Papalotes, Horsetrap Spring

Ashy sandmat is a prostrate perennial growing from woody taproots, which forms grayish mats to 2 feet across. The stems are slender, at times branched or forked in the upper reaches, and normally coated with close-pressed white hairs.

The opposite leaves are grayish green, egg-shaped, to 2/5 inch long, with short, hairy petioles and blunt tips, and a thick cloak of short white, matted hairs, especially underneath.

This spurge blooms opportunistically from spring to fall, with flower clusters (cyathia) single at nodes near the stem tips. One female pistil with three deeply split, cherry-red styles protrudes from the center of a cluster, surrounded by ten to twenty male stamens. The flowers are enclosed in a turbinate, calyx-like floral cup, with four fleshy oblong glands around the rim. The dark red glands may bear white, petal-like appendages much narrower than the glands, or the appendages may be absent. The entire flower cluster is whitehaired, except for the hairless stamens.

The fruit is an oval but sharp-angled, trilobed capsule, to 1/12 inch long, with short bristly hairs, and three four-angled white seeds. Fuzzyfruit is an apt term for this sandmat, and helpful in plant identification.

Botanical notes: The Latin species name *cinerascens* (ashen) refers to the silvery leaves. George Engelmann described this species in 1859, and the first specimen was collected by Josiah Gregg in 1847, at Bishop's Hill near Monterrey, Mexico. Gregg, frontier trader on the Santa Fe and Chihuahua Trails, authored *Commerce of the Prairies* (1844–45) to document his travels.

122. David's Spurge (*Euphorbia davidii*)
(Other common names: David's toothed spurge, toothed spurge, David's poinsettia)

Habitat: xeric, mesic communities, fields, rocky mountain terrain, woodlands; near creeks, springs, roads, trails, waste grounds, on varied soils.

Geographic range: Texas: Trans-Pecos; much of Panhandle; small parts of Edwards Plateau and north-central Texas. US: contiguous forty-eight (excluding northwest, deep south). Elsewhere: scattered in Mexico south to Chiapas; Honduras, Argentina, Paraguay, Uruguay.

Park locations: Panther Canyon, upper Arroyo Mexicano, Ojo Chilicote, Contrabando Creek

This spurge is an annual to 2 feet high, with a single pale green to maroon, erect main stem growing from taproots. Typically, the main stem forms pairs of leafy ascending branches, each pair at right angles to the pair above. Distinctively, the stems are blanketed with a mix of short and more scattered long white hairs.

Usually paired, the leaves are up to 2 inches long or more, broadly elliptic, with a blunt or pointed tip, wedge-shaped base, and bluntly toothed margins. On stout petioles, the green blades are sparsely hairy above, thickly whitehaired below.

Mostly in summer and fall, tight clusters of bell-shaped floral cups (cyathia), with bright white, threadlike fringes, appear at the stem tips, above long leaflike bracts. The bracts may be marked with dark spots or streaks, and chalky-white hues at the base. Each light green floral cup contains up to twenty-five stamens; a stalked, gradually elongating pistil, with three white styles divided near the base; and on the rim, a cupped, greenish-yellow gland without petal-like appendages.

Dangling from the floral cup is the depressed-spheric green fruit, a trilobed capsule to 1/5 inch across, with three tuberculate dark brown seeds.

Botanical notes: In 1984, in the journal *Kurtziana*, Argentine botanist and spurge specialist Rosa Subils named this species for range management ecologist David Lee Anderson. Anderson, coordinator of Argentine range research, collected the specimen. Two varieties of *E. dentata*, var. *gracillima*, described by Charles Millspaugh in 1890 and collected in Arizona by Marcus Jones in 1884, and var. *lancifolia*, described by Oliver Farwell in 1923, are now considered synonymous with *E. davidii*.

123. Squareseed Spurge (*Euphorbia exstipulata*)
(Other common names: Clark Mountain spurge)

Habitat: rocky slopes, sandy draws, clayey flats, grasslands; limestone, also igneous substrates.

Geographic range: Texas: Trans-Pecos (excluding Val Verde County); Crockett, Deaf Smith Counties. US: Texas to southern California; Oklahoma, Utah, Wyoming. Mexico: Sonora, scattered sites in Chihuahua, Coahuila, Durango.

Park locations: Chorro Canyon, Road to Nowhere, Contrabando Creek, Contrabandista Spur Trail

Growing from slender taproots, this spurge is an erect annual to 8 inches high. The main stem forks into flower-bearing branches, each with one flower cluster. The stems may be strigose, with stiff, straight, close-pressed hairs, or scurfy, with enlarged bases of dead leaves at the nodes.

Two varieties are recognized, the widespread var. *exstipulata*, and var. *lata* from Brewster and Presidio Counties and Chihuahua, Mexico. Both varieties have opposite, short-stemmed leaves with cartilaginous, often toothed margins, mostly near the tips. The leaves of var. *exstipulata* are linear to lance-shaped, to 2 3/10 inches long, tapered to the base, with teeth pointing forward. The leaves of var. *lata* are oval to lance-shaped, to 1 3/10 inches long, abruptly narrowed to the base, with teeth outward-pointing.

From late spring to fall, flower clusters (cyathia) form on the forked branches. Each cluster contains up to fourteen stamens and a protruding pistil with three two-branched white styles. The blooms sit in a bell-shaped floral cup with four oblong glands, yellow green to maroon, on the rim. A white or pink appendage, often notched at the tip, attaches to each gland. The clusters may be hairy, or scurfy, with grain-like scales.

The green to maroon fruit is a trilobed ovoid capsule to 1/7 inch long, often strigose or scurfy. Grayish white to brown, the tuberculate seeds are ovoid but four-angled, with two transverse ridges.

Botanical notes: In 1859, George Engelmann named this species *exstipulata*, from Latin words meaning "without stipules," apparently alluding to the tiny glandlike stipules. German immigrant and naturalist Augustus Fendler, trained by Engelmann and sent by Asa Gray to collect in New Mexico, found the first specimen in 1847, on "gravelly banks of the Rio del Norte [Rio Grande] near Santa Fe."

124. Muleear Spurge (*Euphorbia indivisa*)
(Other common names: royal sandmat, royal or pine spurge, pine broomspurge)

Habitat: rocky, alluvial flats, arroyos, disturbed sites, grasslands, mountain slopes and drainages, mid- to high elevations, volcanic soils.

Geographic range: Texas: Trans-Pecos (Jeff Davis, Presidio, Brewster Counties). US: Texas to Arizona. Mexico: Sonora east to Coahuila, south to Durango, San Luis Potosí, into central Mexico.

Park locations: Tascate Hills

This spurge is prostrate, with an annual taproot and leafy, often maroon stems. Long curling hairs are borne on the upper stems. The stem nodes are closely spaced above, causing crowded leaves and flower clusters near the stem tips and on short branchlets. Mostly hairless, the dark- or yellow-green, short-petioled leaves, to 1/2 inch long, are egg-shaped, pointed or blunt at the tip, asymmetrical at the base, with thick, toothed, often red margins.

In summer, flower clusters (cyathia) form singly at the upper stem nodes. Each cluster has one pistil with three styles; five to fifteen stamens around the pistil; and a bell-shaped floral cup enveloping the flowers, with four cupped, dull yellow glands on the rim, each with a petal-like appendage. Oddly, two glands are small and two are much larger. The petal-like appendages differ in size and shape, with two small symmetrical appendages on the smaller glands, and two large asymmetrical appendages on the larger glands. The petal-like structures are white, pink, or red.

The minute shorthaired fruit is an ovoid, trilobed capsule. The capsule contains three oblong, four-sided, buff seeds, with wrinkled surfaces and transverse ridges.

Botanical notes: In 1859, George Engelmann described this species as a variety, *Euphorbia dioica* var. *indivisa*, noting that the plant was distinguished by the "undivided styles." Charles Wright collected the lectotype in 1851, on "stony hills near the Copper Mines," in Grant County, New Mexico. In 1935, Ivar Tidestrom, formerly range plants specialist with the USDA, published new plant names as part of his *Keys to the Flora of Arizona*. In the list, Tidestrom gave this variety species status as *E. indivisa*.

125. Perennial Sandmat (*Euphorbia perennans*)
(Other common names: perennial spurge, perennial broomspurge, Terlingua spurge)

Habitat: arid desert flats, low gravelly, rocky hills, sandy, silty, clayey hills and mounds with sparse vegetation; calcareous, gypseous soils.

Geographic range: Texas: Big Bend (southeast Presidio, south Brewster Counties). US: Texas. Mexico: northern Chihuahua.

Park locations: West Contrabando Trail, Buena Suerte Trail

Of conservation concern: see Appendix A

Once regarded as a rare Big Bend endemic, perennial sandmat is now known to range into adjacent Chihuahua and can be locally common in the southern Big Bend. The hairless plant is a stout perennial to 18 inches tall, which grows from woody roots or sturdy horizontal rhizomes. Tan to grayish black, older stems are stiff, erect, with forking branches in the upper reaches. Younger branches are often maroon-red, fleshy, and flexible.

The opposite, short-stemmed, glaucous leaves are blue gray to gray green. To 2/3 inch long and nearly as broad, the blades are variable in shape—ovate, elliptic, triangular, nearly round, and rounded or blunt at the tip and base.

In summer, flower clusters (cyathia) appear at the stem nodes on erect stalks. A cluster consists of one protruding pistil with three two-branched styles, surrounded by up to forty-five stamens. The flowers rest in an urn-like floral cup with four fleshy, nectar-bearing glands on the rim. Erect, cuplike, the oblong glands lack petal-like appendages. Given the intricate beauty of the flower clusters, the pseudo-petals are scarcely missed.

The fruits are trilobed, three-seeded, egg-shaped capsules, 1/8 inch long, with ovoid, three- to four-angled, buff seeds.

Botanical notes: In 1956, this species was described by Lloyd Shinners, herbarium director at Southern Methodist University, as *Chamaesyce perennans*. The specimen was collected by Barton Warnock in 1937 near Kit Mountain and Chisos Pens in BBNP. The term *perennans* means "perennial." In 1960, in *Southwestern Naturalist*, Barton Warnock and Marshall Johnston transferred the plant to the genus *Euphorbia*. With Donovan Correll, Johnston coauthored the *Manual of the Vascular Plants of Texas* in 1970.

126. Threadstem Sandmat (*Euphorbia revoluta*)
(Other common names: threadstem spurge or broomspurge, rolled leaf or linearleaf spurge)

Habitat: rocky slopes, grassland, sandy washes; limestone, igneous soils.

Geographic range: Texas: Trans-Pecos (excluding Reeves, Pecos, Terrell Counties). US: southwest. Mexico: northeast Chihuahua, western Coahuila.

Park locations: La Iglesia, Outer Loop Trail beyond Righthand Shutup, slopes west of Paso al Solitario, Cienega Mountain, Cienega Creek

Although this sandmat is a wiry, hairless annual to 9 inches long, it has a short, erect main stem growing from slender taproots. The stem divides into delicate maroon, repeatedly forking, branches, which are thinner at each ascending node and threadlike in the upper reaches. The opposite, short-petioled leaves are linear, to 1 inch long, with acute or abruptly pointed tips, obvious midribs, and often strongly revolute margins.

In summer and fall, flower clusters (cyathia) form singly at the upper stem nodes, seemingly in tight clusters, because the nodes are so closely spaced. A cyathium consists of up to eight stamens around a central pistil. On an elongating stalk, the pistil supports three thick white styles. All flowers rest in a bell-shaped floral cup, with four disklike, slightly cupped glands on the rim. Attached to each frequently reddish gland is a minute, usually white, petal-like appendage.

The fruit is a trilobed, three-angled capsule, egg- or globe-shaped, 1/17 inch long. The capsule contains three brown egg-shaped, three-angled seeds, with two or three transverse ridges.

Botanical notes: The Latin epithet *revoluta* (revolute) refers to the often rolled under leaves. This species was described in 1859 by St. Louis botanist George Engelmann. Engelmann cited collections by John Bigelow (near Rock Creek, in Jeff Davis County, Texas, 1852), Charles Wright (at the Santa Rita del Cobre copper mines, Grant County, New Mexico, 1851), and August Fendler (in New Mexico, "between Santa Fe and the Moro River in the mountains," 1847). German immigrant Fendler collected around Santa Fe in 1846 and 1847, and later botanized in Mexico, Panama, Arkansas, and Venezuela.

127. Yuma Spurge (*Euphorbia setiloba*)
(Other common names: Yuma or bristle-lobed sandmat, Yuma broomspurge, fringed or Salton Sink or starfish spurge)

Habitat: sandy, gravelly flats, dry creek beds, roadsides, disturbed sites, desertic mountains; igneous, limestone substrates.

Geographic range: Texas: Trans-Pecos (El Paso, Presidio, Brewster Counties). US: southwest (excluding Colorado). Mexico: Baja California, Sonora, Sinaloa; scattered in Chihuahua, Coahuila.

Park locations: Vista de Bofecillos, west of Rancho Viejo, Agua Adentro, Bofecillos Canyon, Rancherías Spring, Yedra Canyon, Cienega Mountain

With bright white, conspicuously fringed, petal-like appendages and zigzag stems, this ephemeral annual is one of the more unusual and attractive prostrate spurges, a pleasure to encounter in the field. To 8 inches long, often reddish, the stems spread from a central point and branch repeatedly. Growing from a taproot, the main stems and branches are villous, coated with long soft hairs that can also be glandular and clammy to the touch.

The paired, light green or yellow-green leaves are variable, broadly oval to nearly round, elliptic or oblong, seldom more than 1/4 inch long. Short-stemmed, the blades are rounded at the tip, markedly asymmetrical at the base, and villous like the stems.

This short-lived herb blooms opportunistically from spring to fall, often after monsoon rains. The flower clusters (cyathia) form singly on short stalks at the upper stem nodes. Each cyathium consists of three to seven stamens, surrounding a protruding pistil with three deeply divided styles. The cyathium includes a hairy, urn-like floral cup encircling the blooms, with four red or brownish yellow, oblong glands on the rim. Each gland bears a white or pink petal-like appendage, which is much larger than the gland and conspicuously fringed or lobed.

The fruit is a trilobed, three-angled capsule, ovoid or spheric, no more than 1/20 inch long, with long soft hairs. Dull yellow brown with a whitish coat, the three seeds are oblong but three- or four-sided.

Botanical notes: The specific epithet *setiloba* denotes "bristle-lobed," a reference to the white, petal-like gland appendages. This species was described by George Engelmann in the *Pacific Railroad Surveys* of 1857, and was collected by Major George Henry Thomas near Fort Yuma, in Imperial County, California.

128. Slimseed Spurge (*Euphorbia stictospora*)
(Other common names: slimseed sandmat, slimseed broomspurge, slim-seed mat-spurge, slim-seeded euphorbia, fuzzy foliage spurge)

Habitat: sandy flats, gravelly washes, disturbed sites, creosote scrub, grasslands, mountains; limestone, igneous, gypseous soils.

Geographic range: Texas: Trans-Pecos; sand counties, Edwards Plateau; scattered in Panhandle, central, north Texas. US: Texas to Arizona, Texas north to Wyoming, North Dakota; and Missouri, Iowa. Mexico: Chihuahua east to Nuevo León, south to Durango, Zacatecas.

Park locations: Madera Canyon River Access, Llano Loop, Los Alamos, Buena Suerte Trail, Contrabando Waterhole Trail, Cienega Creek

Barton Warnock suggested that this herb might be named "dirty spurge," because the annual is so typically blanketed with dirt and sand particles (Warnock 1977). A dense layer of villous hairs on most plant parts attracts the dirt. This spurge develops prostrate stems to 10 inches long, with many short, lateral branchlets with overlapping leaves and congested flower clusters. The short-stemmed leaves are rounded to oblong, with a blunt tip, lopsided base, and finely serrate edges, or merely a minutely toothed tip.

In summer and fall, flower clusters (cyathia) form singly at the stem nodes and on short branchlets with progressively more close-spaced nodes. Each cyathium contains three to seven stamens, surrounding an elongated and recurved pistil with three red-tipped styles. All flowers are enclosed in an inversely conic floral cup, with stiff, bristly hairs outside, and four oblong or round, usually red glands on the rim. Each gland bears a white or pink petal-like appendage commonly larger than the gland.

The fruit is an ovoid, three-angled capsule, 1/17 inch long, often thickly coated with hairs. The three tan seeds are narrowly oblong, four-angled, sometimes pitted or mottled.

Botanical notes: The specific epithet *stictospora*, from the Greek *stiktos* (spotted) and *spora* (seed), refers to the dotted or pitted seeds. George Engelmann described this species in 1859, and the type specimen was collected by Augustus Fendler in 1847 on the Pawnee River in southwest Kansas.

129. Hairy Spurge (*Euphorbia villifera*)
(Other common names: hairy sandmat, villous spurge, villous rock-spurge, hairy euphorbia)

Habitat: wooded, rocky slopes, mountain canyons, grasslands; calcareous, igneous soils, mid- to high elevations.

Geographic range: Texas: Trans-Pecos (Jeff Davis, Presidio, Brewster, Pecos, Val Verde Counties), Edwards Plateau. US: Texas. Mexico: Chihuahua east to Tamaulipas, south to Oaxaca, Yucatán; Guatemala.

Park locations: Sauceda Road to Solitario turnoff, Cienega Gorge, upper Cienega Trail

Hairy spurge is usually perennial, to 10 inches high, with upright stems growing from persistent taproots. The widely branching, forking stems bear long, spreading hairs.

Short-petioled, the paired, dark or dull green leaves are as much as 1/2 inch long or more, broadly egg-shaped, oblong or triangular, blunt or rounded at the tip, lopsided at the base. The blades are adorned with jagged teeth and erect or spreading, thin white hairs.

From spring to fall, flower clusters (cyathia) form singly at the nodes. Each cyathium is a collection of numerous stamens surrounding a protruding pistil with three white or pink, deeply divided styles. The blooms reside in a bell-shaped floral cup with four fleshy, greenish- or brownish-yellow glands on the rim. Each cuplike gland, oblong or rounded, bears a colorful petal-like attachment that attracts potential pollinators. White, pink, or red, the appendages are relatively large and showy.

The fruit is a trilobed, three-angled, hairless capsule, broader than long, 1/9 inch wide. Orangish but white-coated, the small seeds are narrowly egg-shaped, weakly angled, and blunt-tipped.

Botanical notes: The Latin epithet *villifera* (hair-bearing) refers to the longhaired foliage. In 1849, George Heinrich Adolf Scheele described this species in the journal *Linnaea*, from a specimen collected by Ferdinand Lindheimer at New Braunfels, Texas, in 1846. From 1848 to 1852, Scheele, pastor and botanist in Heersum, Germany, published seven articles in *Linnaea* on plants collected by Lindheimer.

130. Leatherstem (*Jatropha dioica* var. *graminea*)
(Other common names: rubber plant, bloodroot, sangre de drago (dragon's blood), limber bush, leatherwood, sangregrado, sangre de grado, tlapalezpatli [Náhuatl])

Habitat: gravelly washes, alluvial flats, rocky cliffs, desert scrub; limestone, igneous substrates.

Geographic range: Texas: Trans-Pecos (Presidio, Brewster, also north Culberson, west Jeff Davis Counties). US: Texas. Mexico: east Chihuahua, west Coahuila, north Zacatecas.

Park locations: Closed Canyon, Cerro de los Burros Loop, Black Hills, Rancherías Spring, Ojo Chilicote, West Contrabando Trail, Cienega Mountains

Leatherstem is a colony-forming subshrub to 2 feet high, with spreading orange roots. The ruddy-brown stems are thick, leathery, yet fleshy and limber. The mostly smooth stems are erect, seldom branched, but with short, stubby spurs. The yellow stem sap turns blood red when exposed to air.

The leaves are tightly bunched on the short spurs but quickly fall in very dry weather. Largely stemless, the linear leaves average less than 1 1/2 inches long. Some leaves may be weakly lobed or two- to three-parted.

In spring and summer, small flower clusters appear near the stubby branchlet tips. Supported by five red sepals, the white blooms are tubular to urn-shaped, often pink-tinged, with five flaring, recurving lobes and a hairy red throat. Male blooms have ten yellow stamens; female blooms have a pistil with one or two divided styles. The fruit is a leathery, beaked oval capsule 3/5 inch long, twelve-lobed, each lobe with one seed. The large globose seeds are a favorite of white-wing doves.

Native Americans used the stem juice to toughen gums, stop blood flow from wounds (Warnock 1974), and remove teeth stains. Roots were chewed to cure toothaches and stems used as whips (Vines 1960).

Botanical notes: The genus name joins the Greek *iatros* (physician) and *tropheia* (mother's milk), alluding to the milky latex. The Latin species name *dioica* (two houses) implies dioecious, since male and female flowers occur on separate plants. The variety name *graminea* (grasslike) refers to the leaves. This species was described in 1794 by Vicente de Cervantes, the first botany professor in New Spain, at the Royal Botanic Garden in Mexico City. The variety was described in 1944 by Mexican flora expert Rogers McVaugh, from a collection by desert plant ecologist Forrest Shreve at Jimulco, Coahuila, in 1939.

131. Terlingua Milkvetch (*Astragalus terlinguensis*)
(Other common names: Emory's milkvetch, Emory loco, peavine, red-stem peavine)

Habitat: low desert, creeks entering Rio Grande, gravelly arroyos, alluvial flats, gypseous clay hills, silty bluffs; calcareous, igneous soils.

Geographic range: Texas: Trans-Pecos (south Presidio, south Brewster Counties, also southwest Culberson County). US: Texas. Mexico: adjacent Coahuila.

Park locations: Lefthand Shut-up, Los Alamos, Fresno Canyon

This milkvetch is a prostrate annual with slender, radiating stems forming mats 1 foot or more across. Commonly, stems and leaves are silvery, blanketed with a mix of hairs: straight and close-pressed, stiff and rough, long and soft.

The compound, short-stemmed leaves are up to 2 inches long or more, odd-pinnate, with as many as nine pairs of small, crowded leaflets and a leaflet at the tip. The leaflets are variable in shape, mostly elliptic to inversely lance- or egg-shaped, blunt at the tip, frequently with margins rolled upward or nearly folded.

In spring, small clusters of pale lavender flowers form in the leaf axils. To 1/2 inch long, each bloom is butterfly-shaped with a broad, reflexed banner petal, two clawed wing petals, and two fused keel petals. The banner sports a fanlike central eye, with white streaks or folds separated by thin, pale lavender lines. Typically, the blooms have ten stamens, nine stamens with united filaments, and one separate.

The rather short, plump, oblong fruits are stalkless pods to 1/2 inch long or more. The pods are straight or slightly incurved, compared to the similar Emory's milkvetch (*Astragalus emoryanus*), with longer, thinner, more incurved pods.

Botanical notes: Victor Cory collected the type specimen in 1936 and described the species in 1937, noting that the plant was found "in the narrow floodplain of Terlingua Creek . . . about eighteen miles on an airline north of Terlingua, Texas. Apparently, it occurs over the watershed of Terlingua Creek, hence its specific name." A prolific botanical collector, Cory worked at several Agricultural Experiment Stations in Texas, helped develop the Texas A&M and later the SMU herbarium, and in 1937 coauthored with H. B. Parks the first *Catalogue of the Flora of the State of Texas*.

132. Wooton's Loco (*Astragalus wootonii*)
(Other common names: halfmoon or Wooton's milkvetch, garbancillo, tronador, rattleweed, Wooton's or western or bladderpod locoweed)

Habitat: desert scrub, grasslands, alluvial flats, arroyos, road shoulders, disturbed sites, sandy, silty, or clayey soils.

Geographic range: Texas: Trans-Pecos (excluding Pecos, Terrell Counties). US: Texas to southeast California, south Colorado. Mexico: Sonora to west Coahuila, south to Hidalgo, Michoacán, Puebla.

Park locations: Arroyo Mexicano, Sauceda, Llano Loop, Los Alamos

Like many *Astragalus*, this legume is known as "locoweed" because it produces swainsonine, an alkaloid that causes "loco disease" in livestock. Wooton's loco is a spreading, usually prostrate annual or short-lived perennial, much branched at the base. The leafy stems are densely hairy when young, more thinly shorthaired with age. The leaves are up to 4 inches long, short-stemmed to stemless, with up to eleven pairs of leaflets and a terminal leaflet. Each leaflet is oblong, mostly blunt-tipped, sharply folded upward on the edges, and minutely hairy.

In spring, tiny butterfly-shaped flowers form clusters, with as many as ten blooms, on stalks from the leaf axils.

The petite blooms are creamy-white, delicately tinged with pink, lavender, or rose. Supporting each bloom is a five-lobed, straw-colored calyx, with slender, tapered lobes and bristly hairs.

The most striking feature of this milkvetch is the large bladder-like, inflated pods, to 1 1/2 inches long. Greenish yellow to reddish brown, the stemless, abruptly beaked pods become papery with age.

Botanical notes: The Greek word *Astragalus* (ankle bone) may have been used by Pliny to describe a legume with ankle-bone-shaped roots. Another theory suggests that *Astragalus* derives from Greek words for star and milk, because of the flower shape and the belief that the plant increased the milk production of goats. In 1894, Edmund Sheldon, *Astragalus* specialist at the University of Minnesota, named this species for New Mexico botanist Elmer Wooton, who collected the plant near Las Cruces, New Mexico, in 1892. Wooton coauthored *Flora of New Mexico* (1915) with his student, Paul Standley.

133. Isely's Feather Duster (*Calliandra iselyi*)
(Other common names: Isely's stickpea, falsemesquite)

Habitat: desert scrub, hot dry, open terrain; on gravel, clay, gyp, caliche, limestone.

Geographic range: Texas: Big Bend (Brewster, Presidio Counties). US: Texas. Mexico: no record.

Park locations: Solitario northwest rim, east leg of Los Alamos Loop, Terlingua Uplift, Buena Suerte Trail

This fairy duster is a low, densely branched, dwarf shrub to 10 inches high, with slender yet woody and stiff, grayish stems and short, gnarly branchlets. The spreading, silky-haired branchlets are sharply diverging, at times zigzagged, often nearly leafless and bedraggled from heavy browsing.

Short-petioled, the compound leaves are no more than 1/2 inch long, with one pair of larger leaflets (pinnae) and up to nine pairs of small oblong leaflets per pinna. The leaflets are thick, overlapping, rounded or blunt at the tip, ciliate on the margins, and sparsely silky-haired below.

In spring and summer, globelike flower heads, with up to nine small blooms, protrude on long silky-haired stalks from the leaf axils. Each bloom consists of five greenish yellow to reddish brown petals, many protruding stamens with white filaments and yellow anthers, and a long slender style with a headlike stigma. A five-lobed, hairy, cuplike calyx supports each bloom. The fruit is a linear, flattened pod to 1 3/10 inches long, with thick margins and papery valves that curl sharply backward after splitting.

Botanical notes: The genus name *Calliandra*, from Greek words *kallos* (beautiful) and *andra* (stamen), refers to the colorful protruding stamens. Billie Lee Turner described this species in 2000, naming it for legume expert Duane Isely. Turner separated this taxon from *Calliandra conferta*, which is now believed to occur farther east in Texas. The type specimen was collected by Billie Lee and Gayle Turner in 2000 in Brewster County, 2 miles west of Terlingua on FM 170. Barton Warnock collected specimens in 1936 and 1937, west of Alpine in Paradise Canyon, and in the Chisos Mountains in BBNP.

134. Velvet Bundleflower (*Desmanthus velutinus*)
(Other common names: sticky bundleflower, hairy desmanthus)

Habitat: grasslands, hillsides, woodlands, creekbanks, roadsides; calcareous, igneous-derived soils.

Geographic range: Texas: Trans-Pecos (excluding El Paso County), Edwards Plateau, parts of South Texas Plains, northcentral Texas; scattered elsewhere. US: Texas, New Mexico (Eddy, Lincoln Counties). Mexico: Coahuila.

Park locations: Terlingua Uplift, 1/2 mile south of Tres Papalotes

Velvet bundleflower is a perennial to 2 feet high, growing from taproots and spreading at the base, with many sprawling, or at times, ascending branches. Lined with fernlike leaves, the stems may be velvet smooth, coated with soft, shaggy hairs.

The bipinnately compound leaves may reach 3 inches long, on petioles to 1/2 inch long. Each leaf consists of up to seven pairs of pinnae, with as many as twenty-two pairs of leaflets per pinna. Commonly, the leaf axis is ridged and lined with soft white hairs. Each leaflet is narrowly oblong, ciliate on the edges, and often, densely velvety below.

In spring, up to twenty tiny flowers appear in globose heads, on stalks from the leaf axils. Each fertile bloom consists of five greenish-yellow sepals, five pale green, oblong petals, ten protruding stamens with white filaments and yellow anthers, and a projecting style. Often, sepals and petals are red-tinged. Blooms with sterile stamens, with long white filaments lacking anthers, form a skirt below the fertile blooms.

Numerous linear pods, to 3 inches long, grow at the tip of long stalks. Leathery, slightly compressed, the pods are green or red at first, dark brown at maturity, with numerous reddish-brown seeds.

Botanical notes: German taxonomist Carl Ludwig von Willdenow, director of the Berlin Botanical Garden, described the genus in 1806. *Desmanthus*, from Greek words, *desmos* (bundle) and *anthos* (flower), highlights the compact flower clusters. The species name, *velutinus* (velvety), alludes to the villous hairs on stems and leaves. George Heinrich Adolf Scheele, pastor-botanist in Heersum, Germany, described the species in 1848, from a collection by Ferdinand Lindheimer in 1846, near New Braunfels.

135. Siratro (*Macroptilium atropurpureum*)
(Other common names: purple bushbean, purple bean, conchito, atro, frijolillo, jícama silvestre)

Habitat: creek sides, sandy, clayey flats, brushy or gravelly arroyos, disturbed sites.

Geographic range: Texas: Big Bend (Presidio County); Cameron, Kenedy, Kleberg Counties in south Texas. US: Texas, Arizona, Florida, Hawaii, Puerto Rico. Elsewhere: Mexico, Central America, much of Caribbean, South America.

Park locations: Bofecillos Canyon, Yedra Canyon, Leyva Creek, Cienega Mountain

Siratro is a fast-growing perennial vine, with long twining and climbing stems, which grows from deep taproots and may root at the stem nodes. Often, the stems are ridged or grooved and coated with long silky hairs.

On petioles to 2 inches long, the leaves are trifoliate, in three distinct leaflets, each to 3 inches long, lancelike, dark green above, gray green below. The lateral leaflets may have one or two basal lobes, the middle leaflet a basal lobe on one side. The leaflets are sparsely hairy above, densely silky-haired below.

In summer, flower clusters with up to fifteen flowers form in the leaf axils, on stalks to 1 foot long or more. The flowers are butterfly-shaped, with an upper banner petal, two wing petals, and two fused keel petals. The blooms are bizarre: dark reddish purple to nearly black, with wing petals larger than the banner petal, and twisted, pinkish keel petals. Each bloom is supported by a tubular, white-hairy calyx, with five triangular lobes.

The dehiscent fruits are cylindric pods to 4 inches long, with a beak-like tip and up to a dozen seeds. The ripe, hairy pods shatter, propelling the oblong seeds.

Botanical notes: The genus name, from the Greek *macro* (large) and *ptilium* (wing), refers to the wing petals. The epithet *atropurpureum*, from the Latin *atro* (dark) and *purpureum* (purple), refers to the flowers. This species was described in 1825 as *Phaseolus atropurpureus* by Augustin Pyramus de Candolle, based on a watercolor drawing from the 1788–1803 Sessé and Mociño expedition to New Spain. The expedition reports were published years later, in *Plantae Novae Hispaniae* (1887–91) and *Flora Mexicana* (1891–97). German botanist Ignatz Urban, curator of the Berlin Botanical Garden, moved this legume to the genus *Macroptilium* in 1928.

136. Texas Mimosa (*Mimosa texana*)
(Other common names: Texas catclaw, Wherry's mimosa, uña de gato, mimosa de Texas)

Habitat: desert scrub, banks of creeks, springs, canyons; limestone, igneous substrates.

Geographic range: Texas: Trans-Pecos (excluding El Paso, Culberson Counties); parts of Edwards Plateau, South Texas Plains. US: Texas. Mexico: Coahuila to Tamaulipas; Durango, San Luis Potosí.

Park locations: lower Fresno Creek, Rancherías Canyon, Las Cuevas, Los Ojitos, Cienega Creek

Texas mimosa is a small or midsized shrub to 6 feet high, densely branched, with slender branches and twiggy branchlets. Often whitish, the stems are armed with small single prickles, just below the leaf petioles. The dark red prickles are typically flattened, broad at the base and recurved at the tip. To 1 1/2 inches long, the alternate, bipinnately compound leaves are divided into up to four pairs of pinnae, with up to seven pairs of leaflets per pinna. Each leaflet is oblong, rounded or blunt at the tip, and ciliate on the margins. Petioles and leaf axes are sometimes hairy.

In spring and summer, tiny, extraordinarily fragrant, short-stalked flower clusters appear in spheric heads 3/5 inch across. The flowers, seemingly white because of the prominent stamens, are distinctly multicolored. Each bloom has five greenish-yellow, at times red- or pink-tipped petals united at least half their length; ten stamens with white filaments and pale-yellow anthers; and five short, red tinged calyx lobes at the base. Distinctively, the blooms are hairless but at times minutely fringed at the tips.

The hairless fruit is a straight or slightly curved, somewhat flattened, pod to 1 1/2 inches long, with sharp, recurved prickles. Dark- to reddish-brown or brick-red, the oblong pods split on the edges at maturity.

Botanical notes: In 1753, Linnaeus named the genus *Mimosa*, presumably from the Greek *mimos* (mimic), due to response of the leaves to touch (rapid folding) in some species. The type specimen is a 1674 illustration of a Brazilian plant by Danzig merchant-botanist Jacob Breyne, in his *Exoticarum* . . . (First Century of Exotic Plants). Texas mimosa was described by Asa Gray in 1852 as *Mimosa borealis* var. *texana*, from an 1849 collection by Charles Wright, at the "edge of woods on pebbly hills, Austin." In 1901, John Small, curator of the New York Botanical Garden, gave the variety species status.

137. Turner's Mimosa (*Mimosa turneri*)
(Other common names: desert mimosa)

Habitat: near Rio Grande, desert scrub, creek banks, rocky mesas; limestone conglomerate, rubble, clay, igneous soils.

Geographic range: Texas: Trans-Pecos (south Hudspeth, west Jeff Davis, southeast Presidio, south Brewster, Pecos Counties). US: Texas, southern New Mexico. Mexico: scattered in Coahuila, Nuevo León; expected in Chihuahua.

Park locations: Blackrock Campground, Closed Canyon, Upper Guale Mesa, Cienega Creek

To 3 feet high, Turner's mimosa is a straggly shrub with stiff, woody, diverging stems and short branches. Often nearly leafless, the dark stems are covered with a white, shredding epidermis and armed with straight or down-curved prickles.

The leaves form in small bundles on short, scaly, knob-like spur branchlets. The blades are bipinnately compound, usually with one pair of pinnae, which are divided into two, rarely three, pairs of leaflets. The petioles are notably flattened. To 1/6 inch long, each hairless leaflet is egg- or lance-shaped to elliptic.

In late spring and early summer, tiny blooms appear in showy pink globes 1/2 inch across, on stalks from the leaf axils. Each bloom consists of a five-lobed, bell-shaped membranous calyx; five partly united oval petals; and ten protruding stamens with pink filaments and yellow anthers.

Tan to dark red, the fruits are hairless oblong pods to 2 3/10 inches long, curving and twisting, usually with straight prickles. Often constricted between the seeds, the mature pods split into papery sacs, which spill up to nine plump ovoid seeds.

Botanical notes: Legume specialist Rupert Barneby, curator of the New York Botanical Garden's Institute of Systematic Botany, described this species in 1986. The holotype was collected by Barneby in 1985, in southeast Presidio County, on the Rio Grande west of Lajitas, in "washes and gullies under cliffs on limestone conglomerate." Barneby named the species for University of Texas botanist Billie Lee Turner, another legume authority, and author of *The Legumes of Texas* (1958). This mimosa was collected by T. L. Steiger in Alpine, Texas, in 1932, but at that time misidentified as *M. zygophylla*.

138. Retama (*Parkinsonia aculeata*)
(Other common names: Jerusalem thorn, horse bean, Mexican palo verde, jelly bean tree, lluvia de oro, chote, guacóporo, bagote)

Habitat: floodplains; road shoulders, flats, canyons; sandy, clayey, alkaline soils.

Geographic range: Texas: Trans-Pecos (El Paso, Presidio, Brewster, Reeves, Pecos, Val Verde Counties); southern half of state. US: southwest (excluding Colorado); Texas to South Carolina. Elsewhere: Mexico to South America; West Indies.

Park locations: River Road from Lajitas to Contrabando Canyon, Madera Canyon Campground, La Cuesta, Hoodoos

Retama is a large shrub or small tree, fast-growing, with a yellow green trunk or stems growing from deep, spreading taproots. The slender branches are long, spreading, and flexuous, with paired spines. The unusual leaves are described as feathery, plumelike, straplike, and ribbonlike, and each adjective is apt. To 16 inches long, the alternate, hairless leaves are twice divided, with one pinnae pair and up to thirty-plus pairs of tiny oblong leaflets per pinna. The leaflets are widely spaced, early falling, but the persistent spine-tipped midribs replace the leaflets in photosynthesis.

From spring to fall, unbranched clusters of fragrant yellow flowers form at the leaf axils. Each bloom has four rounded petals with wrinkled margins and a banner petal, which develops an orange blotch or turns red with age. Within the petals are ten stamens with yellow filaments and brown anthers and a slender style. The hairless fruit is a linear brown pod to 4 inches long, constricted between the seeds, with up to six oblong seeds.

In tropical America, a leaf infusion is applied to treat diabetes and epilepsy (Standley 1920–26). The wood has been used in papermaking, and Native Americans made bread from the seeds (Vines 1960).

Botanical notes: The genus is named for English herbalist John Parkinson, apothecary to King James I and author of the famous herbal, *Theatrum Botanicum* (1640). The Latin species name *aculeata* (thorny) refers to the thorns. Linnaeus described the species in 1753, citing tropical America as habitat. The lectotype is an illustration in *Hortus Cliffortianus* (1737), a catalog of plants in George Clifford's Hartekamp gardens, written by Linnaeus. Clifford was governor of the Dutch East India Company.

139. Western Honey Mesquite (*Prosopis glandulosa* var. *torreyana*)
(Other common names: western or honey or Torrey's mesquite, algaroba, mezquite, chachaca)

Habitat: Rio Grande, creeks, springs, desert and alluvial flats, grasslands, rocky slopes, woodlands, varied soils.

Geographic range: Texas: Trans-Pecos (El Paso, Hudspeth, Culberson, Jeff Davis, Presidio, Brewster Counties); Loving County. US: Texas to California, Utah, Nevada. Mexico: Baja California to Coahuila; Sinaloa, Zacatecas, San Luis Potosí.

Park locations: Fresno Canyon, Agua Adentro, Llano Loop, Solitario northwest rim, Lower Shutup, South Fork of Alamo de Cesario Creek

This mesquite is a shrub or tree to 30 feet high, with deep taproots or shallow, spreading roots. Often, the tree forms short trunks with dark fissured bark, twisting branches, and hairy branchlets. Needlelike spines, single or paired, and gnarled spurs, are borne on the branches. The deciduous, alternate leaves are compound, usually with one pair of pinnae and up to twenty leaflet pairs per pinna. The pinna, to 5 inches long, is mostly hairless, with relatively close-spaced, oblong leaflets to 1/2 inch long.

In spring and summer, cylindrical clusters to 5 inches long, with many crowded blooms, emerge from the leaf axils. The fragrant blooms have five greenish-yellow, oblong petals; ten projecting yellow-white stamens; a white style with a short stigma; and a five-lobed calyx. The fruits are slender, indehiscent pods to 9 inches long, straight or curved, with a beak-like tip. Green, yellow, or brown, the leathery pods are slightly flattened, and at times constricted between the numerous brown seeds.

The wood is used to make firewood, charcoal, fence posts, and furniture (Benson 1941; Vines 1960). Native Americans pulverized the beans into meal (pinole), which was made into bread or an alcoholic drink (Vines 1960). The sap is made into gum, candy, and glue (Vines 1960; Powell 1998).

Botanical notes: *Prosopis* is a Greek name for burdock (*Arctium lappa*), a prickly aster, but why this term was applied to mesquite is obscure. The species name, *glandulosa* (glandular), refers to the leaflets, and the variety is named for John Torrey. Torrey described the species in 1827, from an 1820 collection by Edwin James on Long's Expedition to the Rocky Mountains. Variety *torreyana* was described by Lyman Benson in 1941, as *P. juliflora* var. *torreyana*, from his 1941 collection at Needles, California. In 1962, Marshall Johnston treated this mesquite as a variety of *P. glandulosa*.

140. Screwbean Mesquite (*Prosopis pubescens*)
(Other common names: tornillo, screwbean, screwpod mesquite, mezquite tornillo, Fremont screwbean, twisted bean)

Habitat: river, tributaries; sandy, gravelly arroyos, creeks, springs, alluvial flats, bajadas; low to mid-elevations.

Geographic range: Texas: Trans-Pecos (El Paso, Hudspeth, Reeves, Presidio, Brewster Counties). US: southwest (excluding Colorado). Mexico: Baja California east to Coahuila.

Park locations: Colorado Canyon River Access, Bofecillos Canyon, Yedra Canyon, Alamito Creek, Cienega Creek

This mesquite is a large shrub or graceful tree to 15 feet high or more, with an extensive root system and one or few trunks with flaky brown bark. The slender branches bear short, straight white spines at the base of the leaf petioles. To 2 inches long, the compound leaves develop one pair of pinnae and up to nine pairs of leaflets per pinna. The alternate leaves are gray green, with glandular-hairy petioles. Well-spaced, the oblong leaflets are blunt or abruptly pointed at the tip, and finely hairy.

In late spring or summer, small flowers appear in dense cylindrical clusters to 3 inches long. Each bloom has five hairy, pale greenish-yellow, united petals; ten projecting yellow stamens with gland-tipped anthers; and a densely hairy pistil with a protruding yellow style. Supporting the bloom is a membranous, bell-shaped calyx with five short lobes. The peculiar, hairy fruits are tightly coiled pods to 2 inches long in shades of yellow, brown, or orange, darker with age. Dangling from the trees in bundles, the pulpy pods contain numerous oval seeds.

Pima Indians treated wounds with the powdered root bark and made syrup from the boiled beans (Vines 1960). The Tewa twisted the pods in the ear to cure earaches (Robbins, Harrington, and Freire-Marreco 1916).

Botanical notes: The Latin species name *pubescens* (hairy) refers to the hairy foliage. British botanist George Bentham described the species in 1846, from a collection by Irish physician-botanical explorer Thomas Coulter in "California between San Miguel and Monterey." Physician for the Real del Monte Mining Company, Coulter collected extensively in Mexico (1824–31) and California and Arizona (1831–32), and returned to Dublin in 1834 with 50,000 specimens for Trinity College.

141. Lindheimer's Senna (*Senna lindheimeriana*)
(Other common names: velvet leaf or showy senna, velvet leaf wild sensitive plant, puppy-dog ears)

Habitat: canyons, arroyos, springs, desert scrub, gravelly bluffs, dry, rocky slopes, open woodlands.

Geographic range: Texas: Trans-Pecos (excluding Reeves, Pecos Counties), Edwards Plateau; Dimmit, Cameron Counties. US: Texas, south New Mexico, southeast Arizona. Mexico: Sonora to Tamaulipas.

Park locations: Monilla Canyon, Agua Adentro, Horse Trap Spring, Ojo Chilicote, Solitario north rim

This legume is an upright, leafy perennial to 4 feet high, with one or a few spreading to erect, copiously hairy stems growing from deep woody roots. The stems are covered with a mix of hairs — long, short, erect, spreading, yellow, velvety.

Often gray green or blue green, the compound leaves are up to 6 inches long, with four to eight leaflet pairs. The long-petioled blades are hairy, with stalked glandular hairs on the midrib at the leaflet nodes. The leaflets are oblong to elliptic, blunt or rounded at the tip, at times bristle-tipped. Each leaflet bears soft fine hairs, especially on the margins. The sensitive blades fold when touched.

The flowers appear in summer and fall in spikelike clusters, on stalks from the upper leaf axils. Each bloom has five oval yellow petals, which are veined, crinkly on the margins, and soon wilting. Within the petals are ten stamens unequal in size, with greenish-yellow filaments and odd reddish-brown, bilobed anthers, and a long, hooked style.

On stout, hairy stalks, in drooping clusters, the fruits are flattened, often curved, linear pods to 2 1/2 inches long. Dull yellow to reddish-brown, the papery pods split on the seams, exposing rows of brown seeds.

Botanical notes: *Senna* is a modern Latin name, from the Arabic *sana* (radiance), presumably because the dried leaves and pods are used as a cathartic. In 1848, German botanist Adolf Scheele named this species *Cassia lindheimeriana*, for Frankfurt-born naturalist Ferdinand Lindheimer, who found the plant near his New Braunfels, Texas, home in 1846. Lindheimer collected Texas plants for his friend, St. Louis botanist George Engelmann, and for Harvard botanist Asa Gray, and in 1852 became editor of the New Braunfels *Herald-Zeitung*. H. S. Irwin and Rupert Barneby moved the species to the *Senna* genus in 1979.

142. Shrubby Senna (*Senna wislizeni*)
(Other common names: Wislizenus' or canyon senna, Wislizenus' wild sensitive plant, hojasén, pinacate)

Habitat: brushy, gravelly, rocky canyon bottoms, bajadas, open grasslands, desertic mountain slopes.

Geographic range: Texas: Trans-Pecos (south Hudspeth, south Culberson, west Jeff Davis, Presidio Counties). US: Texas to southeast Arizona. Mexico: Sonora east to Tamaulipas, Durango southeast to San Luis Potosí, Querétaro.

Park locations: lower Monilla Creek, Tapado Canyon, Arroyo Mexicano, Leyva Creek, Sauceda, Tascate Hills

This senna is a handsome shrub to 6 feet high, a recommended ornamental, with diverging branches and short leafy branchlets. Hairless with age, the dark brown branches are slightly thorny. Young branchlets may bear corky lenslike ridges (lenticels).

The compound leaves are up to 2 inches long, on petioles to 2/5 inch long. Larger leaves form singly on young branchlets, and smaller leaves cluster on knobby spurs. Often, larger leaves have three leaflet pairs, smaller leaves two pairs. The light green leaflets are inversely egg-shaped, rounded or abruptly pointed, with a spinelike tip.

In spring and summer, this shrub is resplendent with crowded clusters of large, golden-yellow blooms. Seldom fully open, the reticent blooms have five broadly spoon-shaped, overarching petals. One petal is larger than the others, and asymmetrical, with unequal sides. Within the petals are ten stamens (three usually sterile) with long red-brown anthers, and a protruding greenish-yellow, hooked style. Supporting each bloom are five yellowish-green sepals. The dangling fruits are straight, purplish-brown, papery pods to 6 inches long, compressed between the seeds, with thick seams.

Botanical notes: In 1852, Asa Gray named this species *Cassia wislizeni*, for German-born, St. Louis physician and botanical collector Adolph Wislizenus. Wislizenus found the plant in 1846 on a trading trip to northern Mexico, at "Carizal and Ojo Caliente, south of El Paso." In 1848, Wislizenus published a record of his travels, *Memoir of a Tour to Northern Mexico*. In 1979, this species was moved to the *Senna* genus by legume authorities H. S. Irwin and Rupert Barneby.

143. Schott's Acacia (*Vachellia schottii*)
(Other common names: Schott's wattle)

Habitat: near Rio Grande; hot desert flats, gravelly arroyos, rocky bluffs; gypseous clay, broken, silty, limestone outcrops.

Geographic range: Texas: Big Bend (south Brewster, southeast Presidio Counties). US: Texas. Mexico: probably adjacent Mexico.

Park locations: West Contrabando Trail, Buena Suerte Trail

Schott's acacia is a small, woody shrub to 4 feet high, with a few stems diverging from the base, and many slender branchlets. Mostly hairless, the maroon branches (gray with age) are armed with straight white, maroon-tipped spines, paired at the nodes.

Alternate or clustered on short spur branchlets, the compound leaves are up to 1 inch long, with one pair of pinnae and as many as six pairs of leaflets per pinna. Threadlike and widely spaced, the leaflets give the foliage a delicate, lacy appearance.

In late spring and summer, tiny florets appear in fragrant yellow globes 2/5 inch across, a major bee attraction. Each bloom consists of five sepals, five petals, up to 100 stamens, and a sometimes glandular, pistil. On 1 1/2 inch long stalks, the globes form singly or clustered on the branchlets. The stalks, with odd little bracts at the middle, may be hairy or gland dotted.

The fruit is a linear pod to 3 inches long, greenish yellow to dark reddish brown, constricted between the seeds and dotted with dark sticky glands. Often curved or twisted, with a short beak, the papery pod splits into two valves, releasing oblong, usually compressed, spotted seeds.

Botanical notes: In 1834, Scottish botanist Robert Wight, naturalist with the British East India Company, and George Arnott Walker-Arnott, University of Glasgow botanist, named the genus *Vachellia* for George Vachell, chaplain of the East India Company in Macau and avid collector of Chinese plants. In 1859, John Torrey named the species *Acacia schottii*, for Arthur Schott, naturalist on the US-Mexico Boundary Survey. The holotype was collected by Charles Parry, the survey's surgeon-botanist, in Brewster County, "near the canon of San Carlos, at the Comanche Crossing [probably Lajitas] of the Río Grande." In 2005, Illinois botanists David Seigler and John Ebinger moved this species to the *Vachellia* genus.

144. Gray Oak (*Quercus grisea*)
(Other common names: shin or scrub or live oak, mountain white oak, encino gris, encino blanco)

Habitat: woodlands, rocky slopes, canyons; riparian sites, bajadas, grasslands; igneous, limestone exposures.

Geographic range: Texas: Trans-Pecos (excluding Reeves, Pecos, Terrell Counties). US: Texas to Arizona. Mexico: Sonora east to Coahuila, south to Durango.

Park locations: Chorro Canyon, Arroyo Mexicano, Yedra Canyon, Skeet Canyon, Tascate Hills, Solitario northwest rim, Solitario Peak, Saltgrass Draw, Cienega Mountains

This white oak is a midsized, much branched tree to 30 feet high, with one trunk, or a large shrub with many stems. Common in the Trans-Pecos, gray oak is mostly evergreen, with robust, spreading, twisting branches, and fissured gray bark. The twigs bear woolly, stellate hairs.

The alternate leathery leaves are variable, often oblong to elliptic, to 3 inches long, on woolly-haired petioles. Stiff, thickish, the blades are pointed or blunt at the tip and smooth on the edges or mucronate, with short, abruptly pointed teeth. The leaves are dull green above, gray green or yellow green below, veined and stellate-hairy, with a felty lower surface.

In spring, male and female flowers form in separate catkins on the same plant. To 3 inches long, the male catkins are dangling clusters of loosely spaced yellow blooms. The female catkins are short, spikelike clusters of up to six blooms. Each male bloom is a cup-shaped calyx with five united lobes, and usually six protruding stamens. The female bloom consists of six united sepals and a three-lobed pistil with three short styles. Single or paired, the hairless fruits are oblong acorns to 3/4 inch long, yellowish green to reddish brown. Each acorn rests in a whitehaired cup of thick overlapping scales.

Botanical notes: *Quercus* is the classical Latin name for oak, possibly from the Celtic *quer* (fine) and *cuez* (tree). The Latin epithet *grisea* (gray) refers to the gray-green leaves. Frederik Liebmann, director of the University of Copenhagen botanical garden, described this species in 1854, and Charles Wright collected the type specimen in West Texas in 1849. Supported by King Christian VIII of Denmark, Liebmann botanized extensively in Mexico (1841–43), but his notable work on oaks, *Chênes de l'Amérique Tropicale* (1869), was published well after his early death in 1856 at the age of forty-three.

145. Hinckley's Oak (*Quercus hinckleyi*)
(Other common names: chaparro)

Habitat: desert scrub, arid, rocky slopes, hilltops, canyon rims; limestone, dolomitic soils at 4,500 feet.

Geographic range: Texas: Big Bend (southeast Presidio, southwest Brewster Counties). US: Texas. Mexico: adjacent.

Park locations: Solitario

Of conservation concern: see Appendix A

The prize of BBR, Hinckley's oak is a relict of an earlier, wetter climate thousands of years ago. Today, the miniature evergreen shrub is limited to the Solitario in BBR, a few sites near the old mining town of Shafter in Presidio County, and in adjacent Mexico. This rare oak is seldom more than 3 feet tall, with many short, rigid, tangled branches and scaly gray bark. Typically, the oak reproduces vegetatively, spreading by rhizomes and forming large patches. Red when young, the mostly hairless twigs quickly turn light brown, then grayish white with age.

Seldom more than 3/5 inch long and broad, the leaves are thick, stiff, round in outline but irregular, heart-shaped or with earlike bulges at the base, and spine-tipped. Short-petioled, the hairless blades are gray- or blue-green and glaucous, with thick, wavy margins and a few long, spiny teeth on each side. Young leaves are especially colorful, in shades of red, pink, and coral.

The flowers of Hinckley's oak are unisexual, petal-less, in separate catkins on the same plant. Appearing in spring on woolly stalks, the male catkins are no more than 1/5 inch long, with few blooms. At the base of each bloom, a reddish-brown, woolly calyx supports protruding stamens with lobed yellow anthers. Female catkins are undescribed. The largely stalkless fruits are reddish-brown ovoid acorns, single or paired, to 3/4 inch long. A shallow cup of tiny close-pressed scales encloses the acorn's bottom third.

Botanical notes: In 1951, in his monograph *The Oaks of Texas*, Cornelius Muller, University of California, Santa Barbara, taxonomist and leading oak authority, named this species for Leon Hinckley, Sul Ross State University botanist. Hinckley collected the oak in 1950, on the "rim of the Solitario Basin." Hinckley is known for his studies of the vegetation of the Sierra Vieja Mountains and the Mount Livermore area, and for his collection of numerous rare plants.

146. Mexican Blue Oak
(Other common names: Sonoran or Arizona blue oak, blue oak, encino azul, encino mexicano azul, bellota de cochi [pig's acorn])

Habitat: igneous canyons, foothills, slopes, oak woodlands, grasslands, with gray oak, cottonwood, ash.

Geographic range: Texas: Trans-Pecos (south Hudspeth, Presidio, Brewster Counties). US: Texas to Arizona. Mexico: Sonora to Coahuila, Baja California Sur to Durango.

Park locations: Chorro Canyon, box canyon west of Cinco Tinajas off Leyva Canyon, South Fork of Alamo de Cesario Creek

In Texas, Mexican blue oak is scarce, known only in a few Trans-Pecos mountains, and three sites at BBR. This evergreen tree may reach 30 feet high, with a thick trunk, scaly gray, furrowed bark, and a widely spreading crown. Young twigs are sparsely to densely hairy, with short, soft hairs, but hairless with age.

The alternate leaves are oblong or elliptic, seldom more than 2 1/2 inches long, rounded or less often pointed at the tip, sometimes with a few small teeth on the margins. Short-petioled, the blades are famously blue-green, or dull green or glaucous above, and paler and glandular-hairy below. A white midrib and network of white veins lines both surfaces.

In spring, as new leaves emerge, the male flowers appear in drooping, spikelike catkins to 1 1/2 inches long. The loosely flowered catkins are greenish yellow but red tinged in age,

with a finely hairy rachis (main axis). Each male bloom consists of a woolly calyx with united lobes, usually surrounding six protruding stamens. Typically, the female flower clusters are less than 3/5 inch long, with up to five blooms on a stiff, thickly hairy stalk with stellate hairs. Each year, this tree produces light brown acorns to 2/3 inch long, which usually ripen in the fall. Single or paired, the acorns are ovoid to oblong, hairless, and thin-shelled. On trees at BBR, a thick, gray-haired, scaly cup may enclose half of the nut or more.

Botanical notes: The Latin species name *oblongifolia*, from *oblongus* (oblong) and *folium* (leaf), applies to the leaves. This species was described by John Torrey in 1853, in *Report of an Expedition down to the Zuni and Colorado Rivers*, by Captain Lorenzo Sitgreaves, Corps of Topographical Engineers. The type specimen was collected by Dr. Samuel Woodhouse, surgeon-naturalist on the Sitgreaves Expedition in western Arizona in 1851.

147. Sandpaper Oak (*Quercus pungens*)
(Other common names: pungent or scrub or shin oak, encino chino, scrub live oak)

Habitat: desertic canyons, gravelly, brushy arroyos, rocky slopes, oak-juniper roughland, limestone, igneous substrates.

Geographic range: Texas: Trans-Pecos (excluding Reeves County), most of west Edwards Plateau. US: Texas to southern Arizona. Mexico: Chihuahua, Coahuila.

Park locations: lower Arroyo Mexicano, Righthand Shutup, Lefthand Shutup, Lower Shutup, Masada Ridge, Saltgrass Draw

Sandpaper oak is a small to large shrub, as much as 10 feet tall, evergreen, and at times thicket-forming. Less often, this oak becomes a small to midsized tree, with thin, flaking, light brown bark. The twigs are grooved, velvety when young, hairless and grayish with age.

Seldom more than 1 1/2 inches long, the leaves are oblong, strikingly wavy on the edges, with teeth or shallow, spine-tipped lobes on each side. Short-petioled, the blades are stiff, yellowish green and often glossy above, with a sharp, abruptly pointed tip. Due to persistent rough hair bases above and a thick coat of hairs below, the leaves feel like sandpaper.

The flowers are unisexual, petal-less, with male and female blooms appearing in spring in separate catkins on the same plant. The male catkins are spikelike, with numerous yellow-red blooms on white-woolly stalks. Female catkins, to 1/5 inch long, have only a few blooms. The fruits are single or paired, light- to reddish-brown, annual acorns, to 2/5 inch long. Hairless, nearly stalkless, the acorns are oblong-cylindric, rounded at the tip. A cup, with thick, tuberculate, whitehaired scales, encloses one-third of the nut.

Botanical notes: The Latin epithet *pungens* (sharply pointed), alludes to the spiny leaves. This species was described by Danish botanist Frederik Liebmann in 1854, and the type specimen was collected by Charles Wright in 1849, on his trek with a military expedition across West Texas to El Paso. Wright traveled 673 miles on foot and brought back 1,400 specimens for Asa Gray at Harvard and cacti for George Engelmann in St. Louis. After tutoring in San Marcos and teaching school in New Braunfels, Wright joined the US-Mexico Boundary Survey as botanist (1851–52).

148. Ocotillo (*Fouquieria splendens*)
(Other common names: coachwhip, Jacob's staff, candlewood, devil's walking stick, vine cactus, desert coral, albarda, barda, ocotillo del corral)

Habitat: hot, open desert flats; arid, exposed, rocky slopes, foothills, alluvial plains; desert scrub, grasslands, to woodlands at 5,000 feet; shallow, rocky, sandy soils, limestone or igneous.

Geographic range: Texas: Trans-Pecos (excluding Reeves County); Crockett, Val Verde Counties. US: Texas to California, Nevada. Mexico: Baja California to Nuevo León, south to Zacatecas, San Luis Potosí.

Park locations: Black Hills, Vista de Bofecillos, Las Cuevas, South Leyva, Encino Loop, Tascate Hills, Solitario northwest rim, Buena Suerte Trail

Ocotillo is a peculiar yet common, widely distributed shrub, with whiplike, thorny stems to 20 feet high, radiating from a central crown. The trunkless shrubs grow from an extensive, shallow root system. The stout, unbranched stems are deeply grooved, with waxy bark and spines.

Ocotillo leaves appear after good rains, and soon fall off until the rains return, so plants are often leafless. Up to five crops of leaves may form each year. When leaves are absent, photosynthesis occurs in the stems. Ocotillo produces two leaf types. First season leaves are long-petioled, alternate, to 2 inches long. When these leaves fall, the petioles harden into spines. Later season leaves, to 1 inch long, grow clustered below the spines. Both leaf types are spatulate and hairless.

In spring, orange-red flowers form fiery clusters, to 10 inches long, at the stem tips. Each waxy tubular bloom has five fused lobes, curled backward at the tip, with up to fifteen protruding red stamens with yellow anthers. Each bloom is supported by five pinkish overlapping sepals with white margins. To 3/5 inch long, the fruits are three-valved capsules with numerous flat, winged seeds designed for wind dispersal.

Botanical notes: Ocotillo, diminutive of the Aztec *ocote* (pine used to make torches), means "little pine," by implication "little torch," for the fiery flowers. In 1823, University of Berlin botanist Karl Kunth named this genus for his teacher Pierre Fouquier (1776–1850), physician to King Louis-Philippe and president of the French Academy of Medicine. The species name *splendens* (splendid) aptly defines the flowers. George Engelmann described this species in 1848, in Adolph Wislizenus's *Memoir of a Tour to Northern Mexico*. The specimen was collected by Wislizenus in 1847, at Santa Cruz, in Chihuahua, Mexico.

149. Scrambled Eggs (*Corydalis aurea*)
(Other common names: golden corydalis, golden or ground smoke, golden fumewort)

Habitat: creeks, woods, disturbed sites; sandy, rocky, gravelly soils; limestone, igneous outcrops.

Geographic range: Texas: Trans-Pecos (excluding Reeves, Pecos, Terrell, Val Verde Counties); scattered in Panhandle. US: west half; Arkansas to Minnesota, Illinois, Great Lakes states, northeast seaboard. Elsewhere: Canada, Mexico (Sonora, Chihuahua south to Sinaloa, Durango).

Park locations: Las Cuevas, dam north of Ojito Adentro, upper Terneros Creek, Cienega Creek

Scrambled eggs is a weak, delicate annual, with slender stems to 14 inches long. The stems are laxly ascending to nearly prostrate, often sprawling on larger plants or rocks for support. The pale green branches are leafy, notably grooved, and glaucous.

To 7 inches long, the leaves are deeply dissected, light green, at times glaucous. Each leaf is compound, with up to seven pairs of pinnae, each pair once or twice divided, with oblong ultimate segments.

Golden blooms appear in spring, in clusters at the branch tips. Each bizarre bloom has four petals varying in size and shape, the two outer petals around the two inner petals. The upper outer petal has an inflated, saclike spur at the base; the other, unspurred outer petal may have a bulging, humpbacked base. Both outer petals are hooded but usually lack a crest found in some species. The inner petals are inversely lance-shaped and united at the tips. Each bloom also consists of two sepals, six stamens, and a style with a bilobed stigma.

The dehiscent fruits are slender, two-valved, cylindrical capsules, to 1 inch long, with many tiny black seeds.

Botanical notes: In 1805, Swiss taxonomist Augustin Pyramus de Candolle named this genus *Corydalis*, from the Greek *korydallis* (crested lark), perhaps a comment on the likeness of the flower's spur to the lark's spur, or the resemblance of the flower's shape to the lark's head. The Latin species name *aurea* (golden) references the flower color. German taxonomist Carl Ludwig von Willdenow described this species in 1809, citing the habitat as Canada, and Muhlenberg as the source. German-educated Henry Muhlenberg, Lutheran pastor of Holy Trinity Church in Lancaster, Pennsylvania, and first president of Franklin and Marshall College, was one of America's pioneering early botanists.

150. Catchfly Prairie Gentian (*Eustoma exaltatum*)
(Other common names: tall prairie gentian; seaside, marsh, beach, bluebell, lesser bluebell, western blue, or catchfly gentian; blue marsh lily, alkali chalice, small bluebell, violeta)

Habitat: riverbanks, creeks entering Rio Grande; alluvial flats, moist grasslands, springs, ephemeral pools; roadside gullies, ditches; sandy, alkaline, gypseous soils.

Geographic range: Texas: Trans-Pecos (excluding El Paso County); parts of South Texas Plains, Coastal Prairies, central Texas, Panhandle. US: California to Florida. Elsewhere: Mexico, parts of Central America, Venezuela; West Indies.

Park locations: Panther Canyon, Rancherías Canyon, Tapado Canyon, Las Cuevas, Alamito Creek, Cienega Creek

Growing from a taproot, this gentian is an erect annual or short-lived perennial to 28 inches high, with one or a few hairless stems. The leafy stems are unbranched or sparsely branched, often glaucous.

The basal leaves are broadly spatulate or oblong, mostly blunt tipped, with short, broad petioles. Lacking petioles, the stem leaves are paired, conspicuously clasping around the stem. Mostly oblong, or oval to lance-shaped, the hairless blades are up to 3 1/2 inches long, thick, and blunt or sharp-pointed at the tip.

From late spring to fall, showy flowers appear singly or in loose, few-flowered clusters near the stem tips, on erect stalks to 4 inches long. The chalice-like blooms are pale violet to deep bluish-purple, white near the base, with a dark purple throat. Seldom more than 1 1/2 inches long, each bloom consists of five deeply divided, oblong to inversely egg-shaped lobes; commonly five stamens with yellow anthers; and a slender style with a bulky, bilobed yellow stigma. The supporting calyx has five tapering linear lobes.

To 3/4 inch long, the fruit is a two-valved, wrinkled oblong capsule, with numerous tiny, pitted seeds.

Botanical notes: The genus name, combining the Greek words *eu* (beautiful) and *stoma* (mouth), refers to the showy flowers. The epithet *exaltatum* (tall, lofty) likely relates to the plant's height. The species was described as *Gentiana exaltata* by Linnaeus in 1762 but moved to the *Eustoma* genus by Scottish botanist George Don in 1837, in his masterwork, *A General History of the Dichlamydeous Plants*. The lectotype is a 1756 illustration in a posthumous work of Charles Plumier. French monk Plumier, royal botanist to King Louis XIV, found this herb on botanical expeditions to the French Antilles and Central America (1689–95).

151. Buckley's Centaury (*Zeltnera calycosa*)
(Other common names: rosita, Buckley's mountain pink, canchalagua)

Habitat: open, arid, barren rocky slopes, on shallow limestone, gypseous soils.

Geographic range: Texas: Trans-Pecos (Presidio, Brewster, Pecos, Terrell, Val Verde Counties), much of Edwards Plateau, South Texas Plains. US: Texas. Mexico: Coahuila east to Tamaulipas.

Park locations: Solitario northwest rim, rim above Righthand Shutup

Buckley's centaury is a small annual, often less than 8 inches high, with one erect stem much branched above. This hairless herb may form a low, rounded mound of short branches or assume a taller, sparsely branched habit to 18 inches high or more.

The basal and lower leaves are oblong, lance-shaped, or spatulate, thickish, often wavy and twisting on the margins. To 1 inch long, the leaves may fall before the flowers bloom. Mid- and upper stem leaves are shorter, narrower. Some basal leaves have petioles, but the paired stem leaves are stalkless. All blades are yellow green.

In spring and summer, slender tubular flowers, with five abruptly flaring lobes, appear in clusters of up to 100 blooms. The blooms are pink, rose pink, or magenta, with lance- or egg-shaped lobes white at the base and a yellow throat. Within the lobes are five protruding stamens with white filaments and yellow anthers, which may be coiled and twisted; and a slender white style, projecting beyond the anthers, with a broad, sometimes two-branched yellow stigma. A green calyx, with five slender lobes, surrounds the floral tube.

The oblong fruits are two-valved, dehiscent capsules to 1/2 inch long, with up to 600 tiny, spheric brown seeds, often with an oily sheen.

Botanical notes: In 2004, Swiss botanist Guilhem Mansion formed the new genus *Zeltnera* from twenty-five North American species, including this one, formerly placed in *Centaurium*. Mansion dedicated *Zeltnera* to Swiss botanists Louis and Nicole Zeltner, for their studies of *Centaurium* and related genera. Naturalist Samuel Botsford Buckley described the species as *Erythraea calycosa* in 1863, from a specimen he collected near Fort Mason, Texas, in 1860 or 1861, while serving as assistant geologist for the Texas Geological Survey.

152. Redstem Stork's Bill (*Erodium cicutarium*)
(Other common names: redstem or cutleaf filaree, crane's bill, pinweed, common stork's-bill or heron's-bill or crowfoot, alfilaree, alfilerillo, alfilaria, aguja del pastor, agujitas, peine de bruja)

Habitat: disturbed sites; low desert to mountains; desert scrub, springs, grasslands; sandy, rocky, clayey, loamy soils.

Geographic range: Texas: Trans-Pecos; most of state (except East Texas, Gulf Coast, scattered in south Texas). US: nearly throughout. Elsewhere: Canada, parts of Mexico to South America; Eurasia, Australia, parts of Africa.

Park locations: Los Ojitos, Los Alamos area

Apparently, Spanish explorers introduced this invasive geranium to California in the early 1700s. Redstem stork's bill is a prostrate to weakly ascending winter annual, branching at the base, with leafy branches arising from a rosette of over-wintering basal leaves. The stems, often reddish, may be hairy and sticky, with glandular and nonglandular hairs. The fernlike, pinnately compound basal leaves are oblong, to 5 inches long, on petioles to 1 inch long or more. The blades are divided into up to thirteen leaflets, each deeply cleft. Stem leaves are smaller.

In spring and summer, dainty flowers form in small clusters, each cluster on a long maroon, bristle-haired stalk from the leaf axils. The blooms have five elliptic, pink or rose petals, each often with three tiny dark streaks, shaped like a bird's foot, at the base. Within the petals are ten rose-purple stamens, five fertile with yellow anthers, five sterile, scalelike, without anthers; and five styles, united to form part of a beak, with rose-purple stigmas. Below the petals are five greenish-yellow, maroon-tinged, oblong sepals.

The fruit is a capsule with five one-seeded, spindle-shaped segments, each attached to one of the united styles that form the beak, or stork's bill. When mature, the segments split but remain attached to the coiling part of the styles. When moist, the coiled style around the seed uncoils, and the unwinding motion plants the seed.

Botanical notes: Descriptions of this genus and species, by Charles L'Héritier, appeared in 1789, in William Aiton's *Hortus Kewensis*, a catalog of plants in the Royal Botanic Garden at Kew. The genus name, from the Greek *erodios* (heron), alludes to the fruit's long beak. The species name joins the Latin *cicuta* (hemlock, also a genus of poisonous plants) and *-arium* (pertaining to), because the leaves resemble *Cicuta* leaves. L'Héritier referred to this plant as "hemlock-leav'd crane's-bill." Linnaeus had described the species in 1753 as *Geranium cicutarium*.

153. Texas Stork's Bill (*Erodium texanum*)
(Other common names: Texas or desert heron's bill, desert or large-flowered stork's bill, pine needle, Texas or false filaree, alfilerilla)

Habitat: desert, alluvial flats, washes, barren hills, desert scrub, low to mid-elevations; rocky, sandy, calcareous, gypseous soils, alluvium.

Geographic range: Texas: Trans-Pecos, much of Texas (nearly absent in Panhandle, east Texas, Gulf Coast). US: southwest (excluding Colorado); Oklahoma; introduced elsewhere. Mexico: Baja California east to Nuevo León.

Park locations: Solitario northwest rim

This herb is a low annual, branching at the base, with prostrate to ascending, long leafy branches. Radiating from central taproots, often trailing, the slender, commonly reddish stems may be thickly coated with short, rough hairs.

To 1 1/2 inches long, the paired leaves are egg-shaped, usually three-lobed, with a large central lobe. The blades are rounded or blunt at the tip, strongly wrinkled and broadly toothed on the margins. Basal leaves are borne on stalks to 4 inches long or more, much longer than the blades. Most leaves have distinct, often reddish veins, and all leaves bear short, close-pressed hairs.

From early spring to summer, five-petal flowers form in small clusters of usually several blooms. Opening in late afternoon, the blooms seem to shy away from light, and the broadly spatulate petals are ephemeral. Within the petals are ten stamens, five fertile with yellow anthers, five much smaller, without anthers, sterile; a silky-haired, five-lobed pistil, and a style column with five fused styles, which ultimately form a long beak. The flowers are supported by five oblong, whitehaired sepals.

The fruit is a schizocarp, a dry, indehiscent capsule that splits at maturity into five one-seeded segments (mericarps). Each hairy mericarp is spindle-shaped, and after separating, remains attached to one of the five united styles, which are spirally coiled at the base. The styles coil and uncoil with changes in humidity, thereby planting or dispersing the seeds in corkscrew fashion.

Botanical notes: The species name *texanum* indicates Texas, where the plant was found. Asa Gray named this species in 1849, in *Genera Florae Americae*... (The Genera of the Plants of the United States Illustrated by Figures and Analyses from Nature), from a collection by Ferdinand Lindheimer in 1846, at New Braunfels, Texas. Isaac Sprague, one of America's foremost botanical and ornithological artists, did the illustrations. Sprague served as a draftsman for John James Audubon on a Missouri River expedition in 1843 and later illustrated botanical works by Asa Gray, John Torrey, and others.

154. Narrowleaf Fendlerbush (*Fendlera linearis*)
(Other common names: stiff fendlerbush)

Habitat: rocky canyons, woodlands, canyon walls, rims; arid slopes; mostly limestone substrates.

Geographic range: Texas: Big Bend (southeast Brewster, south Presidio Counties). US: Texas. Mexico: Chihuahua to Nuevo León.

Park locations: Los Portales, Fresno Peak, Lefthand Shutup, Lower Shutup

Of conservation concern: see Appendix A

This hydrangea is a rigid, upright shrub to 3 feet high, intricately branched, with short, stiff branchlets. Often, the reddish-brown stems are coated with gray bark. Young branchlets are densely hairy, with thorny tips. Mostly stemless, the leathery linear leaves, 3/5 inch long, are blunt at the tips and curled under on the margins. Sparsely hairy, the leaves are paired on the stems or clustered on short branchlets.

Blooming in spring, at times later, the flowers appear singly or in small groups on short gray-haired stalks at the branchlet tips. The unusual white blooms have four wide-spaced, spade-like petals with a stalk-like claw. Often, the white petals are gnawed on the margins and flushed with red or purple colors underneath. Below the petals are four pale green, triangular sepals with soft feathery hairs. Within the petals are eight white stamens with flat filaments, each with two slender protrusions extending beyond the yellow anthers; and a four-segment pistil with four hairy styles shorter than the stamens.

The fruit is a dehiscent, egg-shaped capsule 1/3 inch long, which splits into four brittle segments (valves), each with up to six red-brown seeds. The valves are tan, with long beak-like tips formed by the styles.

Botanical notes: In 1852, in *Plantae Wrightianae*, George Engelmann and Asa Gray dedicated the genus *Fendlera* to "Mr. Augustus Fendler who, next to Wislizenus, was the earliest botanical explorer in New Mexico." Named for the linear leaves, species *linearis* was described in 1920 by Alfred Rehder, curator of the Arnold Arboretum herbarium, from a collection by Cyrus Pringle in 1889, in the Sierra Madre near Monterrey, Mexico. Horticulturist, botanical explorer, and Vermont Quaker, Pringle collected for twenty-eight years in Mexico and Arizona for the Gray Herbarium and the Smithsonian Institution.

155. Bristly Nama (*Nama hispidum*)
(Other common names: rough or hispid nama, sand bells, purple mat, rough fiddleleaf, flor morada, moradita, campanitas de arena)

Habitat: desert arroyos, dry creekbeds, alluvial flats, sandy, gypseous dunes, gravelly slopes, roadsides, trail-sides; limestone, igneous soils.

Geographic range: Texas: Trans-Pecos, much of Texas (excluding east Texas, upper Panhandle, upper Gulf Coast. US: southwest; Oklahoma. Mexico: Baja California east to Tamaulipas, south to Zacatecas, San Luis Potosí.

Park locations: Panther Canyon, Black Hills Creek, Arroyo Mexicano, Los Alamos, Fresno Canyon

Bristly nama is a low, bristle-haired annual to 16 inches high, frequently much less, with mostly upright stems and spreading branches. Commonly, the stems and leaves are covered with short, stiff hairs, some with swollen, pustulate bases. The alternate, stemless leaves are oblong, to 2 inches long. Fleshy and thickish, the blades are often curled under on the edges and strongly scented.

Blooming in early spring and opportunistically into fall, this nama forms low mounds blanketed with purple-pink blooms. Bell-shaped blooms, with five rounded lobes, appear singly on short stalks from the leaf axils, or in small clusters at the branch tips. The flowers have a white ring at the base, above a yellow or greenish-yellow throat. Supporting each flower are five hairy, united sepals. Within each bloom are five stamens with yellow anthers and a two-segment pistil with a short, two-branched style.

The fruit is a partially segmented, globose capsule 1/6 inch long, containing many minuscule yellow seeds.

Botanical notes: The generic name, from the Greek *nama* (stream, spring), likely refers to the habitat of the first species described by Linnaeus in 1759. The Latin species name *hispidum* (hispid) refers to the foliage. In 1861, Asa Gray described this species, citing collections in "Texas and the Mexican borders of the Rio Grande" by Jean Louis Berlandier, Ferdinand Lindheimer, and Thomas Drummond. The type specimen was collected by Swiss-born naturalist Berlandier, probably in south Texas from 1828 to 1834. From 1827 to 1829, Berlandier served as botanist with the Mexican Boundary Commission and later became a physician in Matamoros, Mexico.

156. Rio Grande Phacelia (*Phacelia infundibuliformis*)
(Other common names: Rio Grande scorpionweed)

Habitat: igneous mountain terrain, rocky slopes, creek banks, canyons, below rock walls, amid boulders, in rock crevices; low to mid-elevations.

Geographic range: Texas: Big Bend (southeast Presidio, south Brewster Counties). US: Texas. Mexico: Chihuahua, Coahuila.

Park locations: South Lajitas Mesa, Monilla Canyon, Bofecillos Canyon below Ojito Adentro, Guale 1, drainage from Sauceda to Cinco Tinajas

In the United States only found in the extreme southern Big Bend, this herb is a sturdy, erect annual, much branched from near the base, seldom more than 16 inches tall. Typically, the stems are cloaked with protective hairs — short, rough, stiff, and glandular.

To 4 inches long, the lower leaves are oblong, usually divided into seven or nine leaflets, each cleft or lobed on its own short stem. Leaves on the stems are compound or merely lobed. Generally, the blades are less glandular than the stems, but still coarsely hairy.

In spring, up to forty flowers appear in congested, spirally coiled, one-sided (scorpioid) clusters at the branch tips. The pale purple, short-stalked blooms are funnel-shaped, with five rounded lobes at the tip. Within the blooms are five protruding stamens with purple filaments and yellow anthers, and a divided, two-branched, purple style. Distinctively, scalelike appendages are present at the base of, and alternating with, the filaments. Below each bloom is a calyx with five oblong lobes, each lobe with prominent stalked glands.

The fruit is a dehiscent oblong capsule 1/5 inch long, with as many as twenty tiny oval seeds.

Botanical notes: In 1789, Antoine Laurent de Jussieu, botanist at the Jardins des Plantes in Paris, named this genus from the Greek *phakelos* (fascicle), emphasizing the distinctive flower clusters. The species name joins the Latin *infundibulum* (funnel) and *forma* (form), alluding to the funnel-shaped bloom. John Torrey described the species in 1859, from specimens collected in 1852 by John Bigelow and Charles Wright, on "overhanging rock on a mountain near Lake Santa Maria, Chihuahua."

157. Pope's Phacelia (*Phacelia popei*)
(Other common names: Pope's scorpionweed)

Habitat: desert scrub, open grassland; clay, sandy loam, gypseous, gravelly, rocky soils of limestone, igneous origin; low to mid-elevations.

Geographic range: Texas: Trans-Pecos (excluding Val Verde County), much of Permian Basin, Panhandle; Comanche, Ellis Counties. US: Texas, east New Mexico. Mexico: Nuevo León.

Park locations: Llano Loop

Probably the most common phacelia at BBR, Pope's phacelia is a low, erect annual, branched at the base, frequently forming a rosette of finely divided leaves. The stems, often maroon, are sparingly glandular but covered with both fine spreading hairs and stiff, bristly hairs. This annual can be strongly aromatic.

The leaves are oblong, to 6 inches long, once or twice divided into usually stemless segments. The ultimate segments are narrowly lance-shaped, with three, deeply dissected, terminal lobes. The blades may bear a dense cover of rough, stiff hairs.

In spring, purplish bell-shaped blooms form in spirally coiled, one-sided clusters at the stem tips. Gradually uncoiling, the clusters are crowded, with each bloom very short-stalked. Each bloom has five rounded, spreading, petal-like lobes. Protruding from the blooms are five stamens with purple filaments and white anthers, and a bilobed, purple style, longer than the stamens. Below each bloom is a green five-lobed, bristle-haired calyx.

The fruit is a spheric, four-seeded capsule to 1/10 inch long, noticeably glandular, with dark brown, honeycombed seeds.

Botanical notes: John Torrey and Asa Gray described the species in 1855, in the *Pacific Railroad Surveys*, from "specimens gathered by Dr. Garrard, as well as by Captain Pope, whose name we desire the species to bear." Captain John Pope's type specimen was collected in 1854, on Texas' Llano Estacado. Pope, with the US Army Corps of Topographical Engineers, surveyed part of a railroad route from the Mississippi River to California (from the Red River to Doña Ana, New Mexico, on the Rio Grande).

158. Little Walnut (*Juglans microcarpa*)
(Other common names: Texas black walnut, Texas, Mexican, river, dwarf, or littleleaf walnut, nogal, nogalito, nogalillo, namboca)

Habitat: canyons, creeks, gravelly arroyos, benches, terraces; limestone, igneous-derived soils.

Geographic range: Texas: Trans-Pecos (excluding El Paso, Hudspeth Counties), part of Edwards Plateau, sites in Panhandle; Uvalde, Taylor Counties. US: Texas, New Mexico, Oklahoma, Kansas. Mexico: Chihuahua east to Nuevo León.

Park locations: Arroyo Mexicano, Leyva Creek, South Fork of Alamo de Cesario Creek, Fresno Canyon

Little walnut is a small tree or large shrub to 20 feet high, often strongly scented, and multitrunked. From a long taproot, this walnut develops robust, spreading branches and a broad, rounded crown. Young reddish-brown branchlets and young leaves may be covered with glandular or nonglandular hairs. As much as 1 foot long, the alternate, compound leaves are yellow green above, gray green below, with seventeen to twenty-five lance-shaped leaflets. The leaflets are slightly sickle-shaped, veined, with a tapered, pointed tip.

In early spring before the new leaves, greenish-yellow male and female flowers bloom in separate, spikelike catkins on the same plant. The male catkins, to 2 3/4 inches long, are densely flowered, usually dangling from one-year-old branches. The female blooms appear singly or in one- to four-flowered spikes on new growth. Each male bloom consists of four sepals enclosing up to twenty-five stamens. Each female bloom consists of four sepals and a short, two-branched style.

The globose fruits, to 1 inch wide, are brown when mature, with a tough, thick husk enclosing the smallest nut of all walnut species. The tan nut has a famously hard shell, which splits into two valves.

Botanical notes: In 1753, Linnaeus named the genus *Juglans*, the classical Latin name for walnut, perhaps from the Latin *Jovis* (the god Jove or Jupiter), and *glans* (nut). Pliny used the term *glans Jovis* for walnut, hence the common name Jupiter's acorn. The epithet *microcarpa* refers to the small fruit. This species was described by Jean Louis Berlandier in 1850, in *Diario de Viaje de la Comision de Limites*. Berlandier collected plants for the Mexico Boundary Commission in northern Mexico and South Texas. Berlandier and Rafael Chovell, a mineralogist, authored the *Diario*, a journal of their travels.

159. Knotted Rush (*Juncus nodosus*)
(Other common names: joint, jointed, tuberous, or knotty rush, smaller round-headed rush)

Habitat: damp, wet grounds, creeks, springs, seeps, pools; often calcareous soils.

Geographic range: Texas: Big Bend (Presidio, Brewster, Terrell Counties). US: scattered throughout (excluding most of South, Hawaii). Elsewhere: Canada, Mexico (Chihuahua to Nuevo León, scattered south to Puebla).

Park locations: Rancherías Canyon, Agua Adentro, Panther Canyon, Cienega Creek

Knotted rush is a colony-forming perennial with slender, upright, unbranched stems (culms) to 20 inches high. Creeping rhizomes form tuber-like thickenings at the base of the smooth round culms. Typically, two to four stem leaves and at times a basal leaf, are borne on each culm, some leaves longer than the culms. The blades are hollow cylinders to 1 foot long. Each leaf is partly enclosed in a membranous sheath.

In summer through fall, globose flower heads with up to thirty blooms appear, usually in small clusters, near the culm tips. Each flower cluster, to 2 inches long or more, is supported by an erect, leaflike basal bract. The blooms are greenish to brownish, often red tinged, with three inner and three outer tepals, each tapered to the tip. The short-stalked blooms have three stamens with flat filaments and oblong anthers, and a small style with three longer stigmas.

The fruit is a dehiscent capsule, brown or reddish brown, to 1/5 inch long. Each capsule is narrowly ovoid, beaked, with valves splitting at maturity to release many yellowish-brown, oblong seeds.

Botanical notes: *Juncus* is the classical Latin name for rush, possibly from the Latin *iungo* (to join), because the stems were used in basketmaking and weaving. The Latin epithet *nodosus* (knotty) refers to enlarged nodes on the underground stems. Linnaeus described the species in 1762. The lectotype was collected by Pehr Kalm in Canada, on a 1748–51 expedition to North America. Kalm, an early student of Linnaeus, searched for plants and seeds on behalf of the Royal Swedish Academy of Sciences. A journal of his travels was translated into English as *Travels into North America* (1770).

160. Allthorn (*Koeberlinia spinosa* var. *spinosa*)
(Other common names: crown of thorns, crucifixion thorn, spiny allthorn, junco, corona de Cristo, abrojo)

Habitat: desert, alluvial flats, creosote scrub, sandy, gravelly, brushy arroyos, grasslands, silty, clayey hills, rocky south-facing slopes.

Geographic range: Texas: Trans-Pecos, much of South Texas Plains; scattered sites in Permian Basin, elsewhere. US: Texas to southern Arizona. Mexico: Coahuila to Tamaulipas, south to San Luis Potosí, Guanajuato, Querétaro, Hidalgo.

Park locations: Agua Adentro, South Leyva Campground, Llano Loop, Jackson Pens, Solitario northwest rim, West Contrabando Trail

Allthorn is a viciously thorny shrub to 5 feet high, mound-like, sprawling, or a short-trunked tree to 15 feet or more. Considered indicative of CD boundaries, this shrub is a maze of stiff, smooth, mostly leafless branches. The branches are yellow green, tipped with black, maroon, or gray thorns. Leaves emerge in spring and after good rains, but the blades are minuscule and fleeting. Quickly dropping leaves to reduce water loss, allthorn carries on photosynthesis in the stems.

Small but prolific white flowers bloom from spring to fall in short, few-flowered clusters. On a short, sometimes hairy stalk, each bloom consists of four oblong petals; four egg-shaped to triangular, greenish yellow sepals; eight stamens with flattened filaments thickened at the middle, and pale yellow anthers; and a two-chamber pistil with a slender, persistent style with a headlike stigma.

The fruit is a rounded, two-chambered berry, 1/6 inch across, shiny black when ripe. Each berry has a small, thin beak formed by the persistent style, and one or two seeds per chamber.

Botanical notes: Munich botanist Joseph Zuccarini described this genus and species in 1832, naming the genus for his friend and fellow student in theology at Erlangen, Christoph Ludwig Koeberlin. Koeberlin was a pastor and student of German flora, especially mosses near Memmingen, where he maintained an herbarium. The Latin species name *spinosa* (spiny) emphasizes the thorny branches. Zuccarini described allthorn from Mexican collections (1826–32) by Hungarian-born, Bavarian resident Baron Wilhelm Friedrich Karwinsky, a botanical explorer and mining engineer.

161. Range Ratany (*Krameria erecta*)
(Other common names: littleleaf or small-flower ratany, Pima rhatany, range krameria, purple heather, heart-nut, mezquitillo, cósahui del norte, tamichíl, guisapol colorado)

Habitat: creosote scrub, sotol-yucca environs; sandy, clay, gypseous flats, gravelly bajadas, rocky, brushy slopes, low to mid-elevations.

Geographic range: Texas: Trans-Pecos (excluding Terrell County). US: southwest (excluding Colorado). Mexico: Baja California east to Coahuila; Sinaloa southeast to San Luis Potosí.

Park locations: Black Hills, Arroyo Mexicano, Solitario northwest rim, Buena Suerte Trail, Cienega Creek

This ratany is a compact, rounded shrub to 2 feet high, semiparasitic, with stiff upright branches and many short branchlets. The stems are spineless but sharp-pointed, the older stems with gray or white bark. Younger branchlets may be densely cloaked with grayish-white and sometimes glandular hairs. Although the shrub photosynthesizes, its roots attach to roots of other plants, stealing moisture and nutrients. The plant may die without a host. The stemless linear leaves are alternate or bunched on short branchlets. To 1/2 inch long, the blades are silvery, blanketed with stiff, sometimes glandular hairs.

Blooming in spring and fall, the odd flowers are single but often closely spaced, on glandular-hairy stalks. Each whitehaired bloom has four or five large, egg-shaped, petal-like sepals, pink, rose, or magenta. The petals are much smaller, different in size, shape, and color. The three upper petals are multicolored, maroon above with pale margins, yellow green and clawlike below. The two lower petals are smaller, fleshy, pink, red, or maroon, with saclike blisters. Protruding from the petals are four long, pale yellow green, upcurving stamens with pinkish tips.

The eye-catching fruit is a one-seeded, indehiscent, globose pod, 1/4 inch across, ridged on each side and whitehaired, with needlelike maroon spines.

Botanical notes: The genus *Krameria* was described by Pehr Löfling, Linnaeus's student, from specimens he collected in Venezuela. Linnaeus published Löfling's notes on Spanish and South American plants in *Iter Hispanicum* (1758), following his student's untimely death. Löfling likely named the genus for Johann Georg Heinrich Kramer, Austrian Army surgeon-botanist, and his son, natural scientist Wilhelm Heinrich Kramer. Wilhelm's 1756 flora and fauna of Lower Austria was an early work adopting Linnaeus's binomial naming conventions. In 1827, Josef August Schultes described *K. erecta* from a manuscript by taxonomist Carl Ludwig Willdenow.

162. White Ratany (*Krameria bicolor*)
(Other common names: Gray's ratany, white or Gray's krameria, crimson beak, chacaté, cósahui, mamelique)

Habitat: sandy, clay, gypseous flats; desert scrub, arroyos, grasslands, gravelly bajadas, rocky slopes, open woodlands; low to mid-elevations.

Geographic range: Texas: Trans-Pecos; Loving, Ward, Crane, Crockett Counties. US: southwest (excluding Colorado). Mexico: Baja California east to Coahuila; Sinaloa, scattered south to San Luis Potosí, Hidalgo, Michoacán.

Park locations: Arroyo Mexicano, Solitario northwest rim, Solitario Peak, Lefthand Shutup

White ratany is a low, mounded or spreading, semiparasitic shrub to 28 inches high, with many upright, slender branches. The stems are gray green or whitish, sharp-tipped, often thickly coated with short gray or long silky hairs.

To 4/5 inch long, the leaves are linear to narrowly lance-shaped, with an abrupt, bristlelike tip. The stemless blades are alternate on the branches, gray green, with a dense cover of short white hairs.

In spring, at times fall, unusual pink-magenta flowers appear singly in the leaf axils. The blooms are borne on gray-haired stalks to 1 inch long, each with a pair of leaflike bractlets at the middle. Each bloom has five showy, lance-shaped, petal-like sepals, which are strongly recurved. The five petals, unequal in size and much smaller, are scarcely recognizable as petals. The three erect upper petals are linear, pale yellow green with red tips. The two lower yellow or orange petals are rounded, cupped, and gland-like, with blisters externally. These two glandular petals produce oil instead of nectar. The oil is harvested by industrious female bees of the genus *Centris* and fed to their larvae. Protruding from the petals are four thickish, pale yellow-green stamens, the outer two longer and upward-arching; and a hairy pistil with an upcurved style.

The rounded, whitehaired fruits are one-seeded, indehiscent, bur-like pods to 2/5 inch across. Maroon spines, with hooked barbs at the tip, protrude from the pods.

Botanical notes: This species was described by Harvard botanist Asa Gray as *Krameria canescens* in 1852. Charles Wright collected the shrub in 1849, "on prairies beyond the Pecos [River]" in Pecos County, Texas. Since the name *K. canescens* had been used previously, the botanical name was changed in 1906 to *K. grayi*, to honor Asa Gray. In 2013, Beryl Simpson, with the University of Texas at Austin Plant Resources Center, recognized that an older name, *K. bicolor*, used by Sereno Watson in 1886 to describe the bicolor flowers, takes priority (Simpson 2013).

163. Leafy Rosemary-Mint (*Poliomintha glabrescens*)
(Other common names: orégano, Mexican orégano)

Habitat: desert scrub, gravelly arroyos, rocky canyons, mountain slopes; on limestone.

Geographic range: Texas: Big Bend (southeast Presidio, Brewster Counties). US: Texas. Mexico: Coahuila.

Park locations: Righthand Shutup, Lefthand Shutup, Masada Ridge, Saltgrass Draw

To 2 feet high, this aromatic mint is a low gray-green shrub with diverging stems, often densely branched with upright branches. The slender, leafy stems are tan or reddish brown with short soft hairs, but gray, brittle, and hairless with age.

Nearly stemless, to 3/4 inch long, the leaves are elliptic, pointed or blunt at the tip, at times sparsely hairy, and almost always dotted with tiny sunken pits.

From late spring to fall, flowers emerge singly or in small clusters, on short stalks in the upper leaf axils. Each tubular bloom is two-lipped, with an erect, often notched upper lip, a broad, trilobed lower lip, and a floral tube with long soft hairs inside. The flowers are white, pale lavender or pink, with purple

spots on the lower lip. Extending from the flower's throat are two fertile stamens with white filaments and bilobed pinkish anthers, and a style split at the tip. A green, five-lobed calyx, maroon-tipped and silky-haired, supports the bloom. The fruit consists of four one-seeded, oblong nutlets to 1/12 inch long.

Resembling sage in taste, this mint is sold as a spice in Mexican markets (Von Reis and Lipp 1982).

Botanical notes: Asa Gray described the genus *Poliomintha* in 1870, stating that "the name, composed of the Greek words for hoary-white, or gray [*polios*] and Mint [*mintha*], is suggested by the silvery canescence." The Latin species name *glabrescens*, from *glaber* (hairless), and *escens* (becoming), refers to the foliage. Gray's description of this species was published in 1882 in William Botting Hemsley's *Biologia Centrali-Americana*. The type specimen was collected in 1880 by Edward Palmer, in Soledad, twenty-five miles southwest of Monclova, Coahuila, Mexico. British taxonomist Hemsley was keeper of the Kew Herbarium, and Palmer's specimen was deposited there.

164. Autumn Sage (Salvia greggii)
(Other common names: Gregg's, cherry, or Texas sage, red Chihuahuan sage, Texas red sage)

Habitat: gravelly desert washes, rocky slopes, canyons; arid, loose, well-drained, calcareous soils.

Geographic range: Texas: Trans-Pecos (Presidio, Brewster, Pecos, Terrell Counties), scattered sites on Edwards Plateau, elsewhere. US: Texas. Mexico: Coahuila east to Tamaulipas, south to Durango, Zacatecas, San Luis Potosí.

Park locations: Righthand Shutup, Lower Shutup, Masada Ridge, Saltgrass Draw, Government Road Trail, Fresno Canyon below Pila Montoya Trailhead

Autumn sage is a shrubby perennial or subshrub to 3 feet high, upright, often mounded, with wiry, leafy stems growing from a woody crown. The woody stems are maroon when young, tan or gray with age, mostly hairless. Uncommon in BBNP, this mint is much more frequent in the Solitario of BBR.

With a strong minty aroma, the leaves are pale or dull green, to 1 inch long, elliptic, and short-petioled. Paired or bundled at the stem nodes, the thickish leaves are blunt at the tip, smooth on the edges, mostly hairless but gland dotted.

Blooming opportunistically from spring to fall, the lush scarlet flowers appear in loose, few-flowered clusters near the branch tips. Each bloom is two-lipped, with a hoodlike upper lip, a broad, four-lobed lower lip, and a sharply contracted throat. Within the throat are two stamens, each with a single anther sac, and a projecting style with a forked stigma. At the base of each bloom is a glandular, two-lipped calyx, green and/or maroon. When the fruit matures, the persistent calyx encloses four tiny, one-seeded brown nutlets.

As a hummingbird and butterfly attraction, this shrublet is a popular ornamental in southwest gardens. The leaves are used as a seasoning in Mexico, and the edible, slightly sweet flowers are added to salads.

Botanical notes: *Salvia* is an old Latin name for sage, from *salvare* (to heal), for the purported medicinal properties of some species. In 1872, Asa Gray named this species for Josiah Gregg, who collected the type specimen in the "Canon above Palomas near Saltillo, Mexico, 1848–1849." From 1831 to 1840, Gregg traveled the Santa Fe Trail and in 1844 published *Commerce of the Prairies*, an account of the geography, geology, human and natural history of the area. During the Mexican-American War, Gregg served as US Army guide and interpreter and later as a freelance newspaper correspondent.

165. Mexican Skullcap (*Scutellaria potosina* var. *tessellata*)
(Other common names: Huachuca Mountains skullcap, checkered skullcap)

Habitat: canyons near Rio Grande to mesic mountains; desert scrub, gravelly flats, rocky slopes, woodlands; varied soils, often limestone.

Geographic range: Texas: Trans-Pecos (Hudspeth, Culberson, Presidio, Brewster Counties). US: Texas to Arizona. Mexico: Sonora east to Coahuila.

Park locations: Monilla Canyon, Panther Canyon, Rancherías Canyon, Cienega Mountain

Growing from a woody crown, Mexican skullcap is a nonaromatic perennial to 16 inches high, with upright, four-angled, pale green stems sparingly branched above. Distinctively, the stems are usually clothed with nonglandular and only sparse glandular hairs. The short-petioled leaves, 1/3 inch long, bear both nonglandular and glandular hairs. The gray- or yellow-green blades are often egg-shaped, with a blunt tip and smooth, thick, upturned margins.

From spring to fall, curious hooded, violet-purple flowers form in small groups on short stalks from the leaf axils, or in paired bracts along the stems. Each two-lipped bloom has a helmet-like upper lip, with a projecting central lobe and down-curved side lobes. The spreading lower lip is notched at the tip, with a white, purple-dotted, central patch, and shoulder-like side lobes. Within the flower's throat are four stamens with hairy anthers and a style with a bilobed stigma. The maroon calyx supporting each bloom is two-lipped, the upper lip with a shield-like structure or "skullcap" that gives the genus its common name.

The fruit consists of four ovoid, one-seeded black nutlets with checkered protrusions.

Botanical notes: The genus name, from the Latin *scutella* (shield), refers to the calyx shape, as does the common name skullcap. The species is named for the Mexican state of San Luis Potosí, where the type specimen was found. The Latin variety name *tessellata* (checkered) refers to the checkered nutlets. The species was described by California botanist Townshend Brandegee in 1911, from a collection by freelance collector Carl Purpus in 1910, at Minas de San Rafael, in San Luis Potosí. Variety *tessellata* was described as a species, *S. tessellata*, in 1939 by University of California botanist Carl Epling and collected by Marcus E. Jones in 1903 in the Huachuca Mountains of Arizona. Jones, a self-taught botanist, collected in the western United States and edited his own journal, *Contributions to Western Botany*.

LAMIACEAE — MINT FAMILY

166. Desert Teucrium (*Teucrium depressum*)
(Other common names: small coastal, dwarf, alkali, or combleaf germander)

Habitat: Rio Grande, canyons draining into river; creeks, arroyos, ponds, mesquite scrub, disturbed sites.

Geographic range: Texas: Trans-Pecos (El Paso, Hudspeth, Presidio, Brewster Counties). US: Texas to California. Mexico: Baja California to Coahuila.

Park locations: Fresno Creek, Monilla Canyon, dam north of Ojito Adentro, Llano Loop, Pila Montoya Road, Cienega Creek

Seldom more than 1 foot high, desert teucrium is a colony-forming annual, the smallest *Teucrium* species in the United States. Growing from creeping rhizomes or a woody crown, the four-angled stems are commonly much branched from the base and strictly upright. The upper stems at least bear long, at times rough, spreading hairs.

The light- to gray-green leaves are oblong or spatulate, sometimes lobed. The upper leaves, to 3/5 inch long, are tightly clustered along the branches and often longhaired like the stems. The early-falling basal leaves, to 1 1/2 inches long, are tapered to distinct petioles.

From spring to fall, small, short-petioled flowers emerge from the leaf axils, all along the stems. Each two-lipped bloom is white with purplish streaks or splotches. The short upper lip is divided into two small lobes, and the large, flattish lower lip has a protruding central lobe and two small lateral lobes. Within the flower's throat are four stamens in two pairs of unequal size, projecting between the two lobes of the upper lip; and a protruding style, bilobed at the tip. Supporting each bloom is a bell-shaped calyx with five bristle-tipped lobes. The fruit consists of four one-seeded oval nutlets 1/13 inch long.

Botanical notes: The origin of the generic name is obscure. The name may be borrowed from the Greek *teukrion*, a term used by Dioscorides for a related plant, likely referring to Teukros or Teucer, the first King of Troy, who reportedly used the plant medicinally. The Latin species name *depressum* (flattened) refers to the plant's low stature. The species was described by John Small, curator of the New York Botanical Garden, in 1899, from specimens collected by George Thurber in 1851 at El Paso, and Charles Wright in 1852, in "low bottom of Rio Grande" near El Paso. Thurber was botanist, quartermaster, and commissary on the US-Mexico Boundary Survey (1850–53).

167. Rock Flax (*Linum rupestre*)
(Other common names: lino)

Habitat: grassy flats, gravelly hillsides, rocky canyons, in rock crevices, below shrubs, low to high elevations, on limestone.

Geographic range: Texas: Trans-Pecos (excluding El Paso, Reeves Counties), parts of Edwards Plateau, northcentral Texas; scattered in south Texas. US: Texas, southeast New Mexico. Elsewhere: Mexico (Chihuahua east to Tamaulipas, south through most of Mexico), Guatemala.

Park locations: Arroyo Mexicano, Lefthand Shutup

Rock flax is a fetching perennial seldom more than 16 inches high, mostly hairless, with widely spreading stems branching at the base and above. The wiry stems are markedly ridged or grooved. This flax is common in the Trans-Pecos, but uncommon at BBR, found mostly in limestone habitats. The sparse leaves are linear, to 3/4 inch long, alternate on upper stems, sometimes opposite below. Usually pressed against the stems, the blades are mostly stemless, and sharp-pointed.

From spring through fall, delicate yellow flowers appear on short stalks, in branched clusters, each bloom with five fragile, overlapping petals. The petals, slightly jagged on the edges, soon perish in blustery desert winds. Within the flower's throat are five stamens with oblong anthers and five styles, each with a globular, bristle-haired stigma. Supporting each bloom are five lance-shaped sepals, sharp-pointed and glandular-toothed. The fruit is an oval capsule to 1/8 inch long, which splits into 10 one-seeded chambers, each with a flat, oily, chestnut-brown seed.

Botanical notes: The Latin genus name *Linum* (flax), from Greek *linon* (flax), is the name used by Theophrastus. The Latin species name *rupestris* (on rocks) refers to the habitat. In 1850, in *Plantae Lindheimerianae*, Asa Gray initially described this species as a variety, *Linum bootii* var. *rupestre*, from a collection by Ferdinand Lindheimer in New Braunfels, Texas, in 1846, "growing sparsely on rocky soil or in crevices of naked rocks." Later, in the same article, Gray indicated that the variety "is certainly a distinct species, as Dr. Engelmann had stated." Gray then cited the name *Linum rupestre* and Engelmann's Latin description.

168. Yellow Rocknettle (*Eucnide bartonioides*)
(Other common names: yellow stingbush, Warnock's rock nettle, rock nettle, meloncito, hierba de la peña)

Habitat: Rio Grande floodplain, creeks, springs, steep rocky or gravelly cliffs, overhangs, talus slopes; limestone, igneous exposures.

Geographic range: Texas: Trans-Pecos (Jeff Davis, Presidio, Brewster, Terrell, Val Verde Counties), Edwards Plateau; Webb, Starr Counties. US: Texas. Mexico: Chihuahua, Coahuila to San Luis Potosí.

Park locations: River Road, Monilla Canyon, lower Fresno Canyon, Rancherías Canyon, Madrid Falls area, upper Terneros Creek

A frequent resident of rockslides and the beds of ephemeral creeks, this handsome rocknettle is a low, spreading annual or perennial to 1 foot high, which may form large mounds, or colonize entire cliffsides. The herb develops much branched, leafy stems, with fleshy or slightly brittle branches. The pale green stems bear a mix of hairs, some needlelike and mildly stinging.

Bright to dark green, the alternate leaves are nearly round but heart-shaped at the base, with a blade to 4 inches long and a slender petiole often longer than the blade. Typically, the blades are weakly lobed, irregularly toothed or jagged on the margins, and cloaked with rough, stiff hairs.

Through much of the year, yellow funnel-shaped flowers, 2 inches or more across, bloom in the leaf axils, on long hairy stalks. The blossoms open in late afternoon, capturing and brilliantly reflecting sunlight at the day's end. Each bloom consists of five spade-shaped petals joined at the base and pointed at the tips; up to 100+ yellow stamens, projecting well beyond the petals; and a single style with a five-lobed stigma protruding beyond the stamens. Supporting each blossom are five lance-shaped, persistent sepals.

The fruit is a dry, many-seeded, dehiscent round capsule, to 1/2 inch long, with up to 1,000+ minuscule seeds.

Botanical notes: The generic name, from the Greek *eu-* (good, well) and *knide* (nettle), implies "very nettle-like." The species name combines *Bartonia*, a genus, and the Greek suffix *eidos* (resembling), thus "like *Bartonia* species." Munich botanist Joseph Zuccarini described this genus and species in 1844, from a plant in the Munich Botanical Garden grown from seeds collected in Mexico. The seeds were likely collected by Baron Wilhelm Friedrich Karwinsky, perhaps on his second trip to Mexico (1840–43). The trip was financed by Russian sponsors in St. Petersburg, who paid him to search for mineral deposits as well as plants.

169. Organ Mountain Blazingstar (*Mentzelia asperula*)
(Other common names: rough-stem or mountain stickleaf)

Habitat: desert scrub, canyons, woodlands; igneous, limestone, gypsum hills.

Geographic range: Texas: Trans-Pecos (El Paso, Jeff Davis, Presidio Counties). US: Texas, southwest New Mexico, southern Arizona. Mexico: Nuevo León west to Sonora, south to Durango, Zacatecas, central Mexico.

Park locations: Yedra Canyon, Solitario Peak

This stickleaf is a weak annual to 20 inches high, growing from a slender taproot, with a few upright or partly reclining stems. Seeking the protection of trees and shrubs, this herb is also protected by a cover of rough, stiff hairs on stems and leaves. The glochidiate hairs are barbed like the glochids on cacti.

The alternate leaves are egg-shaped, or lancelike on the upper stems, often strongly lobed or toothed, with earlike basal lobes and/or narrow incised segments. On slender petioles, the blades may reach nearly 3 inches long, with barbed hairs on both petioles and blades.

From late spring to early fall, five-petal orange flowers form in the leaf axils. Each petal is inversely egg-shaped, usually with a short, abrupt, bristlelike tip and a narrow base. Within the petals are ten-plus slender stamens, red or orange, with cream-colored anthers and a threadlike style, shorter than the petals. Supporting each bloom is a light green calyx with five narrow, bristle-haired lobes, each with a long-tapered tip. Below the calyx is a cylindrical ovary, pale green but whitehaired, which becomes the fruit. The attractive blooms open in early morning and may close by noon. Short-stalked or stalkless, the fruit is a dehiscent cylindric capsule to 3/5 inch long, with seven to twelve pear-shaped seeds.

Botanical notes: Charles Plumier, French monk and royal botanist to King Louis XIV, described the genus in 1703, from a plant found on his West Indies journeys (1689–95). Plumier named the genus for German botanist Christian Mentzel, physician to the Elector of Brandenburg and father of the first King of Prussia. In 1753, Linnaeus adopted Plumier's genus. Elmer Wooton and Paul Standley described the species in 1913, from a 1904 collection by O. B. Metcalfe, on Trujillo Creek in Sierra County, New Mexico. The Latin species name, from *asper* (rough) and the suffix *-ula* (somewhat), likely refers to the stems, described by Wooton and Standley as "at first scabrous but becoming smooth below."

170. Mexican Stickleaf (*Mentzelia mexicana*)
(Other common names: Mexican blazingstar)

Habitat: often desert scrub near Rio Grande; silty, clayey, gypseous hills and mounds, alluvial flats, washes, limestone bluffs; low elevations.

Geographic range: Texas: Big Bend (south Presidio, south Brewster Counties). US: Texas. Mexico: eastern Chihuahua, Coahuila.

Park locations: Big Hill, Tapado Canyon, gypseous clay hills north of Buena Suerte Trail

Mexican stickleaf is a low perennial, seldom more than 1 foot high, densely branched from the base, often with whitish stems, brittle with age. Stems and leaves are usually rough, from a mix of straight hairs and odd, pagoda-shaped (glochidiate), barbed hairs.

The distinctive leaves have a wide midrib and relatively small, rounded lobes, commonly shorter than the midrib's width. These characteristics help to distinguish this species from the more widespread long-lobed species, many-flowered stickleaf (*Mentzelia longiloba*). Seldom more than 2 inches long, the stemless blades are oblong, roundly or bluntly toothed on the margins.

From spring to fall, in late afternoon, this stickleaf may be blanketed with ten-petal yellow flowers, each with many glistening yellow stamens. The blossoms appear in the leaf axils or in clusters at the branch tips. The petals are variable, lance- or egg-shaped or spatulate, with a pointed tip and clawlike base. Within the petals are twenty-plus stamens, the outer at times much longer than the inner, and some outer stamens with broad, petal-like filaments; and a three-branched, threadlike style as long as the longer stamens. Supporting each bloom is a calyx with five slender, sharp-pointed lobes.

The fruits are cuplike or short-cylindric capsules to 3/5 inch long, with calyx lobes persistent on the rim. Yellow green, tan, or white, each capsule contains many seeds, ovoid but compressed and winged.

Botanical notes: This herb is primarily a Mexican species, therefore the epithet *mexicana*. The species was described in 1968 by UCLA Loasaceae specialist Henry Joseph Thompson and his student Joyce Zavortink, from a 1940 collection by plant ecologist Forrest Shreve and Canadian-born entomologist E. R. Tinkham, near La Rosa, Coahuila, Mexico. Shreve worked at the Carnegie Institute's Desert Laboratory in Tucson, Arizona, and became one of the foremost authorities on desert plant life. His *Vegetation and Flora of the Sonoran Desert*, coauthored with Ira Wiggins, was published posthumously in 1964. Tinkham published articles on the Cicadidae (1941) and Lepidoptera (1944) of the Big Bend.

171. Coahuila Blazingstar (*Mentzelia pachyrhiza*)
(Other common names: thickroot stickleaf)

Habitat: near Rio Grande; creosote scrub, gravelly hills, rocky cliffs, river canyons; igneous, limestone, gypseous soils.

Geographic range: Texas: Big Bend (southeast Presidio, south Brewster Counties). US: Texas. Mexico: Chihuahua, Coahuila, Durango.

Park locations: South Lajitas Mesa, Las Cuevas

Of conservation concern in Texas: see Appendix B

Like the more widespread chicken-thief stickleaf (*Mentzelia oligosperma*), Coahuila blazingstar is a low, rounded, leafy perennial to 1 foot high, with only one or a few erect or reclining stems, branched at the base and above. The sprawling, strikingly whitish, at times brittle branches arise from a spindle-shaped, fleshy taproot, the nutrient storehouse.

Distinctively, the alternate leaves are borne on relatively long petioles, to half the length of the blades, which are egg-shaped or triangular, to 3/4 inch long. Commonly, the leaves are shallowly lobed, broadly and irregularly toothed on the edges, and rough, with barbed, conspicuously sticky hairs.

Opening in early morning from late spring to fall, five-petal, short-stalked orange flowers appear singly or in few-flowered clusters near the branch tips. The petals are elliptic, with a short, abrupt tip. Within the petals are up to thirty-five stamens, with threadlike orange filaments and yellow anthers, and a style slightly shorter than the stamens. A five-lobed calyx, covered with barbed hairs, supports each bloom.

As much as 1/2 inch long, the stalkless yellow green fruit capsules are cylindrical, with a few small, flattened and wrinkled seeds. Compared to chicken-thief stickleaf, Coahuila blazingstar has smaller leaves but longer petioles, whiter stems, smaller flowers, and smaller seeds with distinctive transverse folds.

Botanical notes: Harvard University botanist and Boraginaceae authority Ivan Johnston described this species in 1940 from a specimen he collected in 1938 in Coahuila, Mexico, 11 miles north of Parras, at the "foot of a steep sandstone slope." Johnston named the species *pachyrhiza*, from the Greek *pachy* (thick) and *rhiza* (root), because "just below the surface of the soil the root becomes abruptly enlarged to form a fusiform or narrowly ellipsoidal mass of fleshy storage tissue. . . . It is unique in the genus." This stickleaf was collected by Josiah Gregg near Saltillo, Coahuila, in 1848, but misidentified as *M. aspera*.

172. California Loosestrife (*Lythrum californicum*)
(Other common names: hierba del cáncer, common or purple loosestrife)

Habitat: damp or wet soil by creeks, springs, tanks, seeps, ditches, gullies; moist grasslands.

Geographic range: Texas: Trans-Pecos (El Paso, Reeves, Presidio, Brewster, Terrell, Val Verde Counties); much of western two-thirds of state. US: southwest (excluding Colorado); Oklahoma, Kansas. Mexico: Baja California east to Tamaulipas.

Park locations: lower Fresno Canyon, Rancherías Canyon, Lower Shutup, Contrabando Waterhole, Cienega Creek

This loosestrife is a leafy perennial to 3 feet high, which spreads by creeping, woody rhizomes. Arising from this underground network are pale green, hairless, four-angled stems. The slender stems are upright or arching, with many diverging and sprawling branches. The mostly stemless, hairless leaves are linear, to 1 inch long, smaller and alternate above, larger and often opposite below. Gray green and glaucous, the blades are smooth on the margins and blunt at the tip.

Small but attractive flowers form in the upper leaf axils from spring to fall, in spikelike clusters, the blooms nearly stalkless, with a pair of small bractlets at the base. Each bloom usually has six fan-shaped petals attached to the rim of the calyx tube. Often wrinkled and curling, the petals are pinkish purple with striking dark midribs. Within the petals are usually six stamens, with pinkish-purple filaments and yellow anthers, and an unbranched white style with a globelike yellow stigma. Supporting each bloom is a yellow green, cylindric calyx, strongly grooved, with short, triangular lobes at the tip.

The yellow-green fruit is a cylindric, ribbed capsule to 1/3 inch long, many-seeded, enclosed by the calyx.

Botanical notes: The genus name *Lythrum*, from the Greek *lythron* (blood), may refer to the flower color, or to the ancient use of some species to stop bleeding and heal wounds. In 1840, this species was described by New York State Botanist John Torrey and his protégé and collaborator Asa Gray, in their milestone work *A Flora of North America*, from a California specimen collected by Scottish botanist David Douglas in 1833. In 1823, Glasgow botanist Sir William Jackson Hooker sent Douglas on a pioneering expedition to California, the Pacific Northwest, and Hawaii, where Douglas died a suspicious death in 1834 at the age of thirty-five, while climbing the Mauna Kea volcano.

173. Hyssopleaf Asphead (*Aspicarpa hyssopifolia*)
(Other common names: hyssop-leaf aspicarpa)

Habitat: limestone hills, slopes above desert drainages, below rocky bluffs, mid-elevations.

Geographic range: Texas: Trans-Pecos (Presidio, Terrell, Val Verde Counties), parts of southern Edwards Plateau, south Texas. US: Texas. Mexico: Chihuahua east to Tamaulipas.

Park locations: Solitario northwest rim, Solitario rim and interior slopes above Righthand Shutup

Hyssopleaf asphead is a shrublet to 1 foot high, woody at the base, with slender stems and short branchlets. Upright, sprawling, or reclining, the grayish-white stems bear a thick cover of silky hairs.

To 1 inch long, the stemless leaves are linear to lance-shaped, at times nearly clasping the stem, with smooth edges and a small, sharp-pointed, often red tip. The blades are paired along the stems, often with each pair at right angles to the pair below. Typically, the leaves are sparsely hairy on the margins and the midribs.

From late spring to fall, this dwarf shrub produces two kinds of flowers: tiny self-fertilizing, petal-less flowers, scarcely noticed, embedded in the leaf axils and never open; and unusual, five-petal, showy blossoms, on stalks from the leaf axils. Each blossom is yellow, aging orange, with wide-spaced, fan-shaped petals, clawed at the base and crinkled on the margins. Within each bloom are five yellow stamens, only two or three fertile with anthers, and a style with a headlike stigma. Supporting each bloom is a yellow-green, silky-haired, five-lobed calyx, four lobes with paired, bulging glands at the base, each with knobby red-brown tips.

The fruit usually consists of two crested nutlets, each with a single seed. The whitehaired nutlets are 1/6 inch long, green but red with age, with swollen sides and a protruding, triangular tip.

Botanical notes: The genus name joins the Greek *aspis* (shield) and *karpos* (fruit), for the shieldlike fruit shape. The species name combines the Greek *hyssopos* (aromatic herb described by Dioscorides) and the Latin *folium* (leaf), for the hyssop-like leaves. Asa Gray described the species in 1850, in his *Plantae Lindheimerianae*, mostly an account of New Braunfels botanist Ferdinand Lindheimer's plant collections in central Texas. Gray listed two 1849 asphead collections by Charles Wright, one on the Rio Grande and the other on "Rio Seco," probably Seco Creek, a tributary of Hondo Creek in Medina County.

MALPIGHIACEAE — BARBADOS CHERRY FAMILY

174. Propellerbush (*Cottsia gracilis*)
(Other common names: slender janusia, propeller-plant, desert or janusia vine, fermina)

Habitat: riverbanks, desert scrub, grasslands, canyons; igneous, limestone slopes.

Geographic range: Texas: Trans-Pecos (excluding Reeves, Pecos, Val Verde Counties). US: Texas, southwest New Mexico, Arizona. Mexico: Baja California east to Coahuila, south to Durango, Zacatecas.

Park locations: Tapado Canyon, Ojito Adentro, Upper Guale Mesa, Chorro Canyon, Cinco Tinajas

Propellerbush is a twining plant, half vine, half shrub, woody at the base, with wiry stems to 16 inches long. The plant drapes over shrubs or twines and clambers through their branches. The pallid stems are coated with malpighian hairs attached to the stems at the middle by a short stalk. The paired leaves, to 1 1/2 inches long, are linear to lance-shaped, nearly stemless, with pointed tips. The blades are green, or red tinged, with malpighian hairs.

From spring to fall, flowers appear in pairs on stalks from the leaf axils, or in small clusters on short branchlets. Each bloom has five propeller-like yellow petals, wide-spaced, clawed at the base, typically fringed, wavy, and wrinkled on the margins. Within the petals are six stamens, some usually sterile, lacking anthers, and a style with a headlike stigma. Supporting each bloom is a glandular calyx with five small lobes, four or five with minuscule oil-producing glands at the base.

Female oil-seeking bees (genus *Centris*) pollinate the blooms while harvesting the oil and take an oil-pollen mix to their nests as larval food. This vine-shrub lacks the self-fertilizing flowers of some Malpighia species. The fruit—green to coral at maturity—consists of one to three propeller-like winged capsules (samaras), to 1/2 inch long, each enclosing a single net-veined nut.

Botanical notes: In 1840, Adrien-Henri de Jussieu named this genus *Janusia*, for Janus, the Roman god of passageways, who was depicted with two faces. Since the plants have paired flowers, and many have two flower types, with and without petals, two-faced Janus was an apt metaphor. The Latin epithet *gracilis* (slender) refers to the stems. Asa Gray described the species in 1852, from a collection by Charles Wright in 1849, in mountains east of El Paso in Culberson County. In 2007, Malpighiaceae specialists William Anderson and Charles Davis placed North American species of *Janusia* in *Cottsia*, a genus described by Marcel Dubard and Paul Dop in 1908 (Anderson and Davis 2007).

175. Dwarf Ayenia (*Ayenia pilosa*)
(Other common names: hairy ayenia)

Habitat: rocky hills, canyons, foothills; mixed desert scrub; limestone, igneous mountains.

Geographic range: Texas: Trans-Pecos (Jeff Davis, Presidio, Brewster, Terrell, Val Verde Counties); parts of South Texas Plains. US: Texas, New Mexico. Mexico: Chihuahua, Coahuila, Tamaulipas, San Luis Potosí.

Park locations: Righthand Shutup

Dwarf ayenia is a seldom seen woody subshrub to 10 inches high, much branched at the base, with ascending to prostrate stems and spreading branches. The branchlets are green and white-haired when young, brown and hairless with age. The plant is much less common than the other two *Ayenia* species in BBR.

Short-petioled, the alternate leaves are oblong to egg-shaped, to 1 inch long. The blades are light green above, paler and strongly veined below, pointed or blunt at the tip, toothed on the edges, and sparingly hairy.

From spring to fall, intricate red to maroon flowers form in small clusters, on slender stalks at the stem nodes. Often, the odd blooms are likened to parasols, hanging baskets, or parachutes. Only 1/8 inch long, each tiny bloom has five fan-shaped, hooded petals, each with a threadlike, pinkish claw curving backward from the petal tip to the central stamen tube below. Collectively, the arching petals form a parachute-like canopy, with the claws resembling parachute cords. Below the petals are ten stamens, with filaments fused into a slender tube. The tube consists of five fertile stamens and five smaller, sterile stamens. At the canopy's center, a short style, with a globular white stigma, scarcely emerges amid the petals. Five pale greenish-yellow sepals, tapered at the tips and stellate-hairy, support each bloom.

The fruit is a round, greenish-yellow capsule to 1/6 inch long, covered with spikelike protrusions. The dehiscent capsule is five-chambered, with each chamber containing one dark, egg-shaped seed.

Botanical notes: Linnaeus described the genus *Ayenia* in 1756, from a cultivated plant grown from Peruvian seeds. The genus was named for Louis de Noailles, the Duc d'Ayen until 1766, when he became Duc de Noailles. A student of botanist Bernard de Jussieu at the Jardin du Roi in Paris, Noailles supposedly influenced King Louis XV's appointment of Jussieu, a friend of Linnaeus, as superintendent of the royal garden at Trianon. The Latin species name *pilosa* (pilose) refers to the leaves. Argentine botanist Carmen Lelia Cristóbal described the species in 1960, from a specimen collected by Carl Purpus in 1910, at Minas de San Rafael, in the Mexican state of San Luis Potosí.

176. Arrowleaf Mallow (*Malvella sagittifolia*)
(Other common names: arrowleaf alkali mallow)

Habitat: ephemeral wetlands, alluvial basins, sandy washes, clayey, loamy, silty flats, bajadas, fields, disturbed sites.

Geographic range: Texas: Trans-Pecos (southeast Hudspeth, Jeff Davis, Presidio, Brewster Counties); sites in Permian Basin, south Panhandle, south Texas. US: Texas to Arizona, Colorado. Mexico: Sonora to Coahuila, Tamaulipas, south to Durango, San Luis Potosí.

Park locations: Llano Loop, drainage off Sauceda Road 3 miles east of Sauceda

Arrowleaf mallow is a low perennial with spreading or trailing, mostly prostrate stems to 16 inches long. Both stems and leaves are gray green to silvery, with shield-shaped, scurfy scales. As much as 1 1/3 inches long, the alternate leaves are narrowly triangular to lance-shaped, with a pointed tip, and two to four arrowhead-shaped lobes at the base. The short-petioled blades, at times lopsided, are conspicuously spotted with scales.

From late spring to fall, single five-petal flowers appear in the leaf axils, on short stalks that are

scaly like the stems. The often inward-arching, fan-shaped petals, to 1/2 inch long, are white, blushing pink, usually pinker or rosier underneath. Within the petals are numerous stamens with white filaments and yellow anthers, the filaments united, attached to the base of the petals; and seven or eight thin style branches, with dot-like red stigmas. Below each bloom is a scaly calyx, with five heart-shaped to triangular lobes.

The fruit is a spheric, but flattened, capsule to 1/4 inch across, which splits into seven or eight finely hairy, one-seeded segments, each with a single pendent seed.

Botanical notes: *Malvella*, diminutive of *Malva*, the genus, means "resembling a small *Malva*." The species name combines the Latin *sagitta* (arrow) and *folium* (leaf), for the "arrowlike leaves." In 1852, Asa Gray described this plant as *Sida lepidota* var. *sagittifolia*, from an 1849 collection by Charles Wright, in a "valley about 80 miles beyond the Pecos," in Pecos County, Texas. In 1974, Paul Fryxell, Malvaceae authority and botanist with the Agricultural Research Service of the USDA, elevated this variety to species status in the genus *Malvella*.

177. Pyramid Flower (*Melochia pyramidata*)
(Other common names: pyramid bush, anglepod melochia, broomwood, huinar, escobilla morada)

Habitat: creek banks, dry creek beds; disturbed sites; sandy, clayey, rocky, alluvial soils, often limestone canyons at low elevations.

Geographic range: Texas: Trans-Pecos (Presidio, Brewster, Terrell, Val Verde Counties), South Texas Plains, Gulf Coast; parts of Edwards Plateau. US: Texas, Louisiana, Florida. Elsewhere: Mexico, Central, South America, West Indies.

Park locations: Monilla Canyon, Contrabando Creek

Pyramid flower is a subshrub to 3 feet high, erect or spreading, with slender, at times reddish, stems, branching at the base. The alternate, long-petioled green leaves, to 2 inches long, are egg- or lance-shaped, with impressed veins above and raised veins below. Blunt or pointed at the tip, the blades are serrate on the edges. The herbage is sparingly hairy, and hairless with age.

From spring to fall, small flowers bloom singly or in loose, few-flowered clusters off short stalks. Each flower has five broadly spatulate petals, pink, pale violet, or lavender, with a yellow throat and tube. Within the flower's throat are five stamens, with pale yellow filaments and bright yellow, oblong anthers and a hairy, branching style. Supporting each bloom is a light green calyx, with five narrowly lance-shaped lobes, the lobes often red tinged and hairy.

The fruit is a five-lobed dehiscent capsule, pyramidal or top-like, to 1/3 inch long. Pale yellow green, with reddish or purplish blotches, each capsule has a long beak and lobes with protruding tips. Each lobe contains two brown, egg-shaped to three-angled seeds.

Botanical notes: Apparently, the genus name *Melochia* was borrowed from an Arabic term (*meluchiye*) for an oriental mallow (*Corchorus olitorius*). Prospero Alpini, physician to the Venetian consul in Cairo and later professor of botany at Padua, described Melochia and other Egyptian plants in 1592. The Latin species name *pyramidata* (pyramid) references the fruit capsules. Linnaeus described the species in 1753, citing an illustration and plant description under an earlier name, *Althaea brasiliana*, by Leonard Plukenet (1692, 1696), and a habitat in Brazil. The lectotype is from Brazil, but the collector is unknown.

178. Narrowleaf Globemallow (*Sphaeralcea angustifolia*)
(Other common names: copper globemallow, narrowleaf desertmallow, hierba del negro, cordón, vara de San José, tlixihuitl [Náhuatl])

Habitat: desert scrub, grassland, woodlands; clayey, loamy, rocky hills, terraces, sandy, gravelly arroyos; calcareous, gypseous soils.

Geographic range: Texas: Trans-Pecos, Permian Basin, west Edwards Plateau; sites in Panhandle; Starr County. US: southwest (excluding Utah); Texas to Nebraska. Mexico: northern Mexico to Mexico City, Puebla.

Park locations: Fresno Canyon, Madera Canyon Campground, Black Hills, Agua Adentro, McGuirk's Tanks, Smith House

Narrowleaf globemallow is an upright perennial or subshrub to 3 feet high, which grows from a broad, woody crown. This common, widespread globemallow produces long, leafy stems, erect and sparingly branched. Both stems and leaves are coated with a woolly down.

To 4 1/2 inches long, the gray-green leaves are narrowly lance-shaped to oblong, blunt-tipped, occasionally with small earlike lobes at the base, and broadly toothed, wrinkled, and wavy margins. Short-petioled, the soft, thick leaves are frequently folded inward toward the midrib.

From early spring to fall, showy, 1-inch-wide blossoms appear in long, leafy, many-flowered clusters. On a short, sturdy stalk, each bloom has five slightly cupped, wedge-shaped petals. In the Big Bend, flower color ranges from lavender-pink to orange or salmon, but other colors are reported. Within the petals are numerous stamens with purple filaments and yellow anthers, the filaments united into a tube around the style; and ten to sixteen style branches, each with a headlike stigma. Supporting the bloom is a stellate-hairy calyx with five lance-shaped to triangular lobes.

The fruit is a capsule that splits into as many as sixteen thin-walled segments to 1/4 inch long. Each segment consists of a dehiscent upper part and indehiscent lower part, with one to three brown or black seeds.

Botanical notes: The generic name combines the Greek *sphaira* (globe) and the genus *Alcea* (mallow), or "globemallow," for the globular fruits. The species name joins the Latin *angustus* (narrow) and *folium* (leaf), for the narrow leaves. In 1786, this species was described by Antonio José Cavanilles as *Malva angustifolia*, from a plant cultivated in the Royal Botanical Garden in Madrid. Garden director and foremost taxonomic botanist, Cavanilles wrote and illustrated a six-volume masterwork, *Icones et Descriptiones Plantarum* (1791–1801), documenting new plants from Latin America and the West Indies. Scottish botanist George Don transferred the species to the *Sphaeralcea* genus in 1831.

179. Juniper Globemallow (*Sphaeralcea digitata*)
(Other common names: slippery globemallow)

Habitat: desert scrub, grasslands, rocky limestone slopes, ravines, canyons; mid-elevations.

Geographic range: Texas: Trans-Pecos (El Paso, Hudspeth, Culberson, Presidio, Brewster Counties). US: Texas to northern Arizona, southern Utah). Mexico: Chihuahua, Coahuila.

Park locations: Solitario northwest rim, Paso al Solitario area, Los Portales, Solitario rim above Righthand Shutup

Juniper globemallow is a perennial or shrublet to sixteen inches high, woody at the base, with slender ascending to nearly reclining stems. On petioles usually shorter than the blades, the leaves, to 3/4 inch long, are deeply cleft into three to five narrow segments. Typically, both stems and leaves bear sparse stellate hairs.

From spring to fall, cup-shaped flowers appear in clusters of up to twenty blooms, each bloom usually single in leaf axils near the branch tips. On a stout stalk, each striking blossom has five often grenadine, wedge-shaped petals. The petals are darker, more saturated in color near the flower's throat. Within the petals are numerous stamens with maroon-red anthers, the filaments united into a tube, and about twelve styles with headlike stigmas. Supporting each bloom is a five-lobed, stellate-hairy calyx, with three minuscule, threadlike bractlets at the base.

The fruit is a spherical but compressed capsule 1/4 inch wide, with about twelve one-seeded segments. Each segment has a dehiscent upper portion and an indehiscent, net-veined lower portion.

Botanical notes: The Latin species name *digitata* (digitate) refers to the lobed leaves. This species was described in 1905, by Edward Lee Greene, as *Malvastrum digitatum*, and collected in 1904 by Orrick Metcalfe at Kingston, New Mexico. Greene, the first botanist at the University of California, Berkeley, published two influential botanical journals, *Pittonia* and *Leaflets of Botanical Observation and Criticism*. Metcalfe, student of Elmer Wooton, collected in the Black Range and Mimbres Valley and authored a thesis on *The Flora of the Mesilla Valley* (1903). In 1913, Per Axel Rydberg reclassified the species as *Sphaeralcea digitata*.

180. Hooked Water Clover (*Marsilea vestita*)
(Other common names: hairy water clover, hairy or hooked pepperwort, clover fern)

Habitat: moist, muddy ground; ephemeral pools, periodically inundated banks, tanks.

Geographic range: Texas: Trans-Pecos (excluding Hudspeth, Culberson, Reeves Counties), Edwards Plateau, much of state (excluding East Texas). US: west half; Florida to Louisiana, Arkansas; Iowa, Minnesota. Elsewhere: Canada, Mexico, Peru.

Park locations: Llano Loop, Tascate Hills

Resembling four-leaf clover, hooked water clover grows from creeping rhizomes that root at the nodes in mud or shallow water. This aquatic fern lacks stems, but the cloverlike leaves rise to nearly 8 inches above the fibrous roots on threadlike petioles. These ferns reproduce sexually by spores, but more often vegetatively by means of branching rhizomes.

The leaf blades are cross-shaped, divided into four wedge-shaped leaflets. Each stemless, hairy leaflet is up to 3/4 inch long, somewhat asymmetrical and hooked, at times with red-brown streaks below.

From spring to fall, this water clover forms nutlike sporocarps resembling peppercorns (hence the name pepperwort). The sporocarps store sporangia, spore-bearing sacs in which the spores develop. To 1/3 inch long, compressed, each sporocarp is borne on a short, upright stalk at the base of the fern's long petiole. The sporocarp is two-toothed, the lower tooth humplike, the upper tooth pointed and often slightly hooked. Megasporangia, with one large female megaspore, and microsporangia, with up to sixty-four male microspores, reside in the sporocarp. When wet, the sporocarps split, releasing spores in a jellylike liquid that carries the spores to water. In water, microspores form male gametophytes, megaspores female gametophytes. Sperm from male gametophytes fertilizes eggs of female gametophytes, and the eggs form new plants.

Botanical notes: The genus is named for Italian Count Luigi Marsigli, who authored a work on fungi (1714) and the first scientific oceanographic study, *Histoire Physique de la Mer* (1725). The Latin species name *vestita* (clothed) refers to the herbage. University of Glasgow botanist Sir William Jackson Hooker and Edinburgh botanist-botanical illustrator Robert Kaye Greville described and illustrated this species in 1830, in *Icones Filicum*. Scottish botanical explorers David Douglas and John Scouler were the collectors, on the Columbia River in the Pacific Northwest, on a Hudson's Bay Company voyage (1824–26).

181. Desert Unicorn Plant (*Proboscidea althaeifolia*)
(Other common names: yellow-flowered, desert, golden, hollyhock, or perennial devil's claw, campanita, aguaro con camote)

Habitat: sandy, gravelly flats, dunes, washes, roadsides.

Geographic range: Texas: Trans-Pecos (El Paso, Hudspeth, Culberson, Presidio Counties). US: Texas to southern California. Elsewhere: Mexico (Baja California east to Coahuila; Baja California Sur, Sinaloa); Peru.

Park locations: just west of Fort Leaton, south of lower Terneros Creek

Desert unicorn is scarce at BBR, encountered once in the Alamito Creek–Terneros Creek Lowlands, and once outside the park near Fort Leaton. To 1 foot high, this clammy-hairy perennial grows from bulky tuber-like roots. At the peak of summer, the roots put forth fleshy, pale yellow-green aerial stems. The stems are widely branched at the base, nearly prostrate to ascending, and coated with glandular hairs. On petioles to 7 inches long, the leaves are less than 3 inches long and wide. The blades are rounded, kidney-shaped, or egg-shaped and often broadly toothed, wavy, or shallowly lobed on the margins. Usually blunt-tipped and heart-shaped basally, the leaves are glandular-hairy like the stems.

In summer, yellow flowers appear in short clusters with up to ten blooms. Each five-lobed blossom is funnel-shaped, to 2 inches long or more. Frequently, the two upper and two lateral lobes are strongly reflexed, the lower lobe protruding, and all lobes ruffled and wavy. Orange nectar guides for pollinators are prominent on the lower lobe. Within the lobes are four fertile stamens in two unequal pairs and one tiny sterile stamen, and a slender style with a bilobed stigma. Supporting each bloom is a membranous, sticky-hairy, five-lobed calyx with two yellow-green, oblong bractlets at the base.

The fruit is a dehiscent woody capsule with a narrow body to 2 3/10 inches long, crested on both sides, and a long-tapered beak. The capsule splits to form two incurved, hooked claws, with oblong black seeds.

Botanical notes: The genus name, from the Greek *proboskis* (snout), alludes to the fruit beak. The species name, from the genus name *Althaea* (hollyhock) and the Latin *folium* (leaf), connotes "with leaves like Althaea." The species was described by English botanist George Bentham in 1844 as *Martynia altheifolia,* in *The Botany of the Voyage of* H.M.S. *Sulphur.* The description was based on a collection by ship surgeon Richard Brinsley Hinds in 1841, at Baja California Sur, on the Bay of Magdalena. Joseph Decaisne, naturalist with the French Museum of Natural History, moved the species to the *Proboscidea* genus in 1865.

182. Doubleclaw (*Proboscidea parviflora*)
(Other common names: small-flower unicorn-plant; red, Wooton's, New Mexico, Arizona, or annual devil's claw, cuernero)

Habitat: sand, gravel, clay, alluvium, near Rio Grande; desert flats, depressions, roadsides.

Geographic range: Texas: Trans-Pecos (El Paso, Hudspeth, Jeff Davis, Presidio, Brewster Counties). US: Texas to southern California, southeast Nevada, southwest Utah. Mexico: Baja California to Chihuahua, south to Sinaloa, Durango.

Park locations: Madera River Access, Arroyo Mexicano

Doubleclaw is a sprawling annual with large leaves and small flowers, to 2 feet tall and 5 feet across. All plant parts are adorned with sticky glandular hairs. Widely branching, the stems are pale yellow green or brown, fleshy, yet rigid. The large, dark green leaves may approach 6 inches long on petioles equally long. Rounded, egg-shaped, or triangular, the blades are blunt at the tip, heart-shaped at the base, wavy and shallowly lobed on the edges. Often, the leaves are bumpy, swollen, and deeply veined.

In summer, up to ten tubular flowers bloom in short clusters. To 1 1/2 inches long, the flowers, with five flaring lobes, are multicolored, white and pink, with a magenta blotch on the upper lobes and a yellow nectar guide from the flower's throat to the broad lower lobe. Within the lobes are four fertile stamens in two unequal pairs, with white filaments and red anthers; a small sterile stamen; and a style with a bilobed flat stigma. Supporting each bloom is a five-lobed calyx with a pair of small bractlets at the base.

The fruit is a fleshy, crested, dehiscent capsule, with a woody body to 4 inches long and an even longer curving beak. The mature capsule splits to form two long, incurving, hooked claws, which snag the legs of passing animals and disperse the seeds.

Botanical notes: The species name, from the Latin *parvus* (small) and *flos* (flower), refers to the flower. In 1898, New Mexico State University botanist Elmer Wooton described this species as *Martynia parviflora*, from his collections in 1897 at the base of the Organ Mountains in Doña Ana County, New Mexico, and at Las Cruces in 1895. Wooton noted earlier collections by George Thurber, in Sonora (1851) and Chihuahua (1852), Mexico. In 1915, Wooton, with the Bureau of Plant Industry, and Paul Standley, with the United States National Museum, placed this plant in the *Proboscidea* genus.

183. Texas Mulberry (*Morus microphylla*)
(Other common names: mountain, Mexican, or littleleaf mulberry, mora, mora cimarrona, apuri [Tarahumara])

Habitat: creeks, springs, canyons; rocky limestone, igneous slopes.

Geographic range: Texas: Trans-Pecos (excluding Reeves, Val Verde Counties); parts of Edwards Plateau, north-central Texas, Panhandle. US: Texas to Arizona; Oklahoma. Mexico: Sonora to Coahuila, south to Durango.

Park locations: Agua Adentro, Las Cuevas, upper Arroyo Mexicano, Ojo Mexicano

This mulberry is a large shrub or tree to 25 feet high, with a stout trunk and scaly gray bark. Often, the bark is marked with corky, lens-shaped horizontal strips (lenticels). Large, dark reddish-brown buds are conspicuous on the hairy young stems and yellow green branchlets. As much as 3 inches long, the alternate leaves are broadly egg-shaped, pointed, often tapered at the tip, coarsely serrate and frequently lobed on the edges. Short-petioled, the blades are rough above and below.

Mostly in March and April, greenish-yellow male and female blooms appear in short catkins on separate plants. Male blooms are borne on hairy stalks in cylindric or spikelike clusters to 3/4 inch long. Female blooms appear on shorter stalks in shorter spikes with fewer flowers. Male blooms consist of four greenish-yellow, finely hairy, egg-shaped sepals and four stamens with threadlike filaments and white anthers. Female blooms have four greenish-yellow sepals, and a style with two white branches.

The edible fruit is a berrylike, aggregate fruit to 3/5 inch long, red to black at maturity. The lustrous fruit consists of tightly clustered achenes (small, indehiscent, one-seeded fruits), each within a fleshy calyx.

Botanical notes: The genus name is the classical Latin *Morus* (mulberry). The species name, from the Greek *mikros* (small) and *phyllon* (leaf), emphasizes the relatively small leaves. Naturalist Samuel Botsford Buckley described the species in 1863, from a plant he collected in "western Texas." Buckley collected botanical and zoological specimens in much of the United States then worked on the Texas Geological Survey (1860–61) and twice served as Texas state geologist (1865–67, 1874–77).

184. African Rue (*Peganum harmala*)
(Other common names: Syrian or wild rue, harmel, harmal, esfand, harmal peganum, alharma, gamarza)

Habitat: roadsides, pastures, alluvial flats, alkaline, clayey, silty, gypseous soils.

Geographic range: Texas: Trans-Pecos, parts of Permian Basin; sites in Edwards Plateau, south Texas, south Panhandle. US: Texas to Nevada, California, north to Washington, Montana. Mexico: expected.

Park locations: Casa Piedra Road

An exotic, invasive plant originally from Central Asia, African rue is a low perennial to 1 foot high. The herb develops spreading or prostrate, much branched, leafy green stems, which grow from a woody crown and deep, branching roots. Stems and leaves are fleshy and largely hairless. The alternate, stemless leaves, to 2 inches long, are dissected into threadlike segments, often with hooked, bristlelike tips.

From spring to fall, clusters of solitary, waxy white flowers appear on the upper half of the stems. At times on curving stalks, the blooms have five oblong petals, each as much as 3/5 inch long. Within the petals

are fifteen protruding stamens, with yellow anthers often longer than the pale green filaments, and a persistent, greenish-yellow style, arrowhead-shaped in the upper half. Supporting each bloom are five green, leaflike linear sepals, typically longer than the petals they surround.

To 2/5 inch long, the fruit is a dehiscent round capsule, usually three-chambered, with a long beak. Pale green but tan and papery with age, the capsule contains up to sixty dark brown, three-angled seeds.

African rue, although toxic to animals and humans, has long been used in traditional medicine to treat illnesses, including depression and heart disease. In Asia, a red dye, "Turkey Red," is obtained from the seeds. In Turkey, dried plant parts are hung in houses to ward off the "evil eye."

Botanical notes: The generic name *Peganum* was borrowed from the ancient Greek *peganon* (rue), which referred to similar fragrant herbs now in the *Ruta* genus. The species name may refer to the Lebanese town Hermel, or to the Arabic *harmal* (an old colloquial name for rue). Linnaeus described this species in 1753, citing Madrid, Alexandria, and Cappadocia and Galatia (both in Turkey) as habitats.

185. Narrowleaf Moonpod (*Acleisanthes angustifolia*)
(Other common names: sparse-leafed moonpod)

Habitat: bluffs above Rio Grande, drainages entering river; calcareous, gypseous, igneous soils.

Geographic range: Texas: Big Bend (Presidio, Brewster Counties). US: Texas. Mexico: Chihuahua, Coahuila, Durango.

Park locations: lower Fresno Canyon, Closed Canyon, Colorado Canyon River Access, Upper Guale Mesa, West Contrabando Trail

This moonpod is a low, erect, woody shrublet to 16 inches tall, branched at the base, with whitish upper branches. Growing from thick taproots, the plants bear a mix of gland-tipped and other hairs. Largely stemless, the paired leaves are linear, to 1 inch long, with thickish blades folded upward toward the midrib. The leaves are dark green, but often whitehaired like the stems, or gray green and maroon tinged.

From spring to fall, fragrant night-blooming, hawkmoth-pollinated flowers form in the axils of upper stems. Opening in late afternoon and closing the next morning, the short-tubed blooms abruptly flare into a five-lobed border (limb). The blooms lack petals, and the tube and limb are formed by a colorful calyx. Each bloom is multicolored, with a yellow-green tube and a brown-orange limb. Within the tube are five projecting stamens and a style with a shield-like stigma protruding above the stamens. This shrublet also develops small self-fertilizing flowers that never open, which saves energy while producing seeds in hostile habitats.

The dangling ornamental fruit is an oblong, compressed anthocarp (achene with the persistent calyx base around it), 1/4 inch long, leathery, ribbed, with five membranous wings and a dried calyx at the tip.

Botanical notes: Asa Gray created the genus *Acleisanthes* in 1853, from the Greek *a* (without), *kleis* (thing that opens and closes), and *anthos* (flower), highlighting the lack of bracts (involucre) below the flower cluster. The species name *angustifolia* joins the Latin *angustus* (narrow) and *folium* (leaf), for the slender leaves. John Torrey described this species as *Selinocarpus angustifolius* in 1859, from a collection by English-born botanist Charles Parry in 1852, on the US-Mexico Boundary Survey, on "gravelly table land near Presidio del Norte." In 2002, Rachel Levin and colleagues moved this plant to the *Acleisanthes* genus.

186. Goosefoot Moonpod (*Acleisanthes chenopodioides*)
(Other common names: none)

Habitat: sandy flats, rocky hillsides, desert scrub, calcareous, gypseous clay, igneous soils.

Geographic range: Texas: Trans-Pecos (excluding Terrell County). US: Texas to Arizona. Mexico: Chihuahua.

Park locations: Cerro de las Burras Loop, lower Arroyo Mexicano, Encino Loop Trail, Sauceda Road near Solitario turnoff

Scattered and uncommon in the Big Bend, once considered rare, this moonpod is a low perennial to 16 inches high, with erect or ascending stems and sturdy, spreading branches. The young gray-green branches and leaves can be densely hairy, with fine white, T-shaped hairs. The fleshy leaves are broadly egg-shaped, to 2 inches long, thick, blunt or pointed at the tip, rounded to heart-shaped at the base, and wavy or curling on the margins. With petioles often approaching the length of the blades, the paired leaves may be notably unequal in size and pale and strongly veined underneath.

From late spring to early fall, morning-blooming flowers appear in rounded clusters of up to twenty-five blooms, each cluster with one or two bracts at the base. On a short stalk, a blossom is no more than 1/4 inch wide, with a short yellow-green tube that flares into a broadly funnel-shaped pink to lavender rim (limb). The limb is divided into five shallow, pleated lobes, often notched at the tip. Within the floral tube are usually two protruding stamens (compared to five in related species) with threadlike pinkish filaments and yellow anthers and a slender projecting style with a shield-shaped stigma.

The fruit is an anthocarp, an achene-like, one-seeded nutlet enclosed in the base of the elongated calyx. The oblong leathery anthocarp is 1/5 inch long, finely hairy, with membranous wings and sharply ribbed grooves between the wings. The oblong seed is light brown, shiny.

Botanical notes: Asa Gray described this species in 1853 as *Selinocarpus chenopodioides*. The epithet *chenopodioides* implies "like plants of the goosefoot genus *Chenopodium*." In Gray's words: "Leaves . . . whitened, at least when young, with an apparently farinose fine pubescence, somewhat as in a *Chenopodium*." Charles Wright first collected the plant in "valleys from Deadman's Pass to the Wells," near Van Horn, in Culberson County, Texas, in 1851. Nyctaginaceae authority Rachel Levin moved this species to the genus *Acleisanthes* in 2002.

187. Angel Trumpets (*Acleisanthes longiflora*)
(Other common names: yerba de la rabia, yerba santa, trompeta de ángel, trompetilla, long-flowered four-o'clock)

Habitat: desert scrub, grasslands, woodlands, creeks, springs, rocky mountain slopes, disturbed sites.

Geographic range: Texas: Trans-Pecos (excluding El Paso County), parts of South Texas Plains, Edwards Plateau, southern Panhandle. US: Texas to southern California. Mexico: Sonora to Tamaulipas, south to Durango.

Park locations: Leyva Canyon, Cinco Tinajas, Fresno Rim Trail, Ojo Chilicote, Solitario northwest rim

Angel trumpets is a low perennial with stems to 3 feet long, which branch extensively, sprawl vinelike on other plants, and trail across the ground. Growing from a woody taproot, the stems are finely hairy. Typically, the leaves are gray green and glaucous, to 1 1/2 inches long, broadly lancelike or triangular, thickish, with wavy, curling, crinkled margins. On petioles to 1/2 inch long, the paired blades are pointed or tapered at the tip, truncate or heart-shaped at the base, and sparsely hairy.

The delicate flowers, with five pleated white lobes and a slender tube to 6 1/2 inches long, are an ideal shape for angelic pronouncements. Mostly solitary, the fragrant night-bloomers appear from spring to fall, opening in late afternoon, then drooping and closing in sunlight the following day. Each bloom has five stamens with yellow anthers, a style with a shieldlike stigma protruding beyond the anthers, and small linear to lance-shaped bracts at the base. This species also produces inconspicuous, self-fertilizing flowers that never open.

The fruit is an oblong anthocarp, a one-seeded nutlet enclosed by the hardened calyx base, to 2/5 inch long, distinctively lacking wings and glandular hairs.

Botanical notes: This four-o'clock perennial is also known as yerba de la rabia and was used, unsuccessfully, in rabies treatment. The meaning of the species name *longiflora* (long flowered) is obvious. This species was described by Asa Gray in 1853, and the type specimen was collected by Charles Wright in 1849, on "roadsides near San Antonio," Texas. Josiah Gregg collected the plant earlier, at Buena Vista (1847), the site of a famous battle in the Mexican-American War, in the Mexican state of Coahuila.

188. Big Bend Ringstem (*Anulocaulis eriosolenus*)
(Other common names: ringstem)

Habitat: rocky limestone bluffs, sandy, silty desert flats, gypseous clay hills, mounds, gravelly arroyos, alluvial fans, bajadas, low elevations, near Rio Grande.

Geographic range: Texas: Big Bend (Presidio, Brewster Counties). US: Texas. Mexico: Chihuahua, Coahuila.

Park locations: South Lajitas Mesa, Terlingua Uplift, Fresno Creek, Fresno Divide Trail

To 3 feet high, this ringstem is a short-lived perennial or annual with erect, ascending stems and sturdy branches. Ringstems are recognized by resinous yellow bands ringing the stems and pinkish-purplish blotches that may decorate the branches. The yellow secretions may protect the plants from aphid infestations (Douglas and Manos 2007).

Mostly basal, the sparse, paired leaves are large, to 5 inches long and broad, round or broadly egg-shaped. On long petioles, the leathery blades are yellow- to dark green above, paler below, often wrinkled or wavy on the margins, lumpy and swollen above. The leaves bear some coarse white hairs, with blister-like protuberances at the base.

From late spring to fall, widely branching flower clusters cover the upper two-thirds of the stems. The white or pink blooms form singly on short branchlets, each bloom with one or two small bracts below. Open near sunset, the 1/3 inch long tubular blooms flare into a five-lobed, funnel-shaped limb. Within the tube are five projecting pink filaments with yellow anthers and a pink style with a shieldlike stigma protruding beyond the anthers. The woolly-haired, short floral tube elongates as the flower wilts.

The fruit is a gray, top-shaped anthocarp, to 1/7 inch long, five-angled, lacking the prominent central ridge of other species.

Botanical notes: The genus name *Anulocaulis*, from the Latin *anulus* (ring) and *caule* (stem), refers to the yellow bands around the stems. Asa Gray described this species as *Boerhavia eriosolena* in 1853, from a collection by Josiah Gregg near Saltillo, Mexico, in 1848. The species name, from the Greek *erion* (wool) and *solen* (tube), relates to the flower tube. Paul Standley, with the United States National Museum, moved this species to the *Anulocaulis* genus in 1909. Standley authored *Trees and Shrubs of Mexico* (1920–26), *Flora of Costa Rica* (1937), and coauthored *Flora of Guatemala* (1946–58).

189. Chihuahuan Ringstem (*Anulocaulis leiosolenus* var. *lasianthus*)
(Other common names: southwestern or hairy-flowered ringstem)

Habitat: creosote scrub, open desert flats, low hills, low elevations; calcareous, gypseous, silty, clay soils.

Geographic range: Texas: Big Bend (Presidio, Brewster Counties). US: Texas. Mexico: Chihuahua.

Park locations: Buena Suerte Trail, Fresno Canyon

Of conservation concern: see Appendix A

Chihuahuan ringstem is a robust perennial to 5 feet high, sparingly branched at the base from thick, gnarly, woody roots, much branched above, with stout ascending stems. Sticky yellow bands around the base of the stems are an identifying feature of the genus and give the plants their name.

Mostly basal, the stiff, long-petioled leathery leaves, to 6 inches long, are broadly egg-shaped to round. Often coated with short hairs with blister-like bases, the blades are dull green above, paler below, wavy and wrinkled, and notched on the margins.

From late spring to early fall, much branched inflorescences form on the upper three-fourths of the stems, on leafless branches or short branchlets. Opening at sundown and closing by midmorning the next day, each trumpet-shaped bloom, to 1 1/2 inches long, abruptly expands into a funnel-shaped, five-lobed limb. The limbs are white or pink, with a yellow throat. Three stamens with magenta filaments and yellow anthers protrude far beyond the limb. A magenta style with a yellow stigma projects beyond the stamens.

The fruit is a gray, top-shaped anthocarp, to 1/4 inch long, with a wavy, skirt-like horizontal wing around the middle and wavy lengthwise ridges on the sides.

Botanical notes: The species name *leiosolenus*, from the Greek *leios* (smooth) and *solen* (tube), refers to the hairless flower tube. This species was described by John Torrey in 1859, as *Boerhavia leiosolena*, and collected by Charles Parry in 1852, "in gypseous soil, Great Canon of the Rio Grande, 70 miles below El Paso." Paul Standley moved the species to the genus *Anulocaulis* in 1909. Variety *lasianthus* was described by Ivan Johnston in 1944, from two Brewster County collections, the first by Barton Warnock at Hot Springs in 1937 and the type specimen by Victor Cory in 1938, 5 1/4 miles east of Terlingua. The varietal epithet *lasianthus*, from the Greek *lasios* (woolly) and *anthus* (flower), refers to the minutely hairy upper floral tube.

190. Velvet Ash (*Fraxinus velutina*)
(Other common names: Arizona, Modesto, desert, or leatherleaf ash, fresno, Uré [Tarahumara])

Habitat: at least semipermanent water sources, creeks, canyons, moist slopes, woodlands; igneous, limestone soils.

Geographic range: Texas: Trans-Pecos (Presidio, Brewster, Jeff Davis, also El Paso and Culberson Counties). US: southwest (excluding Colorado). Mexico: Baja California east to Nuevo León; Nayarit.

Park locations: Chorro Canyon, lower Arroyo Mexicano, Fresno Creek, Cienega Creek

Velvet ash is a fast-growing deciduous tree to 40 feet high, with trunks to 1 foot wide and a shallow root system. The young maroon branches develop gray, furrowed bark with age and spread to form a rounded crown. Twigs and branchlets may be cloaked with velvety hairs. The compound leaves are paired, to 6 inches long or more, with three to seven lancelike or elliptic leaflets, each with pointed or blunt tips and smooth or finely serrate margins. The short-stalked leaflets may be densely hairy below with a coating of soft fine hairs.

Blooming in spring before the new leaves, yellow male flowers and green female flowers appear on separate trees. The flowers emerge on short branchlets or in leaf axils, in crowded clusters to 5 inches long. The tiny flowers lack petals but have four-lobed, bell-shaped calyces. Male flowers develop two or three stamens with oblong anthers. Female blooms form a short style with two stigma branches. To 1 2/5 inches long, the papery pale-yellow fruits (samaras), dangling in tight clusters from the branches, are more conspicuous than the flowers. Each samara, with a long thin wing, contains a single oblong seed.

Velvet ash wood has been used to make wagons and ax handles (Vines 1960). Native Americans made walking sticks and bows from the wood (Watahomigie 1982).

Botanical notes: *Fraxinus* is the ancient Latin name for ash, and the Latin *velutina* (velvety) refers to the leaves. Linnaeus described the genus in 1753. John Torrey, America's first professional botanist, named the species in 1848, in Lieutenant Colonel William Emory's *Notes of a Military Reconnaissance*. Emory collected the lectotype on the reconnaissance, in 1846, in Sierra County, New Mexico. Later, Emory led the US-Mexico Boundary Survey, and surveyed the Gadsden Purchase.

191. Longcapsule Suncup (*Eremothera chamaenerioides*)
(Other common names: fireweed or willowherb suncup, long-capsuled or willow-herb primrose, long fruit suncup)

Habitat: sandy desert flats, creek beds, arroyo bottoms, alluvium; arid, rocky foothills, lower elevations, calcareous, also igneous substrates.

Geographic range: Texas: Trans-Pecos (Brewster, Presidio, also El Paso, Hudspeth Counties). US: southwest (excluding Colorado). Mexico: Baja California, Sonora, Chihuahua near El Paso.

Park locations: South Lajitas Mesa, lower Fresno Creek, Madera Canyon Campground, lower Arroyo Mexicano, Skeet Canyon, Cienega Creek

Uncommon to rare in the Trans-Pecos, longcapsule suncup is a weak but erect annual, branched from the base, as much as 20 inches tall. Often covered with glandular hairs, the slender ascending stems are pale green when young, pink or maroon with age. The alternate leaves are elliptic, to 3 inches long, pointed at the tip, tapered to the base, at times shallowly toothed on the edges. Like the stems, the dark green leaves often turn red or develop red or purplish blotches.

In spring, tubular flowers form in the upper leaf axils, in few-flowered, nodding clusters. Opening at sundown, each bloom has four white or pinkish-white petals, to 1/8 inch long, which turn red with age. Within the petals are eight stamens and a style with a headlike stigma. Sepals below the petals are reflexed and ephemeral. Flower parts may bear straight, stiff hairs, and may also be glandular-hairy like the stems.

The fruit is a many-seeded, slender cylindric capsule to 2 inches long, with tiny gray seeds arranged in a single row in each fruit chamber.

Botanical notes: Peter Raven, director of the Missouri Botanical Garden, coined *Eremothera* in 1964 to name a new section of the genus *Camissonia*. Raven joined the Greek *eremia* (desert) and *thera* (short for *Oenothera*) to identify desert-dwelling plants formerly in genus *Oenothera*. The epithet *chamaenerioides*, likely from the Greek *chamai* (low-growing), *Nerium* (oleander genus) and *-oides* (resembling), means "like a dwarf *Nerium*." The species was described by Asa Gray in 1853 as *O. chamaenerioides*, from a collection by Charles Wright in 1852, below old Fort Quitman, in Hudspeth County, Texas. Onagraceae specialists Warren Wagner and Peter Hoch elevated *Eremothera* to generic status in 2007, renaming this species *E. chamaenerioides*.

192. Boquillas Lizardtail (*Oenothera boquillensis*)
(Other common names: Rio Grande beeblossom, Boquillas gaura, Rio Grande butterfly-weed)

Habitat: desert scrub, canyons, low rocky bluffs; sandy, gravelly arroyos, edges of springs, often near Rio Grande, lower elevations, limestone soils.

Geographic range: Texas: Big Bend (south Brewster, south Presidio Counties). US: Texas. Mexico: Chihuahua east to Nuevo León.

Park locations: South Lajitas Mesa, Rio Grande Corridor

Of conservation concern: see Appendix A

In the United States restricted to the southern Big Bend, this scarce beeblossom is a clumped perennial to 3 feet high, with wiry ascending stems growing from a woody crown. The stems are coated with nonglandular and glandular hairs. To 2 1/2 inches long, the alternate, nearly stemless leaves are smooth or shallowly toothed and wavy on the edges. The basal blades are inversely lance-shaped, the upper blades smaller, more elliptic to lancelike.

In spring and summer, flowers appear in spikelike clusters on few-leaved branches. Opening at night, each bloom has four clawed white petals aging pink or red. The floral tube and ephemeral sepals approach the length of the petals. Within the petals are eight stamens and a style with a four-lobed stigma. The rapidly wilting blooms are pollinated by night-flying owlet moths (Noctuidae).

The fruits are four-chamber capsules to 1/2 inch long, spindle-shaped, slightly curved and tapered at the ends, with several tan or yellow seeds.

Botanical notes: The etymology of *Oenothera* is unclear. Theophrastus used the Greek *oinotheras*, perhaps from the Greek *oeno* (wine) and *thera* (seeking), for a species of *Epilobium*. Sowerby's *English Botany* (1863–72) explains: "the roots having a vinous scent when dried; they were also formerly eaten as incentives to wine drinking, as olives are; hence the name was changed from *onagra*, the ass food, to *oenothera*, the wine trap." The epithet *boquillensis* refers to Boquillas Canyon in BBNP. The species was described as *Gaura boquillensis* by Peter Raven, director of the Missouri Botanical Garden, and David Gregory, Onagraceae specialist, in 1972. Gregory collected the type specimen at Boquillas Canyon in 1960. Warren Wagner and Peter Hoch moved this species to the genus *Oenothera* in 2007.

193. Lizardtail Gaura (*Oenothera curtiflora*)
(Other common names: velvety, willow, downy, velvetleaf, or smallflower gaura, velvetweed, linda tarde)

Habitat: sandy riverbanks, creek beds, washes, seasonal pools, ditches, wet depressions, roadsides.

Geographic range: Texas: Trans-Pecos, much of Texas (excluding east Texas). US: most of US (excluding northeast). Mexico: Baja California to Nuevo León, south to Sinaloa, Durango, Zacatecas. Elsewhere: adventive in South America, Asia.

Park locations: west of Panther Canyon mouth, Colorado Canyon River Access, Cienega Creek

This gaura is a tall, erect annual to 6 feet high, with long stems nearly leafless in the upper reaches. Growing from thick taproots, the stems are densely hairy, with long soft hairs, and short glandular hairs underneath. Velvety like the stems, the alternate leaves are elliptic to lance-shaped, with pointed tips and smooth or shallowly scalloped margins. The soon wilting basal leaves are up to 5 inches long.

From spring to fall, the smallest gaura blooms in the Trans-Pecos appear in slender, crowded terminal spikes to 18 inches long. The towering spikes curl like a lizard's tail. The night-blooming flowers open over many days, from the bottom to the top of the spikes. The minuscule blooms have four clawed petals, white but pink to red with age. Protruding from the floral tube are eight stamens with yellow anthers and a long style with a four-lobed, pale yellow stigma. Below the petals are four slender, strongly reflexed green sepals.

The fruit is an indehiscent, four-angled woody capsule, spindle-shaped but broadest at the base, to 1/4 inch long. Each hairless, ribbed capsule contains up to four yellow or tan seeds.

Botanical notes: The epithet *curtiflora*, from the Latin words *curtus* (short) and *floris* (flower), connotes "short-flowered." German botanist Johann Georg Christian Lehmann, director of the Hamburg Botanical Garden, named the species *Gaura parviflora* in 1830, from a collection by David Douglas on Walla Walla River in Washington. Botanist Edwin James had previously discovered the plant in Colorado in 1819 or 1820, on the Long Expedition to the Rocky Mountains. In 2007, Warren Lambert and Peter Hoch moved this species to the *Oenothera* genus as *O. curtiflora* (the epithet *parviflora* had already been used).

194. Giant Helleborine (*Epipactis gigantea*)
(Other common names: stream or giant stream or chatterbox orchid, chatterbox, giant hellebore, stream orchis, heleborina gigante)

Habitat: springs, seeps, waterfalls; moist, shady banks, seeping canyon slopes.

Geographic range: Texas: Trans-Pecos (Culberson, Presidio, Brewster, Val Verde Counties), much of Edwards Plateau; scattered sites. US: west half. Elsewhere: Canada (British Columbia), Mexico (Baja California, south to Hidalgo), Asia, Australia.

Park locations: Arroyo Primero, Chorro Canyon, Smith Canyon

The only native American *Epipactis* species, giant helleborine is an erect perennial to 3 feet high, with a single stout, hairless stem. The plant grows from creeping rhizomes, reproduces vegetatively by offshoots as well as seeds, and forms colonies in suitable habitats. Like other orchids, this helleborine relies on a symbiotic relationship with fungi to obtain nutrients for seed germination.

As many as fourteen leaves, to 8 inches long, often pleated or veined, are spirally arranged up the stem. The clasping blades are lancelike, with a tapering, pointed tip and smooth, upturned margins.

From spring into summer, up to fifteen blooms appear in a well-spaced, one-sided, spike, each bloom supported by a long, lancelike green bract. On a downcurved stalk, each nodding flower has three spreading, lance-shaped, pale yellow-green sepals, and three small petals, yellow green with reddish-brown streaks and blotches. The protruding, three-lobed, hinged lower lip petal is pouch-like at the base, tonguelike at the tip, with an orange flap that jiggles in the wind like a "chatterbox." The short, pale yellow central column, with one stamen and style united, and pale green anther, loom over the lower lip. The yellow green fruit is a ribbed ovoid capsule to 1 inch long, with thousands of minuscule seeds.

Botanical notes: In 1757, Johann Gottfried Zinn, director of the Göttingen botanical gardens, borrowed the generic name from the Greek *epipaktis*, Theophrastus's term for a plant used to curdle milk. The Latin species name *gigantea* (gigantic) refers to the plant's height. A two-line Latin species description from a David Douglas manuscript, with Sir William Jackson Hooker's comments, appears in Hooker's *Flora Boreali-Americana* (1839), a pioneering work on North American flora. The plant was found by Scottish botanist Douglas "on the subalpine regions of the Blue and Rocky Mountains," and by Dr. John Scouler on the "Columbia River, about Fort Vancouver."

195. Texas Purplespike (*Hexalectris warnockii*)
(Other common names: Texas crested or purple-spike coralroot, Warnock's cockscomb)

Habitat: leaf litter, decaying logs, below oaks in shady woodlands, mesic canyons, creek sides, pour-offs, igneous rock above 4,000 feet.

Geographic range: Texas: Trans-Pecos (Culberson, Jeff Davis, Presidio, Brewster, Terrell Counties); sites in Edwards Plateau, Taylor, Dallas Counties. US: Texas, southeast Arizona (Cochise County). Mexico: Baja California Sur, Sonora, Coahuila.

Park locations: moist protected wetlands at higher elevations

Of conservation concern: see Appendix A

While studying Solitario flora, Jean Hardy Pittman first encountered this lovely orchid in 1995. After many unsuccessful efforts to relocate the plant, on one rewarding day years later, I stumbled on a new BBR site, with thirty-five spectacular Texas purplespikes. This imperiled perennial is erect, with a slender, unbranched, maroon stalk to 16 inches high. The leafless, hairless stalk emerges from a thin ringlike rhizome. All *Hexalectris* species lack chlorophyll and rely on a symbiotic relationship with fungi to extract nutrients from leaf mulch. The absent leaves are replaced by two to four small purple to tan, scalelike bracts.

From June to September, up to ten nodding blooms appear on the stem in a long loose cluster. About 1 inch across, each flower has three spreading, lance-shaped sepals, and three spatulate to oblong maroon petals, waxy and shiny. Prominent is the three-lobed lower lip petal, which is decorated with yellow, ruffle-like crests on the middle lobe, and maroon-tipped. Lateral lobes are white with maroon veins. The central column, with stamen and style united, is slender, whitish, arched above the lower lip petal; the anthers are maroon. Lance-shaped, sharp-pointed floral bracts support the blooms.

The fruit is a narrowly ovoid capsule 3/5 inch long, with many minuscule seeds.

Botanical notes: The name *Hexalectris* joins the Greek *hex* (six) and *alectryon* (rooster) and refers to six (the number varies) wavy crests on the orchid's lip petal, which are reminiscent of a cock's comb. Harvard orchid authority Oakes Ames and his student Donovan Correll described the plant in 1943: "This species is named in honor of Barton H. Warnock, who for many years has been a diligent collector of botanical specimens in the Glass and Chisos Mountains of Texas." Warnock found the type specimen in 1937 in Blue Creek Canyon of the Chisos, but the orchid had been collected earlier in the Chisos, by John Moore and Julian Steyermark in 1931, and oak specialist Cornelius Muller in 1932. Ames published a seven-volume work on orchids, *Orchidaceae* (1905–22), which was illustrated by his wife, Blanche.

196. Woolly Paintbrush (*Castilleja lanata* ssp. *lanata*)
(Other common names: woolly, sierra woolly, whitefelt, or white-woolly Indian paintbrush, hierba de conejo)

Habitat: creosote scrub, desert arroyos, gravelly bajadas, arid, rocky slopes; low to mid-elevations, limestone, igneous, gypseous soils.

Geographic range: Texas: Trans-Pecos (excluding Culberson County); Crockett, Val Verde, Kinney, Upton Counties. US: Texas to southern Arizona. Mexico: Sonora east to Nuevo León, south to Zacatecas, San Luis Potosí.

Park locations: Guale Mesa Road, upper Arroyo Mexicano, Cinco Tinajas, Ojo Chilicote, Fresno Rim Trail, drainage entering Lefthand Shutup

This paintbrush is easily identified by the thick cloak of shaggy white hairs covering all plant parts — stems, leaves, even flowers. To 3 feet high, this erect perennial develops clumped branching stems, often woody at the base, growing from a stout, woody taproot. Like other species, this paintbrush produces chlorophyll and manufactures carbohydrates, but is also parasitic, attaching to and drawing nutrients from the roots of other plants. The alternate, stemless leaves, to 3 inches long, are slender, linear, with blunt tips and thick, upturned margins. The blades are only slightly less hairy than the stems.

From spring to fall, flowers form within congested, spikelike terminal clusters to 4 inches long, each bloom partly hidden by floral bracts more colorful than the blooms. The three-lobed, oblong bracts are red or orange red, occasionally pinkish at the tips. The four-lobed calyx around the base of each bloom is similar in color to the bracts. Each slender tubular flower is strongly two-lipped, with an elongated, hoodlike upper lip and a much smaller, toothed lower lip. The blooms are greenish, with reddish color accents. Within the blooms are two pairs of stamens of unequal length, and a club-like stigma.

The fruit is a two-chamber, egg-shaped dehiscent capsule, to 2/5 inch long, with small brown seeds.

Botanical notes: In 1781, Spanish botanist José Celestino Mutis, physician to the viceroy of New Granada (Colombia), described the genus *Castilleja* from a plant he found in his adopted country. Mutis, head of the Royal Botanical Expedition in New Granada (1783–1808), named the genus for Domingo Castillejo, his former botany professor in Cadiz, Spain. The epithet *lanata* (woolly) refers to the shaggy hairs on stems and leaves. Asa Gray described the species in 1859, from collections by Charles Wright in 1849, at Las Moras Springs in Kinney County, and in Limpia Valley in Jeff Davis County; and in 1851, "on "Piedra Pinta [Pinto Creek]" in Kinney County.

197. Downy Paintbrush (*Castilleja sessiliflora*)
(Other common names: downy, Great Plains, or yellow Indian paintbrush; downy painted cup, downy yellow painted cup)

Habitat: Solitario rim, hilltops, rocky slopes; on limestone, gypsum, with cliffrose, oak, guayule.

Geographic range: Texas: Trans-Pecos (excluding Jeff Davis County), sites in Panhandle, western Edwards Plateau, Permian Basin. US: Texas to Arizona, north to Montana, North Dakota, Minnesota, Wisconsin. Elsewhere: Canada, Mexico (uncommon in Chihuahua, Coahuila, Nuevo León).

Park locations: Solitario northwest rim, Solitario rim above Righthand Shutup

Downy paintbrush is a perennial to 1 foot high, often with a few unbranched stems atop a stout, woody crown. Long soft hairs, at times mixed with glandular hairs, cover the stems. This herb is partly parasitic like all paintbrushes, with specialized appendages (haustoria) on its roots to attach to and extract nutrients from the roots of host plants.

The stems are lined with slender linear leaves to 3 inches long, some split into linear segments. The alternate, yellow green blades are stalkless, blunt tipped, with involute margins and fine hairs.

In spring and summer, flowers protrude from crowded terminal spikes, each bloom supported by lancelike floral bracts. The bracts are yellow green and pink tipped. A four-lobed, yellow green calyx, with slender, glandular-hairy lobes, supports each bloom. To 2 inches long or more, the flowers are pale yellow to white, often tinged pink or green at the tips. The two-lipped bloom has a hoodlike upper lip, a shorter lower lip with three flaring lobes, and an arching tube projecting beyond the bracts. Within the bloom are four stamens in two pairs of unequal length.

The fruit is an ovoid, two-chambered dehiscent capsule, to 3/4 inch long, with many tiny brown, pitted seeds.

Botanical notes: The species name *sessiliflora*, from the Latin *sessilis* (sessile), and *floris* (flower), alludes to the stemless flowers. Frederick Traugott Pursh described the species in 1814, in his *Flora Americae Septentrionalis*, from a collection by British botanist John Bradbury "in Upper Louisiana." Pursh obtained Bradbury's specimens and, without permission, included descriptions of forty-one new plants in an appendix to his *Flora*. In 1817, Bradbury chronicled his journeys in *Travels in the Interior of America in the Years 1809, 1810, and 1811*.

198. Chisos Pricklypoppy (*Argemone chisosensis*)
(Other common names: Chisos Mountain or pink prickly poppy)

Habitat: desert flats, alluvial plains, arroyos, roadsides, rocky hillsides.

Geographic range: Texas: Trans-Pecos, Crockett County, parts of Permian Basin. US: Texas. Mexico: Chihuahua, Coahuila, Durango.

Park locations: 1 1/4 miles west of Lajitas, Monilla Canyon, Las Cuevas, Lefthand Shutup

The most common pricklypoppy in the Trans-Pecos, and the only one at BBR, Chisos pricklypoppy is a robust biennial or perennial to 30 inches high, with one or several sparsely branched, leafy stems. From a fleshy but sturdy taproot, the stems bear vicious prickles and are filled with acrid yellow sap. This thistlelike poppy is remarkably glaucous, with a white, waxy coating on stems and leaves. Alternate, stemless, often clasping, the green leaves are chalky blue on veins above, and prickly, especially on veins underneath. The lower leaves, at least, are pinnately divided into oblong lobes with wavy, spiny margins.

In spring and summer, fragile cuplike flowers, to 4 inches across, appear in clusters at the stem tips. Typically, each bloom, with one or two leafy bracts at the base, has six rounded, white to pinkish petals in two whorls of three. Each showy petal is crinkled, flimsy, and early falling. Below the petals are three sepals, each with a hollow, hornlike, spine-tipped appendage underneath. Within the petals are 150+ stamens, with pale yellow filaments and golden-yellow anthers and a style with a multilobed red stigma.

With notably stout spines, the fruits are many-seeded, dehiscent capsules to nearly 2 inches long. The capsules split into three to five segments, releasing many brown or black pitted globose seeds.

Botanical notes: *Argemone* was borrowed from the Greek *argemon* (cataract) and applied by classical authors to poppy-like plants whose sap was used to treat cataracts. University of Minnesota *Argemone* specialist Gerald Ownbey named this species *chisosensis* in 1958, because the type specimen was collected in Juniper Canyon of the Chisos Mountains in 1921, by Roxana Ferris and Carl Duncan. Ferris later became curator of the Dudley Herbarium at Stanford and authored books on California wildflowers. Duncan taught botany and entomology at San José State College.

199. Fringed Monkeyflower (*Erythranthe chinatiensis*)
(Other common names: toothed, toothpetal, southwest, or sharpwing monkeyflower)

Habitat: moist, shady sites, springs, waterfalls, creeks, seeps; igneous, limestone substrates.

Geographic range: Texas: Presidio County. US: Texas. Mexico: expected in Chihuahua.

Park locations: Bofecillos canyons

Of conservation concern: see Appendix A

Fringed monkeyflower is only known from a handful of sites in the Chinati and Bofecillos Mountains of Presidio County. This rare, newly described perennial was formerly placed within *E. dentiloba*, a Mexican plant classified as globally imperiled. To 8 inches tall, this monkeyflower is mat-forming, with fleshy, often maroon stems. Prostrate, creeping, leafy, the branching stems grow from horizontal roots and may root at the nodes.

The petite leaves are egg-shaped to round, to 3/5 inch long, pointed or blunt at the tip, shallowly toothed or merely bumpy on the thick margins. Dull- or olive-green above, sometimes maroon below, the blades may bear nonglandular or gland-tipped hairs above.

From spring to fall, flowers form in small clusters near the branch tips. On slender, maroon stalks, the blooms rise above the leafy mats. With a funnel-shaped throat and two-lipped, deeply fringed limb, the tubular flowers are yellow with orange spots in the throat and on the broad, reflexed lower lip. Within the throat are four stamens in two unequal pairs and a long style with a bilobed, platelike stigma. Unlike *E. dentiloba*, in this species the anthers of the stamen and the stigma are close, enabling self-fertilization. A five-lobed, pale green or maroon calyx supports each bloom.

The fruit is a two-chamber, many-seeded, dehiscent capsule, enclosed by the expanded calyx.

Botanical notes: In 1840, French botanist Édouard Spach, keeper of the Herbarium at the National Museum of Natural History in Paris, named this genus for a red-flowered species, *E. cardinalis*. Spach formed the name from the Greek *erythros* (red) and *anthos* (flower). The species name *chinatiensis* refers to the Chinati Mountains, the type locality. Guy Nesom, editor of the online botanical journal *Phytoneuron*, described the species in 2012, from a 2004 collection by Emily Lott, S. P. Rankin, Mary Butterwick, and Patty Manning, at the Chinati Mountains State Natural Area in Presidio County.

200. James's Monkeyflower (*Erythranthe inamoena*)
(Other common names: roundleaf or smooth monkeyflower)

Habitat: perennial water sources: canyon bottoms, springs, seeps, waterfalls; igneous, limestone soils.

Geographic range: Texas: Trans-Pecos (Culberson, Jeff Davis, Presidio, Brewster, Val Verde Counties); Edwards Plateau, sites in north-central, north Texas. US: Texas. Mexico: Chihuahua, Coahuila.

Park locations: Fresno, Rancherías, Tapado Canyons, Cinco Tinajas, upper Terneros Creek

James's monkeyflower is a hairless annual to 1 foot high, with fleshy, hollow, often creeping stems growing from fibrous roots, at times rooting at the nodes. The stems are prostrate, in tangled mats, but erect in the flowering branches. The opposite leaves are egg-shaped, or broadly so, to 1 3/10 inches long. Lower leaves are long-petioled, while upper leaves may be stemless and clasping. The blades are shallowly toothed on the margins, blunt-tipped, and flat or heart-shaped at the base.

From spring to September, flowers to 3/4 inch long appear, sometimes all along the stems. Each bloom is tubular, with a yellow, red-spotted, two-lipped limb. The limb has an upright, or recurving, two-lobed upper lip, and a spreading or downcurved, three-lobed, bearded lower lip. The flower's throat is partly obstructed by two bulky ridges. Within the throat are four stamens

and a style with a bilobed stigma. The anthers of the stamens, and the stigma, are close, enabling self-fertilization. A five-lobed green calyx, often maroon-tinged, supports each bloom. The calyx, distinctively, is open in fruit.

The dehiscent fruit is a hairless, two-chamber, many-seeded capsule to 1/4 inch long.

Botanical notes: The species name joins the Latin *in-* (not) and *amoena* (pretty), for the plant's presumed unattractive appearance. In 1903, in his journal *Pittonia*, Edward Lee Greene described this species as *Mimulus inamoenus*, "a coarse plant, growing in mud." The type specimen was collected in 1902 by Samuel Tracy and Franklin Earle, in Limpia Canyon in Jeff Davis County. Tracy, later director of the Mississippi Experiment Station, often collected in Texas and donated his collections to Texas A&M University. His specimens formed the nucleus of the Tracy Herbarium. In 2012, based on new molecular-phylogenetic studies, Guy Nesom moved this and other *Mimulus* species to the *Erythranthe* genus.

201. Trans-Pecos Maidenbush (*Phyllanthopsis arida*)
(Other common names: unknown)

Habitat: rocky limestone slopes, ridgetops, with lechuguilla, sotol, false agave, yucca, and the rare Hinckley's oak; Santa Elena, Glen Rose formations.

Geographic range: Texas: Big Bend (southeast Brewster, southeast Presidio Counties). US: Texas. Mexico: one site in Chihuahua, one in Coahuila.

Park locations: Solitario

Of conservation concern: see Appendix A

This seemingly delicate small shrub is rare, known at a few locations in a limited range, and only one site in BBR. A relict of a cooler, wetter climate, Trans-Pecos maidenbush is up to 2 feet high, with a few dark stems and many leafy branchlets. To 4 inches long, the grayish-white, hairy branchlets are slender yet rigid, with flaking, peeling bark.

The alternate leaves, to 2/5 inch long, are elliptic to nearly round, with round or bluntly pointed tips and short maroon petioles. The blades are pale green with yellowish midribs and obvious hairs.

In summer and fall, male and female flowers appear on separate plants, singly or in small clusters, on short stalks from the upper leaf axils. The male blooms consist of five spreading green sepals with white margins; five greenish-yellow petals smaller than the sepals; a flat, fleshy, ringlike disk; five stamens with white filaments and yellow anthers; and three white style-like branches. The female blooms have five greenish-yellow sepals with transparent margins; five brown, papery petals much smaller than the sepals; a dark brown, five-part disk; and three prominent, two-branched brown styles.

The fruit is a dehiscent capsule, spheric but compressed, 1/7 inch high. Each trilobed capsule contains two wedge-shaped seeds per lobe.

Botanical notes: *Phyllanthopsis* joins the generic name *Phyllanthus* and the Greek suffix, *-opsis* (resembling), meaning "like plants in the genus *Phyllanthus*." The epithet *arida* refers to the arid habitat. This rare plant was collected by Barton Warnock and Marshall Johnston in 1958, in crevices of limestone slopes in the Dead Horse Mountains of BBNP. Warnock and Johnston described the species in 1960 as *Savia arida*, but the shrub was reclassified by Euphorbiaceae specialist Grady Webster in 1967 as *Andrachne arida* and again reclassified in 2008 as *Phyllanthopsis arida*, by Maria Vorontsova and Petra Hoffmann, with the Royal Botanic Gardens at Kew.

202. Rouge Plant (*Rivina humilis*)
(Other common names: pigeonberry, blood or coral berry, baby peppers, dog blood bush, coralito, coralillo, baja tripa)

Habitat: cooler, wetter canyons, creek banks, hillsides, woodlands, varied soils, low to high elevations.

Geographic range: Texas: Trans-Pecos (excluding El Paso, Culberson, Reeves Counties); Crockett County; much of south, central, north Texas; scattered sites. US: Louisiana west to Arizona; Arkansas, Florida. Elsewhere: Mexico, Central, South America, West Indies, Asia, Africa.

Park locations: Arroyo Mexicano, Cinco Tinajas

Nestled below trees or shrubs in shade, this perennial is less than 20 inches high, with upright or spreading, vinelike stems growing from stout roots. Often, the leafy stems are densely branched. Leaf shape is variable, lance- or egg-shaped, triangular, or oblong. To 6 inches long, the long-petioled, alternate blades are green or dark green above, paler below, wavy on the edges and tapered to the tip. Both stems and leaves are mostly hairless.

From spring to fall, pinkish-white flowers form in axillary or terminal spikes as much as 6 inches long. A spike, on a long sturdy stalk, contains up to fifty blooms. Each bloom has four petal-like, oblong sepals, four stamens with white or yellow anthers, and a short style with a headlike stigma. The fruit is a lustrous red berry to 1/5 inch across, with one red to black, broadly lens-shaped seed.

Although poisonous like the leaves and roots, the berries are eaten by the Tarahumara in Chihuahua and used as food and dye in Sonora (Austin 2010). Use of the juice as a beauty aid was documented in a British journal, *The Gardeners' Chronicle*, in 1877: "In the West Indies, from whence it [*Rivina humilis*] was introduced in 1699, it is called the Rouge plant, its berries being used as a cosmetic."

Botanical notes: In 1703, Charles Plumier named this genus for Augustus Quirinus Rivinus, University of Leipzig botanist. Plumier praised Rivinus's 1690 work on plant classification, in which Rivinus proposed that no plant name contain more than two words. A novel idea! In 1753, Linnaeus adopted Plumier's generic name and named the species *humilis* (small) for the plant's modest size. The species was collected by Sir Hans Sloane during his 1687–89 *Voyage to Jamaica*.

203. Water Hyssop (*Bacopa monnieri*)
(Other common names: herb of grace, coastal water hyssop, brahmi, baraima, thyme-leaved gratiola, moneywort, lágrima de bebé, verdolaga de puerco)

Habitat: periodically flooded riverbanks; creek beds, shallow water at springs and ponds.

Geographic range: Texas: Trans-Pecos (Presidio, Brewster, Pecos, Terrell, Val Verde Counties), much of Edwards Plateau, Gulf Coast, South Texas Plains. US: Maryland to Texas; Oklahoma, Arizona, California. Mexico: Chihuahua, Coahuila, some Gulf Coast, Pacific states.

Park locations: Colorado Canyon River Access, Cienega Creek

Water hyssop is a terrestrial or aquatic perennial, with fleshy, hairless, at times floating stems to 1 foot long. Creeping, mat-forming, ascending only at the tips, the yellow-green stems may spread and grow from rhizomes and root at the nodes. To 1 inch long, the stemless leaves are spatulate or oblong, rounded at the tip and smooth on the edges. The thick, fleshy, yellow green blades are often gland dotted.

From spring to fall, small bell-shaped blooms form singly in the leaf axils. The five-lobed flowers are white to pale lavender, with greenish-yellow throats. At the base of each bloom are five sepals, unequal in size and shape, the inner smaller, narrower. Below the sepals are two linear bractlets. Scarcely protruding from the flower's throat are four stamens, with white filaments and anthers with two purple pollen sacs and a style with a headlike or bilobed, green stigma. The fruit is a two-chamber, dehiscent capsule 1/4 inch long, with numerous pitted seeds.

Known as brahmi in Asia, water hyssop has been used in Ayurvedic medicine to treat ulcers, asthma, tumors, depression, and more. A plant extract is sold worldwide as a brain tonic to enhance memory.

Botanical notes: The generic name *Bacopa* was used by French botanist Jean Baptiste Aublet in 1775 to describe a plant he collected in French Guiana (1762–64). Linnaeus described the species in 1756 as *Lysimachia monnieri*, from a now lost specimen from Spanish America. The neotype is an illustration in Patrick Browne's *The Civil and Natural History of Jamaica* (1756). Apparently, Linnaeus named the herb for French botanist Louis Guillaume Le Monnier, head of the Trianon botanical garden at Versailles. In 1891, Austrian botanist Richard Wettstein moved the species to the *Bacopa* genus.

204. Balloonbush (*Epixiphium wislizeni*)
(Other common names: Wislizenus' snapdragon, net-cup, giant, or dune snapdragon vine)

Habitat: deep sandy soil, gypseous dunes, near Rio Grande; open, hot sites, low to mid- elevations.

Geographic range: Texas: Trans-Pecos (excluding Jeff Davis, Reeves, Terrell Counties), west Permian Basin. US: Texas, southern New Mexico, Arizona (Greenlee County). Mexico: northern Chihuahua.

Park locations: 1 1/4 miles west of Lajitas

Balloonbush is a fast-growing annual vine with outsized snapdragon-like flowers and attractive woody fruits often featured in floral arrangements. This hairless vine climbs over other plants with twining leaf petioles. Growing from taproots, the stems become woody at the base with age.

The long-petioled, fleshy leaves, to 2 1/2 inches long, are arrowlike, with basal lobes turned outward. The blades are blunt or abruptly white-pointed at the tip, with a network of veins above.

From spring to fall, solitary two-lipped, tubular flowers, to 1 inch long, appear on short, thick stalks from the leaf axils. The large, open-throated blooms are blue, violet, and white, with conspicuous whitish folds on the lower lip. Within the flower's throat are two pairs of stamens, with hairy filaments and oblong anthers, and a swordlike style with a conic stigma. Supporting the bloom are five lance-shaped sepals, slightly hooked at the tips.

Balloonbush is named for the ballooning, saclike sepals, which expand to enclose the fruit. The woody fruit is an egg-shaped, flattened, dehiscent capsule to 3/5 inch long, with a persistent style. The black seeds are egg-shaped but compressed and winged.

Botanical notes: In 1859, George Engelmann and Asa Gray described this species as *Maurandya wislizeni*, the only species in subgenus *Epixiphium*. In 1797, Casimiro Gómez Ortega, professor at the Royal Botanical Garden of Madrid, named the genus for "Catherina Pancratia Maurandy, wife of Don Augustin Juan, Professor in the Royal Botanic Garden of Carthagena—a learned lady, a sharer, if not indeed a leader in her husband's botanical labors." The subgenus name, *Epixiphium*, joining the Greek *epi* (on) and *xiphos* (sword), refers to the flower's swordlike style. The species is named for Adolf Wislizenus, who collected balloonbush in 1846 near Val Verde, New Mexico. In 1926, Pomona College botanist Philip Munz elevated *Epixiphium* to genus status, with *E. wislizeni* as its only member.

205. Cochise Beardtongue (*Penstemon dasyphyllus*)
(Other common names: thickleaf, purple, or Cochise penstemon, thickleaf or Gila beardtongue)

Habitat: Yucca-sotol-beargrass-mesquite communities, grasslands, gravelly, rocky hills, mesas, slopes; low to mid-elevations, limestone soils.

Geographic range: Texas: Trans-Pecos (Presidio, Brewster, Pecos, Terrell Counties); Crockett County. US: Texas, southwest New Mexico, south Arizona. Mexico: Chihuahua, Coahuila, northern Durango.

Park locations: Solitario northwest rim, Fresno Peak

Cochise beardtongue is a seldom seen perennial to 18 inches high, recognized by the deep blue flowers and thick coat of short white hairs blanketing stems and leaves. Usually erect, the stems are seldom branched, woody at the base, and shorthaired.

The paired leaves are linear, to 2 1/2 inches long, pointed or blunt at the tips and smooth on the upturned margins. Stemless, typically ascending, the stiff gray-green blades are often curved at the tips.

From spring to fall, two-lipped tubular flowers form in spikelike clusters at the stem tips. The clusters are covered with stalked glandular hairs. Blue to purplish or pinkish blue, each bloom has a two-lobed upper lip, three-lobed lower lip, and often colored lines at the base of the swollen throat. Within the throat are four fertile stamens in two pairs, and one hairless, sterile stamen. Many *Penstemon* species have hairy sterile stamens, hence the name beardtongue. The anther sacs, twisted and toothed on the margins, are borne in pairs at the top of the throat. The style is slender, with a flat stigma. Supporting the bloom are five lancelike sepals.

The fruit is a four-chamber, dehiscent capsule to 3/5 inch long, with many small angled, net-veined seeds.

Botanical notes: The generic name *Penstemon* was coined by Virginia physician-botanist John Mitchell in 1748. The term may be derived from the Greek *pente* (five) and *stemon* (stamen), referring to the flower's five stamens, or to the distinctive, sterile fifth stamen. The species name, from the Greek *dasys* (woolly) and *phyllon* (leaf), emphasizes the leaves. Asa Gray described the species in 1859, from 1851 collections by Charles Wright on "stony hills of the Pecos" in Texas, and "near Cooke's Spring" in Luna County, New Mexico, and by Captain Edmund Kirby Smith, John Bigelow, and George Thurber, in the "valley of the Santa Cruz river on mountain sides, and in the valleys of the San Pedro, Sonora."

206. Desert Indianwheat (*Plantago ovata*)
(Other common names: blond, woolly, or desert plantain, psyllium, blond psyllium, Indian wheat, isabgol, pastora, mumsa)

Habitat: near Rio Grande; creosote scrub, sandy, silty dunes, flats, creeks, arroyos, low clay hills.

Geographic range: Texas: Big Bend (Presidio, Brewster Counties). US: southwest. Elsewhere: Mexico (Baja California to Nuevo León); native to Mediterranean.

Park locations: 1 1/4 miles west of Lajitas, Arenosa Campground

This plantain is a stemless low annual that produces many leafless, unbranched scapes to 10 inches long, with flowering spikes at the tip. The herb is blanketed with long silky and woolly white hairs. In basal rosettes, the green or gray-green leaves are linear, as much as 6 inches long. The stemless blades are pointed at the tip and inrolled on the margins.

Mostly in spring, densely flowered cylindrical spikes, to 1 inch long or more, form atop the scapes. The small papery blooms are translucent, with four broadly egg-shaped lobes, each with a conspicuous brown spot at the base. Supporting each bloom are four hairy oval sepals, with green or brown midribs and membranous margins. Below and much like the sepals are whitehaired bracts. Within each bloom are four stamens with lavender filaments and white or pale-yellow anthers and a threadlike style with a long stigma.

The fruit is a dehiscent, egg-shaped capsule to 1/6 inch long, which splits to release two narrowly oblong, orange or yellow seeds.

This annual is a source of psyllium seed husks, which have been used for centuries as a dietary fiber and natural laxative. The husks are also consumed to lower cholesterol and blood pressure, and to treat stomach ailments. Sprouted seeds are eaten in salads.

Botanical notes: *Plantago*, from the Latin *planta* (sole of the foot), and *-ago* (resembling), is a name used by Pliny for plantain, presumably because the leaves of many plantains lie flat and hug the ground. As early as 1542, plantain (*P. major*) was named and illustrated in German herbalist Leonard Fuchs's *De historia stirpium commentarii insignes* (Notable Commentaries on the History of Plants). In 1753, Linnaeus formally described the genus and numerous species, including the type species (*P. major*). Pehr Forsskal, Linnaeus's young student, collected *P. ovata* on a Danish expedition to Egypt and South Arabia (Yemen) from 1761 to 1763, but he died of malaria on the trip, and his plant description was published posthumously, in 1775, in *Flora Aegyptiaco-Arabica*.

207. Coahuila Twintip (*Stemodia coahuilensis*)
(Other common names: Coahuila stemodia)

Habitat: sandy, silty creek beds, seasonally wet washes; below mountain slopes in creosote scrub.

Geographic range: Texas: Trans-Pecos (Jeff Davis, Presidio, Brewster Counties). US: Texas. Mexico: southeast Chihuahua, Coahuila, northern Durango, northern Zacatecas.

Park locations: mouth of Contrabando Canyon, Cinco Tinajas, Skeet Canyon side drainage

Of conservation concern in Texas: see Appendix B

Rare in the United States, found only in the Big Bend, this Mexican transient is a low, often prostrate annual, at times spreading or weakly ascending, with widely branching, vinelike stems growing from slender roots. The maroon stems and the leaves are covered with long stalked, glandular hairs.

Typically, the paired, long-petioled leaves are divided into three spatulate segments, with each segment also lobed. Less commonly, the fleshy, ciliate blades are deeply cleft into opposite lobes, or only shallowly lobed, or undivided.

From spring to fall, often in the monsoon season, blooms form in small clusters in the leaf axils, or in interrupted spikes at the stem tips. Each tubular bloom is two-lipped, with an erect, notched upper lip and a three-lobed lower lip. The lobes are violet purple, the tube and broad throat yellowish white inside and dark purple outside. The throat is ridged, pleated, and bearded. Within the throat are two pairs of stamens of unequal length, with threadlike filaments and white anthers. The long style is slender, with a large stigma. At the base of the bloom are five lance-shaped sepals.

To 1/4 inch long, the hairless fruit is an egg-shaped, dehiscent brown capsule with white cylindric seeds.

Botanical notes: According to Asa Gray in *Synoptical Flora of North America* (1878–86), *Stemodia* is a "name shortened by Linnaeus from Patrick Browne's *Stemodiacra*, meaning stamens with two tips." In 1756, Irish historian and physician-botanist Browne, who had lived in Jamaica for many years, published *The Civil and Natural History of Jamaica*, which described *Stemodiacra* and 103 other new genera. The species name *coahuilensis* refers to the Mexican state of Coahuila, where German-born botanical collector Carl Purpus collected the plant in 1910. James Henrickson, authority on CD flora, described the species in 1989 as *Leucospora coahuilensis*. Billie Lee Turner, professor emeritus at the University of Texas at Austin, transferred the herb to the *Stemodia* genus in 1993.

208. Water Speedwell (*Veronica anagallis-aquatica*)
(Other common names: brook or water pimpernel, blue or sessile water speedwell, blue speedwell, bérula)

Habitat: shallow water, moist sandy creek banks, muddy spring margins, wet fields, depressions.

Geographic range: Texas: Trans-Pecos (El Paso, Presidio Counties), central Texas, sites in north Texas, Panhandle. US: naturalized throughout. Elsewhere: naturalized from Eurasia in Canada, Mexico, Central and South America.

Park locations: 1 mile west of Lajitas, Cienega Creek, Cienega Gorge

Water speedwell is a semiaquatic, mostly hairless perennial to 3 feet high, with long, fleshy but sturdy, stems. The stems are often prostrate, at times rooting at the nodes, or ascending and sometimes branching. The herb grows from rhizomes and reproduces vegetatively as well as sexually. To 4 inches long, the paired, fleshy leaves are largely stemless, clasping at the base, with pointed tips and smooth or finely serrate margins. Upper blades are more oblong, lower blades more rounded and short-petioled.

In spring and summer, many petite flowers appear in loose, spreading clusters on long stalks from the leaf axils. Each flower rests on its own slender stalk, with a small green bract at the base. To 1/4 inch across, the bloom has a short tube abruptly expanded into four flattish, petal-like lobes unequal in size, the upper two lobes united. The delicate blossoms are lavender with deep purple veins, a white ring at the base of the lobes, and a yellow-green throat. Within the throat are only two developed stamens, each with two blue anther sacs and a single style with a white headlike stigma. Four lance-shaped green sepals support each bloom.

The fruit is a round dehiscent capsule 1/6 inch long, slightly compressed, with tiny, slightly flattened, seeds.

Botanical notes: In 1753, Linnaeus named the genus *Veronica* for a European speedwell. The word may combine the Latin *verus* (true) and *iconica* (image). Legend has it that on the road to Calvary, a woman offered Jesus a handkerchief, and his facial imprint, a "true image," was left on the cloth. The woman was elevated to sainthood and given the name St. Veronica. Supposedly, markings on the flowers of some *Veronica* species resemble the image on the cloth. The name *speedwell* refers to presumed medicinal qualities. Linnaeus described this species in 1753, joining *anagallis* (a Primulaceae genus) and *aquatica* (aquatic), implying "an aquatic *Anagallis*."

209. Giant Reed (*Arundo donax*)
(Other common names: giant, river, or wild cane, giant reed grass, carrizo, carrizo gigante, carrizo grande)

Habitat: sandy riverbanks, creek sides, dunes, alluvial flats, ditches, roadside depressions; moist, wet soils.

Geographic range: Texas: Trans-Pecos (Presidio, Brewster, Terrell, Val Verde Counties, scattered elsewhere); parts of South Texas Plains, central, north, southeast Texas, Gulf Coast. US: southern half. Elsewhere: native to East Asia.

Park locations: Rio Grande east of Contrabando Canyon, Grassy Banks, Hoodoos

A highly invasive perennial, giant reed is one of the world's largest grasses, forming impenetrable stands many feet across. This interloper outcompetes native plant species and provides much less useful habitat or food for wildlife. To 1 1/2 inches wide, the stems (culms) are hollow but thick-walled, partitioned at the leaf nodes like bamboo. The culms reproduce vegetatively from creeping rhizomes.

The broad, rigid leaves, to 3 feet long, are tapered to the tip and rough on the edges. Usually hairless, the alternate blades are gray green, glaucous, often folded, with a notable tuft of hairs at the base.

Flowers appear in summer and fall, in plumelike spikes to 2 feet long, at the ends of the culms. The densely branched spikes, in shades of tan, white, silver, and purple, contain countless spikelets, each with two to four tiny florets. To 3/5 inch long, each compressed spikelet consists of paired bracts at the spikelet base (glumes); a lower bract (lemma) and upper bract (palea) enclosing each floret; florets (each with three anthers); and rachilla (main axis). The membranous glumes are boatlike, unequal in size, tapered to the tip.

The fruit is an oblong, tan caryopsis (dry, one-seeded, indehiscent, with the seed coat fused to the fruit wall) 1/7 inch long.

The Bible notes giant reed's use as a measuring instrument. The oldest pipe organ, the Pan pipe, was made from the culms. In Egypt, mummies were wrapped in the leaves. Giant reed is now used to make reeds for woodwind instruments.

Botanical notes: *Arundo* is an old Latin name for a reed, and *donax* is derived from a Greek word for a reed. Linnaeus described this grass in 1753, citing *Florae Leydensis Prodromus* (1740) by Adriaan van Royen as a source. Dutch botanist van Royen was director of the Botanic Garden at Leyden, and the *Florae* was his account of the garden's cultivated plants. The lectotype is stored in Van Royen Herbarium, the National Herbarium of the Netherlands.

210. Sideoats Grama (*Bouteloua curtipendula*)
(Other common names: tall grama, mesquite grass, banderilla, navajita, banderita)

Habitat: desert to mesic mountains, gravelly flats, alluvial basins, grasslands, woodlands.

Geographic range: Texas: Trans-Pecos, much of Texas. US: nearly throughout. Elsewhere: Canada, Mexico, parts of Central, South America.

Park locations: Hoodoos, Arroyo Mexicano, Llano Loop, north of Buena Suerte Trail, Cienega Creek

The state grass of Texas, sideoats grama is a native, warm season perennial to 3 feet high, with erect, hairless, unbranched culms. Two varieties occur in the Trans-Pecos. In variety *curtipendula*, the light green culms grow singly or in small, well-spaced clusters arising from creeping rhizomes. In variety *caespitosa*, the culms may form dense tufts from knotty crowns.

The alternate leaves, to 8 inches long, are linear, flat or folded, and mostly hairless. The ligules (strap-shaped appendages at the leaf-leaf sheath junction) are ciliate scales.

From summer to fall, slender inflorescences to 1 foot long, with up to fifty short spikelike branches, form at the top of the culms. The branches may be one-sided, or in two rows on one side of the axis. Each branch bears up to nine spikelet clusters, each 1/3 inch long. Commonly, a spikelet has one fertile floret and one or two smaller, sterile florets. Usually, the branch axis extends, awn-like, beyond the spikelets. Blooming spikelets assume a rainbowlike array of colors — purple, bronze, green, red, brown, and tan. Anthers of fertile florets may be red, orange, or yellow. The plumelike stigmas are white. The fruit is a tan caryopsis to 1/8 inch long.

Botanical notes: Spanish botanist Mariano Lagasca y Segura named the genus in 1805, for the brothers Claudio and Estéban Boutelou y Soldevilla. Claudio was chief gardener at the Royal Botanical Garden of Madrid; Estéban was chief gardener at the Aranjuez royal garden. Claudio later designed the Seville botanical gardens; Estéban became a leading wine expert. The species name *curtipendula*, from Latin words *curtus* (short) and *pendulus* (pendent), refers to the spikelike branches. In 1803, French explorer André Michaux described this species as *Chloris curtipendula*, from his 1795 Illinois collection. Royal botanist to King Louis XVI, Michaux led an American expedition (1785–95) to collect plants of economic interest. In 1853, John Torrey moved the species to the *Bouteloua* genus.

211. False Grama (*Bouteloua erecta*)
(Other common names: grama china, cathestecum)

Habitat: desert scrub, arid bluffs near Rio Grande; clay flats, bajadas; rocky, gravelly mountains at mid-elevations; loose, broken, igneous, limestone soils.

Geographic range: Texas: Trans-Pecos (south parts of Hudspeth, Presidio, Brewster Counties). US: Texas, south Arizona. Elsewhere: Mexico (south to Oaxaca), Guatemala, Honduras, El Salvador.

Park locations: Hoodoos, Papalote Alto, Javelin 4WD Trail, Madrid Falls

Rare in Texas, sometimes locally common in the Big Bend, false grama is a patch-forming perennial with erect, unbranched culms to 1 foot high. Each patch, a tight cluster of hairless culms, is likely a new clonal plant formed by spreading runners (stolons). The stolons are creeping stems that reproduce vegetatively by rooting at the nodes.

In basal clumps, the leaves are linear, to 5 inches long, flat but commonly incurved on the edges. The upper leaf surfaces bear long straight hairs, often with swollen bases. The blades are sheathed, the sheaths open, loose, the lower ones papery and woolly-haired. The ligule is a minute fringe of hairs at the leaf base.

From late spring to fall, flowers appear in short, terminal clusters near the culm tips. Seldom more than 1 inch long, each cluster has four to eight short, spikelike branches in two rows on the flowering stem. On a shorthaired, downcurved stalk, each branch supports three tightly clustered spikelets. The branches are either all male or bisexual, mixed on the same or on separate plants. On bisexual branches, the glumes at the spikelet base are often maroon-tinged and shaggy-haired. The lemmas are lobed at the tip, with veins projecting as needlelike bristles. The fruit is a caryopsis.

Botanical notes: In 1884, George Vasey and Eduard Hackel described this species as *Cathestecum erectum*, from specimens collected by Valery Havard and Edward Palmer. Havard found the lectotype in 1883, on "arid bluffs of Rio Grande, from El Paso to Presidio." The syntype was collected earlier, by Palmer in Sonora, Mexico, in 1869. The meaning of the generic name *Cathestecum*, from the Greek *kathestekos* (stationary), is unclear. The species name *erectum* denotes "erect." James Travis Columbo, with the Rancho Santa Ana Botanic Garden, transferred the species to the *Bouteloua* genus in 1999.

212. Black Grama (*Bouteloua eriopoda*)
(Other common names: woollyfoot or hairyfoot grama, grama negra, navajita negra)

Habitat: desert to mountains; grasslands, rocky, grassy slopes; igneous, limestone soils, low to high elevations.

Geographic range: Texas: Trans-Pecos (excluding Val Verde County), much of Permian Basin, Panhandle, part of Edwards Plateau. US: southwest; Oklahoma, Kansas, Wyoming. Mexico: Sonora east to Coahuila, south to Durango.

Park locations: Upper Guale Mesa, Arroyo Mexicano, Yedra Canyon, east Solitario

Black grama is a widespread native perennial and nutritious forage grass to 2 feet tall, arising from a fast-growing, shallow root system. The unbranched culms may be tightly clustered, with sprawling or abruptly bent stems with hard, knotty bases. Less often, the culms are prostrate runners (stolons), which root at the nodes. This grass is recognized by the white-woolly hairs that may cover lower portions of the culms.

The slender leaves, to 4 inches long, are clustered near the base of the culms. Flat or wavy, the gray-green blades are folded upward on the edges and pointed at the tips. The ligule is a tiny fringe of hairs.

From late spring to fall, especially in response to summer rains, flowers appear in slender clusters to 4 inches long, near the culm tips. Each cluster has three to eight spikelike branches, with each branch supporting up to eighteen comblike spikelets. The branches are white-woolly-bearded near the base. The stemless spikelets are aligned in two rows along the edges of the branch, often close-pressed to the branch stalk and overlapping. Each spikelet bears one lower bisexual floret and usually, one or two upper male or sterile florets. Commonly, the lemmas are hairy at the base, with three veins projecting as rough bristles, the midvein bristle the longest. The fruit is a slender caryopsis to 1/8 inch long.

Botanical notes: The epithet *eriopoda* joins the Greek *erio-* (wool) and *poda* (foot), alluding to the white-woolly lower stems. The species was described by John Torrey as *Chondrosum eriopodum* in 1848, in Lieutenant William Hemsley Emory's *Notes of a military reconnaissance from Fort Leavenworth, in Missouri, to San Diego, in California*. . . . Torrey's description was likely based on an 1846 specimen collected by Lieutenant Emory in New Mexico, between the Rio Grande and Gila River.

213. Fluffgrass (*Dasyochloa pulchella*)
(Other common names: low woollygrass, desert or false fluffgrass, zacate borreguero)

Habitat: creosote scrub, sandy flats, gravelly hills, bajadas; limestone, igneous, gypeous soils.

Geographic range: Texas: Trans-Pecos, sites in Permian Basin, southern Panhandle, west Edwards Plateau. US: southwest. Mexico: Baja California east to Tamaulipas, south to Mexico City.

Park locations: lower Contrabando Creek, Colorado Canyon River Access, Agua Adentro, Arroyo Mexicano, Outer Loop Trail

Only a few inches high, this fluffy, matlike "bunchgrass" is a perennial with tufted culms, often connected by wiry, creeping runners (stolons). Erect and unbranched, the culms produce leafy clusters at the tips, which may bend to the ground and take root, forming new stolons. Up to 2 1/2 inches long, the slender, stiff leaves are flat or involute, sharp-pointed, and rough underneath. Leaf sheaths are finely grooved and membranous on the margins. The ligule is a fringe of thin hairs.

From late spring to fall, flowers form in compact, headlike clusters to 1 inch long, at the stem tips, within the bundled leaves. On a short stalk, each flower cluster—straw-colored, pale green, or purple-tinged—is composed of short, spikelike branches with two to four spikelets. The flattened spikelets bear four to ten tiny florets, most fertile, at least one bisexual. The hairless glumes at the spikelet base are tan to purplish, with a bristlelike tip, papery edges, and a green vein. Frequently purple-tinged, the three-veined lemmas are divided into two lobes, with the midvein protruding between the lobes as a needlelike bristle.

Often, the plant is blanketed with white fuzz or fluff, which gives the grass its name. The hairlike fibers are excreted and evaporated water-soluble crystals (Valdes-Reyna 2003). The fruit is an oval translucent caryopsis to 1/17 inch long.

Botanical notes: The generic name, from the Greek *dasys* (hairy) and *chloe* (grass), alludes to the "hairy" inflorescences. The epithet *pulchella* connotes "pretty." In 1815, German taxonomist Karl Kunth described the species as *Triodia pulchella*. Alexander von Humboldt and Aimé Bonpland collected the type specimen between Guanajuato and Mina de Belgrado, Mexico, about 1803. Herbarium curator at the New York Botanical Garden, Per Axel Rydberg, moved this species to genus *Dasyochloa* in 1906.

214. Tanglehead (*Heteropogon contortus*)
(Other common names: black or bunch spear grass, twisted beardgrass, pili or tangle grass, zacate colorado, barba negra, zacate aceitillo)

Habitat: by Rio Grande, arroyos, creeks, springs; grasslands, hillsides, woodlands; varied soils.

Geographic range: Texas: Trans-Pecos (excluding Reeves County), much of South Texas Plains; parts of Edwards Plateau; Harris County. US: Texas to southern California; Florida. Elsewhere: native to Eurasia, Australia, Africa.

Park locations: Hoodoos, Agua Adentro, Rancherías Springs, lower Arroyo Mexicano, Cinco Tinajas

Tanglehead is a stout clump-forming perennial to 30 inches high, with slender, erect culms. Broomlike, the hairless culms are much branched above, with many upright branches. Rust-red fall colors and black culm nodes typify this attractive species. To 6 inches long, the leaves are flat or folded, abruptly pointed, often bristle-haired at the base. The leaf sheaths are compressed and ridged. The ligules are tiny ciliate membranes.

Mostly in summer and fall, unbranched flower clusters, to 3 inches long, form at the culm tips or upper leaf axils. In each cluster, spikelets are arranged in twelve to twenty-two pairs, with one stemless spikelet and one stalked spikelet in each pair. In the lower part of the cluster, the paired spikelets are sterile or male, and awnless. In the upper part, stemless spikelets are bisexual and awned, stalked spikelets sterile or male, and awnless. Stemless spikelets are brown, with a hard, thick, red-bearded base (callus) and long awn. The bristle-haired awn is bent and often twisted. Typically, stemless spikelets bear two florets, the lower sterile, the upper fertile. The three anthers are yellow or red, the stigmas purplish.

The fruit is a white cylindric caryopsis 1/6 inch long. The awn attached to the fruit responds to moisture changes by twisting up when dry, straightening when wet, in a motion that buries the seed in moist soil.

Botanical notes: The generic name *Heteropogon*, from the Greek *heteros* (different) and *pogon* (beard), alludes to the difference in the stemless and stalked spikelets in the upper part of the inflorescence. The Latin epithet *contortus* (twisted) refers to the twisted awns. Linnaeus described this species in 1753 as *Andropogon contortus*, citing India as the location. In 1817, French naturalist Ambroise Marie François Joseph Palisot, Baron de Beauvois moved the species to the *Heteropogon* genus.

215. Needleleaf Gilia (*Giliastrum acerosum*)
(Other common names: blue bowls, prickleleaf gilia, spiny blue bowls)

Habitat: creosote scrub, desert flats to mountains; on limestone, gypsum, novaculite.

Geographic range: Texas: Trans-Pecos; much of Permian Basin, Plains Country, Panhandle. US: Texas to Arizona, north to Kansas, Colorado. Mexico: Chihuahua east to Nuevo León.

Park locations: Solitario northwest rim, near Paso al Solitario, Lefthand Shutup

A delight to find on barren desert slopes, this gilia is a low perennial to 6 inches high, typically compact, rounded, at times strongly woody at the base, resembling a dwarf shrub. Growing from taproots, the spreading or ascending stems are often densely branched and covered with stalked glandular hairs.

Obscured below a copious floral display, the alternate leaves are short, to 3/4 inch long, dissected into stiff, prickly, needlelike segments. Each segment is abruptly and sharply tipped, and glandular.

From spring to fall, deep blue to lavender flowers appear in loose clusters, on stalks to 1 inch long at the stem tips. Each bloom is wheel-shaped or bowl-like, with five spreading oval lobes, a short tube, and a yellow throat with a jagged white rim. Protruding from the flower's throat are five stamens with large yellow anthers and a slender style with three white stigma lobes elevated above the anthers. At the base of each bloom is a bell-shaped, glandular-hairy, five-lobed calyx, each dark green lobe with wide membranous margins.

The fruit is a three-chamber, broadly egg-shaped capsule 1/4 inch long, with two to six seeds per chamber.

Botanical notes: *Giliastrum* joins *Gilia* (the genus name) and the Latin suffix *-astrum* (incompletely similar), meaning "something like" *Gilia* plants. In 1794, Hipólito Ruiz López and Jose Pavón, botanical explorers sent to South America by King Carlos III of Spain, named *Gilia* for Filippo Luigi Gilii, Italian astronomer and director of the Vatican Observatory. Gilii, with Argentine botanist Gaspar Xuarez, coauthored a study of plants brought to Rome from South America. In 1870, Asa Gray described this species as a variety, *Gilia rigidula* var. *acerosa*, collected by Augustus Fendler, Alexander Gordon, and Charles Wright. Gordon found the plant at Raton, New Mexico, in 1848; Wright on the Rio Grande in Texas, and near Del Rio in 1849. In 1917, Per Axel Rydberg formed the new genus *Giliastrum* in his *Flora of the Rocky Mountains* and elevated this plant to species status as *G. acerosum*.

216. Iron Ipomopsis (*Ipomopsis laxiflora*)
(Other common names: iron skyrocket, iron standing-cypress, nodding ipomopsis, loose-flowered gilia)

Habitat: sandy soils, grasslands, fields, roadsides, dry ground near seasonal pools and creeks.

Geographic range: Texas: Trans-Pecos (Presidio, Reeves, Pecos, also El Paso, Culberson, Jeff Davis Counties), parts of Permian Basin, Panhandle. US: Texas to Arizona, north to Kansas, Colorado. Mexico: not reported.

Park locations: Llano Loop

This phlox is a delicate annual to 16 inches high, with wiry, pale green stems. Erect or widely spreading and branching, the stems are nearly hairless, glandular-hairy, or coated with curling hairs.

To 1 inch long, the alternate leaves are clustered along the branches and gradually smaller upward. The blades are divided into linear segments, but upper leaves may be undivided. The dark green blades may bear short, crinkled hairs like the stems.

From spring to early fall, trumpet-shaped flowers form in loose, leafy clusters, single or paired on stalks at the branch tips. Each bloom is white, lavender-tinged, with a slender tube and five abruptly spreading lobes. Within the flower's throat are five stamens with yellow anthers and a style with three stigma branches. Supporting the blooms are five linear green sepals, with white, membranous margins.

The cylindric fruit is a three-chamber dehiscent capsule to 2/5 inch long, with brown spindle-shaped seeds.

Botanical notes: In 1803, French explorer André Michaux formed the genus *Ipomopsis* from the generic name *Ipomoea* and the Greek suffix *-opsis* (resembling), to indicate "like *Ipomoea*," for the similar flowers. Royal botanist to King Louis XVI, Michaux traveled through the United States and Canada (1785–96) collecting plants for his South Carolina garden and for shipment to France. The epithet *laxiflora* refers to the lax flower clusters. In 1890, University of Chicago botanist John Coulter described this plant as a variety, *Gilia macombii* var. *laxiflora*, from a collection by botanical explorer Greenleaf Nealley in 1889, at Camp Charlotte, in Irion County, Texas. Nealley collected mostly forage plants for the USDA. Coulter also collected in Texas and authored *Botany of Western Texas* (1891–94). In 1956, University of Texas at Austin botanist Verne Grant placed the plant in the genus *Ipomopsis* and gave it species status.

217. Paleflower Ipomopsis (*Ipomopsis longiflora* var. *neomexicana*)
(Other common names: flaxflowered or New Mexico flaxflowered ipomopsis, blue or white-flowered gilia, blue or pale blue trumpets, white-flower skyrocket, pitillo)

Habitat: dunes, sandy, gypseous desert flats, low gravelly hills, bajadas; creosote scrub, grasslands.

Geographic range: Texas: Trans-Pecos (excluding Terrell, Val Verde Counties). US: Texas to eastern Arizona, southern Colorado, southeast Utah. Mexico: Chihuahua, Coahuila.

Park locations: 1 1/4 miles west of Lajitas, Black Hills, Los Alamos

Seemingly out of place in the brutal desert sun, paleflower ipomopsis is an annual or biennial to 20 inches high, flimsy in appearance, with thin yellow green stems. Upright or sprawling, the leafy stems branch in the lower half. Both stems and leaves are hairless to sparsely hairy. The alternate leaves are up to 1 1/2 inches long, dissected into slender, often stiff and abruptly tipped, linear segments, or undivided. The fleeting basal leaves may wilt before the flowers bloom.

From late spring to early fall, delicate, early falling, trumpetlike flowers appear in lax clusters at the stem tips. Each bloom has a slender

lavender tube with five flaring, pinwheel-like lobes. The lobes are pale lavender, white, or pink. Attached to the upper floral tube are five stamens with yellow anthers. The style is trilobed, with stigma branches in the flower's throat. Although self-pollinating, the blooms attract night-feeding hawk moths. Supporting each bloom are five linear green sepals with white membranous margins. The fruit is a three-chamber, narrowly oval, dehiscent capsule to 3/5 inch long.

Botanical notes: The species name *longiflora*, from Latin *longus* (long) and *floris* (flower), refers to the long-tubed flowers; the varietal name, *neomexicana*, to the New Mexico habitat. The species was described by John Torrey in 1827 as *Cantua longiflora*, from a collection by Edwin James "on the Canadian," on the 1820 Long Expedition. In 1956, University of Texas botanist Verne Grant moved the species to the *Ipomopsis* genus. In 2001, Dieter Wilken, Santa Barbara Botanic Garden botanist, described this variety as a subspecies, *I. longiflora* subsp. *neomexicana*, from a specimen he collected in Sierra County, New Mexico, in 1985. Billie Lee Turner treated the subspecies as a variety in 2003.

218. Blue Milkwort (*Hebecarpa barbeyana*)
(Other common names: Barbey's or slenderlobe or narrowleaf milkwort)

Habitat: desert scrub, creek banks, grasslands, gravelly bajadas, rocky canyons, slopes; low to high elevations, varied soils.

Geographic range: Texas: Trans-Pecos (excluding Reeves, Val Verde Counties); Crockett County. US: Texas, southern New Mexico, Arizona. Mexico: Sonora east to Tamaulipas, south to Hidalgo, Guerrero.

Park locations: Las Cuevas, Arroyo Mexicano, Cinco Tinajas, Tres Papalotes, Cienega Creek, Cienega Mountain

Blue milkwort is a small perennial to 20 inches high, with a few upright green stems branched below from a woody base. Typically, both stems and leaves are shorthaired and fuzzy to the touch. As much as 1 3/4 inches long, the alternate leaves are variable, the lower more oblong and early falling, the upper more narrowly lance-shaped. The blades are largely stemless, with pointed tips and smooth margins.

Flowers appear from spring to fall, in loose clusters at the stem tips or in the leaf axils. The blooms resemble butterfly-shaped flowers of the legume family. Each bloom has five sepals, the two lateral sepals large, oval, petal-like, called "wings," and three petals, the lower "keel" petal boat-shaped, with a rounded or beak-like tip. The two upper petals are tiny and strap-shaped. The blooms are multicolored, the wings pale lavender or white, the petals purplish but the keel yellow or green at the protruding tip. Within the petals are eight stamens with yellow anthers and a style with a bilobed stigma.

The pale green fruit is a two-chamber, flattened dehiscent capsule to 2/5 inch long, with hairy black seeds.

Botanical notes: The genus name *Hebecarpa*, from the Greek *hebe* (pubescent) and *carpus* (fruit), likely refers to the fruits of other species (not ours). In 1893, University of Geneva botanist Robert Chodat named this species *Polygala barbeyana*, for William Barbey, from collections by Charles Parry and Edward Palmer in 1878, and by Johann Schaffner in the Mexican state of San Luis Potosí. Swiss botanist-philanthropist Barbey, and his wife Caroline Barbey-Boissier, collected for their Barbey-Boissier Herbarium in Geneva. In 1856, Schaffner moved to San Luis Potosí, where he practiced pharmacy, collected plants, and built a large herbarium. In 2011, milkwort specialist J. Richard Abbott placed this species in his new genus *Hebecarpa*, formerly a subgenus of *Polygala*.

219. Smallflower Milkwort (*Rhinotropis nudata*)
(Other common names: littleleaf milkwort)

Habitat: desert scrub, rocky seasonal creeks, limestone canyons, slopes, ledges.

Geographic range: Texas: Big Bend (Presidio, Brewster Counties). US: Texas. Mexico: Chihuahua east to Nuevo León, Tamaulipas.

Park locations: Los Portales, Solitario rim above the Righthand Shutup, Lower Shutup

Rare in the United States, this milkwort is a rock-crevice-loving perennial or subshrub to 16 inches tall, growing from woody roots and woody at the base. The green stems are nearly leafless, broomlike, mostly erect, and minutely hairy. To 3/5 inch long, mostly absent or early falling, the alternate, stemless leaves are narrowly lance-shaped or linear. Persistent blades are scalelike.

From spring to fall, tiny blooms with an intriguing architecture form in loose, few-flowered, terminal clusters. Each flower has five sepals, two large lateral sepals, the "wings," petal-like, oval; and three petals, the two upper petals minuscule, earlike, and the lower protruding petal, the "keel," with a short conical beak. The blooms are multicolored, the wings white, tinged greenish yellow, with green or purple veins; the earlike upper petals with maroon markings; the keel green and/or yellow; the beak yellow or white. Typically, the keel is minutely hairy. The seven or eight stamens are united much of their length, and the style is slender, bent, with a bilobed stigma.

To 1/6 inch in length, the fruit is a two-chamber, dehiscent green capsule, oval, egg-shaped, or oblong, compressed, and sparingly hairy.

Botanical notes: The generic name *Rhinotropis*, from the Greek *rhinos* (nose) and *tropis* (keel), refers to the beaked keel petal. The species name is the Latin *nudata* (nude), for the nearly leafless stems. Botanist Townshend Brandegee described the species as *Polygala nudata* in 1911, from a collection by botanical explorer Carl Purpus in 1910, at Sierra de la Paila in Coahuila, Mexico. Brandegee and his wife Catherine, curator of botany at the California Academy of Sciences in San Francisco, were leading California botanists and published their own botanical journal *Zoe*. In 2011, milkwort authority J. Richard Abbott moved the species to his new genus *Rhinotropis*, formerly a subgenus of *Polygala*.

220. Havard's Wild Buckwheat (*Eriogonum havardii*)
(Other common names: Havard's buckwheat)

Habitat: barren desert flats, creosote-mesquite scrub, sotol-yucca grassland, rocky slopes, open woodlands; on silt, clay, limestone, gypsum.

Geographic range: Texas: Trans-Pecos (excluding El Paso, Jeff Davis, Reeves Counties); Winkler County. US: Texas, New Mexico. Mexico: not recorded.

Park locations: Los Portales, Solitario rim above the Righthand Shutup, West Contrabando Trail, Contrabando Dome Trail, Buena Suerte Trail

Havard's wild buckwheat is a robust perennial to 2 feet high, with a mound-like basal rosette of leaves, leafless stems, and a rounded, knobby crown growing from a woody taproot. Towering above the leaves, the slender stems are erect, mostly hairless, but woolly within the leaf rosette.

The mat-forming leaves are elliptic to inversely lance-shaped, to 2 inches long, on petioles half as long. Pointed at the tips, smooth on the margins, grooved, the silvery blades are densely coated with soft hairs.

From spring to fall, clusters of tiny blooms appear in open, branching inflorescences to 16 inches long. On a slender, erect stalk, each small flower cluster is supported within a bell-shaped arrangement of three scalelike green bracts. To 1/8 inch long, the tiny blooms are bright yellow with greenish midribs and cloaked with white-woolly hairs. Each bloom is composed of petal-like, lance-shaped tepals fused below, in two rings of three. Within the tepals are nine protruding stamens with oval yellow anthers.

The fruit is a brown, hairless achene (small, dry, indehiscent, one-seeded) to 1/10 inch long.

Botanical notes: French botanist André Michaux named the genus *Eriogonum* in 1803, combining the Greek *erio-* (woolly) and *gonu* (joint), emphasizing the hairy stem nodes of the type species. In 1883, Sereno Watson, curator of the Gray Herbarium at Harvard, named the species for French-born botanist and US Army surgeon Valery Havard, who collected the herb in the Chinati Mountains of Presidio County in 1882. Stationed at nearby military posts, Havard collected throughout West Texas, and in 1885 authored *Report on the Flora of Western and Southern Texas*. Havard also wrote articles on the economic uses of plants by Native Americans and early settlers.

221. Bushy Wild Buckwheat (*Eriogonum suffruticosum*)
(Other common names: bushy or shrubby buckwheat, shrubby wild buckwheat)

Habitat: barren desert flats with creosote scrub; silty, clayey, gypseous hills and mounds, often with loose, broken, limestone rock, debris; rocky limestone outcrops, low to mid-elevations.

Geographic range: Texas: Trans-Pecos (southwest Brewster, southeast Presidio Counties). US: Texas. Mexico: no record.

Park locations: Solitario Peak, Fresno Peak, Los Alamos, Contrabando Dome, Buena Suerte Trail

Of conservation concern: see Appendix A

To 8 inches high, bushy wild buckwheat is an artful, highly photogenic miniature shrub, treelike or bonsai-like, growing from a stout, branching woody crown. The stems produce many spreading, woody branchlets, often tangled, twisted, and gnarled, with peeling bark. The short flowering stems are erect, to 1 inch long or more, and woolly-haired.

Small leaves form in tight bundles at the tips of short branchlets. At most 1/3 inch long, the short-petioled leaves are elliptic, with thick revolute margins. More finely silky-haired above, the blades bear short, matted hairs below.

From March to May, small profusely blooming flower clusters appear at the branch tips. Each bloom, to 1/4 inch long, is composed of six tepals in two rings of three each, the outer fan-shaped, the inner inversely lance-shaped. The tepals are white with a prominent red midrib. Protruding from each small bloom are nine stamens with hairy threadlike filaments and three yellow styles with headlike stigmas.

The hairless fruits are light brown, three-angled achenes 1/8 inch long.

Botanical notes: The Latin epithet *suffruticosum* denotes "somewhat shrubby." Sereno Watson described this species in 1885, from a collection by US Army surgeon Valery Havard in 1883, "on the foothills of the Bofecillos Mountains." Stationed at Fort Duncan and San Antonio, Havard traveled with Captain William Livermore's expeditions (1883–84) to the upper Rio Grande, including present-day BBR. As an assistant to Asa Gray and curator of the Gray Herbarium at Harvard, Watson named many species newly discovered in the American West.

222. Fiddle Dock (*Rumex pulcher*)
(Other common names: fiddleleaf or fiddleneck dock)

Habitat: Rio Grande, periodically inundated sites, intermittent creeks, disturbed sites.

Geographic range: Texas: Trans-Pecos (El Paso, Hudspeth, Presidio, Val Verde Counties); parts of east half of Texas. US: much of east and west coasts, southern states west to New Mexico, Missouri. Elsewhere: native to Mediterranean.

Park locations: 1 mile west of Lajitas, Grassy Banks, Colorado Canyon head, Cinco Tinajas

Introduced and weedy, fiddle dock is a coarse perennial to 3 feet high, with erect, hairless stems branched above the base, growing from thick roots. The stems, fruits, and leaves turn deep ruddy red with age. To 6 inches long, the alternate, mostly basal leaves are oblong, or occasionally fiddle-shaped. Thick, stiff, yet wavy and curling, the blades are often red tinged and blunt-tipped. Often long-petioled below, the leaves are smaller, largely stemless above.

From spring to early summer, flowers form on the upper two-thirds of the stems in loose, many-branched inflorescences, each branch with numerous separate, compact flower clusters. The bell-shaped green flowers, red with age, consist of six sepal-like tepals in two rings of three, the outer lance-shaped tepals small, and the inner tepals much larger. Within each flower are six stamens with yellow anthers and three spreading styles with shield-like stigmas. Expanded after blooming, the triangular inner tepals (valves at this stage) become hard and veiny with spiny wings and a central warty tubercle.

The fruits are three-angled achenes to 1/9 inch long, ruddy-red or black, enclosed by the enlarged valves.

Botanical notes: The Latin genus name *rumex* is a term used by Pliny for a sorrel, possibly from Latin *rumo* (to suck), because of the ancient Roman practice of sucking the leaves to quench thirst. However, *rumex* also connoted "spear," and perhaps alluded to the daggerlike leaves of some sorrels. The species name is the Latin *pulcher* (pretty). In 1753, Linnaeus described this species when he named the genus and cited Swiss botanist Jean Bauhin, physician to Duke Frederick of Württemberg, as a source. Bauhin, apparently thinking the sorrel "pretty," named it *Lapathum pulchrum* in his *Historia plantarum universalis* (1650–51), a monumental work describing 5,200+ plants.

223. Sinkerleaf Purslane (*Portulaca halimoides*)
(Other common names: silkcotton or dwarf purslane, desert portulaca, mañanitas)

Habitat: desert flats, dunes, arroyos; gravelly banks, bluffs above creeks, springs; loose, sandy, alluvial soils.

Geographic range: Texas: Trans-Pecos (El Paso, Hudspeth, Presidio, Brewster Counties); Ward, Edwards, Llano Counties. US: southwest; Oklahoma. Elsewhere: Mexico, south to Brazil; West Indies.

Park locations: Leyva Canyon, Skeet Canyon, Cerro Chilicote Loop

Uncommon in BBR, sinkerleaf purslane is a fleshy annual growing from slender, fibrous roots. Only a few inches high, the branching stems reach 10 inches in length. Except for large clumps of cottony hairs in the leaf axils, the mostly prostrate stems are hairless. In hot desert environs after summer rains, the stems may turn red or maroon. The seemingly swollen leaves are cylindric or slightly flattened, to 3/4 inch long. Stemless, with rounded or blunt tips, the green, often red-tinged blades may form tight clusters at the branch tips.

From late spring to fall, small flowers appear at the branch tips, single or in small headlike clusters. The flowers arise from bract-like rings of leaves filled with strands of hairs. Each bloom consists of five oblong yellow petals; two triangular sepals, often red or pink; up to eighteen yellow stamens; and a yellow style with multiple stigma branches. The blooms remain open only briefly, usually midmorning to early afternoon.

The fruit is a membranous, many-seeded capsule, which splits below the middle so that the top comes off like a lid. The tiny egg-shaped capsule contains gray or black, coiled seeds.

Botanical notes: *Portulaca*, an old Latin name for purslane, likely combines the Latin *portula* (little gate) and *-aca* (pertaining to), alluding to the lid of the fruit capsule. However, *Portulaca* may signify "milk carrier," from the Latin *portare* (to carry) and *lac* (milk), for the stems' milky sap. The species name joins the genus name *Halimium* and the suffix *-oides* (resembling), for the likeness to *Halimium* plants of the rock rose (Cistaceae) family. Linnaeus described the species in 1762, citing *Voyage to . . . Jamaica* (1707) by Sir Hans Sloane and *The Civil and Natural History of Jamaica* (1756) by Irish botanist-historian Patrick Browne as sources.

224. Shrubby Purslane (*Portulaca suffrutescens*)
(Other common names: copper or subshrubby purslane, shrubby portulaca)

Habitat: sandy, gravelly flats, grasslands, rocky ridgetops; igneous, limestone substrates.

Geographic range: Texas: Trans-Pecos (El Paso, Jeff Davis, Presidio, Brewster Counties). US: Texas west to Arizona. Mexico: Baja California Sur, Sonora, Chihuahua, Sinaloa, Durango, Guerrero.

Park locations: Agua Adentro, Llano Dome near Papalote Llano Nuevo, Tascate Hills

Shrubby purslane is a stout perennial, often woody at the base, with thick tuberous roots and erect or ascending stems to 1 foot long. The succulent, much branched herb is largely hairless, except for tufts of white-woolly hairs at the leaf nodes. The alternate, stemless leaves are cylindric, to 1 inch long, and tapered to a pointed tip. Bractlike leaves form bundles at the branch tips, just below the flowers.

Mostly from late spring to late summer, flowers appear at the branch tips in small headlike clusters. The flowers are orange, copper, or bronze, and darker-colored or red in the throat. Each bloom has

five petals, spatulate to broadly oblong, and notched at the tip; two yellow or red, triangular sepals; up to thirty-five stamens with red filaments and yellow anthers; and a long red style with up to eight stigma branches. The blooms may open midmorning and close by midday. Like all *Portulaca*, this purslane is self-pollinating, and cross-pollination is infrequent (Matthews et al. 1994).

The circumscissile fruit is a many-seeded round capsule to 1/5 inch across. The upper half of the capsule comes off like a lid to release tiny, iridescent, blue-gray to black seeds.

Botanical notes: The epithet *suffrutescens*, from the Latin prefix *suf-* (somewhat) and *frutescens* (shrubby), characterizes the plant habit. George Engelmann described the species in 1881, from collections by Charles Wright, "in western New Mexico, at the copper mines [Santa Rita del Cobre]," in 1851; Elliott Coues and Edward Palmer, at "Fort Whipple, northern Arizona, 1865"; and Engelmann himself, in 1880, "in the Santa Rita Mountains, southern Arizona." US Army surgeon-naturalist Coues was stationed at Fort Whipple; Palmer served as assistant surgeon at Fort Lincoln in the Verde Valley.

225. Wingpod Purslane (*Portulaca umbraticola* var. *lanceolata*)
(Other common names: Chinese hat, wingpod portulaca, crownpod purslane)

Habitat: by creeks, tanks, pools, sandy flats, grasslands, igneous mountains.

Geographic range: Texas: Trans-Pecos (El Paso, Hudspeth, Jeff Davis, Presidio, Brewster Counties); much of east half of state. US: Texas to Arizona, Oklahoma, east to Mississippi, Arkansas, Missouri. Elsewhere: variety *umbraticola*, Mexico to Argentina.

Park locations: Skeet Canyon, Tascate Hills

Wingpod purslane is a low succulent annual from fleshy, fibrous roots, fast-growing, prolific, ephemeral. The brownish-green or reddish stems, to 10 inches long, are prostrate to ascending. Mostly stemless, to 1 1/2 inches long, the alternate leaves are flat, not cylindric. Often red tinged, the fleshy blades are oblong or inversely lance-shaped and rounded at the tip. Several leaves form bract-like rings below the flowers.

From late spring to early fall, flowers form in small clusters at the branch tips. The flowers are yellow, suffused with red, orange, or pink at the tips. Each bloom has five broadly spatulate petals, two egg-shaped sepals with membranous margins, up to thirty stamens with red filaments and yellow anthers, and one short style with numerous yellow stigma branches.

The fruit is a many-seeded, top-like capsule to 1/5 inch wide, which splits above the middle, with the top coming off like a lid. A thin, translucent wing, formed from the persistent calyx base, encircles the capsule, and the upper capsule (lid) is depressed. Together, the lid and wing resemble a Chinese hat.

Botanical notes: The epithet *umbraticola* combines the Latin *umbro* (to shade) and the suffix *-icola* (inhabitant of), for the purslane's shady habitat. The variety name *lanceolata* (lanceolate) refers to the upper leaves, which may be more oblanceolate. Karl Kunth described the species in 1823, from a collection by Baron von Humboldt and Aimé Bonpland near Cumaná in Venezuela, likely in 1799 or 1800. George Engelmann described the variety as a species, *Portulaca lanceolata*, in 1850, from a collection by New Braunfels botanist Ferdinand Lindheimer in 1848, in the "granite region of the Liano [Llano River]." In 2003, Billie Lee Turner and colleagues treated the plant as a variety of *P. umbraticola*.

226. Maidenhair Fern (*Adiantum capillus-veneris*)
(Other common names: southern, black, or common maidenhair fern, Venus's-hair fern, culantrillo, cilantrillo, adianto)

Habitat: full or partial shade at perennial water sources; springs, creeks, waterfalls, seeps, moist canyon walls, limestone, also igneous rock, Rio Grande to higher, wooded mountains.

Geographic range: Texas: Trans-Pecos (excluding El Paso, Reeves, Pecos Counties), Edwards Plateau, parts of Panhandle, north, east Texas. US: mostly southern half. Elsewhere: Mexico to Venezuela, Peru; West Indies; parts of Eurasia, Africa.

Park locations: Fresno and Panther canyons, Ojito Adentro, Madrid Falls, Mexicano Falls, Smith Spring

Unlike some desert-dwelling ferns, maidenhair fern grows only in cool, wet habitats. To 2 feet long, the fronds dangle gracefully from steep rock walls or seeps. The rhizomatous stems are wiry, short creeping, at times branching, with shiny gold or brown scales. Often, the fronds are tightly clustered and overlapping, with lance-shaped blades. To 18 inches long, the hairless pale green blades are two or three times divided into smaller leaflet pairs (pinnae), and ultimately undivided segments (pinnules). The blades form on slender, dark purple to black petioles (stipes), which are grooved above.

The fan-shaped pinnules are stalked, with the stalk's dark color extending into the pinnule base. The edges of fertile pinnules are lobed; margins of sterile pinnules are toothed. In many ferns, outgrowths known as indusia cover and protect the sori, clusters of sporangia (spore-bearing sacs) on the leaf surfaces. In this fern, the apical margins of the fertile pinnules curl under to form "false" indusia. Sporangia are borne along the ends of the veins beneath the flaplike folds. The yellow or tan spores mature in spring or summer.

Botanical notes: Since water does not adhere to the fronds, Pliny named this genus *Adiantum*, from the Greek *a* (not) and *diantos* (wet). A waxy coat on the leaves causes water to roll off. The epithet *capillus-veneris*, combining *capillus* (hair) and *veneris* (of Venus), is based on a Roman myth of Venus, goddess of love, who supposedly emerged from the sea with dry hair. Linnaeus described the genus and species in 1753, citing southern Europe as the location. The lectotype was collected in France by Pierre Magnol, professor at Montpellier, the first botanist to develop the concept of plant families, in 1689.

227. Hairy Bommer (*Bommeria hispida*)
(Other common names: copper fern, dancing or hairy bommeria, helecho terciopelo)

Habitat: igneous mountain slopes, on rocks, ledges, in rock crevices, at base of cliffs and boulders, under overhangs, on talus and scree; partial shade.

Geographic range: Texas: Trans-Pecos (El Paso, Jeff Davis, Presidio, Brewster Counties). US: Texas to Arizona. Mexico: Baja California Sur east to Coahuila, south to Hidalgo, Jalisco.

Park locations: Skeet Canyon, Cienega Creek

Hairy bommer forms extensive patches across the ground, beneath boulders and nestled in rock crevices at moist, shady sites. This captivating fern grows from spreading horizontal roots and forms widely creeping and branching stems with chaffy brown scales. The erect fronds are scattered along the stems, with the blades parallel to the ground.

Hairy bommer is distinguished by highly divided leaves with a thick cloak of hairs and scales. The blades change colors with age, from pale to dull or dark green, or often copper or red at maturity. To 3 inches long, the pentagonal blades are divided into three primary divisions, which are further dissected into smaller (pinnate-pinnatifid) segments. The upper blade surface is lined with short, straight hairs; the lower surface bears a mix of hairs and scales. Tan to chestnut brown, the slender stipes are also hairy.

Unlike most ferns in the maidenhair fern family (*Pteridaceae*), hairy bommer develops no true or false indusia. The margins of the ultimate segments of the fronds are not recurved or modified, leaving the sporangia (spore cases) unprotected. The sporangia form along the veins on the lower leaf surface, near the margins, and cover much of the segment. The globular brown spores mature in summer and fall, sixty-four in each sporangium.

Botanical notes: In 1876, French fern authority Eugène Pierre Fournier named the genus *Bommeria* for University of Brussels pteridologist Jean Edouard Bommer. The Latin epithet *hispida* (hispid) refers to the leaves. This species was described as *Gymnogramma hispida* by Leipzig fern specialist Georg Mettenius in 1869, from a collection by Charles Wright in 1849, in "Jeff Davis County, near present Fort Davis, Pass of the Limpia, top of mountain." Columbia University fern expert Lucien Underwood moved the fern to the *Bommeria* genus in 1902.

228. Helecho de La Llave (*Llavea cordifolia*)
(Other common names: Llave's fern, cordate-leaf llavea, llavea)

Habitat: shady, moist, rocky, wooded slopes, canyons, on rock walls, limestone soil.

Geographic range: Texas: Big Bend (southeast Presidio County). US: Texas. Elsewhere: parts of Mexico, Guatemala, Costa Rica.

Park locations: moist limestone bluff

Of conservation concern in Texas: see Appendix B

Photo: Carlos Velazco-Macias

Found only once in the United States, in 1992 at BBR, helecho de la Llave (Llave's fern) is a mostly Mexican fern, ranging from the Sierra Madre Oriental in Nuevo León south to Guatemala and Costa Rica. Efforts to relocate the fern in BBR have been unsuccessful so far, but the search continues for this rare, morphologically isolated species, with two leaf types.

Llave's fern grows from short, stout roots with blackish scales. The fronds are up to 3 feet long, with egg-shaped blades two or three times divided into smaller pinnae pairs. The long, grooved petioles, with white or greenish-yellow scales at the base, may approach the length of the blades.

The lower part of each frond consists of sterile, egg-shaped segments, while the upper part is a branching cluster of podlike fertile segments. To 1 inch long or more, the stalked sterile segments are bright green, often minutely toothed on the edges, and glaucous underneath.

Borne at the frond tips, the stalked fertile segments are linear, to 3 inches long, pale green but reddish brown with age. The edges of these segments curl under and partly hide the sporangia, which line veins on the segment undersides. The granular spores are spheric but four-sided.

Botanical notes: In 1816, Mariano Lagasca y Segura, director of the Madrid Botanic Gardens, named this genus for Pablo de la Llave, who discovered the type species, *Llavea cordifolia*, in Mexico. La Llave, Mexican naturalist, priest, and Veracruz state senator, later became director of the Natural History Museum in independent Mexico. The Latin epithet *cordifolia* connotes "heart-shaped leaves," a reference to the sterile segments. In 1992, R. J. O'Kennon and Barton Warnock collected this fern at BBR on a moist limestone bluff.

229. Bonaire Lip Fern (*Myriopteris aurea*)
(Other common names: golden or slender lip fern, calawala)

Habitat: desert scrub, woodlands, springs, creeks, igneous slopes.

Geographic range: Texas: Trans-Pecos (El Paso, Jeff Davis, Presidio, Brewster Counties). US: Texas, southwest New Mexico, Arizona. Elsewhere: Mexico (northern Mexico south to Chiapas); Central America to Argentina; West Indies.

Park locations: Ojito Adentro, La Iglesia, Yedra crossover trail to Old Abandoned Road Trail, Skeet Canyon, Tascate Hills

The stout rhizomatous stems of Bonaire lip fern are short-creeping or densely clustered, often much branched and adorned with lancelike, dark scales with pale brown margins. To 2 feet long, mostly erect or ascending, the fronds are green to dark green above, with shiny hairs.

The long blades are narrowly linear, only 1 1/2 inches wide. Each blade consists of up to forty pinnae pairs, and each pinna up to eight pairs of lobes (ultimate segments). The oblong pinnae bear stiff white hairs above and thicker short, woolly hairs below, which are often white with age. The dark or reddish-brown, whitehaired rachises are rounded above.

False indusia are formed by the light green or whitish recurved margins of the segments. The slender margins may or may not partly conceal the sporangia. Dark red sporangia are borne on the undersides of the segments, along the margins. In summer and fall, each sporangium produces thirty-two spores.

Botanical notes: The genus name *Myriopteris*, joining the Greek *myrios* (myriad) and *pteris* (fern), refers to the fern's many leaflets. French clergyman-botanist, Abbe Jean-Louis Poiret, described this fern as *Pteris aurea* in 1804, from a Peruvian collection of

Joseph de Jussieu, brother of famous botanists Antoine and Bernard de Jussieu. According to Poiret, the epithet *aurea* (golden) refers to the "nearly golden down" of the upper leaflet surfaces. In 1953, Caribbean taxonomist George Proctor placed the fern in the *Cheilanthes* genus as *C. bonariensis* (the epithet *aurea* was already in use). In 2013, based on new phylogenetic studies, Amanda Grusz and Michael Windham of Duke University transferred forty-seven *Cheilanthes* species to the genus *Myriopteris*. Our fern was moved with the older epithet, as *M. aurea*.

230. Prickly Lip Fern (*Myriopteris scabra*)
(Other common names: rough, warty, or spinose lipfern)

Habitat: protected sites in dry, hot desert, rock crevices, cliff bases, ledges, rocky canyon bottoms, ravines, on limestone, also igneous soils.

Geographic range: Texas: Trans-Pecos (excluding El Paso, Culberson, Reeves Counties), parts of Edwards Plateau, South Texas Plains; scattered sites. US: Texas, Oklahoma. Mexico: Chihuahua to Tamaulipas, Durango southeast to San Luis Potosí, Querétaro.

Park locations: East Rancherías Trail, Rancherías Canyon, Panther Canyon, Chorro Canyon, Cinco Tinajas, Los Portales

At home in the driest desert habitats, this fern is the only *Myriopteris* with bristly, prickle-like hairs. The plants grow from sturdy, short-creeping rhizomatous stems, which are covered with oblong brown scales. Erect or spreading, the stems form small clusters to 1 foot long or more. The leaves rest on dark brown or reddish-brown, chaffy, scaly stipes equaling the length of the blades. The rachises are similar in color, chaffy and sparsely scaly, with stiff hairs below and close-pressed hairs above.

Oblong or lance-shaped, less than 2 inches wide, the blades are green to dark or dull green above, light green below, once or twice divided into smaller pairs of pinnae that may also be cleft or lobed. Unlike the beadlike segments of many *Myriopteris* species, the ultimate segments of the blades are mostly oblong or egg-shaped. Each segment is rough, covered with short, stiff hairs, often with enlarged bases.

False indusia are formed by the recurved segment margins. Green or white, minutely toothed on the edges, the false indusia scarcely conceal the sporangia. The lustrous tan to black sporangia are clustered along the segment margins. Each sporangium produces sixty-four spores from summer to fall.

Botanical notes: Glasgow botanist Sir William Jackson Hooker described this species as *Cheilanthes aspera* in 1852, from an 1849 Charles Wright collection on Turkey Creek in Uvalde County. However, that name was illegitimate, previously used for another fern. The species was described as *Pellaea scabra* in 1905 and has been known since 1918 as *Cheilanthes horridula*. In 2013, when Amanda Grusz and Michael Windham transferred forty-seven *Cheilanthes* species to the genus *Myriopteris*, this fern was moved with the older epithet, as *M. scabra*. The Latin epithet *scabra* (rough) refers to the upper leaf surfaces.

231. Villous Lip Fern (*Myriopteris windhamii*)
(Other common names: scaly or hairy lipfern)

Habitat: rock niches, shady ledges, boulder-filled canyons, rocky slopes; limestone, also igneous substrates, low to high elevations.

Geographic range: Texas: Trans-Pecos (excluding Reeves County). US: Texas to Arizona. Mexico: Chihuahua, Coahuila, Durango, Zacatecas.

Park locations: Leyva Canyon, drainage from Sauceda to Cinco Tinajas, Los Portales, Lefthand Shutup, Lower Shutup

The rhizomatous stems of villous lip fern are short, sturdy, clumped, and coated with bicolored scales. The scales have a broad, dark central region and brown or rust-colored margins. To 1 foot long, the clustered fronds are stiffly erect to spreading.

Typically, the blades are oblong or lance-shaped, to 2 inches wide, tapered to the tip, three or four times divided into smaller pinnae pairs. The blades rest on wiry black or dark brown stipes that are rounded above, often nearly the length of the blades, and coated with white or tan scales.

The ultimate segments of the fronds are beadlike, densely cloaked with long, coarse, crinkled hairs above, less hairy but copiously scaly below. Nearly colorless scales, chaffy and overlapping, obscure the midveins on the segment undersides and may extend beyond the segment margins.

The recurved edges of the ultimate segments form false indusia. Concealed beneath overlapping scales, the sporangia are borne along the segment margins. In summer and fall, thirty-two spores develop in each sporangium. The plant is apogamous, forming sporophytes asexually without fertilization.

Botanical notes: In 1918, this species was described by Massachusetts self-taught botanist and fern enthusiast George Davenport as *Cheilanthes villosa*. The fern was col-

lected by John Lemmon, self-taught botanist for the California State Board of Forestry, in the Chiricahua Mountains of Arizona in 1881. Lemmon, often with his wife Sarah, a botanical illustrator, collected widely in the western states and authored popular works on natural history. In 2013, in a revision of *Cheilanthes* taxonomy with fern authority Michael Windham, Amanda Grusz of Duke University moved this species to the *Myriopteris* genus as *M. windhamii* (the specific epithet *villosa* was already used). The species name honored Windham "for his lifelong dedication to the study of cheilanthoid ferns."

232. Foreign Cloak Fern (*Notholaena aliena*)
(Other common names: Mexican cloak fern)

Habitat: desert scrub, arid, rocky bluffs, talus slopes; in rock fissures, below boulders; volcanic substrates.

Geographic range: Texas: Big Bend (southeast Presidio, south Brewster Counties). US: Texas. Mexico: Chihuahua, Coahuila, Tamaulipas.

Park locations: Agua Adentro Mountain, Chorro Canyon

Of conservation concern in Texas: see Appendix B

Foreign cloak fern is scarce in the southern Big Bend of Texas and absent elsewhere in the United States.

Residing in arid desert habitats, this fern grows from short-creeping, short-branching, horizontal rhizomes. The chaffy rhizomes are coated with dark brown or black scales with ciliate margins. Fronds to 6 inches high arise on slim, arching stipes nearly the length of the blades. The chaffy stipes are pale to dark brown, sparingly farinose, with some glandular hairs and scales.

Coated with long, loose hairs, the upper blade surface may also be sparsely glandular and farinose. The lower surface is caked with farina, and midribs and midveins below are lined with tan or white, often ciliate scales. Slightly recurved, the segment margins scarcely hide the sporangia.

The sporangia occur near the vein tips on the segment undersides, partly below the margins and partly protected by layers of wax and scales. This fern produces sixteen spores per sporangium from early spring to fall, but apparently also reproduces asexually without fertilization.

Botanical notes: The genus name, from the Greek *notho* (false) and *chlaena* (coat), refers to the revolute segment margins or false indusia partly protecting the sporangia. The Latin species name *aliena* (foreign) was likely chosen because the fern was found in Mexico and not known to occur in the United States. William Maxon, editor of the *American Fern Society Journal*, described the species in 1916, from a collection by Edward Palmer in 1880, at Soledad, in Coahuila, Mexico. Palmer traveled frequently in Mexico (1878–1910), collecting for the Gray Herbarium and Peabody Museum at Harvard and for the Smithsonian Institution.

233. Scaled Cloak Fern (*Notholaena aschenborniana*)
(Other common names: Aschenborn cloak fern, helecho de capa falsa de Aschenborn)

Habitat: rocky slopes, shady canyons, among shrubs; mesquite scrub, oak-juniper roughland, mostly limestone at low to mid-elevations.

Geographic range: Texas: Trans-Pecos (Culberson, Presidio, Brewster, Pecos, Terrell, Val Verde Counties); Real County. US: Texas, south Arizona. Mexico: northern Mexico south to Oaxaca.

Park locations: Ojo Escondido, Los Portales, Solitario rim above Righthand Shutup

Of conservation concern in Texas: see Appendix B

Not previously known in BBR or Presidio County, this species is a limestone-loving, mostly Mexican fern, rare in West Texas and Arizona, with stout, stiffly erect to ascending, rhizomatous stems. The stems bear needlelike black scales with thick ciliate margins. Tight clusters of dark green to gray-green fronds, to 14 inches high, grow from the stems.

The fronds are supported on short black or dark brown stipes, which are rounded above and thickly cloaked with scales. Seldom much more than 1 inch wide, the blades are lance-shaped, elongate, pointed at the tip, twice divided into ten-plus pairs of leaflets (pinnae), which are divided into smaller ultimate segments.

Largely stemless, the segments are oblong, blunt-tipped, bulging, closely arranged along, and partly fused to, the pinna midrib. The segment upper surface bears stellate hairs and white, stellate scales. The segment lower surface is coated with white or yellow farina, commonly hidden beneath overlapping brown, stellate scales. Only slightly recurved, the segment margins scarcely hide the sporangia.

Sporangia are borne near the vein tips, partly below the segment margins, often within the dense farina layer. This fern, which produces thirty-two spores in each sporangium from late spring to fall, also reproduces asexually without fertilization.

Botanical notes: German botanist Johann Klotzsch, keeper of the Royal Herbarium in Berlin, named this species in 1847 for German physician and botanical explorer of Mexico, Alwin Aschenborn. On his 1840s travels, Aschenborn collected many new species in the valley of Mexico and mountains to the west, including scaled cloak fern, at Chapultepec, in the Federal District of Mexico.

234. Gray's Cloak Fern (*Notholaena grayi* ssp. *grayi*)
(Other common names: none)

Habitat: rocky, grassy slopes, foothills; loose soils; granite, limestone.

Geographic range: Texas: Big Bend (Brewster, Jeff Davis, Presidio Counties); Edwards Plateau (Burnet, Llano, Uvalde Counties). US: Texas, New Mexico, Arizona. Mexico: northern Mexico.

Park locations: foothills of the Llano

Of conservation concern in Texas: see Appendix B

The first Texas fern list (Bush 1903) cited early collections of Gray's cloak fern in Presidio and Jeff Davis Counties by two famous explorer-collectors, Valery Havard in the Chinati Mountains in 1880, and Greenleaf Nealley in the Chinati Mountains and Limpia Canyon circa 1889. Known at only one site in BBR, the fern grows from ascending or short creeping, rhizomatous stems with dark brown scales. To 8 inches long, the gray-green, clustered fronds are oblong to lance-shaped, twice divided into six or more pairs of pinnae, which are divided into ultimate segments. The long broad stipes are reddish brown, rounded above, somewhat hairy and scaly, and farinose, with waxy-mealy-powdery farina grains.

Oblong, fused to the midrib, the ultimate segments are rounded or blunt at the tip, shallowly lobed or smooth on the thick margins. The upper segment surface is largely hairless and sparsely farinose, but the lower surface is blanketed with bright white farina and some tan needlelike scales. Generally, the scales are smooth, not ciliate (like those of *N. aliena*) on the edges. Segment margins are only slightly curled under and scarcely hide the sporangia. Sporangia are borne near the vein tips, partly below the margins and often hidden within the farina. This fern produces sixteen spores per sporangium from spring to fall, but is apogamous, reproducing asexually without fertilization.

Botanical notes: George Davenport described this species in 1879, in Daniel Cady Eaton's *Ferns of North America*, and again in 1880, naming the fern for Harvard botanist Asa Gray. Mining engineer William Courtis collected the type specimen in 1880, in the Chiricahua and Oro Blanco Mountains of southeast Arizona. Surveyor Arthur Schott found the fern earlier, on the US-Mexico Boundary Survey, "on the boundary of Sonora, 5000 ft, Sierra del Pajarito, 1855," but his specimen was not correctly identified until 1899.

235. Gregg's Cloak Fern (*Notholaena greggii*)
(Other common names: marshmallow cloakfern)

Habitat: near Rio Grande, sheltered by rocks on steep, open mountain slopes; limestone, occasionally gypseous, rarely igneous outcrops.

Geographic range: Texas: Big Bend (south Presidio, south Brewster Counties). US: Texas. Mexico: Chihuahua, Coahuila, Durango, site in Nuevo León.

Park locations: Solitario rim above Righthand Shutup, Terlingua Uplift

Of conservation concern in Texas: see Appendix B

In the United States, only found within 15 miles of the Rio Grande in south Brewster and south Presidio Counties, Gregg's cloak fern grows on limestone or gypseous soils in arid mountainous terrain. The fern forms erect or sprawling clusters of gray-green or whitish fronds to 8 inches high. Growing from thick rhizomatous stems, the fronds are supported on reddish brown stipes, grooved or flattened above.

Narrowly triangular to lance-shaped, the blades are up to 4 inches long, two or three times divided into pinnae pairs that are also divided. The rachises are grooved like the stipes, waxy but not scaly. Usually fused to the midrib, the thick ultimate leaf segments are egg-shaped to triangular, rounded or blunt at the tips, and smooth on the margins. Lacking scales, the segments are protected by a white glandular layer above and a dense layer of white farina below.

In this species, the pinnule margins are conspicuously rolled under, overlapping much of the lower surface and at least partly covering the sporangia. Near the margins and vein tips, the sporangia may be obscured by white farina as well as the recurved margins. This fern produces sixty-four spores per sporangium in summer and fall.

Botanical notes: When German pteridologist Georg Mettenius, director of the Botanic Garden of Leipzig, died an untimely death from cholera, his colleague, Friedrich "Max" Kuhn, edited and published many of his fern descriptions posthumously, in the European journal *Linnaea* (1868–69). This fern was described as *Pellaea greggii*, for Josiah Gregg, who collected the plant in 1847 on a rocky hill northwest of Mapimí, in Durango, Mexico. In 1916, William Maxon, curator at the United States National Herbarium, transferred the species to the *Notholaena* genus.

236. Standley's Cloak Fern (*Notholaena standleyi*)
(Other common names: star cloak fern, northern desert star cloakfern)

Habitat: rocky slopes, canyons; in rock pockets, below rock walls, overhangs, desert scrub to woodlands; low to high elevations, igneous, limestone formations.

Geographic range: Texas: Trans-Pecos (excluding Reeves County); Edwards Plateau (Edwards, Gillespie, Llano Counties). US: Texas to southern Arizona, Oklahoma, southeast Colorado. Mexico: Baja California east to Tamaulipas, Baja California Sur east to Durango; San Luis Potosí south to Oaxaca.

Park locations: Ojito Adentro, Upper Guale Mesa, Chorro Canyon, Cinco Tinajas, Cienega Mountain

Standley's cloak fern is a common, widespread fern with pentagonal, oddly dissected blades. The fern grows from short, sturdy, branching rhizomes that support frond clusters to 1 foot high. Narrowly lance-shaped, bicolored scales, with a black mid-region and thin, dark brown edges, cover the stems. The blades are thickish, leathery, to 4 inches wide, on wiry maroon stipes as long or longer than the blades.

Each blade has an upper triangular, short-stalked pinna, with often ten-plus pairs of lobed or unlobed segments, and a larger pair of lateral pinnae, similarly divided, except with the lowest basal segments expanded to form finlike lower pinnae. Without scales and hairs, the blades are farinose below, with white or yellow exudate.

In dry conditions, the blades curl into whitish balls, with the lower surface exposed. Oblong, fused to the pinna midrib, the segments are rounded or blunt at the tips, with smooth or minutely toothed margins. Only slightly curled under, the segment margins scarcely obscure the sporangia, which are borne below the margins near the vein tips. Commonly, this fern produces thirty-two spores per sporangium from spring to fall.

Botanical notes: The species was described by Sir William Jackson Hooker in 1864 as *Notholaena candida* var. *quinquefidopalmata*, from a collection by John Bigelow at La Cuesta, New Mexico, in 1853. In 1879, Yale University botanist Daniel Eaton treated the variety as a species, *N. hookeri*, naming it for Hooker. However, that epithet had been used for another fern, so fern specialist William Maxon in 1915 renamed the plant *Notholaena standleyi*. The name honored Paul Standley, assistant curator of the United States National Herbarium, for his *Flora of New Mexico* (1915), coauthored with New Mexico botanist E. O. Wooton.

237. Wright's Cliff Brake (*Pellaea wrightiana*)
(Other common names: none)

Habitat: below boulders, rock walls; on shady ledges, bluffs above canyon bottoms; mostly igneous exposures.

Geographic range: Texas: Trans-Pecos (excluding Reeves, Pecos, Terrell, Val Verde Counties), parts of Edwards Plateau, scattered sites in north, southeast, south Texas. US: Texas to Arizona, Oklahoma, Colorado, Utah; North, South Carolina. Mexico: Baja California east to Coahuila.

Park locations: Skeet Canyon, Llano Dome, Tascate Hills

Infrequent but widespread in the southwest United States and Mexico, Wright's cliff brake grows from robust, ascending rhizomatous stems. The stems bear needlelike black scales with thin tan, toothed margins. To 16 inches long or more, the clustered fronds are mostly erect, with green, gray-green, or blue-green, oblong blades that turn orange red with age. The blades rest on brown to maroon petioles, which are slightly flattened or grooved above.

The blades are once or twice divided into smaller leaflet pairs (pinnae), with the pinnae near the tip sometimes divided into only three leaflets (ultimate segments), and the pinnae near the base with one to four pairs of leaflets and a terminal, short-stalked leaflet. Although lacking hairs and scales, the blades may be sparsely farinose underneath.

The leathery leaflets are oblong, abruptly sharp-pointed, and smooth on the margins. The margins are rolled underneath to form slender false indusia with white, minutely toothed edges. Partly hidden by the revolute margins, the brown to black sporangia often form a continuous band along the marginal flaps. Normally, sixty-four spores are produced in each sporangium from spring to fall.

Botanical notes: Johann Heinrich Link, director of the Royal Botanical Garden of Berlin, named this genus *Pellaea* in 1841, but the meaning is unclear. The name may originate from the Greek *pellaios* (dark), referring to the leaf stalks or the leaves. In 1858, in his *Species Filicum* documenting nearly 2,500 ferns, Sir William Jackson Hooker named this species for Charles Wright. Wright collected the fern in Grant County, New Mexico, in 1851, in mountains about the Santa Rita del Cobre copper mine.

238. Golden Columbine (*Aquilegia chrysantha* var. *chrysantha*)
(Other common names: yellow, southwestern yellow, golden spur, or canary columbine; aquileña, aguileña, colombina dorada)

Habitat: mesic mountains, waterfalls, springs, seeps; moist bluffs in shady canyons, mid- to high elevations.

Geographic range: Texas: Big Bend (Presidio, south Brewster Counties). US: Texas to Arizona, south-central Colorado, southwest Utah.

Park locations: Arroyo Primero spring, Chorro Canyon, Madrid Falls, Mexicano Falls, Smith Spring

A treasure to find at moist nooks and hidden desert seeps, this columbine is a robust perennial to 3 feet high, growing from woody rhizomes. The plant is bushy, with leafy yellow-green stems branching above. The basal leaves are up to three times divided into three leaflets, which are also lobed or parted. On slender petioles to 8 inches long, the blades are up to 18 inches long. The leaflets are thin, hairless, and at times glaucous.

Mostly from late spring to midsummer, large yellow flowers form at the tips of long flowering stems, single or in clusters of up to ten blooms. On a stalk to 4 inches long, each bloom has five oblong or spatulate petals, each with a long,

tapering hollow tube (spur), three times the petal's length, which projects backward and curves outward. A gland at the spur's tip produces nectar. Within the petals are many partly fused stamens, the innermost sterile and scalelike, the outer fertile, with long filaments and oval anthers; and five long, slender styles. Below the petals are five yellow, spreading, lance-shaped sepals.

The fruits are beaked cylindrical follicles to 1 inch long, with numerous shiny black seeds.

Botanical notes: Linnaeus described the genus *Aquilegia* in 1753, but Casper Bauhin used the name as early as 1623 for a European columbine. *Aquilegia* may be formed from the Latin *aquila* (eagle), for the likeness of the flower's spurs to an eagle's talons; or from *aqua* (water) and *legere* (to collect), for the wet habitat or the liquid nectar in the spurs. The epithet *chrysantha* joins the Greek *chrysos* (golden) and *anthos* (flower). The species was described by Asa Gray in 1873, and collected by George Thurber in Mabibi, Sonora, Mexico, in 1851. In 1854, in *Plantae Novae Thurberianae*, Asa Gray documented many Thurber specimens obtained in New Mexico and Sonora on the US-Mexico Boundary Survey.

239. Old Man's Beard (*Clematis drummondii*)
(Other common names: virgin's or Texas virgin's bower, Drummond's clematis, goat's beard, barba de chivato, barba de chivo, barba de viejo)

Habitat: desert scrub, arroyos, creeks, canyons, woodlands, roads, low to high elevations.

Geographic range: Texas: Trans-Pecos, much of Edwards Plateau, South Texas Plains; sites in Panhandle, north, north-central Texas, elsewhere. US: Texas to California, Oklahoma, Colorado. Mexico: Sonora to Tamaulipas, Baja California Sur to San Luis Potosí, scattered south.

Park locations: Fresno Canyon, Agua Adentro, Arroyo Mexicano, airstrip, Lower Shutup, Los Alamos

Old man's beard is a semiwoody perennial vine with stems many feet long, which sprawls or climbs over host shrubs with twisting, tendril-like leaf petioles. The compound leaves are usually divided into five or seven leaflets, which are often trilobed or parted, and incised on the margins. The blades rest on hairy, wiry petioles to 4 inches long. Each soft-haired leaflet is triangular or egg-shaped in outline.

From spring to fall, flowers form in small clusters in the leaf axils, on stalks to 3 inches long. Male and female blooms appear on separate plants. The flowers lack petals but have four narrowly oblong, petal-like sepals. The sepals are creamy-white, silky-haired, curling and ciliate on the edges. Within the male blooms are up to ninety slender stamens with yellow anthers, which give the blooms a yellow aura. Within the female blooms are up to ninety pistils, each with a long white style and many sterile stamens.

The fruit is a silky-haired, orangish-brown achene to 2 inches long, oval or egg-shaped. In fruit, female vines are strikingly bearded, with ashy white fluff from the long fruit beaks (persistent styles).

Botanical notes: *Clematis* is a name used by Dioscorides for a climbing vine, from the Greek *clema* (tendril). Linnaeus described the genus in 1753, but the type species, *Clematis vitalba*, was documented in 1586, as *Clematis tertia*, by Italian herbalist Pietro Andrea Mattioli. In 1838, John Torrey and Asa Gray named this species for Scottish botanist Thomas Drummond, who collected the vine by the Colorado River in central Texas in 1833–34. This was Drummond's last trip to America for the Glasgow Botanical Society: he died mysteriously in Cuba on his way back to Scotland.

240. Leather Flower (*Clematis pitcheri* var. *dictyota*)
(Other common names: bluebill, Pitcher's or purple or bellflower clematis, purple or Pitcher's leather flower, bluebell, barbas de viejo)

Habitat: rocky, wooded canyons, creeks, springs, mesic hillsides, woodlands, mid- to high elevations.

Geographic range: Texas: Trans-Pecos (El Paso, Culberson, Jeff Davis, Presidio, Brewster Counties). US: Texas, New Mexico. Mexico: Coahuila.

Park locations: Yedra Canyon, route from Yedra Trailhead to Old Abandoned Road Trail

Leather flower is a deciduous perennial vine, often woody at the base, that sprawls or climbs on host shrubs or trees. Extending to 13 feet by means of twisting, tendril-like leaf petioles, the slender stems are reddish brown and ribbed.

The opposite leaves may be simple or compound, with three to five leaflets that may be lobed or parted. Compared to the other variety (*pitcheri*), the leaflets are small, to 2 inches long, thick, and lobed. Each lance- or egg-shaped leaflet is green above, paler and often densely hairy below, with netlike veins. The terminal leaflet may clasp objects for support and serve as a climbing mechanism.

From spring to fall, usually single, nodding, urn-shaped flowers form on long stalks from the leaf axils. The curious leathery blooms lack petals but have four thick petal-like sepals united much of their length. The ribbed sepals are lancelike, with recurving tips. Flower color is deep reddish purple, bluish purple, or paler shades. Within the sepals are many distinctively hairless stamens and many pistils with long curving styles.

The fruits are disklike achenes, 1/3 inch wide, arranged in tight globose clusters, with a curving, often silky-haired, tail (persistent style) to 1 inch long.

Botanical notes: The species is named for Zina Pitcher, US Army surgeon, botanical collector, and later mayor of Detroit. John Torrey and Asa Gray described the species in 1838, from collections by Dr. Pitcher, and by Thomas Nuttall, on the Red River in Arkansas. The Latin varietal name *dictyota* (netted) refers to the leaves. This variety was described as a species, *Clematis dictyota*, in 1903, by Edward Lee Greene in his journal *Pittonia*, from a 1902 collection by botanist-agronomist Samuel Mills Tracy and Franklin Sumner Earle in Limpia Canyon, in Jeff Davis County. *Clematis* specialist W. Michael Dennis reclassified the vine as a variety of *C. pitcheri* in 1979.

241. Desert Spikes (*Oligomeris linifolia*)
(Other common names: lineleaf or flaxleaf whitepuff, desert cambess, narrowleaf oligomeris, hierba de la pulga)

Habitat: desert, alluvial flats, dunes, arroyos, silty, clayey low hills, disturbed sites, calcareous or alkaline soils.

Geographic range: Texas: Trans-Pecos (excluding El Paso, Culberson Counties), parts of Lower Rio Grande Valley. US: southwest (excluding Colorado). Elsewhere: Mexico (Baja California east to Nuevo León, south to Durango, Zacatecas); Africa, Europe, Middle East, India.

Park locations: Monilla Canyon mouth, Rancherías Canyon south of Highway 170, Smith House

Desert spikes is an inconspicuous annual to 14 inches high, with fleshy, erect stems branching at the base. The leafy stems are hairless, ribbed, pale or yellow green, red or orange red with age. To 1 1/2 inches long, the linear leaves are alternate or bundled along the stems. The stemless blades are hairless, with blunt, abruptly pointed, or curving tips. The blades may also turn red or grayish red and glaucous.

From early spring to summer, minuscule white flowers form in narrow, congested spikes to 6 inches long. The blooms have two egg-shaped petals and four lance-shaped green sepals with thin margins. Within each bloom are three stamens with white filaments and yellow anthers, and four recurved styles. Leaflike linear bracts, roughly the length of the sepals, support the bloom.

The tan fruit is a dehiscent capsule to 1/8 inch long, spheric but compressed, with four beaks. The membranous capsule splits open at the tip, releasing many glossy black seeds.

Botanical notes: French botanist Jacques Cambessèdes described the genus *Oligomeris* in 1844, in an account of rare plants collected by French botanist Victor Jacquemont in East India (1828–32). The generic name joins the Greek *oligos* (few) and *meros* (part) since the few petals and stamens distinguish this genus from the related genus *Reseda*. The species name, from the Latin *linea* (line) and *folium* (leaf), refers to the leaves. University of Copenhagen botanist Martin Vahl described this annual posthumously in 1815, as *Reseda linifolia*. The lectotype, collected in Egypt in 1813, is attributed to Jens Hornemann, Vahl's successor. James Macbride, with the Gray Herbarium at Harvard, moved the plant to the *Oligomeris* genus in 1918.

242. Warnock's Snakewood (*Condalia warnockii* var. *warnockii*)
(Other common names: Warnock's condalia, Mexican crucillo, crucillo, squawbush, Mexican buckthorn, mesquilillo)

Habitat: desert, alluvial flats, gravelly, brushy arroyos, rocky, wooded hillsides, limestone, igneous substrates.

Geographic range: Texas: Trans-Pecos (excluding Culberson County); Crane, Edwards Counties. US: Texas to New Mexico. Mexico: Chihuahua, Coahuila, northern Durango, northern Zacatecas.

Park locations: near Guale 1 Campsite, Arroyo Mexicano, Solitario north rim, Buena Suerte Trail, upper Alamito Creek

To 6 feet high, Warnock's snakewood is an upright, often compact shrub, intricately branched, or widely spreading and treelike. The rigid branches are spiny and velvety, with gray to black furrowed bark. Branchlets, tipped with dark red thorns, extend at right angles from the branches. In stormy weather, the shrub can be foreboding — more suitable in the Scottish moors than the CD.

To 1/4 inch long, the leaves alternate along the stems or form tight bundles on short branchlets. Thick, slightly folded, the blades are elliptic, pointed at the tip, tapered to the base, and smooth on the edges. The yellowish-green leaves bear short rough hairs above and below.

In summer, or in spring or fall after rains, yellow blooms appear singly or in small clusters on the branchlets. Each tiny bloom, to 1/8 inch across, consists of five triangular sepals on the rim of a bell-shaped floral cup. The sharp, abruptly pointed sepals are coated with fine, stiff hairs. Five stamens alternate with the sepals on the rim. Within the floral cup, a pistil supports a purple style with a headlike stigma.

The fruit is a dark maroon to black, edible drupe to 1/4 inch long, short-cylindric, with one egg-shaped seed.

Botanical notes: In 1799, Spanish botanist Antonio Cavanilles named the genus *Condalia* for Spanish physician Antonio Condal. Condal served on Swedish botanist Pehr Löfling's naturalist team, on the José Iturriaga expedition to the Orinoco River in Venezuela (1754–61). In 1962, Marshall Johnston named this species for Barton Warnock, Sul Ross State University botanist and author of three popular works on Trans-Pecos flora. Warnock and O. C. Wallmo collected the type specimen in 1955, between Alamo Spring and Burro Spring in BBNP.

243. Lotebush (*Ziziphus obtusifolia*)
(Other common names: graythorn, clepe, gumdrop tree, Texas buckthorn, white crucillo, abrojo)

Habitat: desert, grasslands, mountains, woodlands; alluvial flats, creeks; limestone, igneous substrates.

Geographic range: Texas: Trans-Pecos; much of South Texas Plains, Edwards Plateau; parts of Panhandle, north Texas. US: southwest (excluding Colorado); Oklahoma. Mexico: Baja California east to Tamaulipas, Baja California Sur, Durango southeast to San Luis Potosí, Veracruz.

Park locations: Madrid Falls, Sauceda, Horse Trap Spring, Cienega Creek

Lotebush is a robust shrub to 10 feet high, with spreading, densely branched, grayish-white stems. The curving, arching branches are armed with right-angled thorn-tipped branchlets to 3 inches long. Stems and leaves are hairless or sparsely hairy. Alternate or clustered on the branchlets, the nearly stemless leaves are oblong or elliptic, to 1 inch long. The pale green to grayish or yellowish-green blades are rounded at the tips and smooth or finely serrate on the margins.

In spring and summer, inconspicuous flowers form in small, compact clusters on the branchlets, or in the leaf axils of new season's growth. To 1/5 inch wide, the minuscule blooms are yellow to white, with five early falling, spatulate, cupped petals. Within each bloom are five short stamen, and a slender, purplish-olive, two-branched style. At the base of the bloom are five triangular, yellow-green sepals.

The fruit is a globose drupe to 2/5 inch long, bluish black to dark reddish purple at maturity. Edible but not tasty, the drupe has a two-chamber stone with one seed per chamber.

Botanical notes: The name *Ziziphus* was used by Pliny, borrowed from the Arabic *zizouf*, an ancient name for the jujube tree (*Ziziphus lotus*). The Latin species name, from *obtusus* (blunt) and *folium* (leaf), refers to the leaf tips. The species was described by William Jackson Hooker in 1840 as *Rhamnus obtusifolia*, from a collection by Scottish naturalist Thomas Drummond in 1835, in Austin County, Texas. From 1830 to 1835, Drummond traveled through the Allegheny Mountains to Missouri, Louisiana, Texas, and Cuba, where he died of unknown causes. Asa Gray moved the species to the *Ziziphus* genus in 1848.

244. Apache Plume (*Fallugia paradoxa*)
(Other common names: poñil, feather rose, feather duster bush, fallugie, plumed arroyo shrub)

Habitat: arroyos, alluvial flats, canyon bottoms, springs, intermittent creeks, rocky hillsides; low to high elevations, limestone, igneous substrates.

Geographic range: Texas: Trans-Pecos; Upton County; parts of Edwards Plateau. US: southwest, Oklahoma. Mexico: mostly Chihuahua, Coahuila, also Sonora, Durango, Zacatecas.

Park locations: Ojo Chilicote, Solitario, upper Alamito Creek

To 6 feet high, Apache plume is an upright, much branched, and leafy shrub that spreads from rhizomes to form small thickets. The slender branches are grayish white or tan, with shredding bark. Young stems may be cloaked with matted hairs. Often, the branches form a tangled jumble, giving the shrub an unruly, disheveled appearance.

Alternate or clustered on the branches, the leaves are up to 3/5 inch long, wedge-shaped in outline, with three to seven linear lobes. The lobes may bear fine hairs above, and rust-colored hairs below.

From spring to fall showy white flowers, to 1 1/2 inches across, appear singly or in small clusters at the branch tips. Each bloom has five rounded petals, fragile and ephemeral. Within the petals are many stamens with white filaments and oblong, four-lobed yellow anthers and many pistils, each with a feathery purple style. Below the petals are five egg-shaped, hairy sepals and five slender bractlets.

The fruit is a spindle-shaped achene to 1/6 inch long, with a persistent, plumelike style. The mature fruits form feathery red "heads," said to resemble the headdress of Apache chiefs.

Botanical notes: Austrian botanist Stephan Endlicher, director of the University of Vienna botanical garden, named the genus *Fallugia* in 1840, for the abbot-botanical author of Vallambrosa, Italy, Virgilio Fallugi (1627–1707).

The species was described by David Don in 1825 as *Sieversia paradoxa*, from Mexican collections in 1788 by the Spaniard Martin de Sessé y Lacasta and Mexican naturalist José Mariano Mociño. Don, secretary of the Linnaean Society of London, regarded the shrub as "paradoxical" since other *Sieversia* species were herbs. Endlicher gave the paradoxical shrub its own genus.

245. Toothed Serviceberry (*Malacomeles pringlei*)
(Other common names: Big Bend serviceberry, southern false serviceberry, membrillo cimarrón, tlaxistle)

Habitat: rocky, brushy, wooded canyons, mountain slopes, sandy arroyos; limestone or igneous soils.

Geographic range: Texas: Trans-Pecos (Presidio, Brewster, Pecos Counties). US: Texas. Mexico: Chihuahua to Nuevo León.

Park locations: Upper Guale Mesa, Arroyo Mexicano, Chorro Canyon, Tascate Hills, Solitario Peak, Righthand Shutup

This serviceberry is an upright, spineless shrub to 8 feet high, rounded, densely branched, with stiff gray branches and short branchlets. Young branchlets are often maroon and densely woolly. The leathery, olive green leaves are rounded or oval, to 3/5 inch long, with well-spaced, spinelike teeth on the upper edges, and grayish white, woolly hairs underneath.

In spring, even before new leaves form, flowers appear in small clusters on young branchlets. The bell-shaped, red-stalked blooms are white or suffused with pink. Each bloom has five rounded, veiny, early falling petals. On the rim of the floral cup are twenty white stamens, with awl-shaped filaments and large bilobed anthers. The three styles are distinct, separate. Below the petals are five egg-shaped, woolly sepals, and oblong bracts. The berrylike fruit is a multichambered red pome to 1/3 inch wide, with one seed per chamber.

In Mexico, the stems were made into canes known as *varitas de apizaco* (Standley 1920–26). The dried or cooked berries were a preferred food of Native Americans (Warnock 1977).

Botanical notes: In 1874, Joseph Decaisne, at the French Museum of Natural History, used the name *Malacomeles* for a section of the genus *Cotoneaster*. Decaisne joined the Greek *malakos* (soft) and *melon* (apple) to distinguish plants with soft fruits from other *Cotoneaster* species. German taxonomist Adolf Engler, director of the Berlin-Dahlem Botanical Garden, gave *Malacomeles* generic status in 1897. In 1890, University of Berlin botanist Bernhard Adalbert Emil Koehne named this species *Amelanchier pringlei*, for botanical explorer Cyrus Pringle, who collected the plant in 1885 in the Santa Eulalia Mountains of Chihuahua. In 2011, Billie Turner placed this shrub in the *Malacomeles* genus.

246. Mat Rock Spiraea (*Petrophytum caespitosum*)
(Other common names: tufted, caespitose, or Rocky Mountain rockmat, rock or dwarf spiraea, rock rose)

Habitat: crevices, rock pockets on steep, exposed, limestone slopes, mountaintops; mid- to high elevations.

Geographic range: Texas: Trans-Pecos (El Paso, Hudspeth, Culberson, Presidio, Brewster, Pecos Counties); Real, Frio Counties. US: west excluding Washington; South Dakota. Mexico: northern Sonora, northern Chihuahua, central Coahuila.

Park locations: peak east of Pico 4522

Mat rock spiraea is a mat-forming dwarf shrub that carpets barren rocks or hangs precariously from rock faces. To 3 feet across but only a few inches high, with woody, maroon stems and roots with blackish bark, this remarkable miniature is a delight to find. The short, leafy branchlets, or "spurs," are stubby, knotty, and gnarled. Leaves, to 1/2 inch long, grow in minirosettes at the spur tips. The silky-haired, silvery blades are largely stemless, spatulate, and leathery.

In summer and fall, small white flowers bloom in crowded cylindrical clusters to 2 inches long, on reddish brown stalks towering above the leaf rosettes. The stalks bear scattered ascending, needlelike bracts. Each flower has five spatulate petals, usually erect, persistent, and often notched at the tip.

Within the petals are about twenty protruding stamens, with white filaments and pale yellow, bilobed anthers, and up to five pistils, with threadlike styles and minute stigmas. Below the petals are five triangular maroon sepals. The dehiscent fruits are spindle-shaped leathery follicles, 1/13 inch long.

Botanical notes: The genus name is formed from the Greek *petros* (rock) and *phyton* (plant) since the plant grows on rocks. The species name, from the Latin *caespitosus* (cespitose), reflects the tufted habit. The term *Petrophytum*, a section of the genus *Spiraea*, appeared in a Thomas Nuttall manuscript and was published in 1840, in John Torrey and Asa Gray's *A Flora of North America*. The flora contained Nuttall's description of *Spiraea caespitosa*, which Nuttall found "on high shelving rocks in the Rocky Mountains, towards the sources of the Platte. July. —A singular dwarf alpine plant." In 1900, when Per Axel Rydberg elevated *Petrophytum* to generic status, this species became *Petrophytum caespitosum*.

247. Heath Cliffrose (*Purshia ericifolia*)
(Other common names: unknown)

Habitat: rock pockets on rocky limestone hills; windswept canyon rims.

Geographic range: Texas: Big Bend (southeast Presidio, south Brewster Counties). US: Texas. Mexico: Coahuila.

Park locations: Solitario northwest rim, Lower Shut-up, Terlingua Uplift, Cienega Creek

Heath cliffrose, a shrub to 3 feet high, is much branched, with reddish brown bark and rigid, tangled branches. Sculptured and contorted on windswept slopes, the woody branches support many woolly-haired spur branchlets, only 3/4 inch long.

Leathery leaves grow in bundles at the tips of the spurs. To 1/3 inch long, the stemless blades are linear, curved at the tips, and densely woolly below. The stems are marked with leaf bases that persist after the leaves have fallen.

From spring to fall, fragrant white flowers appear singly at the branchlet tips. On a short, woolly-haired stalk, each bloom is 3/5 inch across, with five rounded, spreading petals, sometimes notched at the tip; many stamens

with pale yellow filaments and yellow anthers; and eight pistils, each with a feathery style. Below each bloom are five elliptic yellow or red sepals, often with woolly hairs. The fruit is a cluster of hairy oblong achenes to 1/4 inch long, with feathery tails that enhance seed dispersal.

Native Americans used the leaves to cleanse wounds and induce vomiting (Warnock 1970).

Botanical notes: In 1818, Augustin Pyramus de Candolle named the genus *Purshia* for Frederick Pursh, who, in 1814, published a controversial work, *Flora Americae Septentrionalis*, on the plants of North America. Pursh's *Flora* included plants collected by John Bradbury and Thomas Nuttall, and 132 plants from the Lewis and Clark Expedition. Pursh had lived in the United States (1799–1811), working as a gardener, curator, and botanical collector. The Latin species name, from *Erica* (heath, the genus) and *folium* (leaf), refers to the heathlike leaves. The species was described as *Cowania ericifolia* by John Torrey in 1853. Torrey cited Charles Parry as collector, in autumn 1852, in "crevices of limestone rocks on the Rio Grande, below Presidio del Norte." University of Texas research fellow James Henrickson combined this and other species into the *Purshia* genus in 1986.

248. Northern Dewberry (*Rubus flagellaris*)
(Other common names: whiplash, American or common dewberry, trailing bramble, trailing blackberry)

Habitat: by water in rocky, wooded canyons; fields, thickets, woodlands, disturbed sites; often sandy soils.

Geographic range: Texas: Trans-Pecos (Presidio County); scattered in East Texas. US: east half; Texas to Nebraska. Elsewhere: Canada, Mexico (Nuevo León).

Park locations: Ojito Adentro

Northern dewberry is a shrub or perennial vine, growing from a woody taproot, with whiplike stems to 3 feet long or more. The young green, first-year stems (primocanes) are often prostrate, creeping, with hooked prickles, at times rooting at the tips but seldom branching or flowering. Second-year stems (floricanes) are more slender, erect, reddish, sometimes branching, with flowers, some hairs, and sharp prickles.

With prickly petioles to 2 inches long or more, the leaves are alternate along the stems and compound, with three to five leaflets. Smaller leaves on flowering stems may be undivided and unlobed. Each leaflet, to 3 inches long, is egg-shaped to elliptic, toothed or lobed, green above and paler below. The terminal leaflet is short-stalked, but lateral leaflets are mostly stemless, frequently with lopsided bases.

From mid-spring to early summer, large white flowers form on the flowering stems (floricanes), single or in small clusters. The showy flowers have five fan-shaped petals, each with a clawlike base. Each petal is flimsy, ephemeral, often wrinkled or wavy on the margins. Within the flower's throat are many white stamens, with threadlike filaments and bilobed anthers. The stamens surround numerous green pistils. Supporting the petals are five egg-shaped green sepals. Each fertile flower produces a cluster of round or short-oblong, one-seeded drupelets, to 1 inch long. Black when ripe, the drupelets are said to be both tart and sweet.

Botanical notes: The generic name *Rubus* is the classical Latin term for a blackberry, borrowed from the Latin *ruber* (red). The Latin species name *flagellaris*, from *flagellum* (whip), alludes to the whiplike canes. German taxonomist Carl Ludwig von Willdenow, director of the Berlin Botanical Garden, described the species in 1809, from material collected in North America and cultivated at the garden. Willdenow was a founder of phytogeography, the study of the distribution of plants.

249. Buttonbush (*Cephalanthus occidentalis*)
(Other common names: common buttonbush, button willow, honeybells, honeyballs, buttonball, riverbush, swampwood, little snowball, pinball, crane willow, rosa de San Juan, jazmín, uvero)

Habitat: riverbanks, creeksides, spring margins, alluvial flats; low-mid elevations.

Geographic range: Texas: Trans-Pecos (Jeff Davis, Presidio, Brewster, Terrell, Val Verde Counties); parts of Edwards Plateau, South Texas Plains, northern Panhandle; much of east half of state. US: east half; Texas to California; Texas to Nebraska. Elsewhere: Canada, Mexico, Central America, West Indies.

Park locations: Cinco Tinajas, drainage from Sauceda to Cinco Tinajas, Horse Trap Spring, South Fork of Alamo de Cesario Creek, Fresno Canyon

Buttonbush is a fast-growing, spreading shrub or small tree to 10 feet tall. This deciduous shrub forms many slender, at times four-angled branchlets, with corky, slit-like grooves. Mostly hairless, the pale green branchlets turn brown or gray with age.

To 8 inches long, the paired, short-stemmed, oblong leaves are tapered or abruptly pointed at the tip and wavy on the margins. Lustrous and hairless above, the blades may bear coarse hairs along the veins below.

From late spring to fall, fragrant flowers, with prominent protruding styles, bloom in pincushion-like heads 1 inch across. On long thick stalks, the heads contain many tiny, four- or five-lobed, tubular blooms, white to pale yellow. Within each bloom are four or five stamens with short filaments and a threadlike style with a headlike stigma. Supporting each bloom is a minuscule four- or five-lobed calyx. The flower heads harden into round clusters of inversely conic, two-seeded orange brown nutlets, each to 1/3 inch long.

Botanical notes: Linnaeus coined the name *Cephalanthus* in 1738, joining the Greek *kephale* (head) and *anthos* (flower) for the globelike flower clusters. The Latin epithet *occidentalis* signifies "western." In 1753, Linnaeus formally described the genus and species, citing North America as the habitat, but the shrub has a significant pre-Linnaean history. Royal Professor of Botany Leonard Plukenet, superintendent of Queen Mary II's Hampton Court exotic gardens, published an account of the plant in his *Phytographia* (illustration, 1691) and his *Almagestum Botanicum* (description, 1696).

250. Arizona Cottonwood (*Populus fremontii* subsp. *mesetae*)
(Other common names: Frémont, plateau, meseta, or alamo cottonwood, chopo, álamo cimarron, álamo)

Habitat: reliable water sources; riverbanks, protected canyons, creek sides, springs; alluvial, sandy, sandy clay soils, low to mid-elevations.

Geographic range: Texas: Trans-Pecos (excluding Terrell, Val Verde Counties); Ward County. US: Texas, New Mexico, probably southeast Arizona. Mexico: south to Valley of Mexico.

Park locations: Palo Amarillo Creek, Agua Adentro, Bofecillos Canyon, Rancherías Spring, Chorro Canyon, Yedra Canyon, Leyva Creek, Ojo Chilicote, South Fork Alamo de Cesario Creek, Fresno Canyon, Upper Alamito Creek, Cienega Creek

This fast-growing, deciduous tree is small for a cottonwood, 20 to 65 feet high, with a spreading crown and a trunk to 4 feet wide. Often forming a few stout branches just above the base, the trunk develops pale, smooth bark when young, which becomes deeply furrowed with age. Young branchlets are brown or tan, and yellow-haired, but paler or white after several seasons. The trees may reproduce vegetatively from stumps or buried branches.

The yellow green leaves, to 3 inches long, rest on petioles nearly as long. Blades appearing early on primary branches tend to be broadly triangular, short-tapered, and flat at the base. Blades appearing late, or early on secondary branches, are more diamond-shaped, tapered, and wedge-shaped at the base. Early blades are coarsely toothed, later blades more sharply serrate. Both blades and petioles bear yellow-tinged hairs. Distinctively, petioles are flattened just below the blades.

In early spring, before new leaves form, tiny flowers appear on young branchlets of the previous season, in drooping catkins to 5 inches long. Male and female catkins occur on separate trees. Each bloom is inserted on a cuplike disk that later encloses part of the fruit. Below each bloom is an early falling, membranous bract. Flowers lack petals and sepals. Male blooms consist of many stamens with yellow or purple anthers. Female blooms consist of a stalkless ovary and a short style with two to four platelike stigmas.

The short-stalked brown fruit is a two- to four-valved dehiscent capsule to 1/3 inch long. The mature fruit releases tiny cottony seeds (which give the tree its name) with long silky hairs to aid wind dispersal. The seeds form a fluffy white blanket beneath the tree.

In New Mexico, the Pueblo used the wood to make drums, and Isletans to make boats and rafts. At some pueblos, catkins were eaten raw or put in meat stews, and bark was made into bone splints (Dunmire and Tierney 1995).

Botanical notes: Some believe Pliny gave poplar the Latin name *populus* (people) because the many leaves, in a wind, resemble people in motion. However, in Rome, poplar was known as *arbor populi* (tree of the people), likely because it was planted in public places. This species is named for explorer and military governor of California John Frémont, who led expeditions to the Rocky Mountains and northern California (1842–46). *Mesetae*, the subspecies name, refers to the tree's main location in Mexico's Meseta Central. The species was described by Sereno Watson in 1875, from a collection by Frémont in 1846 "on Deer Creek at Lassens in the Upper Sacramento Valley." *Populus* expert James Eckenwalder described the subspecies in 1977, from Cyrus Pringle's 1886 collection in a "valley near Chihuahua."

251. Goodding's Willow (*Salix gooddingii*)
(Other common names: Goodding's or southwestern or western black willow, black or Dudley or valley willow, sauce, sauz)

Habitat: near water; creeks, springs, sites subject to flooding.

Geographic range: Texas: Trans-Pecos (excluding Reeves County), parts of Permian Basin. US: southwest; Oklahoma. Mexico: Baja California east to Coahuila; northwest Sinaloa.

Park locations: Las Cuevas, Bofecillos Creek, Leyva Canyon, Horsetrap Spring, upper Alamito Creek

This willow is a deciduous tree, usually with a single trunk, or a large shrub with numerous stems to 30 feet high. Growing from large fibrous roots, the tree develops slender grayish-brown branches and yellowish-gray young branchlets. The rough, thick bark is often deeply furrowed. To 5 inches long, the drooping, alternate leaves are lancelike to oblong, yellow green above and below. Mostly hairless, the leaves hang from short, often longhaired petioles, which are gland dotted at the base. The blades are tapered to a sharp tip, wedge-shaped at the base, and finely serrate along the margins.

In spring, flowers appear in elongated catkins to 3 inches long or more, with male and female catkins on separate plants. The catkins often form on leafy stalks on short leafy branchlets. The flowers lack petals, sepals, and floral disks. Each male bloom consists of four to eight stamens, with pale yellow, hairy filaments, orange or purple, bilobed anthers, and a scalelike bract at the base. Each female bloom consists of a pear-shaped ovary, a style with two often bilobed stigmas, and a scalelike bract below. The fruit is a two-segment lance- or egg-shaped capsule, 1/4 inch long, with tiny cottony seeds.

Hippocrates described medicinal properties of willow trees in the fifth century BCE, when chewing the bark was recommended to reduce fever and inflammation. Native Americans used the bark as an aspirin substitute long before 1897, when salicylic acid in the bark was used to make aspirin.

Botanical notes: The genus name is the classical Latin *Salix* (willow). In *Theatrum Botanicum* of 1640, John Parkinson suggested that the willow is called "*Salix a saliendo* in Latine, because it groweth with that speed that it seemeth to leape." In 1812, in *The British Flora*, Robert Thornton added that Salix is "from salio, L. to leap or spring, from the quickness of its growth." In 1905, USDA agrostologist Carlton Ball named the species for Leslie Goodding, who collected the type specimen in 1902, on a tributary of the Virgin River in southeast Nevada.

252. Western Soapberry (*Sapindus saponaria*)

(Other common names: wingleaf, winged, tropical, or Florida soapberry, soapnut, (wild) chinaberry, (wild) chinatree, Indian soap plant, amole, jaboncillo, amolillo, abolillo, palo blanco)

Habitat: creeks, springs, arroyos, grasslands, woodlands, roadsides.

Geographic range: Texas: Trans-Pecos (excluding Hudspeth, Reeves Counties), much of Edwards Plateau, Panhandle, scattered elsewhere. US: Louisiana west to Arizona, north to Missouri; Texas to Kansas, Colorado. Elsewhere: Mexico to South America, West Indies.

Park locations: Agua Adentro, Leyva Creek, Pila Montoya Trail, South Fork of Alamo de Cesario Creek

This soapberry is a shrub to 15 feet high, or a small to midsize tree, with erect or ascending branches and a broad, rounded canopy. Western soapberry grows in small clusters or forms large groves. The trees have scaly trunks and rough, splitting bark. Young branchlets are grayish brown to yellow, and often densely hairy. The alternate leaves, to 15 inches long, are divided into up to eighteen paired leaflets, each to 3 inches long. Each leaflet is lancelike, curved, hooked at the tip, green above, paler below, golden in autumn.

In spring and summer, fragrant white flowers appear in crowded pyramidal clusters. Each short-stalked bloom consists of five egg-shaped petals, clawed at the base, ciliate on the margins; eight protruding stamens with bilobed yellow anthers; and a style with a lobed stigma. Below the petals are five sepals, round to egg-shaped, with hairs along the margins. The waxy fruit is a translucent amber berry, 1/2 inch across, with a hard, black seed. The fruit pulp contains poisonous saponin, which lathers when wet.

In Mexico, the pulp is used to wash clothes and hair (Warnock 1974). The fruit is used to treat kidney problems and rheumatism, and the seeds to make buttons and necklaces (Vines 1960; Powell 1998).

Botanical notes: The generic name *Sapindus*, from the Latin *sapo* (soap) and *Indus* (Indian), refers to the berries' use as soap by Native Americans. The redundant species name combines the Latin *saponis* (soap) and *-aria* (resembling), for the soap-like fruit. Linnaeus described the genus and species in 1753, citing habitats in Brazil and Jamaica, and Leonard Plukenet as a source. The lectotype is an illustration in Plukenet's *Phytographia* of 1692.

253. Big Bend Silverleaf (*Leucophyllum minus*)
(Other common names: lesser Texas silverleaf, Big Bend cenizo, Big Bend barometerbush, littleleaf rain sage)

Habitat: gravel, clay flats, desert scrub, sandy arroyos, rocky hills, grasslands.

Geographic range: Texas: Trans-Pecos (excluding El Paso, Reeves, Terrell Counties). US: Texas, New Mexico. Mexico: Chihuahua, Coahuila.

Park locations: Las Cuevas, Upper Guale Mesa, Rancherías Spring, Solitario northwest rim, Solitario rim above Righthand Shutup, Buena Suerte Trail

Big Bend silverleaf is a rounded shrub to 3 feet high, with many leafy branches. Stems and leaves are covered with woolly, branching, stellate hairs, giving this cenizo (ashy) a distinct silvery appearance. The shrub is spineless, but old branches become thornlike with age. Crowded on the branches or short spur shoots, the alternate, short-stemmed gray leaves are spatulate, to 2/5 inch long and rounded at the tip.

In summer and fall, often after soaking rains, pale flowers — lavender, violet, blue, or white — appear on short, hairy stalks in the leaf axils. Each bloom is funnel-shaped, with a slender tube expanded into five hairy lobes. A white blotch with orange or yellow dots fills the flower's throat. The rounded lobes are spreading or reflexed, notched at the tips, and ciliate. Within each bloom are four stamens, in two unequal pairs, with white or yellow anthers, and a style with a bilobed, flattened white stigma. Supporting the bloom is a five-lobed calyx. The profusely blooming flowers soon fall, leaving a memorable carpet at the shrub's base. Cenizos, or "barometerbushes," are believed to bloom before rains in response to humidity changes. More often, cenizos bloom soon after rains.

The fruit is a dehiscent brown capsule to 1/5 inch long, egg-shaped, with small brown, egg-shaped seeds.

Botanical notes: The genus name *Leucophyllum*, from the Greek *leucos* (white) and *phyllon* (leaf), emphasizes the leaf color. French botanist Aimé Bonpland, who traveled with Prussian explorer Alexander von Humboldt to South America, Mexico, and Cuba (1799–1804), described the genus in 1809 in *Voyage de Humboldt et Bonpland*, their expedition account. In 1859, the species was described by Asa Gray, from collections by Charles Wright (1849, 1851) "at hills on and near the Pecos [River]," and by John Bigelow and Charles Parry "between Van Horn's Wells [southwest Culberson County] and Muerta [Jeff Davis County?]." Gray gave this silverleaf the Latin name *minus* (less), commenting that the plant is "a low spreading shrub, only two feet high."

254. Flower of Stone (*Selaginella lepidophylla*)
(Other common names: resurrection fern, dinosaur plant, rose or false rose of Jericho, stone flower, flor de piedra, siempre viva, doradilla)

Habitat: rocky outcrops near Rio Grande to north-facing mountain slopes; often on limestone.

Geographic range: Texas: Trans-Pecos (excluding Jeff Davis, Reeves Counties); Crockett County. US: Texas, New Mexico. Elsewhere: Mexico (Baja California east to Tamaulipas, south to Oaxaca), Central America.

Park locations: below South Lajitas Mesa, near Panther Canyon mouth, Rancherías Spring, Cinco Tinajas, Buena Suerte Trail, Cienega Creek

Spikemosses are primitive, spore-producing plants, with ancestors that evolved 400 million years ago. A "resurrection" fern, flower of stone is a perennial spikemoss with leafy, profusely branching stems and prostrate leaf rosettes to 10 inches across. The rosettes grow flat when moist but become dormant when dry, curling into reddish brown, nestlike balls to preserve moisture. Even after years of drought, the plants "resurrect" with rain, unfurl, and resume photosynthesis. All leaves are thick, triangular or egg-shaped, green above, brown below, with white, ciliate margins.

Overlapping fertile leaves bearing spore cases (sporangia) form a conelike structure, to 1/2 inch long, at the branchlet tips. Megasporangia are borne mostly in the lower part of the cone with up to four large female megaspores, and microsporangia mostly in the upper part, with many tiny male microspores. Megaspores form female gametophytes producing egg cells, microspores male gametophytes producing sperm cells. Sperm and egg unite to produce a sporophyte, which becomes a new spikemoss.

Botanical notes: The generic name *Selaginella* connotes "small *Selago*," a name used by Pliny for clubmoss (*Lycopodium*). The species name, from the Greek *lepis* (scale) and *phyllon* (leaf), refers to the scalelike leaves. This fern was described in 1830 as *Lycopodium lepidophyllum*, by Scottish botanists William Jackson Hooker and Robert Greville, and collected by Henry Dundas, captain in the Royal Navy, at Tepic, near San Blas, Mexico. Belgian botanist Antoine Spring, University of Liège spikemoss authority, moved the species to the *Selaginella* genus in 1840.

255. Peruvian Spikemoss (*Selaginella peruviana*)
(Other common names: flor de piedra)

Habitat: igneous slopes, base of rock walls, boulders; mid- to high elevations.

Geographic range: Texas: Trans-Pecos (excluding Reeves County), part of Edwards Plateau; Kinney, Comanche Counties. US: Texas, New Mexico, Oklahoma. Elsewhere: Mexico (Sonora east to Nuevo León, south to Oaxaca), Peru, Bolivia, Argentina.

Park locations: Guale 1 Campsite, Skeet Canyon, Tascate Hills

This perennial spikemoss forms creeping, often crowded mats, rather than rosettes. The slender stems are green or brown, to 5 inches long, with many leafy branches, commonly with upturned tips. Dorsiventral, the stems seem different when viewed from above or below. Spreading leaves on the lower stem surface give the stems a flattened appearance. Rhizophores, downward-growing, leafless shoots, form threadlike roots on touching the soil.

Arranged in vertical rows, the scalelike leaves are close-pressed or ascending, and overlapping. The leaves are upcurved, with a long white, persistent bristle-tip, and ciliate margins. Upper-side blades are green, linear or lance-shaped; underside blades are yellow brown or tan, narrower, slightly longer, and decurrent on the stem.

Producing spores in summer and fall, the overlapping fertile leaves are borne in four-sided, clublike cones to 3/4 inch long, at the branchlet tips. The fertile leaves are egg-shaped to triangular, with a relatively short bristle-tip, and cilia on the edges. In the fertile leaf axils are megasporangia, with up to four large female megaspores, mostly in the lower part of each cone, and microsporangia, with numerous male microspores, usually in the upper part of the cone.

Botanical notes: The Latin species name *peruviana* (Peruvian) refers to the habitat. In 1867, German pteridologist Carl August Julius Milde described this species as *Selaginella rupestris fo. peruviana*, a form of the *rupestris* species. The holotype was collected by Spanish botanist Hipólito Ruiz López at Huanuco, in the Andean mountains of Peru, during the Ruiz and Pavón expedition to Peru and Chile (1777–88). In 1900, German botanist Georg Hans Emmo Wolfgang Hieronymus, curator of the Berlin Botanical Garden, elevated the plant to species status.

256. Greenleaf Fives Eyes (*Chamaesaracha coronopus*)
(Other common names: green false nightshade)

Habitat: open flats, bajadas, grasslands, open woodlands, roadsides; clay, sandy, rocky soils, low-high elevations.

Geographic range: Texas: Trans-Pecos (excluding Terrell, Val Verde Counties); Ward, Andrews Counties. US: Texas to Arizona, Colorado, Utah. Mexico: Chihuahua, San Luis Potosí.

Park locations: Llano Loop

Greenleaf five eyes is a seemingly delicate perennial to 1 foot high with slender stems branching from the base. The herb is variable, with nearly prostrate, sprawling, or ascending stems spreading from rhizomes. To 2 1/2 inches long, the leaves are linear, usually alternate and stemless, rounded or pointed at the tips, and smooth to shallowly toothed or lobed on the margins. Scattered scurfy hairs on the leaves collect dust and dirt.

From spring to fall, saucer-shaped flowers, mostly single, form in the leaf axils, on slender, hairy stalks. The flowers are pale greenish yellow or greenish white, to 3/4 inch across, with five thin lobes. The flaring lobes are purplish near the throat, wavy and fringed on the margins. Clogging the flower's throat are five fluffy, white-woolly pads. Within the throat are five stamens with white filaments and large, oblong yellow anthers, and a style with a globular or bilobed green stigma. Supporting each bloom is a bell-shaped calyx with five thin, hairy lobes.

The fruit is a yellow berry to 1/3 inch long, partly within the enlarged calyx, with tiny, orange brown seeds.

Botanical notes: In 1874, using the Greek prefix *chamae* (low), Asa Gray proposed *Chamaesaracha* as a section of *Saracha*, an American genus named by Ruiz and Pavón in 1794. Ruiz and Pavón honored Spanish Benedictine monk Isidore Saracha, pharmacy director at Santo Domingo de Silos monastery, and instructor of botanist Luis Née. George Bentham and William Jackson Hooker elevated *Chamaesaracha* to generic status in 1876. The species was described in 1852 by Michel Dunal as *Solanum coronopus*, from a Jean Louis Berlandier collection between Laredo and San Antonio in 1828. In 1876, Asa Gray moved the species to the *Chamaesaracha* genus. The epithet, from the Greek *koronopous* (crowfoot), was used by herbalists for plants with crowfoot-shaped leaves or branching stems.

257. Oakleaf Datura (*Datura quercifolia*)
(Other common names: Chinese or oakleaf thorn apple, oak-leaved angel's trumpet, oakleaf jimsonweed, toloache)

Habitat: disturbed sites, clay, alluvial flats, bajadas, creeks, arroyos; limestone, igneous soils.

Geographic range: Texas: Trans-Pecos (Culberson, Jeff Davis, south Presidio, Brewster Counties). US: Texas, New Mexico, Oklahoma. Mexico: Sonora east to Nuevo León, south to central Mexico.

Park locations: Leyva Creek, Sauceda, Alamito Creek, Cienega Creek

Oakleaf datura is a foul-smelling, leafy annual to 3 feet high. The plant contains alkaloids toxic to humans and animals, which cause hallucination, delirium, and death. From a shallow taproot, this annual develops stout branching stems. The stems are sparsely hairy, but the young, reddish-brown branches may be coated with downy hairs.

The alternate leaves are up to 6 inches long, egg-shaped in outline but deeply lobed. The stiff lobes are wavy, curling, and toothed. Often on long maroon petioles, the blades are pointed at the tips, wedge-shaped at the base, and hairy below.

From late spring to fall, tubular flowers, white or pale violet, bloom singly at the branch forks. The delicate blooms seem reclusive, nestled amid menacing leaves. To 3 inches long, each bloom has a funnel-shaped border with five pleated, sharp-tipped lobes. Within the tube are five stamens with purple anthers and a style with a bilobed, greenish-yellow stigma. The tubular calyx is five-lobed and ribbed.

The fruit is a four-chamber, dehiscent capsule, egg-shaped, to 1 3/4 inches long, with calyx remnants forming a basal skirt. Each capsule, with bulky, broad-based spikes, splits to release numerous black seeds.

Botanical notes: The genus name *Datura*, from the Sanskrit *dhatturah*, Hindu *dhatura*, and Arabic *tatorah*, is an ancient name for jimsonweed. In the eleventh century, Persian physician-philosopher Avicenna recounted the medicinal and hallucinogenic properties of a plant later named *Datura metel* by Linnaeus. The epithet *quercifolia* combines *Quercus* (the oak genus), and the Latin *folium* (leaf), for the oak-like leaves. The species was described by German taxonomist Karl Kunth in 1818, from a collection about 1803 by Baron von Humboldt and Aimé Bonpland in the Mexican state of Guanajuato.

258. Sacred Datura (*Datura wrightii*)

(Other common names: sacred or southwestern thorn-apple, western or Wright's jimsonweed, Wright's or sweet-scented datura, moonflower, angel's or devil's trumpet, Indian whiskey, Indian apple, tolguacha, toluaca, toloache, belladona)

Habitat: sand, gravelly floodplains, alluvial flats, banks of creeks, springs, openings in mountains, canyons, disturbed sites; igneous, calcareous soils.

Geographic range: Texas: Trans-Pecos; parts of Edwards Plateau, widely scattered elsewhere. US: native Texas to California, now most of US. Mexico: native in northern Mexico.

Park locations: Casa Piedra Road, Palo Amarillo Creek, Agua Adentro, Leyva Creek, Sauceda

Sacred datura is a strong-smelling annual or perennial, erect or spreading, to 3 feet high and wider, with thick stems and stout tuberous roots. The stems are much branched, gray-haired, often glaucous. To 6 inches long, the alternate leaves are egg-shaped, on broad, often maroon petioles. Wavy, at times toothed or weakly lobed, the blades are blunt or abruptly pointed at the tip, dark green above, yellow green and velvety below.

From late spring to fall, funnel-shaped flowers, to 6 inches long, appear singly at the branch forks. Tinged with violet or yellow, the white night-blooming flowers produce a narcotic scent that attracts hawk moths. Each bloom opens to a spreading, pleated border with five fused lobes, each pointed and hooked at the tip. Within the flower's throat are five stamens with cream-colored, linear anthers and a style with a bilobed stigma. In fruit, remnants of the green, five-lobed calyx form a basal skirt.

The fruit is a four-chamber, dehiscent green capsule to 1 1/2 inches long. The hairy, spheric capsule, with small, flexuous prickles, contains numerous light brown, discoid seeds.

Datura species have been used for 3,000 years as sacred visionary plants. The Aztecs employed toloache (probably *D. inoxia*) as a painkiller in ritual sacrifices. Native Americans used the plant to induce visions. The Tarahumara of northern Mexico took peyote as an antidote to datura poisoning (Boyd 2013).

Botanical notes: In 1859, with an illustration in the magazine *Gartenflora*, German botanist Eduard August von Regel named this species for Charles Wright. The illustration depicted a plant grown from seeds collected by Wright. Wright likely collected *D. wrightii* on the Rio Grande in Texas in 1849.

259. Downy Wolfberry (*Lycium puberulum* var. *puberulum*)
(Other common names: downy desert-thorn)

Habitat: gravelly, sandy, clay-gyp flats, creosote scrub, rocky bluffs; limestone, igneous soils.

Geographic range: Texas: Trans-Pecos (El Paso, Hudspeth, Culberson, Jeff Davis, Presidio, Brewster Counties). US: Texas. Mexico: Chihuahua, Coahuila, Durango.

Park locations: Solitario rim above Righthand Shutup, Terlingua Uplift, Fresno Canyon, West Contrabando Trail, Cienega Creek

Downy wolfberry is a surprisingly thorny shrub to 5 feet high, with diverging, slightly zigzagged branches. This thornbush is armed with needlelike spines and covered with gray, chocolate, or black bark. Young branchlets may be thickly shorthaired. The gray-green, spatulate leaves are alternate or bundled along the stems, with some gland-tipped hairs. To 1 1/2 inches long, the stemless blades are rounded or blunt at the tip, smooth on the edges, with a distinct midvein.

In spring, single or paired, short-stalked flowers emerge from the leaf axils. Each tubular bloom is up to 1/2 inch long, with a reflexed, five-lobed border. The flowers are white, the tube at times lavender-tinged, and the lobes and throat often pale green or yellowish green. Within the tube are five stamens, with filaments attached to the tube for much of their length, and short anthers, and a style with a headlike or bilobed stigma. Often densely glandular, the persistent calyx is bell-shaped, with five oblong lobes.

The fruit is a hardened, blue-gray berry to 1/3 inch long, globose but contracted below the middle. Each berry is two-chambered, with a hard, inner layer (endocarp) enclosing one or two seeds per chamber.

Botanical notes: According to Asa Gray, Linnaeus's genus name *Lycium* was borrowed from the "ancient Greek *lycion*, a prickly shrub growing in Lycia [country in Asia Minor, present-day Turkey]." The Latin species and variety name *puberulum* (puberulent), from *puber* (downy) and the suffix *-ulum* (slightly), alludes to the slightly hairy leaves. Gray described the species in 1862, from a collection by Charles Wright on the "western borders of Texas, near El Paso." Based on Wright's field notes, the specimen was likely obtained in 1852, below old Fort Quitman in southern Hudspeth County (Shaw 1987).

260. Netted Globecherry (*Margaranthus solanaceus*)
(Other common names: netted globeberry, totomache)

Habitat: sandy, gravelly, rocky, alluvial soils, creeks, springs, canyons; mid- to high elevations.

Geographic range: Texas: Trans-Pecos (Hudspeth, Culberson, Jeff Davis, Presidio, Brewster Counties), Lower Rio Grande Valley; Bandera, Fayette Counties. US: Texas to Arizona. Elsewhere: Mexico to Guatemala, Honduras; West Indies.

Park locations: Arroyo Mexicano, Leyva Creek, Cinco Tinajas, Jackson Pens, upper Alamito Creek

Resembling a groundcherry (*Physalis*), but with a much different flower, this singular nightshade is an erect or spreading annual to 2 feet high, with multibranched yellow-green or brown stems. Growing from a taproot, the leafy stems are mostly hairless. On slender petioles, the leaves are thin, to 2 inches long, green above, yellow green below. The blades are lancelike to egg-shaped, with a pointed or tapered tip.

In summer and fall, flowers appear singly in the leaf axils on short, curving, white-haired stalks. To 1/6 inch long, the iridescent, urnlike blooms are dark blue purple, constricted at the five-toothed mouth. Within the flower's throat are five stamens with bluish anthers and a threadlike style with a disklike stigma. A bell-shaped, densely hairy calyx supports each bloom. The

fruit is a thin-walled dry berry to 1/3 inch wide, enclosed in the bladderlike fruiting calyx. The greenish-yellow calyx is membranous, ribbed, and veined. Each berry contains twenty to thirty disklike, pitted seeds.

In the Mexican state of Puebla, aerial plant parts are boiled and eaten to treat diarrhea (Hernández et. al. 2003) and made into a tea to treat diabetes and bile (Canales et al. 2005).

Botanical notes: The genus name may derive from the Greek *margarite* (pearl) and *anthos* (flower), for the shiny flowers. German botanist Diederich von Schlechtendal, director of the University of Halle botanical garden, described the genus and species in 1838. The type specimen was cultivated in the garden from seeds collected in Mexico by German explorer Carl Ehrenberg. Ehrenberg, mining superintendent in the 1830s at Real del Monte in Hidalgo, Mexico, made extensive collections, often cultivating and shipping living plants to Halle and Berlin.

261. Tree Tobacco (*Nicotiana glauca*)
(Other common names: wild or glaucous tobacco, mustard tree, tronadora, tabaquillo, tabaco silvestre, tabaco cimarrón, virginio, gigante, gigantón)

Habitat: Rio Grande, river canyons, creeks, springs, disturbed sites; sandy, gravelly, clayey, alluvial, rocky substrates, low elevations.

Geographic range: Texas: Trans-Pecos (near Rio Grande from El Paso to Del Rio; Pecos County); parts of South Texas Plains, central Texas; scattered sites. US: Florida to Alabama, Texas to California, Utah; other states. Elsewhere: native to South America; naturalized north to Mexico; introduced worldwide.

Park locations: River Road, Fresno Canyon, Madera Canyon Campground, Ojito Adentro

Tree tobacco was introduced in the United States in the late 1800s, later naturalized in the southwest, and is now prolific along the Rio Grande. This evergreen shrub or small tree is seldom more than 10 feet high, with slender, sparingly branched, erect or arching stems. The fast-growing branches are hairless but glaucous, and all plant parts are poisonous, producing nicotine and other toxic alkaloids.

The alternate leaves are glaucous, powdery, often more blue than green. Broadly egg-shaped, or up-turned and scoop-like, the long-stemmed leaves are up to 8 inches long, with pointed or blunt tips and smooth margins.

Blooming most of the year, greenish yellow tubular flowers appear in loose clusters at the branch tips. To 2 inches long, each bloom is abruptly expanded at the tip into a five-angled, pleated border with five short, fused lobes. The flowers are shorthaired externally. Within each bloom are five stamens, with greenish white filaments and short-oblong, yellow-green anthers and a white style with a green, headlike stigma. The light green calyx supporting each bloom is tubular, with five triangular lobes. The brown fruit is a two-chamber, many-seeded, dehiscent egg-shaped capsule, 2/5 inch long.

Botanical notes: Linnaeus described the genus in 1753, but the name was used much earlier. In 1623, Caspar Bauhin described *Nicotiana major latifolia*, a plant Linnaeus named *Nicotiana tabacum* 130 years later. The genus is named for Jean Nicot, secretary to King Francis II of France and ambassador to Portugal, where Portuguese botanist Damião de Goes introduced him to tobacco. In 1560, Nicot introduced tobacco and snuff to the French court and convinced Queen Catherine de Medici of its healing powers. The species name *glauca*, from the Greek *glaucos* (glaucous), refers to the leaves. Scottish botanist Robert Graham described the species in 1828, noting that the plant was "raised in 1827 from seeds communicated . . . to the Royal Botanic Garden, Edinburgh, by Mr. Smith at Monkwood, whose son sent them from Buenos Ayres."

262. Lanceleaf Groundcherry (*Physalis angulata* var. *lanceifolia*)
(Other common names: cutleaf groundcherry)

Habitat: moist, disturbed sites, floodplains, fields; roads, ditches, dumps; sandbars, marshes.

Geographic range: Texas: Trans-Pecos (Presidio, Brewster Counties); Llano, Travis, Refugio, Cameron Counties. US: Texas to California; Oklahoma, Florida. Mexico: cited in Campeche, Sinaloa, Sonora.

Park locations: 1 mile west of Lajitas

Rare in the Trans-Pecos, this groundcherry is a sprawling or upright annual to 3 feet high, with notably angled and grooved, diffusely branching stems. Stems and leaves may be sparsely hairy, with short, thin hairs. On long petioles, the mostly alternate leaves are lance-shaped, to 5 inches long. The blades are sharp-pointed or tapered at the tip, wedge-shaped at the base, irregularly incised, and wavy on the margins.

In late spring and summer, dainty whitish-yellow flowers bloom singly in the leaf axils. The bell-shaped flowers, to 1/4 inch across, dangle from slender, finely hairy stalks. Each bloom has a spreading, sometimes reflexed, five-lobed border, which is darker yellow at the base. Within the flower's throat are five stamens, with threadlike filaments and oblong yellow, often blue-tinged anthers and a slender style with a headlike or bilobed stigma. The yellow-green, five-lobed calyx is bell-shaped, with maroon-tinged ribs.

In fruit, the ten-ribbed calyx becomes inflated, with maroon ribs and veins. The calyx encloses a two-chamber berry 2/5 inch across, with many tiny yellow seeds.

Botanical notes: The generic name *Physalis*, from the Greek *physa* (bladder), refers to the bladderlike fruits. The Latin species name *angulata* (angled), according to Solanaceae specialist W. G. D'Arcy, "is derived from angles on the branches which are indeed conspicuous." The Latin variety name *lanceifolia* connotes "lanceolate leaves." University of Breslau botanist Christian Nees described this variety as a species, *Physalis lanceifolia*, in 1831, citing habitats in Peru and Mexico. In 1958, Oklahoma State University botanist Umaldy Theodore Waterfall treated Nees's groundcherry as a variety of *P. angulata*, a widespread species described by Linnaeus in 1753.

263. Heartleaf Groundcherry (*Physalis hederifolia*)
(Other common names: ivyleaf or ivy-leaved groundcherry, ivy leafed tomatillo, hillside groundcherry, rama de amores)

Habitat: sandy flats, desert scrub, gravelly arroyos, springs, grasslands, woodlands, rocky slopes, mesas, canyons; low to high elevations, often calcareous soils.

Geographic range: Texas: Trans-Pecos (excluding Val Verde County), parts of Edwards Plateau, Permian Basin, northern Panhandle, South Texas Plains, scattered elsewhere; US: southwest; Oklahoma to South Dakota, Wyoming, Montana. Mexico: northern Mexico south to Michoacán.

Park locations: Arroyo Mexicano, Yedra Canyon, Ojo Chilicote, Lower Shutup, Cienega Mountain

Heartleaf groundcherry is a low perennial to 18 inches high, with erect or ascending gray-green stems and spreading branches. The stems and leaves may be densely hairy, with a mix of long, branching, or gland-tipped hairs, often sticky to the touch.

To 2 inches long, on thick, hairy petioles, the leaves are broadly egg-shaped to triangular, rounded or pointed at the tip, truncate or heart-shaped at the base, and broadly toothed on the margins. The blades are gray green above, paler below, with white gland-tipped hairs outlining the margins.

From spring to fall, yellow flowers arise singly from the leaf axils, on stalks to 1/3 inch long. The dangling blooms are bell-shaped, to 3/5 inch across, with five fused, spreading or reflexed lobes. Commonly, the lobes are marked with dark, often reddish-brown, blotches and radiating streaks at the base. Within each bloom are five stamens, with short filaments and tapering, tubular yellow anthers and a protruding, pale yellow style with a rounded, greenish-yellow stigma. The supporting, five-lobed calyx is bell-shaped, with a mix of hairs.

The fruit is a round, many-seeded berry, enclosed by the enlarged fruiting calyx. The light green calyx, to 1 inch long, is thin, ten-ribbed, inflated like a Chinese lantern.

Botanical notes: The species name, from *hedera* (ivy, the genus *Hedera*) and *folium* (leaf), implies "with ivylike leaves." Asa Gray described the species in 1874, from an 1849 collection by Charles Wright on the "western borders of Texas," most likely on mountain slopes near El Paso. Josiah Gregg obtained a specimen even earlier, at Bishop's Hill near Monterrey, Mexico, in 1847.

264. Watermelon Nightshade (*Solanum citrullifolium*)
(Other common names: melonleaf nightshade)

Habitat: grasslands, basins, canyons, mountains, disturbed sites; igneous-derived or calcareous soils.

Geographic range: Texas: Trans-Pecos (Hudspeth, Jeff Davis, Presidio, Brewster Counties); parts of Edwards Plateau. US: Texas, New Mexico, Kansas, Iowa, Florida. Mexico: Sonora east to Nuevo León, south to San Luis Potosí.

Park locations: lower Monilla Canyon, Black Hills, Yedra Canyon, Cinco Tinajas, Solitario Basin, Cienega Mountain

Growing from a taproot, this nightshade is a low, intricately branched, leafy annual to 30 inches high. The yellow-green stems turn white with age. Most plant parts are armed with straight white prickles. Stems and leaves bear gland-tipped hairs. Like watermelon leaves, the leaves are broadly egg-shaped in outline, but deeply, irregularly cleft, with wavy, curling lobes. To 5 inches long, on thick, prickly petioles, the blades are often prickly along the primary veins and glandular-hairy above, with stellate hairs below.

From spring to fall, large purple flowers appear in clusters usually arising between the stem nodes. On stout stalks, the flowers are short-tubed, with a star-shaped, five-lobed border. The lobes are triangular, tapered, and curved to the tip. Within the lobes are five protruding stamens, with short greenish-yellow filaments and long bulky anthers. The four upper anthers are yellow; the lower anther is longer, reddish brown with age, with a curved violet tip. The long pale green style is also curved, with a small stigma.

The fruit is a two-chamber berry to 1/2 inch across, enclosed by a spiny calyx, with up to sixty dark brown seeds.

Botanical notes: Pliny, in his *Natural History*, used the Latin name *solanum* for a nightshade known to the Greeks as *strychnos*, but the meaning is unclear. The name may be derived from the Latin *solamen* (solace), for the plant's medicinal properties; or the Latin *sol* (sun), for the likeness of the flowers to the sun's rays, or the plant's sunny habitat. The species name, from the Latin *Citrullus* (watermelon genus), and *folium* (leaf), suggests "with leaves like *Citrullus*." German plant morphologist, Alexander Carl Heinrich Braun, described the annual in 1849, from a plant grown from seed in the Freiburg Botanical Garden. Apparently, the seeds came from central Texas, supplied by Ferdinand Lindheimer.

265. Buffalobur Nightshade (*Solanum rostratum*)
(Other common names: buffalo bur(r), Texas or Kansas or Mexican thistle, prickly or spiny or horned nightshade, buffalo berry, mala mujer, duraznillo, hierba del sapo, ayohuiztle, soíwari [Tarahumara])

Habitat: open disturbed sites, pastures, fields, sites subject to flooding; sandy, rocky soils.

Geographic range: Texas: Trans-Pecos (excluding Hudspeth, Terrell Counties), Edwards Plateau, Panhandle, north Texas; scattered elsewhere. US: throughout; native to Great Plains. Elsewhere: Canada, Mexico; invasive worldwide.

Park locations: Colorado Canyon River Access, Hoodoos

This nightshade is a taprooted annual to 28 inches high, with ascending, diverging branches. Sharp, noxious, yellow or white thorns, spiny yellow burs, and yellow stellate hairs on most plant parts give the annual a characteristic yellow cast. In pioneer times the herb shared mud wallows with bison, where its spiny burs lodged in the hides of the bison, and awaited dispersal. On long prickly petioles, the leaves are egg-shaped in outline, to 6 inches long, once or twice lobed or cleft, with wavy and wrinkled margins. Dull green above, yellow-green below, the blades bear prickles along the major veins.

From spring to fall, yellow star-shaped flowers form in long clusters on lateral stalks between the stem nodes. Each bloom is short-tubed, with five thin, spreading, fused lobes. The lobes are egg-shaped, curling, and slightly jagged. Projecting from the flower's throat are five stamens, with short filaments and long tubular, tapering yellow anthers. The four upper anthers are yellow; the lower anther is longer, downturned, curved at the tip, dark reddish brown with age. Also downturned, and hooked at the tip, is the long slender, pale yellow style. The five-lobed calyx bears spinelike prickles and stellate hairs.

The fruit is a berry to 1/2 inch long, enclosed by the enlarged green calyx, with many dark brown seeds.

Botanical notes: The species name, from the Latin *rostratus* (beaked), refers to the bloom's largest, beaklike anther. Michel Dunal, chair of medical natural history at Montpellier, described the species in 1813, from a specimen cultivated in the Montpellier Botanical Garden. A plant (then identified as *Solanum bejarense*) was collected by Jean Louis Berlandier in 1828, at San Antonio, Texas.

266. Pink Baby's-Breath (*Talinum paniculatum*)
(Other common names: jewels of Opar, big talinum, panicled flameflower, rama del sapo, quelite de monte)

Habitat: canyon walls, below dams in igneous mountains; moist clay, sandy soils.

Geographic range: Texas: Trans-Pecos (El Paso, Jeff Davis, Presidio, Brewster Counties), parts of Edwards Plateau, South Texas Plains, Gulf Coast. US: Louisiana to Arizona; North Carolina to Florida; Kentucky. Elsewhere: Mexico, Central, South America, West Indies.

Park locations: drainage from Sauceda to Cinco Tinajas, Tascate Hills

Pink baby's-breath is a stout perennial to 3 feet high, with erect stems and tuberous roots. The maroon stems are succulent, with ascending, wand-like branches. Up to 5 inches long, the fleshy leaves are short-stemmed or stemless, broadly elliptic to inversely egg-shaped. The blades are rounded or blunt at the tip, smooth on the margins.

In summer and fall, blooms form in loose clusters on the flowering branches. On a slender stalk, each bloom is up to 1/4 inch across, with five oval petals. Within the petals are numerous stamens with yellow filaments and oblong, bilobed, golden anthers, and a green ovary and a style with three linear stigmas. Below the petals are two tan sepals. Typically, this herb produces yellow blooms in the Trans-Pecos, but often pink or red blooms elsewhere. Pink baby's-breath is also known as jewels of Opar, a name taken from a Tarzan novel, because the amber fruits resemble lustrous jewels.

The fruit is a three-valved dehiscent capsule to 1/5 inch long, globular to three-angled. The capsule splits lengthwise from the tip to expose many shiny black seeds.

Botanical notes: The Latin species name *paniculatum* (paniculate) denotes the shape of the flower clusters. The species was described in 1760 by Austrian botanist Nicolaus Joseph von Jacquin as *Portulaca paniculata*. Sent by Austrian Emperor Francis I to the West Indies (1755–59) in search of flora for the Schoenbrunn Palace gardens, Jacquin returned with botanical collections and illustrations eventually documented in *Hortus Schoenbrunnensis* (1797–1804). Jacquin was the first German author to implement Linnaeus's binomial system. In 1791, German botanist Joseph Gaertner transferred the species to the *Talinum* genus.

267. Yellow Flameflower (*Talinum polygaloides*)
(Other common names: narrowleaf flameflower, polygala-like talinum)

Habitat: desert flats, arroyos, hillsides, disturbed sites; sandy, gravelly soils; mid- to low elevations.

Geographic range: Texas: Trans-Pecos, parts of Permian Basin, southern Panhandle, Rolling Plains, Edwards Plateau, South Texas Plains. US: Texas to Arizona. Elsewhere: Mexico (Chihuahua to Nuevo León, south to Durango, San Luis Potosí), South America (Bolivia, Chile, Argentina, Paraguay).

Park locations: Llano Loop, Tascate Hills, Tres Papalotes, Buena Suerte Trail, Cienega Gorge

This flameflower is a low perennial to 20 inches high, at times weak and spindly, leaning on larger plants, or shrubby, woody at the base, with woody tuberous roots. Mostly hairless, the slender stems are sprawling to erect, and woodier with age.

The leaves seem nearly as long and slender as the stems. To 2 1/2 inches long, the green leaves are linear, stemless, flat or revolute, with edges tucked under. The alternate blades have a short, pointed tip, smooth margins, and a prominent midvein above.

From late spring to early fall, yellow flowers appear on thin stalks from the leaf axils, or in few-flowered clusters. Each cluster is short-stalked, with two bracts at the base. To 3/5 inch across, each flower has five inversely egg-shaped petals, which may turn red with age. Within the flower's throat are up to thirty stamens, with bilobed, oblong yellow anthers; a green ovary; and a yellow style with three linear, curving, yellow stigma branches. Two early falling, often chaffy and membranous, sepals support each bloom.

The fruit is a globose, three-chamber, dehiscent capsule to 1/5 inch long, with many shiny black seeds.

Botanical notes: French naturalist Michel Adanson named the genus in 1763, in a pioneering work, *Familles des Plantes*, but the meaning is unclear. According to Asa Gray, *Talinum* is "probably an unmeaning name." Others noted that *Talinum* is the name of a Senegalese plant, and that Adanson collected flora in Senegal and wrote dictionaries of the native languages. The epithet *polygaloides*, from *Polygala* (the genus) and *-oides* (similar), means "*Polygala*-like *Talinum*." Scottish physician John Gillies, resident of Mendoza, Argentina, for some years, collected the species in 1822, "in the Jarillal, and along the foot of the mountains near Mendoza." Gillies's species account was published in 1831.

268. Athel (*Tamarix aphylla*)
(Other common names: athel or leafless tamarisk, athel tree, athel pine, pinabete, taraje, pino salado)

Habitat: cultivated or escaped, naturalized at local sites near Rio Grande; sandy, silty riverbanks, towns, farms.

Geographic range: Texas: Big Bend (Presidio, Brewster Counties), South Texas Plains. US: southwest (excluding Colorado). Elsewhere: Mexico; native to North Africa.

Park locations: 1 mile west of Lajitas, Redford, Rio Grande below Presidio, Botella

Like other tamarisks, athel drains water from streams and pushes out native species. This Old World evergreen tamarisk occurs in the Big Bend mostly where cultivated (Redford, Presidio), but it sometimes escapes and has become naturalized at several sites along the river. Although fast-growing from extensive roots, athel is less invasive than other salt cedars, reproducing mostly vegetatively rather than by seed.

A handsome tree to 30 feet high, athel has stout, widely spreading stems with reddish brown bark, slender smooth branches and drooping, needlelike young branchlets. Wrapped in tiny, close-spaced, sheathing leaves, the branchlets appear jointed. Salt-secreting glands, on the blue-green, linear blades, form crusty white joints along the twigs.

In summer, white flowers, also bearing salt-secreting glands, bloom prolifically in spikelike clusters to 3 inches long. Each bloom has five oblong petals to 1/10 inch long. Within the bloom's throat are five stamens with threadlike white filaments and maroon anthers, a pistil with three short styles with headlike stigmas, and a red nectary disk below the ovary. The mature fruit is a brown capsule to 1/5 inch long, which splits into three valves, releasing tiny brown bristle-haired seeds.

Botanical notes: The genus name is the Latin *Tamarix* (tamarisk), likely referring to River Tamaris in Galicia. The epithet *aphylla*, from the *Greek a* (without) and *phyllon* (leaf), refers to the scalelike leaves. Linnaeus's student Abrahamus Juslenius described the species as *Thuja aphylla* in 1755, with Egypt cited as the habitat and a 1738 work by Reverend Thomas Shaw as a source. However, the plant described turned out to be a cypress. Shaw, a chaplain, traveled extensively in North Africa. In 1882, German botanist Hermann Karsten placed this species in the *Tamarix* genus.

269. Salt cedar (*Tamarix chinensis*)
(Other common names: five-stamen or Chinese or China tamarisk, Chinese salt cedar, tamarisco chino, taray de China)

Habitat: Rio Grande, tributary canyons, springs, creeks, tanks, disturbed sites; often saline soils.

Geographic range: Texas: Trans-Pecos; scattered elsewhere. US: southwest, much of northwest; Oklahoma, Arkansas, some eastern states. Elsewhere: naturalized Canada to Mexico.

Park locations: 1 mile west of Lajitas, Rio Grande between stables and movie set, Madera Canyon Campground, Colorado Canyon River Access, Arenosa Campground, Fresno Canyon

Salt cedar, a shrub or tree with a trunk to 20 feet tall, is much branched, with erect or spreading branches, and many wiry branchlets. The hairless branches are covered with reddish-brown to blackish bark. Native to China and Korea, this tamarisk was introduced in the United States in the early 1800s and has become highly invasive in the southwest. The fast-growing shrubs develop an extensive root system that consumes large amounts of water and lowers water tables, driving out native species.

The common name is salt cedar, since the leaves resemble cedar (juniper) leaves and leaf glands secrete salts, and because the shrub grows in saline environs. The alternate, stemless green leaves are scalelike, not sheathing, but clasping and sometimes overlapping.

Tiny fragrant flowers bloom profusely from spring to fall, in crowded clusters at the branch tips. Each bloom has five white or pink, oblong petals. Inside the flower's throat are five stamens, with threadlike white filaments and red anthers; a red, conical pistil with three clublike white styles with headlike white stigmas; and a red nectary disk below the ovary. At the base of each bloom are five egg-shaped sepals and a small green bract. The fruit is a brown, three-valved, dehiscent capsule to 1/6 inch long, with many minute seeds.

Botanical notes: The Latin species name *chinensis* (Chinese) indicates the native habitat. In 1790, João de Loureiro, a Portuguese Jesuit missionary, mathematician, and naturalist who spent thirty-five years in Cochinchina (southern Vietnam), described this species in his *Flora Cochinchinensis*, listing the Canton province of China as the habitat. Loureiro's flora described 672 plant genera from Vietnam, China, and Mozambique, with information on medicinal, economic, and other human uses.

270. Southern Cattail (*Typha domingensis*)
(Other common names: tule, Dominican or slender cattail, espadaña, masa de agua, cola de gato)

Habitat: riverbanks, periodically flooded, perennially wet sites.

Geographic range: Texas: Trans-Pecos (Hudspeth, Culberson, Presidio, Brewster, Val Verde Counties), much of state. US: southern three-fourths. Elsewhere: Mexico to South America, West Indies; invasive in tropics worldwide.

Park locations: Rio Grande between stables and movie set, lower Rancherías Canyon, Ojito Adentro, Chorro Canyon, Smith Canyon

Southern cattail is an aquatic perennial to 10 feet high, with erect, fibrous, unbranched stems arising from creeping rhizomes. This cattail spreads vegetatively to form often dense colonies. The hairless, jointless green stems are smooth and slim but 2/3 inch wide in bloom.

A flowering stem bears up to ten erect, linear leaves, which sheath the stem base. To 3/5 inch wide, the stalkless leaves are nearly as long as the flowering spikes. The hairless yellow-green blades are ribbonlike, spongy but firm, tapered at the tip.

From spring through summer, thousands of tiny male and female blooms form on the same stalk, in separate cylindric spikes. The tan male spike, up to 14 inches long, is tapered to the tip. Below the male spike, the cinnamon-brown female spike is shorter but thicker. The male spike withers, but the female spike persists into winter. A male bloom consists of two to four stamens, with white filaments and linear yellow anthers. The smaller female bloom consists of a pistil and a style with a linear white stigma. Long hairlike bractlets are intermingled with the female blooms. The fruit is a thin-walled, spindle-shaped follicle, dehiscent with the long stalk and one tiny seed.

Botanical notes: The Greek *typhe* was Theophrastus's term for cattail, related to the Greek *typhein* (to smoke), and *typhos* (marsh). Linnaeus's genus name, *Typha*, may be derived from *typhein*, because blowing seeds resemble smoke, flower spikes make smoky fires, or the fruiting spikes' color is smoky; or from *typhos*, for the marshy habitat. The species name *domingensis* signifies "Dominican Republic." Taxonomist Christiaan Persoon described the species in 1807 in *Synopsis Plantarum*, his massive effort to describe all flowering plants.

271. Beebrush (*Aloysia gratissima*)
(Other common names: whitebrush, common beebrush, common beebush, privet lippia, cedrón, cedrón del monte, palo amarillo, vara dulce, jazminillo, chaparro blanco, quebradora)

Habitat: sandy, gravelly creeks, rocky canyons; desert flats, grasslands, mountains, woodlands; igneous, limestone soils.

Geographic range: Texas: Trans-Pecos (excluding El Paso County), much of South Texas Plains, Edwards Plateau; scattered sites. US: Texas to Arizona. Elsewhere: Mexico (Sonora east to Tamaulipas, Durango south to Oaxaca); South America.

Park locations: Tapado Canyon, Agua Adentro, Madrid Falls Trail, upper Terneros Creek, Horsetrap Spring, Solitario northwest rim, Fresno Canyon

Beebrush is an upright, intensely fragrant, much branched shrub to 10 feet high. The thicket-forming shrub has slender, yet stiff, woody stems, with diverging, sharp-tipped branches.

Short-stemmed or stemless, the paired deciduous leaves are oblong or elliptic, to 1 inch long. The blades are pointed or blunt at the tip, wedge-shaped at the base, smooth on the margins, minutely hairy above, resin-dotted and hairy below.

From spring to fall, often after rains, white flowers appear on upper branches in crowded or more open clusters to 3 inches long. Often violet-tinged, each bloom is less than 1/8 inch across, with a slender tube that flares into a four-lobed border. The vanilla-scented flowers produce a much-sought-after honey. Within the flower's throat are four stamens, in two unequal pairs, with yellow anthers; a two-chamber ovary; and cylindric style with a headlike stigma. Below each bloom is a four-lobed, bristle-haired calyx, which encloses the mature fruit.

The tiny bilobed fruit consists of two one-seeded oblong nutlets.

Botanical notes: In 1784, Spanish botanists Casimiro Gómez Ortega and Antonio Palau y Verdera named this genus for Maria Louisa Teresa (*Aloysia* was formed from letters of her name), wife of King Charles IV of Spain. The Latin epithet *gratissima* signifies "most pleasing." The species was described as *Verbena gratissima* in 1830 by Scottish botanist John Gillies and William Jackson Hooker, director of the Royal Botanic Gardens at Kew, from a specimen collected by Gillies near Mendoza, Argentina. In 1962, Argentinian Verbenaceae specialist Nélida Sara Troncoso moved the plant to the *Aloysia* genus.

272. Oreganillo (*Aloysia wrightii*)
(Other common names: Wright's beebrush, Mexican oregano, spicebush, Wright's or mintbush lippia, vara dulce, high mass, altamisa)

Habitat: desert scrub, arroyos, springs, canyons, woodlands; igneous, limestone substrates.

Geographic range: Texas: Trans-Pecos (excluding Reeves, Val Verde Counties); Crockett, Tom Green Counties. US: southwest (excluding Colorado). Mexico: Sonora east to Nuevo León, south to Durango, Zacatecas.

Park locations: Agua Adentro, Arroyo Mexicano, Pico de las Aguilas, above Madrid Falls, Yedra Canyon

Oreganillo is a densely branched, sweetly fragrant, often rounded shrub to 4 1/2 feet high, with slim, brittle stems and thin, peeling bark. Older woody stems, with reddish-brown bark and yellow wood, become hairless with age; younger stems often bear short, thin hairs.

Short-stemmed or stemless, the paired leaves are egg-shaped or rounded, with broad teeth on the edges. Deeply veined and wrinkled, the blades bear short, stiff hairs and blister-like swellings on the upper sides and longer gray hairs and resinous glands underneath. The crushed leaves smell like oregano, and dried leaves are sold in Mexican markets as a food seasoning.

Mostly in summer and fall, tiny white flowers form in many-flowered clusters, on stalks from the upper branches. A bee favorite, the blooms have a slender tube that flares into a two-lipped, four-lobed border (limb). Within the flower's throat are four stamens, in two unequal pairs, with anthers with two pollen sacs; a two-chamber ovary; and a style with a headlike stigma. The four-lobed calyx is tubular, with bristly and glandular hairs. To 1/17 inch long, the bilobed fruit splits into two one-seeded nutlets.

Botanical notes: In 1858, Asa Gray named this species *Lippia wrightii*, from collections by Charles Wright, John Bigelow, Charles Parry, and George Thurber. The holotype is two 1851 Wright collections, one near the Chiricahua Mountains in Cochise County, Arizona, one at Guadalupe Pass, on the Arizona–New Mexico border. California botanist A. Arthur Heller transferred the plant to the *Aloysia* genus in 1906. Heller collected in the western states, edited his botanical journal, *Muhlenbergia*, and in 1895, authored *Botanical Explorations of Southern Texas*.

273. Groovestem Bouchea (*Bouchea linifolia*)
(Other common names: flaxleaf bouchea)

Habitat: desert scrub, gravelly creeks, brushy arroyos, springs, rocky canyons, open hillsides; limestone soils.

Geographic range: Texas: Trans-Pecos (Presidio, Brewster, Pecos, Terrell, Val Verde Counties); Crockett, Sutton, Kinney, Uvalde Counties. US: Texas. Mexico: Coahuila.

Park locations: Contrabando Waterhole, Lower Shutup, Contrabandista Spur Trail

Groovestem bouchea is a shrubby perennial to 3 feet high, with upright, sparingly branched stems, woody at the base. Mostly hairless, the stiff, leafy branches are notably grooved.

The stalkless leaves are opposite or clustered in rings of three at branch nodes. To 2 inches long, the blades are linear to narrowly lance-shaped, sharp-pointed at the tip, and tapered to the base. Leaf margins are sometimes revolute, with scattered hairs.

Showy flowers appear from spring to fall in loose clusters to 6 inches long, at the stem tips or upper branches. To 1 inch long or more, the flowers are funnel-shaped, with a long, often curving or arching tube that spreads to form a broad five-lobed border. Each bloom has a projecting lower lip, smaller reflexed upper lobes, and curling margins. Reported flower colors are lavender, purple, blue violet, magenta, and white. Within the flower's throat are four stamens, in two unequal pairs, with anthers with two pollen sacs; a two-chamber ovary; and a threadlike style with a bilobed stigma — one lobe large, clublike, the other tiny, toothlike. Supporting each bloom is a green or reddish-brown, tubular calyx, with five slender lobes. Needlelike bracts, and two small, inner bractlets, are usually present below each bloom.

The bilobed fruit splits into two one-seeded nutlets, enclosed in the calyx at maturity. To 1/3 inch long, each black nutlet is linear, ridged, with a shaggy-haired beak.

Botanical notes: In 1832, Adelbert von Chamisso, keeper of the Royal Botanical Garden in Berlin, named the genus *Bouchea* for his peers, Carl David Bouché and his father Peter Carl Bouché. Peter was scion of a famous family of gardeners. Carl worked at and was later inspector of the Royal Botanical Garden. The Latin species name, from *linea* (line) and *folium* (leaf), refers to the linear leaves. Asa Gray described the species in 1859, from collections by Arthur Schott and Charles Wright in "Western Texas." According to Wright's field notes, Wright collected the lectotype on the "pebbly bed of a creek between Piedra Pinta [Pinto Creek in Kinney County] & San Felipe [Creek in Val Verde County]" in 1851.

274. Christmas Mistletoe (*Phoradendron leucarpum* ssp. *tomentosum*)
(Other common names: American or hairy mistletoe, injerto)

Habitat: parasitic mostly on hackberry, mesquite, acacia; sandy, alluvial flats, desert scrub, grassland to woodlands.

Geographic range: Texas: Trans-Pecos (excluding Reeves, Pecos Counties), much of state. US: Texas to Kansas; Louisiana, Arkansas. Mexico: Chihuahua east to Tamaulipas, Durango southeast to San Luis Potosí.

Park locations: upper Rancherías Canyon, West Fork of Comanche Creek, Cienega Creek

Christmas mistletoe is a rounded evergreen shrub to 3 feet long, with intricately branched stems. The stems are thick, brittle, smooth but enlarged at the nodes. The herbage is hairless to densely stellate-hairy. Although the shrubs contain chlorophyll and produce some nutrients, they parasitize woody plants with rootlike organs that penetrate the host and attach to conducting tissue to steal water and minerals. Mistletoes seldom kill their host.

The short-petioled, opposite leaves are inversely egg-shaped, oval or rounded, to 1 3/10 inches long. Gray green or yellowish, the blades are thick, leathery, rounded at the tip, tapered to the base, smooth on the edges, and at times densely short-haired.

Yellow male and female flowers bloom on separate plants from October to March, in segmented spikes to 1 1/2 inches long. The spikes consist of two or three fertile segments, male spikes with up to forty flowers in three rows per segment, female spikes with up to ten flowers per segment, clustered at the tip. Embedded in the segments, all blooms have three triangular, scalelike petals, the male with a bilobed anther at the base of each petal, the female with a style with a headlike stigma.

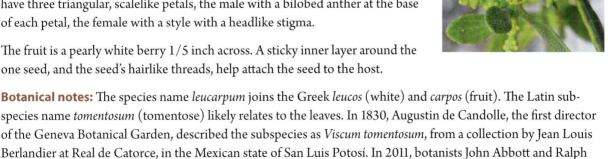

The fruit is a pearly white berry 1/5 inch across. A sticky inner layer around the one seed, and the seed's hairlike threads, help attach the seed to the host.

Botanical notes: The species name *leucarpum* joins the Greek *leucos* (white) and *carpos* (fruit). The Latin subspecies name *tomentosum* (tomentose) likely relates to the leaves. In 1830, Augustin de Candolle, the first director of the Geneva Botanical Garden, described the subspecies as *Viscum tomentosum*, from a collection by Jean Louis Berlandier at Real de Catorce, in the Mexican state of San Luis Potosí. In 2011, botanists John Abbott and Ralph Thompson gave this mistletoe its current botanical name.

275. Bigleaf Mistletoe (*Phoradendron macrophyllum*)
(Other common names: Colorado Desert or broadleaf mistletoe, muérdago)

Habitat: Rio Grande, tributaries; parasitic on cottonwood, ash, willow, walnut; moist woodlands, creeks, lower elevations.

Geographic range: Texas: Trans-Pecos (El Paso, Hudspeth, Presidio, Brewster Counties). US: Texas to California. Mexico: Baja California to Coahuila.

Park locations: Cienega Creek, Cienega Gorge

This mistletoe is like Christmas mistletoe but more robust, with larger leaves and less hairy. To 3 feet across, the semiparasitic evergreen shrub is round or broadly oblong, with long thick branches. The yellow-green branches are often dangling, fleshy yet woody with age, and brittle at the nodes. Like other species, the shrub extracts water and nutrients from hosts by means of modified roots but also contains chlorophyll and performs photosynthesis. To 2 inches long, the short-petioled, paired, yellow-green leaves are oval, oblong, or nearly round. Often the blades are thick, stiff, and brittle, with noticeable veins.

Male and female flowers form on separate plants. From November to March, tiny yellow flowers appear in segmented, often clustered spikes to 1 1/2 inches long. The stout spikes grow on short stalks from the leaf axils. Male spikes consist of four or five fertile segments, each to 1/2 inch long, with up to forty flowers in three rows. Female spikes usually have three to five fertile segments, each to 3/5 inch long, with up to fifteen flowers. The minute blooms are embedded in the segments. All blooms have three triangular, scalelike petals, the male blooms with a bilobed, stemless anther at the base of each petal, and the female blooms with a one-chamber ovary and a short style with a headlike stigma.

The fruit is a globe-shaped, one-seeded berry, pearly white and hairless, to 1/4 inch long. Maturing in September, the berry contains a mucilaginous pulp surrounding a single seed.

Botanical notes: The genus name combines the Greek *phor* (thief) and *dendron* (tree) because mistletoe steals nutrients from trees. The epithet, *macrophylum*, from the Greek *makros* (large) and *phyllon* (leaf), refers to the leaves. University of Halle botanist Curt Sprengel described the species in 1825, and Italian physician Carlo Guiseppe Luigi Bertero collected the plant in Puerto Rico in 1820. Bertero, ship's doctor and botanical explorer of the West Indies, settled in Chile, but in 1831, on a trip to Tahiti, his ship was lost at sea. Sprengel is best known for his botanical history, *Historia Rei Herbariae* (1807–8), which assessed older plant descriptions in the new Linnaean framework.

276. Canyon Grape (*Vitis arizonica*)
(Other common names: Arizona or Arizona canyon or gulch grape, parra, parra or uva del monte, uva cimarrona)

Habitat: springs, creeks, canyons; sandy, clayey, loamy soils; igneous, limestone substrates.

Geographic range: Texas: Trans-Pecos (excluding Reeves, Pecos, Terrell Counties). US: Texas to Arizona, Utah, Nevada. Mexico: Baja California to Tamaulipas; Durango.

Park locations: lower Monilla Canyon, Las Cuevas, Agua Adentro, Bofecillos Canyon, Ojito Adentro, upper Rancherías Canyon, Arroyo Mexicano, Chorro Canyon, Yedra Canyon

This grape is a deciduous, much branched, woody vine, with slender stems to 20 feet long. The vine scrambles over rocks, shrubs, and trees and climbs with twining tendrils. The stems, often with gray or red, shredding bark, grow from a shallow, weakly developed root system. Young stems may be hairless or ashy gray and woolly-haired.

The alternate green leaves, to 5 inches long, are broadly heart-shaped in outline. On maroon or pink petioles, the blades are weakly lobed or trilobed, coarsely toothed, and densely hairy to hairless. Young leaves are cottony, older leaves longhaired above, thickly shorthaired below.

From spring to summer, tiny fragrant, greenish-white flowers bloom in drooping clusters to 4 inches long, with each cluster opposite a leaf. Male and female blooms usually occur on separate plants, the male plants far outnumbering the female. Each bloom, to 1/8 inch long, has five incurving petals united at the tips. Within the petals are five protruding stamens, with slender filaments and inturned anthers; a nectary disk at the two-chamber ovary base; and a short conic style with a bilobed or headlike stigma. To 2/5 inch wide, the mature grape is purple to black, juicy, tasty but tart, with two to four large bony seeds.

Botanical notes: The genus name is the classical Latin *Vitis* (vine). The species name is the Latin *arizonica* (from Arizona). George Engelmann described this grape in 1868 in *American Naturalist*. Collections by Elliott Coues, Edward Palmer, George Thurber, and Charles Wright were later designated as type specimens. Thurber collected the vine in Sonora, Mexico, in 1851; Wright at Santa Cruz, in Sonora, in 1851; and Dr. Coues and Palmer at Fort Whipple, Arizona, in 1865. In 1872, US Army surgeon-naturalist Coues authored *Key to North American Birds*, an early contribution to bird taxonomy.

277. Orange Caltrop (*Kallstroemia grandiflora*)
(Other common names: Arizona, Mexican, desert, or summer poppy, Arizona caltrop, mal de ojo, abrojo de flor amarillo, baiburín)

Habitat: river, creeks, arroyos, desert scrub, alluvial flats, bajadas, hillsides; limestone, igneous environs.

Geographic range: Texas: Trans-Pecos (excluding Reeves, Terrell, Val Verde Counties); Loving County. US: Texas to California. Mexico: Sonora to Coahuila, Durango; coast (Sinaloa to Chiapas).

Park locations: below South Lajitas Mesa, river from Lajitas to west of Panther Canyon, Cerro de las Burras Loop, Vista de Bofecillos, Palo Amarillo Creek, Las Cuevas, Arroyo Mexicano

Orange caltrop is an annual to 3 feet high, with nearly prostrate to ascending stems. The stems bear a dense mix of silky, rough, sometimes yellow hairs. To 3 inches long, the paired leaves are elliptic, with up to eight pairs of oblong leaflets. The leaves are unequal in size, with the smaller leaf alternating from side to side along the stem. The leaflets bear stiff hairs and silky hairs on the margins.

Blooming profusely after monsoon rains, this caltrop forms extensive displays of showy flowers. From late spring to fall, large solitary flowers, to 1 1/2 inches across, bloom on stalks from the axils of the smaller leaves. Each bloom has five fan-shaped, shallowly scalloped petals, orange with red-orange veins and a yellow throat. Within the throat are ten stamens, with yellow filaments and oblong yellow anthers; a ten-lobed ovary; and a bristle-haired style with a clublike dark stigma. Below the petals are five hairy sepals.

The fruit is a ten-lobed, egg-shaped schizocarp, to 1/5 inch long, with an elongated beak. The bristly fruit splits into ten one-seeded nutlets (mericarps), but the beak remains.

Botanical notes: In 1777, Giovanni Antonio Scopoli, University of Pavia, Italy, professor, named the genus for his friend, Andreas Kallström, Swedish gardener at the royal botanical gardens in Stockholm. The epithet *grandiflora* connotes "large-flowered." In 1852, Asa Gray published a description of this species by John Torrey, citing a collection by Major William Emory in 1846, on the Gila River in Graham County, Arizona. Emory, US Army Corps of Topographical Engineers cartographer, later directed the US-Mexico Boundary Survey.

278. Warty Caltrop (*Kallstroemia parviflora*)
(Other common names: few-flowered or small-flowered caltrop, small-flowered carpetweed, contrayerba)

Habitat: desert, alluvial flats, creeks, canyons, hillsides, disturbed sites; varied soils.

Geographic range: Texas: Trans-Pecos; scattered in much of Texas. US: southwest; Texas to Kansas, east to Mississippi, Missouri, Illinois. Elsewhere: Mexico (Sonora to Nuevo León, Durango south to Guerrero); South America.

Park locations: Contrabando Creek, Fresno Canyon, East Rancherías Trail, Palo Amarillo Creek, Agua Adentro

This caltrop is a prostrate, creeping annual, only 2 inches high. Profusely branching from a central taproot, the stout stems usually bear coarse and silky hairs. Short-petioled, the leaves are oblong, to 2 1/2 inches long, with up to five pairs of elliptic leaflets. The paired leaves are unequal in size, with the smaller leaf alternating from side to side along the stems. The leaflets may be asymmetrical, and hairy.

From spring to fall, single flowers form on stalks from axils of the smaller leaves. Each bloom has five orange or orange-yellow, fanlike petals and a yellow-green throat. The petals are rounded or slightly notched at the tips, darker orange at the base. Within the throat are ten stamens in inner and outer rings, the inner stamens large, fertile, the outer stamens much smaller, sterile. Fertile stamens have orange filaments and elliptic orange anthers with two pollen sacs. Within the stamens is a ten-lobed green ovary and a conic, shorthaired style with a ten-lobed stigma. The style and stigma turn orange red with age. Below the petals are five pale green, narrowly lance-shaped, strongly ciliate sepals.

The fruit is a ten-lobed schizocarp to 1/6 inch long, broadly egg-shaped, with a conic orange-red beak twice as long. The fruit splits into ten one-seeded nutlets, each with a "warty" back, but the beak remains.

Botanical notes: The Latin epithet *parviflora* combines *parvus* (small) and *flos* (flower). In 1898, the species was described by John Bitting Smith Norton, with the Missouri Botanical Garden, from collections by Charles Pollard at Agricultural College, Mississippi, in 1896, and by E. H. Wilkinson at San Antonio, Texas, in 1897. Pollard, assistant curator at the US National Museum, collected in the southeastern states and edited the journal *Plant World*. Wilkinson is known only from plant collections around San Antonio.

279. Perennial Caltrop (*Kallstroemia perennans*)
(Other common names: perennial desert poppy)

Habitat: exposed desert flats, mounds, barren hills, shallow gullies; gypseous clay, limestone rubble.

Geographic range: Texas: Trans-Pecos (Presidio, Brewster, Terrell, Val Verde Counties). US: Texas. Mexico: not recorded.

Park locations: Contrabando Lowlands

Of conservation concern: see Appendix A

Perennial caltrop is a rare endemic known only from a handful of sites on limestone and gypseous clay soils. One of the rarest plants at BBR, this caltrop is a copiously hairy, low perennial to 10 inches high, with several stems growing from stout, semiwoody roots. Ascending or erect, the stems are coated with stiff, sharp, close-pressed hairs.

The paired leaves are elliptic, to 2 inches long, with up to five pairs of oblong or egg-shaped leaflets. Leaf pairs are unequal in size, with the smaller leaf alternating from side to side along the stems. The leaflets may be asymmetrical, with coarse hairs.

From late spring to early fall, often in September, single flowers bloom on broad, densely hairy stalks in the axils of the alternately smaller leaves. To 1 1/2 inches across, each bloom has five orange, fan-shaped petals. Thin, with darker veins, the petals are rounded or shallowly notched at the tips, and reddish orange at the base. Within the flower's throat are ten orange-red stamens, with tapering filaments and large egg-shaped anthers, and a long, thick, dark orange red cylindric style, with a ten-ribbed, oblong, gray-haired stigma. Below the petals are five slender, persistent sepals with white, bristly hairs.

The fruit is a ten-lobed, hairy schizocarp, broadly egg-shaped, to 1/4 inch long, with a longer, hairy cylindric beak. At maturity, the schizocarp splits into ten one-seeded, ridged and pitted, nutlets.

Botanical notes: The species name is the Latin *perennans* (perennial). Louis Williams, Washington University botanist, described the species as *Kallstroemia hirsuta* in 1935, from a 1913 collection by Charles Orcutt at Langtry in Val Verde County. Orchid specialist Williams authored *The Orchidaceae of Mexico* in 1965. San Diego naturalist and publisher of *The West American Scientist*, Orcutt collected plants and shells in the southwest, Mexico, Jamaica, and Haiti. In 1950, on discovery that the name *K. hirsuta* had been used long ago in 1890 to describe another species, University of Texas botanist Billie Turner renamed the species *K. perennans*.

280. Creosote Bush (*Larrea tridentata*)
(Other common names: greasewood, gobernadora, hediondilla, guamis, hedionda, chaparral, falsa alcaparra)

Habitat: desert flats, bajadas, low hills; well-drained soils of sand, gravel, alluvium.

Geographic range: Texas: Trans-Pecos, much of Permian Basin, South Texas Plains. US: southwest (excluding Colorado). Mexico: south as far as San Luis Potosí, Querétaro.

Park locations: Rancherías Canyon, Closed Canyon, Black Hills, Encino Loop Trail, Llano Loop, Contrabando Lowlands

To 6 feet high, creosote is a dominant shrub of southwest deserts, which forms gray to black, leafy stems with enlarged nodes. Radiating from a central base, the slender, limber stems ascend, then often arch and droop. The shrub is known as hediondilla (little stinker) since the resinous foliage is odorous when wet. The resin reflects light, limits water loss, and deters predators.

Much branched, widely spreading, the shallow root system gathers moisture adroitly and produces a toxin to ward off competing plants. Creosote is known as *gobernadora* (governess), due to this ability to control its environs. The world's longest-living shrub, creosote may form large clonal rings, some in the Mojave Desert more than 11,000 years old.

Almost stemless, the yellow-green leaves consist of a pair of resinous leaflets fused basally. To 2/5 inch long, each leaflet is oblong to egg-shaped, slightly hooked and pointed at the tip, and lopsided at the base.

From spring to fall, single yellow flowers form on stalks from the leaf axils. About 1 inch across, each bloom has five inversely egg-shaped petals, rounded or notched at the tip, slim at the base, and oddly twisted. Within the petals are ten stamens with threadlike filaments and oblong, arrowlike anthers; a five-lobed, hairy ovary; and cylindric style with a simple or lobed stigma. The five hairy yellow sepals are egg-shaped, unequal in size, ephemeral. The fluffy fruit is a five-lobed round capsule, 1/4 inch wide, woolly-haired, tipped by the style. The capsule splits into five one-seeded nutlets with many oblong seeds.

Botanical notes: In 1800, José Cavanilles, director of the Madrid botanical gardens, named this genus for Spanish clergyman Juan Antonio Hernández Pérez de Larrea. Larrea, dean of Zaragosa Cathedral and Bishop of Valladolid (1802–3), was a leading Enlightenment figure. The epithet *tridentata* refers to a three-toothed scale at the base of each filament. The species was described in 1824 by Geneva botanist Augustin de Candolle, as *Zygophyllum tridentatum*, based on an illustration from the Sessé and Mociño New Spain expedition (1787–1803). Frederick Coville, first director of the US National Arboretum, transferred the species to the *Larrea* genus in 1893.

281. Goathead (*Tribulus terrestris*)
(Other common names: puncture vine, puncture weed, caltrop, bullhead, Mexican or Texas sandbur, devil's thorn, tackweed, abrojo, cadillo, torito, toboso)

Habitat: sandy flats, roads, disturbed sites.

Geographic range: Texas: Trans-Pecos, much of Texas. US: nearly throughout. Elsewhere: Canada, Mexico, South America; native to southern Europe, Mediterranean.

Park locations: Contrabando Movie Set, lower Contrabando Creek, Sauceda, Llano Loop

Goathead is a prostrate, profusely branched, creeping annual to 3 feet long, with brown or maroon stems radiating from a deep central taproot.

This naturalized, highly invasive annual spreads rapidly to form extensive mats. Both stems and leaves bear long silky and shorter, rough hairs. To 2 inches long, oval in outline, the leaves are compound, with three to six leaflet pairs. One of each pair is alternately smaller along the stems. Often, the dark green, oblong leaflets are slightly lopsided at the base.

Blooming from spring to fall, small yellow, five-petal single flowers emerge from the axils of the smaller leaves on hairy stalks. Often lighter in color near the edges, the yellow petals are spatulate and rounded at the tips. Within the petals are ten yellow stamens, in two rings of five each; a five-lobed, thickly hairy green ovary; and a style with a five-branched, pale green stigma. Below the petals are five pale green, egg-shaped sepals, white on the margins, and fuzzy-haired.

The fruit is a woody bur to 2/5 inch wide, five-lobed, green, tan, yellow brown or reddish brown with age. The mature bur splits into five wedge-shaped, three- to five-seeded nutlets, each with rigid, hornlike spines, and small prickles, warts, and tubercles. Goathead's common names characterize these injurious burs. Goathead is now sold as a supplement to increase testosterone and sexual performance.

Botanical notes: The Latin word *tribulus* (three-pointed) refers to a caltrop, an iron ball with spikes, used on battlefields to obstruct the enemy. The term was applied to plants with spiny fruits. The species was named *terrestris* (terrestrial) for its prostrate habit. Linnaeus described goathead in 1753, indicating a habitat in southern Europe and citing Casper Bauhin, Flemish botanist Mathias de l'Obel, and others as sources. L'Obel, physician to William the Silent, Prince of Orange, later botanist to King James I of England, had named and illustrated *T. terrestris* as early as 1581, in his *Plantarum seu stirpium icones*.

APPENDIX A

Plants on Texas Parks and Wildlife Department Rare, Threatened, and Endangered Species list

July 2019

#	Common Name	Botanical Name	Global Rank	State Rank
1	Littleleaf moonpod	*Acleisanthes parvifolia*	vulnerable	vulnerable
2	Chihuahuan ringstem	*Anulocaulis leiosolenus* var. *lasianthus*	vulnerable	imperiled
3	Creeping rockvine	*Bonamia repens*	vulnerable	imperiled
4	Turner's thistle	*Cirsium turneri*	vulnerable	vulnerable
5	Texas claret-cup cactus	*Echinocereus coccineus* var. *paucispinus*	vulnerable	vulnerable
6	Graybeard cactus	*Echinocereus viridiflorus* var. *canus*	critically imperiled	critically imperiled
7	Mariposa cactus	*Echinomastus mariposensis*	vulnerable	imperiled
8	Boke's button cactus	*Epithelantha bokei*	vulnerable	vulnerable
9	Bushy wild buckwheat	*Eriogonum suffruticosum*	imperiled	imperiled
10	Fringed monkeyflower*	*Erythranthe chinatiensis*	critically imperiled	critically imperiled
11	Desert pincushion cactus	*Escobaria dasyacantha*	vulnerable	vulnerable
12	Duncan's pincushion cactus	*Escobaria duncanii*	critically imperiled	critically imperiled
13	Silverlace cactus	*Escobaria sneedii* var. *albicolumnaria*	imperiled	imperiled
14	Perennial Sandmat	*Euphorbia perennans*	vulnerable	vulnerable
15	Mimicking matspurge	*Euphorbia simulans*	vulnerable	vulnerable
16	Narrowleaf fendlerbush	*Fendlera linearis*	vulnerable	critically imperiled
17	Hairy hedeoma	*Hedeoma mollis*	vulnerable	vulnerable
18	Texas purplespike	*Hexalectris warnockii*	imperiled	imperiled
19	Havard's ipomopsis	*Ipomopsis* (now *Dayia*) *havardii*	vulnerable	vulnerable
20	Warnock's justicia	*Justicia warnockii*	vulnerable	vulnerable
21	Perennial caltrop	*Kallstroemia perennans*	critically imperiled	critically imperiled
22	Boquillas lizardtail	*Oenothera boquillensis*	vulnerable	imperiled
23	Desert night-blooming cereus	*Peniocereus greggii* var. *greggii*	imperiled	imperiled

#	Common Name	Botanical Name	Global Rank	State Rank
24	Slimlobe rockdaisy	*Perityle dissecta*	imperiled	imperiled
25	Trans-Pecos maidenbush	*Phyllanthopsis arida*	imperiled	critically imperiled
26	Havard's plum	*Prunus havardii*	vulnerable	vulnerable
27	Hinckley's oak	*Quercus hinckleyi*	imperiled	imperiled
28	Durango yellowcress	*Rorippa ramosa*	imperiled	critically imperiled
29	Lyreleaf twistflower	*Streptanthus carinatus* ssp. *carinatus*	vulnerable	vulnerable
30	Texas thelypody	*Thelypodium texanum*	vulnerable	vulnerable
31	Fleshy tidestromia	*Tidestromia carnosa*	vulnerable	imperiled
32	Gyp daisy	*Xylorhiza wrightii*	vulnerable	vulnerable

**E. chinatiensis* is a new species, segregated from *E. dentiloba*, only found in Presidio County in the United States. Global and state ranks are set to the state rank of *E. dentiloba*, formerly *Mimulus dentilobus*.

Note

All plants above appear on TPWD's list of *Species of Greatest Conservation Need in Texas*. The list was developed for TPWD's revised Texas Conservation Action Plan through expert consultation and public feedback. Species are ranked using a conservation status system established by NatureServe. NatureServe ranks are based on multiple criteria including range extent, known occurrences, abundance, and threats. The global and state ranks are available online at explorer.natureserve.org/.

Definition of Global Rank

Critically imperiled—At very high risk of extinction or elimination due to very restricted range, very few populations or occurrences, very steep declines, very severe threats, or other factors.

Imperiled—At high risk of extinction or elimination due to restricted range, few populations or occurrences, steep declines, severe threats, or other factors.

Vulnerable—At moderate risk of extinction or elimination due to a fairly restricted range, relatively few populations or occurrences, recent and widespread declines, threats, or other factors.

Definition of State Rank

Critically Imperiled—At very high risk of extirpation in the jurisdiction due to very restricted range, very few populations or occurrences, very steep declines, severe threats, or other factors.

Imperiled—At high risk of extirpation in the jurisdiction due to restricted range, few populations or occurrences, steep declines, severe threats, or other factors.

Vulnerable—At moderate risk of extirpation in the jurisdiction due to a fairly restricted range, relatively few populations or occurrences, recent and widespread declines, threats, or other factors.

APPENDIX B
Other Plants of Possible Conservation Concern in Texas

#	Common Name	Botanical Name	Global Rank	State Rank
1	Mexican gumweed	*Grindelia oxylepis*	not ranked	not ranked
2	Helecho de la Llave	*Llavea cordifolia*	not listed	not listed
3	Coahuila blazingstar	*Mentzelia pachyrhiza*	not ranked	not ranked
4	Foreign cloak fern	*Notholaena aliena*	apparently secure	critically imperiled
5	Scaled cloak fern	*Notholaena aschenborniana*	apparently secure	imperiled
6	Gray's cloak fern	*Notholaena grayi*	apparently secure	imperiled
7	Gregg's cloak fern	*Notholaena greggii*	apparently secure	imperiled
8	Coahuila twintip	*Stemodia coahuilensis*	not listed	not listed
9	Gypsum tansy aster	*Xanthisma gypsophilum*	not ranked	not ranked

*Global and state ranks are from NatureServe.

Mexican gumweed: In the United States, Mexican gumweed is limited to southern Presidio County in Texas and Doña Ana County in New Mexico. In Mexico, its distribution ranges from Chihuahua east to Coahuila and south to Durango, Zacatecas, and San Luis Potosí. The species is not yet ranked by NatureServe. Mexican gumweed was first collected in the United States by J. H. Russell in 1946, near Presidio, Texas, and not identified in the United States again until recently. Richard Spellenberg, botanist at New Mexico State University, discovered the plant in Doña Ana County, New Mexico, in 2014. I have collected and photographed the plant at several locations in BBR since June 15, 2009, but the specimens were not clearly identified, with mature cypselae, as *Grindelia oxylepis* until September 2015. These plants are in the Rio Grande floodplain, between Lajitas and Redford, at periodically inundated sandy or sandy-gravelly sites.

Helecho de la Llave: R. J. O'Kennon and Barton Warnock collected this fern at BBR in 1992, on a moist limestone bluff. The fern has not been relocated in the United States, although many others and I have searched for it extensively at BBR. However, helecho de la Llave ranges through much of Mexico to Guatemala and Costa Rica. The BBR search continues.

Coahuila blazingstar: This stickleaf is only known from southeast Presidio County and southern Brewster County in the United States. Apparently, the plant is uncommon in BBR but more common and widespread in southern Brewster County, mostly at BBNP and Black Gap Wildlife Management Area. In Mexico, Coahuila blazingstar occurs in parts of Chihuahua, Coahuila, and Durango. The species is not yet ranked by NatureServe.

In *Ferns and Lycophytes of Texas* (Diggs and Lipscomb 2014), the following four ferns are considered species of conservation concern in Texas, based on their rareness and limited distribution in the state:

Foreign cloak fern: In the United States, this fern is only known from about six collections in Brewster County (BBNP, Black Gap WMA, Lower Canyons of the Rio Grande), and a few sites in Presidio County (BBR, Chinati Mountains). In BBR, I collected the fern in Chorro Canyon, and on east-facing slopes of Agua Adentro Mountain. In Mexico, foreign cloak fern occurs in Chihuahua, Coahuila, and Tamaulipas.

Scaled cloak fern: In Texas, this fern has a sparse but widely scattered distribution, mostly on limestone, in Culberson, Presidio, Brewster, Pecos, Terrell, Val Verde, and Real Counties. Scaled cloak fern also occurs in southern Arizona and ranges across Mexico, from Sonora east to Tamaulipas, and south to Oaxaca.

Gray's cloak fern: In Texas, the distribution of Gray's cloak fern is scattered and disjunct, across Jeff Davis, Presidio, and Brewster Counties in the Trans-Pecos, and Burnet, Llano, and Uvalde Counties in the Edwards Plateau. Its range extends across New Mexico into southern Arizona, and into Mexico, from Chihuahua east to Nuevo Leon, and south as far as Hidalgo and Jalisco. After ten years of hiking, I have found this fern at only one site in BBR.

Gregg's cloak fern: This mostly Mexican fern is restricted to limestone and gypsum substrates in southern Presidio and southern Brewster Counties in the United States. At BBR, I have collected this species on the slopes of the Solitario rim, on Terlingua Uplift, and in the Contrabando Lowlands near Terlingua Uplift. In Brewster County, the plant has been found in Boquillas Canyon and on Mesa de Anguila in BBNP, and at Black Gap Wildlife Management Area. In northern Mexico, Gregg's cloak fern occurs in Chihuahua, Coahuila, and Durango, and at one site in Nuevo Leon.

Coahuila twintip: A mostly Mexican species, this plantain is a worthy addition to the Texas flora, with relatively recent collections in Presidio, Jeff Davis, and Brewster Counties in the United States. Apparently, the plant was first discovered by Richard Worthington in 1995, in Presidio County, in the "vicinity of Sauceda Ranch," in what is now BBR. In 2004, Billie Turner located the species in silty limestone soils at one site in extreme western Jeff Davis County (Turner 2005). In Brewster County, Coahuila twintip was first identified by Wendy Weckesser at Reed Plateau and later collected by others and me in BBNP, below Stuart's Peak and near Ernst Tinaja. Subsequently, from 2009 to 2012, I encountered Coahuila twintip at three Presidio County locations in BBR. Coahuila twintip is much more widespread in northern Mexico, documented in southeastern Chihuahua, Coahuila, northern Durango, and northern Zacatecas.

Gypsum tansy aster: The range of this aster in Texas is restricted to southeast Presidio County, along or near the Rio Grande, from Rancherías Creek near the head of Colorado Canyon to Big Hill. Numerous specimens have been collected, but they are concentrated at only two general locations. From 2009 to the present, I have encountered a very small number of plants, although they seem to bloom regularly after good rains from late spring to fall. My understanding is that this species was once much more plentiful. In New Mexico, gypsum tansy aster has been collected in Sierra and Socorro Counties, at White Sands Missile Range. In northern Mexico, this species ranges from Chihuahua east to Coahuila and south to Durango and San Luis Potosí.

Other plants may be under some threat in Texas, but secure or apparently secure globally.

All thirty-two BBR plants on the TPWD Rare, Threatened, and Endangered Species List (Appendix A) are ranked by NatureServe as plants of conservation concern globally. NatureServe rates another thirty-seven BBR plants of conservation concern in Texas, but these plants are regarded as secure or apparently secure globally. Four of those are the ferns listed above, one with a state rank of critically imperiled and three with a state rank of imperiled. Of the thirty-three other species, nineteen have a state rank of vulnerable, six have a state rank of imperiled, and eight have a state rank of critically imperiled. Unfortunately, many of these rankings have not been reviewed in recent years.

APPENDIX C

Additions or Revisions to Floral Inventory of Big Bend Ranch.

#	Common Name	Botanical Name	Conservation Status (from NatureServe)	Treated or Possibly Treated Previously As:	Powell Herbarium	Warnock Specimen Inventory, 1992–1993
1	Littleleaf moonpod	Acleisanthes parvifolia	vulnerable globally and in TX		P	
2	Big Bend ringstem	Anulocaulis eriosolenus	apparently secure globally		S, P	WC (Desert Garden)
3	Bear Mountain milkweed	Asclepias scaposa	globally secure		S, P	
4	Big saltbush	Atriplex lentiformis ssp. lentiformis	not ranked globally, not listed in TX		S, P	
5	Dwarf ayenia	Ayenia pilosa	globally secure		S, P	
6	Cosmopolitan bulrush	Bolboschoenus maritimus ssp. paladosus	globally secure		S, P	
7	Creeping rockvine	Bonamia repens	vulnerable globally, imperiled in TX		P	
8	Lemmon's brickellbush	Brickellia lemmonii var. lemmonii	vulnerable globally and in TX		P	
9	Isely's feather duster	Calliandra iselyi	vulnerable globally and in TX	Calliandra conferta	P	
10	Greenleaf five eyes	Chamaesaracha coronopus	globally secure		P	
11	Turner's thistle	Cirsium turneri	vulnerable globally and in TX		P	
12	Watermelon	Citrullus lanatus	exotic		P	
13	Leather flower	Clematis pitcheri var. dictyota	not ranked globally or in TX		P	
14	Sea urchin cactus	Coryphantha echinus var. echinus	apparently secure globally		P	WC (Desert Garden)
15	Big-needle pincushion cactus	Coryphantha macromeris var. macromeris	globally secure		P	WC (Desert Garden)
16	Big Bend croton	Croton bigbendensis	not listed		P	WC (Croton dioicus)
17	Talayote	Cynanchum racemosum var. unifarium	globally secure		P	
18	Yellow nut-grass	Cyperus esculentus var. macrostachyus	globally secure		S, P	
19	New Mexico tickflower	Desmodium neomexicanum	globally secure		P	
20	Thickleaf drymary	Drymaria pachyphylla	globally secure		P	WC (Desert Garden)
21	Earlobe mustard	Dryopetalon auriculatum	apparently secure globally		S	
22	Mariposa cactus	Echinomastus mariposensis	imperiled globally and in TX; THREATENED species		P	
23	Warnock's cactus	Echinomastus warnockii	apparently secure globally		P	

#	Common Name	Botanical Name	Conservation Status (from NatureServe)	Treated or Possibly Treated Previously As:	Powell Herbarium	Warnock Specimen Inventory, 1992–1993
24	Balloonbush	*Epixiphium wislizeni*	apparently secure globally, vulnerable in TX		S, P	
25	Ferriss' scouring rush	*Equisetum xferrissii*	hybrid without conservation value	*Equisetum laevigatum*	S, P	
26	Havard's wild buckwheat	*Eriogonum havardii*	apparently secure globally, vulnerable in TX		P	
27	Texas stork's bill	*Erodium texanum*	globally secure		P	WC (Desert Garden)
28	Fringed monkeyflower	*Erythranthe chinatiensis*	not listed	*Mimulus dentilobus*	S, P	
29	James's monkeyflower	*Erythranthe inamoena*	globally secure	*Mimulus glabratus*	S, P	
30	Duncan's pincushion cactus	*Escobaria duncanii*	critically imperiled globally and in TX		S, P	
31	Abrams' spurge	*Euphorbia abramsiana*	apparently secure globally, not yet listed in TX		S, P	
32	Pointed sandmat	*Euphorbia acuta*	globally secure		S, P	WC (Desert Garden)
33	Muleear spurge	*Euphorbia indivisa*	globally secure		S, P	
34	Spurca spurge	*Euphorbia spurca*	not listed	*Euphorbia theriaca*	P	
35	Hairy spurge	*Euphorbia villifera*	globally secure		P	
36	Powell's heliotrope	*Euploca powelliorum*	not listed	*Heliotropium torreyi*	P	
37	Cluster flaveria	*Flaveria trinervia*	globally secure		P	
38	Red dome blanketflower	*Gaillardia pinnatifida*	globally secure		P	
39	Davis Mountains mock vervain	*Glandularia pubera*	apparently secure globally	*Glandularia wrightii*	P	
40	Mexican gumweed	*Grindelia oxylepis*	not ranked globally or in TX		S, P	
41	Arizona cockroach plant	*Haplophyton cimicidum* var. *crooksii*	globally secure, critically imperiled in TX		P	
42	Swollenstalk sneezeweed	*Helenium amphibolum*	not ranked globally or in TX	*Helenium elegans* var. *elegans*	S, P	
43	Longleaf false goldeneye	*Heliomeris longifolia*	apparently secure globally		S, P	
44	Arizona goldenaster	*Heterotheca arizonica*	not ranked globally or in TX	*Heterotheca fulcrata*	P	

#	Common Name	Botanical Name	Conservation Status (from NatureServe)	Treated or Possibly Treated Previously As:	Powell Herbarium	Warnock Specimen Inventory, 1992–1993
45	Torrey's Tievine	Ipomoea cordatotriloba var. torreyana	apparently secure globally		P	
46	Ivyleaf morning-glory	Ipomoea hederacea	globally secure		S, P	
47	Iron ipomopsis	Ipomopsis laxiflora	globally secure		P	
48	Paleflower ipomopsis	Ipomopsis longiflora var. neomexicana	variety not listed	I. longiflora ssp. longiflora	S, P	
49	Helecho de la Llave	Llavea cordifolia	not listed (not yet relocated in park)			
50	Toothed serviceberry	Malacomeles pringlei	not listed (formerly regarded as M. denticulata)	Amelanchier denticulata	P	
51	Little mallow	Malva parviflora	exotic		S, P	
52	Arrowleaf mallow	Malvella sagittifolia	globally secure		S, P	
53	Hooked water clover	Marsilea vestita	globally secure		S, P	
54	Pyramid flower	Melochia pyramidata	globally secure		P	
55	Mexican stickleaf	Mentzelia mexicana	apparently secure globally, imperiled in TX		S, P	
56	Emory's mimosa	Mimosa emoryana	apparently secure globally		S, P	
57	Texas mimosa	Mimosa texana	globally secure, apparently secure in TX		P	WC
58	Foreign cloak fern	Notholaena aliena	apparently secure globally, critically imperiled in TX		S, P	
59	Scaled cloak fern	Notholaena aschenborniana	apparently secure globally, imperiled in TX		S, P	
60	Gregg's cloak fern	Notholaena greggii	apparently secure globally, imperiled in TX		S, P	
61	Texas toadflax	Nuttallanthus texanus	apparently secure globally		S, P	
62	Mexican navelwort	Omphalodes alienoides	not listed	Omphalodes aliena	P	
63	Diploid purple prickly pear	Opuntia azurea var. diplopurpurea	not listed	Opuntia macrocentra	P	
64	Big Hill prickly pear	Opuntia azurea var. discolor	not listed		P	

#	Common Name	Botanical Name	Conservation Status (from NatureServe)	Treated or Possibly Treated Previously As:	Powell Herbarium	Warnock Specimen Inventory, 1992–1993
65	Comanche prickly pear	*Opuntia camanchica*	vulnerable globally and in TX	*Opuntia phaeacantha*	P	
66	Sweet prickly pear	*Opuntia dulcis*	not listed (regarded as synonym of *O. phaeacantha* var. *phaeacantha*)	*Opuntia phaeacantha*	P	
67	Palmer's cat's eye	*Oreocarya palmeri*	apparently secure globally, vulnerable in TX		S, P	WC (Desert Garden)
68	Rio Grande palafox	*Palafoxia riograndensis*	apparently secure globally, imperiled in TX		S, P	WC
69	African rue	*Peganum harmala*	exotic		P	
70	Creeping cliff brake	*Pellaea intermedia*	globally secure, vulnerable in TX		P	
71	Rio Grande phacelia	*Phacelia infundibuliformis*	globally secure		S, P	
72	Rock phacelia	*Phacelia rupestris*	apparently secure globally		P	
73	Slimleaf bean	*Phaseolus angustissimus*	apparently secure globally		P	
74	Date palm	*Phoenix dactylifera*	not ranked globally, not listed in TX		P	
75	Bigleaf mistletoe	*Phoradendron macrophyllum*	globally secure		P	
76	Lanceleaf groundcherry	*Physalis angulata* var. *lanceifolia*	variety not listed		S, P	
77	Thurber's stemsucker	*Pilostyles thurberi*	globally secure, vulnerable in TX		S, P	
78	Desert indianwheat	*Plantago ovata*	apparently secure globally		P	
79	Redseed plantain	*Plantago rhodosperma*	not ranked globally or in TX		S, P	
80	Silversheath knotweed	*Polygonum argyrocoleon*	exotic		S, P	
81	Prostrate knotweed	*Polygonum aviculare*	exotic		S, P	
82	Parry's holdback	*Pomaria melanosticta*	apparently secure globally, vulnerable in TX		P	
83	Wingpod purslane	*Portulaca umbraticola* var. *lanceolata*	not ranked globally or in TX		S, P	
84	Screwbean mesquite	*Prosopis pubescens*	globally secure		P	WC
85	Red-tip rabbit-tobacco	*Pseudognaphalium luteoalbum*	exotic, not listed in TX		S, P	

#	Common Name	Botanical Name	Conservation Status (from NatureServe)	Treated or Possibly Treated Previously As:	Powell Herbarium	Warnock Specimen Inventory, 1992–1993
86	Annual bastard cabbage	Rapistrum rugosum	exotic		P	
87	Durango yellowcress	Rorippa ramosa	vulnerable globally, critically imperiled in TX		S, P	
88	Hooded arrowhead	Sagittaria montevidensis ssp. calycina	globally secure		P	
89	Rocky Mountain sage	Salvia reflexa	globally secure		P	
90	Common threesquare	Schoenoplectus pungens	globally secure		P	
91	Guajillo	Senegalia berlandieri	apparently secure globally		P	
92	Western sea purslane	Sesuvium verrucosum	globally secure		P	
93	West Indian nightshade	Solanum ptycanthum	globally secure		P	
94	Buffalobur nightshade	Solanum rostratum	globally secure		P	
95	Coahuila twintip	Stemodia coahuilensis	not listed	1995 Worthington collection (SRSC), "vicinity of Sauceda Ranch"	S, P	
96	Pink baby's-breath	Talinum paniculatum	exotic		S, P	
97	Yellow flameflower	Talinum polygaloides	not listed	Talinum aurantiacum	P	WC (Talinum angustissimum)
98	Stemmy four-nerved daisy	Tetraneuris scaposa	apparently secure globally		P	
99	Fleshy tidestromia	Tidestromia carnosa	vulnerable globally, imperiled in TX		S, P	
100	Huisache	Vachellia farnesiana	globally secure		S, P	
101	Water speedwell	Veronica anagallis-aquatica	globally secure		S, P	
102	Gypsum tansy aster	Xanthisma gypsophilum	not ranked globally or in TX	Machaeranthera blephariphylla	P	
103	Threeflower goldenweed	Xylothamia triantha	apparently secure globally, critically imperiled in TX		P	WC (Desert Garden)
104	Buckley's centaury	Zeltnera calycosa	apparently secure globally		S, P	

Note: This table lists 104 species, mentioned in the "Tour" or "Flora" sections of this work, that constitute additions or revisions to two previous floras, the last complete BBR flora (Worthington 1995) and the subsequent Solitario flora (Hardy 2009). Of these species, 103 were photographed by the author in BBR. Helecho de la Llave (*Llavea cordifolia*) was photographed by Carlos Velazco-Macias in Mexico. This plant was collected in BBR but has not been relocated.

Powell Herbarium: "P" indicates that a Morey plant photo, with date and location, has been deposited at the A. Michael Powell Herbarium at Sul Ross State University in Alpine, Texas. "S" indicates that a Morey plant specimen, with date and location, has been deposited at the herbarium.

Warnock Specimen Inventory: In 1992–1993, Jean Hardy Pittman compiled an inventory of Barton Warnock specimens housed at that time at Sauceda. Plants listed in the inventory, and noted here as WC (Warnock Collection), were never incorporated into the subsequent 1995 and 2009 floras. Some collections were from the Desert Garden at Lajitas, and some of these plants were brought to the garden from locations outside BBR.

GLOSSARY OF BOTANICAL AND GEOLOGICAL TERMS

achene	Dry, one-seeded, indehiscent fruit, with seed attached to fruit wall at one point
acorn	One-seeded, thick-walled oval nut seated in cuplike base; fruit of oak
agrostologist	Specialist in scientific study of grasses
alkaline soil	Soil with pH above 7, often with limited nutrients and solubility
alkaloid	Usually alkaline, nitrogenous organic compound, some toxic, such as caffeine, nicotine, morphine
alluvial	Derived from alluvium
alluvial fan	Triangular, fanlike deposits of alluvium
alluvium	Loose soil or sediment deposited by flowing water
alternate	With only one leaf or other structure borne at a node, not opposite or whorled
annual	Plant that completes life cycle in single year
anther	Upper, pollen-bearing part of the stamen
anthocarp	Structure in which part of perianth remains attached to fruit, as in Nyctaginaceae
apogamous	In ferns, forming a sporophyte asexually without fertilization
appressed	Flat against a surface, pressed closely
areole	On cacti stems, a specialized, often raised or depressed area from which hairs, spines, branches, and flowers may grow
attenuate	Tapering gradually to a slender point
awl-shaped	Slender, tapered from base to sharp point
awn	Bristlelike or hairlike projection
axil	Angle between two organs; e.g., upper angle formed by leaf or branch with stem
beak	A firm, pointed, usually tapering, terminal projection, as in a fruit
bearded	With long, stiff hairs
bedrock mortar	Circular depression in solid rock, used to grind food
bentonite	Absorbent clay formed from aging of volcanic ash, composed mostly of smectite
berry	Small, fleshy, pulpy, multiseeded, indehiscent fruit
biennial	Plant that completes life cycle within two years
bipinnate	Refers to pinnately compound leaf, with primary leaflets also pinnately divided
bisexual	With both stamens and pistil(s) functional in the same flower
blade	Of a leaf, the flat, expanded part, excluding the stem

bract	Modified leaf subtending a flower or flower cluster
bractlet	A small or secondary bract
bristle	Stiff, slender hair or hairlike projection
bunchgrass	A grass that grows in clumps
bur	Structure with rough or prickly covering; e.g., the seeds or fruits of some plants
calcareous	Rich in calcium carbonate; a soil type often derived from limestone
caldera	Crater formed by collapse of a volcanic structure, usually caused by eruption
caliche	A layer of calcium carbonate on or in soils of arid regions
calyx	Usually green outer whorl of a flower; collectively, the sepals
capsule	Dry fruit composed of more than one carpel, splitting open at maturity
carpel	female ovule-bearing sex organ of a flower; a simple pistil or member of compound pistil
caryopsis	Dry, one-seeded, indehiscent fruit, with seed coat fused to the fruit wall
catkin	Usually slender, drooping spike of unisexual flowers that lack petals
central spine	Usually longer, broader, stouter spine from center of the areole
cheilanthoid	Relating to ferns in Cheilanthoideae subfamily of Pteridaceae, common in southwest US and other arid environs
chert	Hard, fine-grained sedimentary rock composed mostly of silica (SiO_2)
chlorophyll	A green pigment in plants, used to absorb energy from light and make carbohydrates
cholla	Cactus with cylindrical stem segments
cilia	Marginal hairs
ciliate	With marginal hairs
clasping	With leaf base surrounding or partly surrounding the stem
claw	Narrowed base of some petals, sepals, or bracts
cleistogamous	Refers to self-pollinating flowers, often small, unopened, or not fully open
clone	Genetically identical, vegetatively propagated plant
column	Columnar structure of united filaments (e.g., Malvaceae), or united filaments and style (e.g., Orchidaceae)
compound	Composed of two or more similar parts
cone	Reproductive structure with axis and spirally organized cluster of sporophylls or scalelike parts (bearing spores in ferns, pollen or seeds in seed plants)
conglomerate	Coarse-grained sedimentary rock, with larger, rounded fragments cemented within finer-grained particles
crest	Ridge on surface or summit of a plant part
crown	Persistent base of plant; also, canopy of tree; or in Asclepiadaceae, the corona
culm	In grasses and sedges, the usually hollow aerial stem
cyathium	In Euphorbiaceae, a cluster of unisexual flowers within a cup-shaped involucre
cypsela	Dry, one-seeded, indehiscent fruit in Asteraceae, with an adherent calyx, derived from an inferior ovary
deciduous	Falling seasonally, or in drought
dehiscent	Opening at maturity, e.g., to release seeds

dike	Hardened intrusion filling fissures in usually softer, existing rock
diploid	With two sets of chromosomes in the nucleus of a sporophyte cell, one from each parent
disjunct	Occurring in two or more widely separated geographic regions
disk	Outgrowth of receptacle; central part of flower head in Asteraceae
disk floret	Usually small flower borne in central part or disk of a flower head in Asteraceae
dolomitic	Containing dolomite, a mineral rich in magnesium and calcium carbonate
dome	Spherical protrusion of rock strata, caused by compression and uplift
dorsiventral	With different upper and lower surfaces, e.g., some stems, leaves
drupe	Indehiscent fruit with fleshy outer layer and hard inner layer around a single seed
drupelet	Small drupe, e.g., in *Rubus*
ellipsoidal	Having a surface with all plane sections ellipses or circles
elliptic	Shaped like an ellipse; oval in outline, widest in middle, tapering equally to ends
endemic	Restricted to or native to a specific geographic region
endocarp	Innermost layer of the fruit wall
epithet	Species, subspecies, or variety name
evergreen	With leaves throughout the year; not deciduous
exocarp	Outermost layer of the fruit wall
exotic	Not native, naturalized, introduced
exudate	Oozing of fluid, e.g., from plant pores
family	Group of related organisms, consisting of one or more genera
farina	Mealy exudate secreted by glandular hairs
farinose	Covered with farina
faulting	Fracture in Earth's crust caused by movement of rock on either side
filament	Stalk of the stamen, supporting the anther; a threadlike or hairlike structure
flatiron	Steep, up-tilted, triangular landform
floral cup	Cuplike perianth; or hypanthium, fused basal parts of sepals, petals, and stamens
floret	Small or diminutive flower, usually in a dense flower cluster, as in Asteraceae
folding	Bending of Earth's crust due to pressure and temperature
follicle	Dry, single-chambered fruit that splits open lengthwise along one seam
frond	Leaf of a fern, including blade and petiole
fruit	Seed-bearing organ of a flowering plant, formed by mature ovary and accessory tissue after flowering
gametophyte	Gamete-producing organism, alternating with sporophyte, spore-producing organism in plant's life cycle
genus (pl. genera)	Grouping of one or more related species
gland	Organ that secretes a fluid
glaucous	With whitish, bluish, or silvery waxy coating or bloom
globose	Spherical
glochid	In Cactaceae, minute, fine, barbed, modified spine or bristle

glochidiate	With a barbed tip
glume	In grasses, one of usually two bracts at spikelet base
granite	Coarse-grained, light-colored igneous rock, composed mostly of feldspars and quartz
gyp	Gypsum
gypsum	A sedimentary rock; or soft, chalklike mineral of hydrated calcium sulfate
habit	General appearance of a plant
head	Dense cluster of stemless or short-stemmed florets
herbaceous	Herblike; not woody; often with stems dying back at end of growing season
herbarium	Collection of preserved plant material; place maintaining such a collection
holotype	Specimen or illustration forming the basis of an original species description
hood	In milkweed, part of the flower's corona, one of five hollow arched structures
horn	In milkweed, a slender hornlike appendage exserted from the hood
hybrid	Cross between two species
igneous	A product of volcanic activity, formed by solidification of molten rock
indehiscent	Not splitting open at maturity
indusia	In ferns, outgrowths covering the sporangia
inflated	Bladdery, swollen, expanded
inflorescence	Flower cluster; or the typical arrangement of flowers on a plant
introduced	Not native; brought into a region
intrusion	Movement of magma beneath the Earth's crust into overlying strata
invasive	Non-native to the ecosystem, capable of economic or environmental damage
involucre	Ring of bracts, below or around a flower cluster
joint	Point of articulation; a node on a stem
keel	Longitudinal ridge, like the keel of a boat; two lower, united petals of papilionaceous flower
laccolith	Igneous intrusion into sedimentary rock, forming domelike structure
lanceolate	Longer than wide, broadest toward the base, tapering to the tip, lance-shaped
latex	The often white, milky liquid or sap produced by some plants
lax	Loose, open, not compact, not rigid
leaflet	Ultimate segment of compound leaf
lectotype	Specimen or illustration designated as the type in the absence of a holotype
legume	Dry fruit or seedpod, typical of Fabaceae; plant in Fabaceae family
lemma	Lower bract enclosing a grass floret
lenticel	Usually lens-shaped pore in bark of woody plant
ligule	In grasses, small straplike appendage at junction of leaf sheath and blade
limb	Upper, expanded part of corolla with fused petals
limestone	Sedimentary rock, mostly calcite, often derived from marine organisms
linear	Very narrow, with parallel sides
lip	Upper or lower division of a two-lipped flower or calyx; larger petal of an orchid

lobe	Often rounded part of an organ; e.g., a leaf segment separated by sinuses extending no more than halfway to the leaf midrib
megasporangia	Plant structures in which only megaspores are formed
megaspore	Larger spore that produces the female gametophyte
mericarp	Usually one-seeded segment resulting from division of a fruit (schizocarp) at maturity
mesic	With moderate moisture
microsporangia	Plant structures in which only microspores are formed
microspore	Smaller spore that produces the male gametophyte
midrib	Of a leaf, the central vein
monocline	Steplike fold or dip in otherwise flat or gently sloping rock strata
morphologist	Student of the form and structure of organisms
mucronate	Ending in a short, sharp point
naturalized	Established in a non-native region
nectar	Sugary, nutrient-rich secretion from the nectary glands of plants
nectary	Nectar-secreting plant gland
net-veined	Reticulated; with a network of veins
node	Point on stem where leaves, branches, or flowers grow
novaculite	In BBR, silica-rich crystalline chert with high fossil content, formed 400 Ma
nut	Hard, dry, one-seeded, indehiscent fruit
nutlet	Small nut; seed covered by a stony layer; in some plants, a one-seeded section of a fruit
oblanceolate	Inversely lanceolate, with broadest part above the middle
oblong	Several times as long as wide with roughly parallel sides
odd-pinnate	Pinnate leaf with single, terminal leaflet
opposite	On opposing sides, as in opposite sides of the stem; two at a node
oval	Broadly elliptic
ovary	Basal ovule-bearing portion of the pistil, which becomes the fruit
ovate	Egg-shaped in outline, widest end at base
ovoid	Egg-shaped, with widest end at base, three-dimensional
pad	Flattened stem of prickly pear cactus
palea	Upper bract enclosing a grass floret
panicle	Compound raceme; any loosely branched, complex inflorescence
papilionaceous	Butterfly-shaped; having a flower with banner, wings, and keel petals
pappus	Modified calyx of an Asteraceae floret, composed of scales, bristles, or hairs, borne at the fruit (cypsela) tip
parasite	Organism growing on or in, and obtaining nourishment from, a host
pepo	Fleshy, many-seeded fruit of Cucurbitaceae, with hard rind
perennial	A plant living for several years or more
perianth	Calyx and corolla together
petal	Usually colorful part of corolla, surrounding stamens and pistils

petiole	Leaf stalk
photosynthesis	Process by which green plants use energy from sunlight to convert carbon dioxide and water into food
phyllaries	In Asteraceae, bracts subtending the flower head
phylogenetic	Relating to the evolutionary history of a group of related species
pictograph	Ancient or prehistoric rock painting or drawing
pila	Water storage tank
pinna (pl. pinnae)	Primary division of a pinnate leaf
pinnate	With divisions arranged on each side of common axis; e.g., with leaflets on common rachis of pinnate leaf
pinnatifid	Pinnately lobed
pinnule	Secondary division of a pinnate leaf
pistil	Female, ovule-bearing flower organ, usually composed of ovary, style, and stigma
pistillate	Female; bearing pistils but no functional stamens
pod	Dry fruit or seedpod that splits open when mature; legume
pollen	Powdery, grainy mass containing male reproductive cells, produced in anthers of flowers or male cones of seed plants
pome	In some Rosaceae, an indehiscent fruit with several seed chambers, and outer fleshy layer formed from base of flower (floral cup)
prickle	Hard, sharp, spiny, often curved outgrowth
prickly pear	Cactus with flattened stem joints
pteridologist	Student of ferns and related plants
pubescence	Covering of short, soft, erect hairs
pustulate	With pustules or blister-like swellings
raceme	Simple flower cluster, with flowers on unbranched lateral stalks (pedicels) off the main stem, the oldest at the base
rachis	Main axis of compound leaf or inflorescence
radial spine	Spine radiating from the edge of a spine cluster, mostly parallel to stem surface
ray floret	Outer, petal-like, strap-shaped corolla (ligule), typically on perimeter of central disk in Asteraceae
recurved	Curving backward or downward
reflexed	Abruptly turned backward or downward
resin	Sticky, gummy substance
resinous	Containing resin
revolute	Rolled downward or backward; e.g., leaf margins curled under toward the midrib
rhizome	Usually horizontal underground stem, producing shoots and roots at the nodes, and scale leaves
rhyolite	Extrusive, silica-rich igneous rock
rib	Longitudinal ridge; e.g., rib on cactus stem, or prominent vein of leaf
rim sill	A rhyolite, the oldest igneous rock exposed in the Solitario (37.5 Ma)
rootstock	Rhizome; root part used in propagation
rosette	Usually circular, crowded basal cluster of radiating leaves

saline	Containing salt
samara	Dry, winged, indehiscent fruit
saponin	Naturally occurring, foam-producing, plant glycoside, often toxic
scale	Small, thin, usually flat, elongate outgrowth; e.g., reduced or vestigial leaf
scape	Leafless flower stalk rising from ground level
schizocarp	Dry fruit splitting into two or more one-seeded chambers (mericarps) when ripe
scorpioid	Of a cymose inflorescence with a coiled axis, blooms opening downward from tip
scrub	Dense shrub vegetation, often stunted by lack of moisture
scurfy	Bearing small, loose, whitish, bran-like scales or crust on the surface
segment	Subdivision of an organ; part of divided leaf or frond
sepal	Usually green, leaflike segment of the calyx, beneath the petals
serrate	With small, sharp teeth pointing forward along the margin
serrated	Saw-toothed, with forward-pointing teeth
serrulate	Finely serrate
shale	A mudstone; fissile, fine-grained, layered rock of compacted silt and claylike sediment
sheath	Thin, tubular structure surrounding an organ; e.g., lower part of leaf enclosing the stem in some grasses and sedges
shrub	Large, woody, perennial plant, smaller than a tree, usually with multiple stems
shrublet	Small or dwarf shrub
silt	Fine, loose, sedimentary particles larger than clay, smaller than sand
slickrock	Smooth, barren, weathered rock polished by wind and water
spatulate	Spatula-shaped; wider and rounded at tip, narrowed to base
species	Division of a genus, with related individuals usually only reproducing among themselves
specimen	Part of an individual used as an example of its species
spike	Unbranched, elongated cluster of stalkless flowers
spikelet	Small spike; part of grass flower cluster, with usually two glumes and one or more florets
spine	Sharp-pointed, stiff, typically woody projection, a modified leaf
sporangia	Plural of sporangium
sporangium	Structure in which spores are formed
spore	Small, usually one-celled, reproductive structure; e.g., of ferns or other nonflowering plants
sporocarp	In some ferns, specialized multicellular structure containing sporangia
sporophyll	Spore-bearing leaf or modified leaf
sporophyte	Spore-producing generation, alternating with gamete-producing generation, in plant's life cycle
spur	Short branch or branchlet; slender extension at base of petal or sepal
stamen	Male flower organ bearing pollen in the anther; filament and anther collectively
staminate	Male; bearing stamens but no functional pistils
stellate	Star-shaped, with branches radiating from a central point
stem	Plant axis, with nodes from which leaves, branches, and flowers grow
sterile	Barren, infertile; unable to develop reproductive organs

stigma	Part of the pistil of a flower that receives the pollen, usually at tip of style
stipe	Stalk, such as petiole of fern leaf
stipule	One of usually a pair of leaflike appendages at the base of a petiole
stolon	Long thin, prostrate stem, forming roots and often shoots at the nodes
strata	Distinct layers of sedimentary rock or soil
strigose	With stiff, straight, appressed hairs
style	Tubular upper portion of pistil above the ovary and below the stigma
subshrub	Small shrub; short woody plant, often with woody base and annual stems
subspecies	Classification intermediate between a species and a variety
substrate	Underlying stratum; surface on which an organism grows
succulent	Fleshy, juicy, containing or storing water
symbiotic	Refers to dependent relation between two organisms in proximity, usually beneficial to both
syntype	Any of two or more specimens cited in an original taxon description if no holotype was designated; or any of two or more specimens originally designated as types
talus	Typically, small rocks or rock fragments at base of cliff
taproot	Main, dominant root, descending vertically, from which secondary roots may arise
taxon	A unit of taxonomy; e.g., family, genus, species, subspecies, variety
taxonomist	Specialist in taxonomy
taxonomy	Study of the identification, naming, and hierarchical classification of life forms
tendril	Slender, coiling structure, modified from stem or leaf, that enables climbing or attachment for support
tepal	A term used to refer to petals and sepals of a flower (perianth), when they are not clearly distinguishable
ternate	In threes; e.g., a leaf with three leaflets
thorn	Stiff, sharp-pointed, woody structure; a modified stem
throat	Opening of a flower with fused or united petals
tinaja	Small surface pocket or depression in bedrock, which may fill with water
tomentose	Densely covered with short, soft, matted hairs
tooth	Small marginal, usually pointed lobe, often referring to edge of leaf or petal
tree	Woody plant usually with one main trunk, usually larger than a shrub
tuber	Enlarged, usually underground, modified stem, often used for storage of nutrients and/or vegetative reproduction
tubercle	Small, rounded, wartlike protrusion on surface of organ
tuff	Cooled, hardened, compacted ash and debris, a product of volcanic eruption
tuft	Cluster, clump
type	Herbarium specimen or illustration to which a taxon name is permanently attached
ultimate segment	Smallest leaf subdivision
umbel	Flat or domelike flower cluster, with flowers on stalks of roughly equal length from a common point
unisexual	Male or female; flowers with only functional stamens or functional pistils
utricle	Small, bladderlike, thin-walled, one- seeded, usually indehiscent fruit

valve	One of the parts into which a dehiscent capsule splits at maturity
variety	Subdivision of a species or subspecies
vein	Threadlike ridge in a leaf; a strand of vascular tissue which transports nutrients
villous	Covered with long, soft, not matted, hairs
viscid	Adhesive; sticky
volcanic	Relating to or product of a volcano
whorl	Circular arrangement of three or more organs borne from a common point; e.g., a whorl of leaves
wing	Thin, membranous extension of a plant part; in some plant families, one of two lateral petals
winged	With a wing
woolly	With long, soft, closely interwoven or matted hairs
xeric	With limited moisture

GLOSSARY OF SPANISH TERMS

Spanish Term	English Translation
Acebuches	Desert olive (*Forestiera angustifolia*), common in Acebuches Canyon
Agarito	Spanish name, probably from *agrito*, diminutive of *agrio* (sour)
Agua Adentro	Water within/inside; may also refer to water in backcountry, away from river
Aguila	Eagle
Aguja	Needle or sharp peak
Alamito	Little cottonwood
Alamo	Cottonwood
Alamo de Cesario Creek	Named for Cesaria, Mexican girl captured by Apaches and tied to a cottonwood tree, later rescued by relatives
Alto	High
Arenosa	Sandy
Arroyo	Small, seasonally dry drainage
Auras	Black vultures
Bandidos	Bandits
Baños de Leyva	Leyva baths
Bebederos	Water troughs
Blanca	White
Bofecillos	Perhaps small lungs; possibly from "bosquecillos," little forests
Boludo	Bald, round
Bosque	Woodland; in arid southwest United States, often applied to river floodplain forests of mesquite, cottonwood, willow, etc.
Botella	Bottle
Buena suerte	Good luck
Burras	Female donkeys
Camino Viejo	Old road; probably older alignment of Marfa-Lajitas road
Candelilla	*Euphorbia antisyphilitica*, little candle
Carrizo	Cane; in Big Bend, usually applied to giant reed (*Arundo donax*)
Casa Piedra	Rock house; community north of park on Alamito Creek, settled 1883
Cerro	Hill

Chilicote	Ranch hands' name for Lindheimer's senna (*Senna lindheimeri*); term also applied to *Cucurbita foetidissima*, likely from chilicayote (squash species)
Chillicothe	Erroneous spelling of chilicote
Chinati	Possibly from Apache word for mountain pass, or from chanate (grackle)
Chiqueras	Baby goat shelters
Chorro	Rapid flow or gush of water; place where water issues
Chupadero	Seep or very small spring; also hummingbird
Cibolo Creek	Bison creek
Ciénaga, ciénega	Marsh
Cinco Tinajas	Five water basins in bedrock
Colorado	Red
Contrabandista	Smuggler
Contrabando	Contraband
Cuesta	Steep grade or inclined mesa
Cueva Larga	Long cave
Cuevas Amarillas	Yellow caves
Damianita	Colloquial name for *Chrysactinia mexicana*; diminutive of damiana (*Turnera diffusa*), a shrub with major distribution in Mexico
Dos Pilas	Two water storage tanks
El Despoblado	The unpopulated place
El Padre al Altar	The priest at the altar
Elephante	Elephant; misspelling of Spanish "elefante"
Encino	Oak
Entradas	Entries; a reference to the arrival of the Spanish in the region
Escobilla	Brush
Escondido	Hidden
Fresno	Ash
Guale	Nickname of Guadalupe Carrasco, who ran sheep on Guale Mesa, early 1900s
Guayacan	Colloquial name for *Guaiacum angustifolium*; tree of life
Guayule	Colloquial name for *Parthenium argentatum*; from Nahuatl word for rubber
Helecho de la Llave	Llave's fern; named for Pablo de la Llave, Mexican naturalist
Iglesia	Church
La Guitarra	The guitar
La Junta de los Rios	The junction of the rivers (Rio Grande and Rio Concho)
La Mota	The mound, hillock
La Posta	The post
Lajitas	Diminutive of lajas, little flat rocks of the Boquillas Formation
Las Cuevas	The caves
Lechuguilla	Colloquial name for *Agave lechuguilla*, diminutive of lechuga (lettuce)

León	Panther
Leyva	Name of family that lived in Leyva Canyon
Llano	Plain
Los Cuates	The twins
Los Hermanos	The brothers
Los Portales	The openings or portals, referring to cavernous overhangs at entrance
Madera	Timber
Manos Arriba	Hands up
Mejorana	Marjoram; mejorana (*Lantana achyranthifolia*) is called Mexican marjoram
Melón Loco	Crazy melon; Valery Havard, *1885 Report on the Flora of Western and Southern Texas*: "the fruit is considered poisonous by the Mexicans"
Mesa de Anguila	Eel mesa, perhaps a gringo error for aguila (eagle)
Monilla	Colloquial name for Mexican buckeye (*Ungnadia speciosa*)
Morita	Little mulberry
Mulato	Mixed race person
Nopalera	Prickly pear patch
Nuevo	New
Ocotillo	Colloquial name for *Fouquieria splendens*; diminutive of ocote (pine), from Nahuatl ocotyl (torch). Wax at base of stem is highly flammable.
Ojito adentro	Little spring within/inside, or in backcountry away from river
Ojitos	Small springs
Ojo	Spring (literally, eye)
Ojo de vibora	Snake's eye, colloquial name for *Evolvolus alsinoides*
Oreganillo	*Aloysia wrightii*, little oregano. Oregano is common name of *Origanum vulgare* and *Lippia graveolens*
Oreja de perro	Dog's ear, colloquial name for *Tiquilia canescens*
Oso	Bear
Palo amarillo	Yellow wood; colloquial name for agarito
Papalote	Windmill
Papalotito	Little windmill
Paso	Pass, also ford or crossing
Paso Lajitas	Lajitas crossing or ford on Rio Grande
Peñasco	Crag, rocky outcrop
Pico	Peak
Pico de Fiero	Peak of iron, a reference to Solitario Peak
Pila	Water storage tank
Pila de Gato	Cat tank; refers to a mountain lion at this site
Polvo	Dust
Presidio	Fortified settlement
Prieto	Black or dark

Primero	First
Rancherías	Small ranches or Native American encampments
Rancho Viejo	Old ranch
Redford	Named after a nearby river ford, Vado Colorado (red ford)
Redondo	Round
Retama	Broom, colloquial name for *Parkinsonia aculeata*
Rincon	Corner, out-of-the-way nook
Sabino	Salt cedar
Santana Mesa	Named for a Comanche chief
Sauceda	Later name for Saucita Ranch
Saucita	Little willow; derived from sauz (willow)
Seco	Dry
Sierra	Mountain range
Sierra Rica	Mountain range named rica (rich) because of mineral riches or piñon
Solitario	Hermit, loner
Sotol	Colloquial name for *Dasylirion*
Tanque	Earthen stock tank
Tapado	Covered, filled, impassable
Tascate	Juniper
Terlingua	Possibly from tres lenguas (three tongues), referring to three languages (English, Spanish, Native American), or to three forks in Terlingua Creek
Terneros	Heifer calves
Tinajas	Water basins in bedrock
Tortolo	Turtledove
Tres Papalotes	Three windmills
Vado Colorado	Red ford; Redford is named for this nearby ford
Yedra	Misspelling of hiedra (poison ivy)

A NOTE ON SOURCES

This work belongs to me, and to those who preceded me in recording the human and natural history of Big Bend Ranch, to the botanical collectors who discovered the plants, and to the botanists who have subsequently described them. I want to acknowledge their contributions.

Introduction and Tour of Big Bend Ranch State Park (BBR) by Physiographic Region

Park Guides

These introductory sections draw heavily on a wealth of resources available from Texas Parks and Wildlife Department (TPWD), many of them cited on the BBR website: tpwd.texas.gov/state-parks/big-bend-ranch:

Big Bend Ranch State Park Road Guide (Nored 2014a) documents major attractions on the main park road to Sauceda, the backcountry park headquarters. *Roads to Nowhere* (Riskind and Sholly 2009) introduces visitors to the 70 miles of challenging, unmaintained 4WD roads. *Big Bend Ranch Biking Guide: The Other Side of Nowhere* (Phillips et al., 2010) is a compendium of biking routes ranging in length from five miles to 107 miles. *Campsites of the Big Bend Ranch State Park* (Nored 2014b) describes fifty-two campsites scattered throughout the park. *El Solitario*, a free TPWD periodical, offers information on BBR visitor services, facilities, and activities, with articles on hiking and biking trails, horseback riding, camping, river access, park attractions, and park history.

Geology

Geology of Big Bend Ranch State Park, Texas (Henry 1998) is my primary source of information on park geology. *River Road Vistas: A Journey along the River Road* (MacLeod 2008) is especially helpful in describing the topography and geology of the Rio Grande Corridor Physiographic Region. *Big Bend Ranch State Park: Geology at the Crossroads* (Hall 2013), by my friend and sometimes hiking companion Blaine Hall, provides a guide to the four major geological trends affecting BBR over the last 500+ million years. *In and Out of Time: Geologic Hike of Los Portales Shutup and the Righthand Shutup, El Solitario, Big Bend Ranch State Park, Texas* (Hall and Hunt 2015), a thrilling introduction to park geology, describes a day-long drive through Fresno Canyon and a hike through two Solitario canyons.

Archaeology

I am indebted to four extensive archaeological surveys of BBR, not only for their contributions to the park's human history but also for their description of physiographic areas, natural resources, geology, and vegetation.

Archeological Reconnaissance on Big Bend Ranch State Park, Presidio and Brewster Counties, Texas, 1988–1994 (Ing, 1996a and b), documents cultural resources throughout BBR and records 179 archaeological sites. *Archeological Reconnaissance of Upper Fresno Canyon Rim, Big Bend Ranch State Park, Texas* (Sanchez 1999) covers an area from Los Alamos to Chorro Canyon and records forty-six archaeological sites. *Archeological Survey of Select Boundary and Power Line Segments: Big Bend Ranch State Park, Presidio County, Texas* (Ohl and Cloud 2001) covers an area along the park's main power line and segments along the park boundary and records thirty-nine archaeological sites. *Trails through Time: Archeological Reconnaissance of Selected Trail Corridors, Big Bend Ranch State Park, Presidio and Brewster Counties, Texas, 2004–2010* (Roberts, Gibbs, and Gibbs 2017) surveys 117 miles of park trails, and records 159 previously unknown archaeological sites.

Natural Area Surveys

In 1976, studies published by the Lyndon B. Johnson School of Public Affairs surveyed, mapped, and photographed four major natural areas of BBR. The studies cover human and natural history, geology, archaeology, flora, and fauna. Although dated, these natural area surveys of *The Solitario* (No. 9), *Fresno Canyon* (No. 10), *Colorado Canyon* (No. 11), and *Bofecillos Mountains* (No. 12) remain relevant and informative today. The historical survey by Bruce Saunders; the geology by Dwight Deal; the vegetational surveys of

Mary Butterwick, Jim Lamb, and Stuart Strong; and the archaeological reconnaissances of Barbara Baskin and William Hudson informed my work.

Other Area Studies

El Camino Del Rio — The River Road, FM 170 from Study Butte to Presidio and through Big Bend Ranch State Park (Alloway 1995) offers a wealth of historical, topographical, and geological information on park locations along the Rio Grande Corridor, from Lajitas to Presidio. *The Upper Canyons of the Rio Grande: Presidio to Terlingua Creek, Including Colorado Canyon and Santa Elena Canyon* (Aulbach and Gorski 2000), a guide for canoeists, includes interesting historical, topographical, and geological notes on attractions along the Rio Grande. *The Recreation Potential of Chorro Canyon, Presidio County, Texas* (McKann 1975), a master's thesis, provides insights into the natural, cultural, and scenic resources of Chorro Canyon not available elsewhere.

Big Bend Ranch Plant Descriptions

Botanical Plant Names

The scientific names (genus + specific epithet) of flowering plants used in this work coincide largely with those in *Flowering Plants of Trans-Pecos Texas and Adjacent Areas* (Powell and Worthington 2018). The scientific names of ferns coincide with those in *The Ferns and Lycophytes of Texas* (Diggs and Lipscomb 2014), but *Cheilanthes* species were moved to the *Myriopteris* genus based on new phylogenetic studies.

Common Plant Names

Common plant names cited in the two works mentioned above are also often cited in this study. Other common names appear in a multitude of other sources, many in the three popular wildflower works of Barton Warnock (*Wildflowers of the Big Bend Country, Texas*, 1970; *Wildflowers of the Guadalupe Mountains and the Sand Dune Country, Texas*, 1974; and *Wildflowers of the Davis Mountains and Marathon Basin, Texas*, 1977), and some from an earlier study of Robert Vines (*Trees, Shrubs, and Woody Vines of the Southwest*, 1960). Other sources, especially of Spanish common names, include the online website SEINet Portal Network (swbiodiversity.org/seinet/index.php), an Arizona–New Mexico gateway to a network of herbaria, museums, and agencies that provide plant data; and the Mexican online website, CONABIO (www.conabio.gob.mx/malezasdemexico/2inicio/home-malezas-mexico.htm), which contains photos and fact sheets for over 700 weed species.

Plant Habitat

The habitats listed for individual plants are a function of my field work at BBR and a review of herbaria data, but numerous other sources were consulted as well. The most important are *Flowering Plants of Trans-Pecos Texas and Adjacent Areas* (Powell and Worthington 2018); the Flora of North America (FNA) online website (www.efloras.org); *Manual of the Vascular Plants of Texas* (Correll and Johnston 1979); *The Flora of the Chihuahuan Desert Region* (Henrickson and Johnston, prepublication copy 2004); and for ferns, *The Ferns and Lycophytes of Texas* (Diggs and Lipscomb 2014).

Geographic Range of Plants: Texas

The primary sources consulted on plant distributions across Texas counties include: *Flowering Plants of Trans-Pecos Texas and Adjacent Areas* (Powell and Worthington 2018); *Atlas of the Vascular Plants of Texas* (Turner et al. 2003); the online USDA PLANTS Database (plants.sc.egov.usda.gov/java/); *Manual of the Vascular Plants of Texas* (Correll and Johnston 1979); and for ferns, *The Ferns and Lycophytes of Texas* (Diggs and Lipscomb 2014).

Geographic Range of Plants: US

The primary sources consulted on plant distributions across the United States include the sources listed above for Texas (except *Atlas of the Vascular Plants of Texas*), and in addition: the Flora of North America (FNA) online website (www.efloras.org); NatureServe Explorer's, *An Online Encyclopedia of Life* (explorer.natureserve.org/); and the online website SEINet Portal Network and associated herbaria (swbiodiversity.org/seinet).

Geographic Range of Plants: Mexico and Elsewhere

The primary sources consulted on plant distributions in Mexico and elsewhere include the sources listed above for the United States (except the online USDA PLANTS Database and NatureServe Explorer's *An Online Encyclopedia of Life*), and in addition, mostly for Mexican plants: *The Flora of the Chihuahuan Desert Region* (Henrickson and Johnston, prepublication copy 2004); the Mexican online website, CONABIO (conabio.gob.mx/malezasdemexico/2inicio/home-malezas-mexico.htm), and *The Pteridophytes of Mexico* (Mickel and Smith 2004).

Park Locations

Nearly all of the park plant locations cited are locations where I photographed plants in the field. In a few cases, the listed sites include locations documented in *A Floral*

Inventory of the Big Bend Ranch State Natural Area, Presidio and Brewster Counties, Texas (Worthington 1995), or in *Flora and Vegetation of the Solitario Dome, Brewster and Presidio Counties, Texas* (Hardy 2009).

Conservation Status

The principal sources of information on conservation status of plants are Texas Parks and Wildlife Department's *Rare, Threatened, and Endangered Species of Texas by County, July 17, 2019* (tpwd.texas.gov/gis/rtest/); and NatureServe Explorer's *An Online Encyclopedia of Life* (explorer.natureserve.org/).

Plant Morphology

Information on plant parts, blooming periods, and flower colors is partly a function of my examination of plants in the field, specimens, and close-up plant photographs. I also consulted a wealth of botanical data from other sources: first and foremost, the many works of Dr. A. Michael Powell on trees, shrubs, cacti, and ferns, culminating in *Flowering Plants of Trans-Pecos Texas and Adjacent Areas* (Powell and Worthington 2018); and second, the Flora of North America (FNA) online website (www.efloras.org).

For an earlier book, *Little Big Bend: Common, Uncommon, and Rare Plants of Big Bend National Park* (Morey 2008), I had relied on some older botanical works: *Trees, Shrubs, and Woody Vines of the Southwest* (Vines 1960); *Manual of the Vascular Plants of Texas* (Correll and Johnston 1979); Barton Warnock's three wildflower books (*Wildflowers of the Big Bend Country, Texas*, 1970; *Wildflowers of the Davis Mountains and Marathon Basin, Texas*, 1977; *Wildflowers of the Guadalupe Mountains and the Sand Dune Country, Texas*, 1974); and *The Flora of the Chihuahuan Desert Region* (Henrickson and Johnston, prepublication copy 2004). I may be attached at the hip to these studies, and I continued to review them for information on BBR plants.

On ferns, I regularly perused *The Ferns and Lycophytes of Texas* (Diggs and Lipscomb 2014), my new fern Bible, as well as the earlier work, *Ferns and Fern Allies of the Trans-Pecos and Adjacent Areas* (Yarborough and Powell 2002). Also helpful were fern descriptions, as well as geographic range data, in *The Pteridophytes of Mexico* (Mickel and Smith 2004).

Grasses of the Trans-Pecos and Adjacent Areas (Powell 1994) was my principal source on grasses. A more recent work with many good close-up plant photos, *Field Guide to Common Texas Grasses* (Hatch, Umphres, and Ardoin 2015), covers many grasses that occur at BBR.

Rare Plants of Texas (Poole et al. 2004) was consulted on BBR plants included in that work.

Plant Uses

The many works of A. Michael Powell, the three wildflower books of Barton Warnock, and Robert Vine's *Trees, Shrubs, and Woody Vines of the Southwest*, all previously mentioned, are the most frequently referenced sources of information on human uses of plants. Countless other references were searched, especially for the early use of plants by Native Americans, and many are listed in the "Literature Cited" section. I often referred to *Native American Medicinal Plants: An Ethnobotanical Dictionary* (Moerman 2009), a wonderful starting point for further research.

Botanical Notes

Origin and Meaning of Botanical Names

The origin and meaning of Latin plant names are often unclear, expert views may strongly differ, and commonly accepted opinions may change with time. Consequently, some authors avoid this subject, but the subject has always intrigued me. I relied most heavily on the work of A. Michael Powell, especially *Flowering Plants of Trans-Pecos Texas and Adjacent Areas* (Powell and Worthington 2018), and the earlier work of Robert Vines (*Trees, Shrubs, and Woody Vines of the Southwest* 1960), for answers. For ferns, *The Ferns and Lycophytes of Texas* (Diggs and Lipscomb 2014) was helpful.

Yet, I reviewed countless other sources for disparate opinions. *The Names of Plants* (Gledhill 2002), *Dictionary of Plant Names* (Coombes 1985), and *A Dictionary of Common Wildflowers of Texas and the Southern Great Plains* (Holloway 2005) were regularly consulted. Online websites, Michael Charters's *California Plant Names: Latin and Greek Meanings and Derivations, A Dictionary of Botanical and Biographical Etymology* (calflora.net/botanicalnames/), and Dave's Garden's *Botanary: The Botanical Dictionary* (davesgarden.com/guides/botanary/), were also informative. Where opinions differed, I went to original sources (Linnaeus, etc.) or often old botanical journals. Some names remain a mystery.

First Descriptions and Collections of Plants

Online herbaria websites were the primary source for information on botanists, botanical collectors, and the locations where type specimens were collected. Websites most frequently consulted were the Missouri Botanical Garden's *Tropicos* (www.Tropicos.org); the Harvard University Herbaria and Libraries' *Collections*: Index of Botanical Specimens (kiki.huh.harvard.edu/databases/specimen_index.html); and the New York Botanical

Garden's *C. V. Starr Virtual Herbarium* (http://sweetgum.nybg.org/science/vh/).

Often, the locations of Charles Wright's plant collections were difficult to determine because of confusion between the field numbers that Wright assigned to specimens and the distribution numbers that were later assigned to the specimens. A painstaking study by Elizabeth Shaw, with the Harvard University Herbaria, *Charles Wright on the Boundary, 1849–1852; or, Plantae Wrightianae Revisited* (Shaw 1987) was a godsend in helping to resolve these issues.

Biographical Information on Botanists and Botanical Collectors

This is my hobby. Over many years, I have compiled a biographical database on botanists and botanical collectors, with research from a wide array of disparate sources. No single source was especially helpful for my purposes, but some entertaining works that I would recommend to readers are *Naturalists of the Frontier* (Geiser 1948); *Pioneer Naturalists: The Discovery and Naming of North American Plants and Animals* (Evans 1993); *One Hundred and One Botanists* (Isely 1994); and *A Wild Flower by Any Other Name: Sketches of Pioneer Naturalists Who Named Our Western Plants* (Nilsson 1994).

LITERATURE CITED

Abrams, L., and R. S. Ferris. 1923–60. *An Illustrated Flora of the Pacific States: Washington, Oregon, and California.* Stanford, CA: Stanford University Press.

Adanson, M. 1763. *Familles des plantes.* Paris: Vincent.

Aiton, W. T. 1812. *Hortus Kewensis; or, A Catalogue of the Plants Cultivated in the Royal Botanic Garden at Kew.* London: Longman, Hurst, Rees, Orme, and Brown.

Alloway, D. 1995. *El Camino Del Rio — The River Road, FM 170 from Study Butte to Presidio and through Big Bend Ranch State Park.* Austin: Texas Parks and Wildlife Department.

Ames, O. 1905–22. *Orchidaceae: Illustrations and Studies of the Family Orchidaceae.* Illustrated by B. Ames. Boston: Houghton, Mifflin.

Anderson, W. R., and C. C. Davis. 2007. "Generic Adjustments in Neotropical Malpighiaceae." *Contributions from the University of Michigan Herbarium* 25: 163, f. 14.

Anthony, M. 1956. "The Opuntiae of the Big Bend Region of Texas." *American Midland Naturalist* 55, no. 1: 225–56.

Arruda da Cámara, M. 1836. "Pamphlets Translated by Koster, H." *Travels in Brazil.* London: Longman, Hurst, Rees, Orme, and Brown.

Aulbach, L. F., and L. C. Gorski. 2000. *The Upper Canyons of the Rio Grande: Presidio to Terlingua Creek, Including Colorado Canyon and Santa Elena Canyon.* Houston: Wilderness Area Map Service.

Austin, D. F. 2010. *Baboquivari Mountain Plants: Identification, Ecology, and Ethnobotany.* Tucson: University of Arizona Press.

Backeberg, C., and F. M. Knuth. 1935. *Kaktus—ABC: En haandbog for fagfolk og amatorer.* Copenhagen: Gyldendal.

Bauhin, J., and J. H. Cherler. 1650–51. *Historia Plantarum Universalis.* Yverdon: D. Chabrey and F. L. Graffenried.

Benson, L. 1941. "The Mesquites and Screw-Beans of the United States." *American Journal of Botany* 28, no. 9:748–54.

Benson, L., and R. A. Darrow. 1954. *The Trees and Shrubs of the Southwestern Deserts*, illustrated by L. B. Hamilton. Tucson: University of Arizona Press.

Bentham, G. 1839–57. *Plantas Hartwegianas.* London: G. Pamplin.

———. 1844. In *The Botany of the Voyage of H.M.S. Sulphur, under the Command of Captain Sir Edward Belcher... during the Years 1836–42*, by R. B. Hinds. London: Smith, Elder.

Berlandier, J. L., and R. Chovell. 1850. *Diario de viaje de la Comision de Limites...* Mexico City: Tip. de J. Navarro.

Boyd, C. E. 2013. In *Painters in Prehistory: Archaeology and Art of the Lower Pecos Canyonlands*, edited by H. J. Shafer. San Antonio: Trinity University Press.

Bradbury, J. 1817. *Travels in the Interior of America in the Years 1809, 1810, and 1811.* Liverpool: Printed by Smith and Galway, published by Sherwood, Neely, and Jones, London.

Brandimarte, C. 2011. "The Madrid House: Squatters Ruled." *El Solitario*, Texas Parks and Wildlife Department, Big Bend Ranch State Park (Winter–Spring).

Breyne, J. 1674. *Exoticarum aliarumque minus cognitarum planatrum centuria prima* [First century of exotic plants]. Danzig: David-Fridericus Rhetius.

Britton, N. L., and J. N. Rose. 1922. *The Cactaceae: Descriptions and Illustrations of Plants of the Cactus Family.* Vol. 3. Washington, DC: Carnegie Institution of Washington.

Browne, P. 1756. The *Civil and Natural History of Jamaica.* London: T. Osborne and J. Shipton in Gray's-Inn.

Brune, G. 1981. *Springs of Texas.* Vol. 1, Part 4: 374. Fort Worth: Branch-Smith.

Bryan, K. B. 2011. *Birds of Big Bend Ranch State Park and Vicinity: A Field Checklist.* Natural Resource Program, Texas Parks and Wildlife Department.

Burman, J. 1737. *Thesaurus Zeylanicus: Exhibens Plantas in Insula Zeylana Nascentes...* Amsterdam: Janssonio Waesbergios and Salomonem Schouten.

Burnett, J. P. 2002. *Trans-Pecos Texas: A Study of Exploration.* Rockport, TX: Christopher Blum / Artworx Graphic Design.

Canales, M., T. Hernandez, J. Caballero, A. Romo de Vivar, G. Avila, A. Duran, and R. Lira. 2005. "Informant Consensus

Factor and Antibacterial Activity of the Medicinal Plants Used by the People of San Rafael Coxcatlan, Puebla, Mexico." *Journal of Ethnopharmacology* 97: 429–39.

Castetter, E. F. 1935. "Ethnobiological Studies in the American Southwest I. Uncultivated Native Plants Used as Sources of Food." *University of New Mexico Bulletin* 4, no. 1: 1–44.

Castetter, E. F., and M. E. Opler. 1936. *The Ethnobiology of the Chiricahua and Mescalero Apache. A. The Use of Plants for Foods, Beverages, and Narcotics.* Ethnobiological Studies in the American Southwest 3. Albuquerque: University of New Mexico Bulletin, Biological Series 4(5).

Cavanilles, A. J. 1791–1801. *Icones et descriptiones plantarum.* Madrid: Typographia Regia.

Charters, M. L. 2019. *California Plant Names: Latin and Greek Meanings and Derivations, a Dictionary of Botanical and Biographical Etymology.* calflora.net/botanical-names/.

Cloud, W. A., and R. J. Mallouf. 1996. "Material Culture: Discussion, Temporal and Cultural Affiliation of Artifacts." *Archeological Reconnaissance on Big Bend Ranch State Park, Presidio and Brewster Counties, Texas, 1988–1994.* Alpine, TX: Center for Big Bend Studies, Sul Ross State University.

Columella, L. J. M. 1st century AD *De Re Rustica.* English translation by H. B. Ash. *Lucius Junius Moderatus Columella on Agriculture.* Cambridge, MA: Harvard University Press, 1960.

CONABIO (La Comisión Nacional para el Conocimiento y Uso de la Biodiversidad). conabio.gob.mx/malezasde mexico/2inicio/home-malezas-mexico.htm.

Cook, W. 2020. *History of DDRS* (Dallquist Desert Research Station). Midwestern State University. msutexas.edu/academics/scienceandmath/ddrs/history.php.

Coombes, A. J. 1985. *Dictionary of Plant Names.* Portland, OR: Timber Press.

Correll, D. D., and M. C. Johnston. 1979. *Manual of the Vascular Plants of Texas.* Richardson: University of Texas at Dallas.

Corry, C. E., E. Herrin, F. W. McDowell, and K. A. Phillips. 1990. *Geology of the Solitario, Trans-Pecos Texas.* Geological Society of America, Special Paper 250.

Cory, V. 1937. "Some New Plants from Texas." *Rhodora* 39: 419–20.

Cory, V. L., and H. B. Parks. 1937. *Catalogue of the Flora of the State of Texas.* Texas Agricultural Experiment Station Bull. 550. College Station.

Coues, E. 1872. *Key to North American Birds.* Salem: Salem Press.

Coulter, J. M. 1891–94. *Botany of Western Texas: A Manual of the Phanerograms and Pteridophytes of Western Texas.* Washington, DC: Government Printing Office.

Coulter, J. M., revised by A. Nelson. 1909. *New Manual of Botany of the Central Rocky Mountains (Vascular Plants).* New York: American Book.

Coville, F. V. 1895. "The Botanical Explorations of Thomas Coulter in Mexico and California." *Botanical Gazette* 20, no. 12: 519–31.

Daugherty, F. W. 1972. "The Terlingua Mercury District." In *Geology of the Big Bend Area, Texas.* West Texas Geological Society Publication, 72–59.

Davenport, G. 1879. In *Ferns of North America*, by D. C. Eaton. Salem: S. E. Cassino, Naturalist's Agency.

Dave's Garden. 2005. *Botanary: The Botanical Dictionary.* davesgarden.com/guides/botanary/.

Davis, W. T., and M. A. Davis. 1984. "Marfa Pays Tribute to Its Pioneers." *Big Bend Sentinel*, February 16.

De Candolle, A. P. 1828. *Prodromus systematis naturalis regni vegetabilis.* Paris: Treuttel and Wurtz.

Deal, D. 1976a. "The Geologic Environment of Colorado Canyon of the Rio Grande, Southeastern Presidio County, Texas and Chihuahua, Mexico." In *Colorado Canyon: A Natural Area Survey, No. 11.* Lyndon B. Johnson School of Public Affairs. University of Texas at Austin.

———. 1976b. "The Geologic Environment of Fresno Canyon, Southeastern Presidio County, Texas." In *Fresno Canyon: A Natural Area Survey, No. 10.* Lyndon B. Johnson School of Public Affairs. University of Texas at Austin.

———. 1976c. "The Geologic Environment of the Bofecillos Mountains, Southeastern Presidio County, Texas." In *Bofecillos Mountains: A Natural Area Survey, No. 12.* Lyndon B. Johnson School of Public Affairs, University of Texas at Austin.

———. 1976d. "The Geologic Environment of the Solitario, Brewster and Presidio Counties, Texas." In *The Solitario: A Natural Area Survey,* No. 9. Lyndon B. Johnson School of Public Affairs, University of Texas at Austin.

Decaisne, J. 1840. In *Plantas Hartwegianas*, by G. Bentham. London: G. Pamplin.

Delesalle, V. A. 1989. "Year-to-Year Changes in Phenotypic Gender in a Monoecious Cucurbit, *Apodanthera undulata.*" *American Journal of Botany* 76, no. 1: 30–39.

Dietrich, D. N. F. 1831–54. *Flora Universalis in Colorirten Abbildungen.* Jena: August Schmid.

Diggs, G. M., Jr., and B. L. Lipscomb. 2014. *The Ferns and Lycophytes of Texas.* Fort Worth: Botanical Research Institute of Texas.

Dioscorides, P. 50–70 CE. *De Materia Medica.*

Don, G. 1837. *A General History of the Dichlamydeous Plants.* London: J. G. and F. Rivington.

Douglas, D. 1839. In *Flora Boreali-Americana; or, The Botany of the Northern Parts of British America*, by W. J. Hooker. London: H. G. Bohn.

Douglas, N. A., and P. S. Manos. 2007. "Molecular Phylogeny of Nyctaginaceae: Taxonomy, Biogeography, and Characters Associated with a Radiation of Xerophytic Genera in North America." *American Journal of Botany* 94, no. 5: 856–72.

Duncan, F., and Daughter. 1946–47. "Collectors and Rockhounds: What Frank Duncan Hasn't Got His Daughter Has." *Harry Oliver's Desert Rat Scrap Book* (Winter): 4.

———. 1947. "Calcite Crystal Impressions in Agate: Perfection of Specimen Determines Price." Box 63. Terlingua, Texas. *Desert Magazine* 10, no. 5: 41. 25 Cents.

Dunmire, W. W., and G. D. Tierney. 1995. *Wild Plants of the Pueblo Province: Exploring Ancient and Enduring Uses*. Santa Fe: Museum of New Mexico Press.

Durst, B. 2015. "Discover Chorro Vista Loop on the West Fresno Rim." *El Solitario*. Texas Parks and Wildlife Department, Big Bend Ranch State Park.

Elliott, S. 1824. *A Sketch of the Botany of South Carolina and Georgia*. Charleston: J. R. Schenck.

Engelmann, G. 1848. In *Memoir of a Tour to Northern Mexico: Connected with Col. Doniphan's Expedition, in 1846 and 1847*, by F. A. Wislizenus. Washington, DC: Tippin and Streeper.

———. 1856. "Synopsis of the Cactaceae of the Territory of the United States and Adjacent Regions." *Proceedings of the American Academy of Arts and Sciences*. Vol. 3. Boston: Metcalf.

———. 1857. In *Pacific Railroad Surveys: Reports of Explorations and Surveys, to Ascertain the Most Practicable and Economical Route for a Railroad from the Mississippi River to the Pacific Ocean*. Washington, DC: Beverley Tucker.

———. 1868. "The North American Grapes." *American Naturalist* 2, no. 6: 320–22.

———. 1882. "*Yucca elata*, n. sp., *Y. macrocarpa*, n. sp." *Botanical Gazette* 7: 17.

Evans, H. E. 1993. *Pioneer Naturalists: The Discovery and Naming of North American Plants and Animals*. New York: Henry Holt.

Felger, R. S., and M. B. Moser. 1974. "Seri Indian Pharmacopoeia." *Economic Botany* 28, no. 4: 414–36.

Ferguson, D. J. 1989. "Revision of the US Members of the *Echinocereus triglochidiatus* Group." *Cactus and Succulent Journal* 61: 217–24.

Fewkes, J. W. 1896. "A Contribution to Ethnobotany." *American Anthropologist* 9: 14–21.

Flora of North America (FNA). 2019. efloras.org/flora_page.aspx?flora id=1.

Forsskål, P., and C. Niebuhr. 1775. *Flora Aegyptiaco-Arabica*. Copenhagen: Möllerus, typographer, published by Heineck and Faber.

Fuchs, L. 1542. *De historia stirpium commentarii insignes* [Notable commentaries on the history of plants]. Basel: Michael Isingrin.

Galeotti, H. 1848. In *Illustrations and Descriptions of Blooming Cacti. Abbildung undBeschreibung blühender Cacteen*, by L. K. G. Pfeiifer and C. F. Otto. N.p.

Gardeners' Chronicle: A Weekly Illustrated Journal of Horticulture and Allied Subjects. 1877, p. 724. G. T. Warren, Balcombe Place. London: Bradbury, Agnew.

Geiser, S. W. 1948. *Naturalists of the Frontier*. Dallas: Southern Methodist University.

Gentry, Howard Scott. 1998. *Gentry's Rio Mayo Plants: The Tropical Deciduous Forest and Environs of Northwest Mexico*. Revised and edited by P. S. Martin, D. Yetman, M. Fishbein, P. Jenkins, T. R. Van Devender, and R. K. Wilson. Tucson: University of Arizona Press.

Gerow, D. 2011. "Today's Madrid House." *El Solitario* (Winter–Spring). Texas Parks and Wildlife Department, Big Bend Ranch State Park.

Gibbs, T. 2016. "Archaic People of the Big Bend: Desert Survivalists." *El Solitario*. Texas Parks and Wildlife Department, Big Bend Ranch State Park.

Gledhill, D. 2002. *The Names of Plants*. Cambridge: Cambridge University Press.

Gray, A. 1849. *Plantae Fendlerianae Novi-Mexicanae: An Account of a Collection of Plants Made Chiefly in the Vicinity of Santa Fe, New Mexico, by Augustus Fendler*. Memoirs of the American Academy of Arts and Sciences 4: 1–116.

———. 1850. In *Plantae Lindheimerianae: An Enumeration of F. Lindheimer's Collection of Texan Plants . . .*, by G. Engelmann and A. Gray. Boston: Freeman and Bolles.

———. 1852. *Plantae Wrightianae texano-neo-mexicanae: An Account of a Collection of Plants Made by Charles Wright . . .* Washington, DC: Smithsonian Institution.

———. 1854. *Plantae Novae Thurberianae: The Characters of Some New Genera and Species of Plants, in a Collection Made by George Thurber, Esq., of the Late Mexican Boundary Commission, Chiefly in New Mexico and Sonora*. Cambridge: Metcalf.

———. 1878–86. *Synoptical Flora of North America*. New York: Ivison, Blakeman, and Taylor.

———. 1882. Cited by W. B. Hemsley in *Biologia Centrali-Americana: Contributions to the Knowledge of the Fauna and Flora of Mexico and Central America*, by F. D. Godman and O. Salvin. London: Published for the editors by R. H. Porter.

Gray, A., and I. Sprague. 1849. *Genera florae Americae boreali-orientalis illustrata: The Genera of the Plants of the United States Illustrated by Figures and Analyses from Nature.* Boston: J. Munroe.

Gregg, J. *Commerce of the Prairies.* 1844. New York: Henry G. Langley.

Greene, E. L. 1899. *Grindelia oxylepis. Pittonia* 4: 42.

———. 1903a. *Clematis dictyota. Pittonia* 5: 133.

———. 1903b. *Mimulus inamoenus. Pittonia* 5: 137.

Grusz, A. 2013. "*Myriopteris windhamii* sp. nov., a New Name for *Cheilanthes villosa* (Pteridaceae)." *American Fern Journal* 103, no. 2: 112–17.

Grusz, A., and M. Windham. 2013. "Toward a Monophyletic *Cheilanthes*: The Resurrection and Recircumscription of *Myriopteris* (Pteridaceae)." *PhytoKeys* 32: 49–64.

Hall, B. R. 2013. *Big Bend Ranch State Park: Geology at the Crossroads.* Austin: Texas Parks and Wildlife Department.

Hall, B. R., and B. B. Hunt. 2015. *In and Out of Time: Geologic Hike of Los Portales Shutup and the Righthand Shutup, El Solitario, Big Bend Ranch State Park, Texas.* Texas Parks and Wildlife Department.

Hampson, J. 2015. *Rock Art and Regional Identity: A Comparative Perspective.* Walnut Creek, CA: Left Coast Press.

Hardy, J. E. 2009. *Flora and Vegetation of the Solitario Dome, Brewster and Presidio Counties, Texas.* Houston: Iron Mountain Press.

Hartung, T. 2016. *Cattail Moonshine and Milkweed Medicine: The Curious Stories of 43 Amazing North American Native Plants.* North Adams: Storey Publishing.

Harvard University Herbaria and Libraries (HUH). 2019. *Collections: Index of Botanical Specimens.* kiki.huh.harvard.edu/databases/specimen index.html.

Hatch, S. L., K. C. Umphres, and A. J. Ardoin. 2015. *Field Guide to Common Texas Grasses.* College Station: Texas A&M University Press.

Havard, V. 1885. "Report on the Flora of Western and Southern Texas." *Proceedings of the US National Museum* 8: 449–533. Washington, DC.

———. 1896. "Drink Plants of the North American Indians." *Bulletin of the Torrey Botanical Club* 23, no. 2: 33–46.

Heil, K. D., and S. Brack. 1988. "The Cacti of Big Bend National Park." *Cactus and Succulent Journal* 60: 17–34.

Heller, A. A. 1895. *Botanical Explorations in Southern Texas during the Season of 1894.* Contributions from the Herbarium of Franklin and Marshall College, No. 1. Lancaster: New Era Printing House.

Henrickson, J., and M. C. Johnston, eds. 2004. *The Flora of the Chihuahuan Desert Region.* Unpublished copy.

Henry, C. D. 1996. "Igneous Geology of the Solitario." In *Geology of the Solitario Dome, Trans-Pecos Texas*, edited by C. D. Henry and W. R. Muehlberger. Bureau of Economic Geology, the University of Texas at Austin.

———. 1998. *Geology of Big Bend Ranch State Park, Texas.* Bureau of Economic Geology, University of Texas at Austin.

Hermann, P. 1698. *Paradisus batavus.* Edited by W. Sherard. Lugduni Batavorum: Impensis Viduae. Apud Abrahamum Elzevier, Academiae Typographum.

Hernández, T., M. Canales, J. G. Avila, A. Duran, J. Cabarello, A. Romo de Vivar, and R. Lira. 2003. "Ethnobotany and Antibacterial Activity of Some Plants Used in Traditional Medicine of Zapotitlan de Las Salinas, Puebla (Mexico)." *Journal of Ethnopharmacology* 88: 181–88.

Hester, J. P. 1939. "New Species of Cacti." *Cactus and Succulent Journal* 10: 179–82.

———. 1941a. *Escobaria albicolumnaria* Hester. *Desert Plant Life* 13: 129.

———. 1941b. *Escobessaya duncanii* Hester. *Desert Plant Life* 13: 192.

———. 1945. *Echinomastus mariposensis* Hester. *Desert Plant Life* 17: 59–60.

Holloway, J. E. 2005. *A Dictionary of Common Wildflowers of Texas and the Southern Great Plains.* Fort Worth: Texas Christian University Press.

Hooker, W. J., and R. K. Greville. 1830. *Icones Filicum . . . Figures and descriptions of ferns, principally of such as have been altogether unnoticed by botanists, or as have not yet been correctly figured.* London: Prostant Venales Apud Treuttel et Würtz, Treuttel Fil. et Richter.

Hooker, W. J. 1858. *Species Filicum.* Vol. 2. London: William Pamplin.

Hoyt, C. 2002. "The Chihuahuan Desert: Diversity at Risk." *Endangered Species Bulletin* 27, no. 2: 16–17.

———. 2015. *Galleries: Student Work; Stark and Beautiful Images of the Chihuahuan Desert.* matadornetwork.com/trips/stark-beautiful-images-chihuahuan-desert/.

Humboldt, A. de, and A. Bonpland. 1808. *Voyage de Humboldt et Bonpland: Sixieme Partie, Botanique. Plantes Equinoxiales.* Paris: F. Schoell.

Hunter, K. L., J. L. Betancourt, B. R. Riddle, T. R. Van Devender, K. L. Cole, and W. G. Spaulding. 2001. Ploidy Race Distributions since the Last Glacial Maximum in the North American Desert Shrub, *Larrea tridentata*." *Global Ecology and Biogeology* 10: 521–33.

Ing, J. D. 1996a. "Environmental Setting." In *Archeological Reconnaissance on Big Bend Ranch State Park, Presidio and Brewster Counties, Texas, 1988–1994.* Center for Big Bend Studies. Alpine, TX: Sul Ross State University.

———. 1996b. "The Reconnaissance." In *Archeological Reconnaissance on Big Bend Ranch State Park, Presidio and*

Brewster Counties, Texas, 1988–1994. Center for Big Bend Studies. Alpine, TX: Sul Ross State University.

Isely, D. 1994. *One Hundred and One Botanists*. Ames: Iowa State University Press.

Jacquin, N. J. 1786–96. *Collectanea ad Botanicam, Chemiam, et Historiam Naturalem Spectantia, cum Figuris*. Vindobonae: Ex Officina Wappleriana.

———. 1797–1804. *Hortus Schoenbrunnensis: Plantarum rariorum horti caesarei Schoenbrunnensis descriptiones et icones*. Vienna: C. F. Wappler.

Jones, D. E. 1968. "Comanche Plant Medicine." *Papers in Anthropology* 9: 1–12. University of Oklahoma.

Justice, J. 1754. *The Scots' Gardener's Director*. Edinburgh.

Kaempfer, E. 1712. *Amoenitatum Exoticarum Politico-Physico-Medicarum Fasciculi V*. Lemgo: Henrici Wilhelmi Meyeri.

Kalm, P. 1770–71. *Travels into North America*. English translation by J. R. Forster. Warrington: W. Eyres.

Katzer, G. 2012. "Gernot Katzer's Spice Pages: Garden cress (*Lepidium sativum* L.), Water cress (*Nasturtium officinale* L.), and Nasturtium (*Tropaeolum majus* L.)." gernot-katzers-spice pages.com/engl/Lepi_sat.html.

Kelley, J. C. 1986. *Jumano and Patarabueye: Relations at La Junta de los Rios*. Anthropological Papers 77. Museum of Anthropology. Ann Arbor: University of Michigan.

Kirkpatrick, Z. M. 1992. *Wildflowers of the Western Plains: A Field Guide*. Austin: University of Texas Press.

Kleiman, R., F. R. Earle, and I. A. Wolff. 1966. "The Trans-3-enoic Acids of *Grindelia oxylepis* Seed Oil." *Lipids* 1: 301–4.

Klepper, E. D. 2007. "A Desert Love Story." *Texas Parks and Wildlife Magazine*, April 2007.

Knuth, F. M. 1930. *Den nye kaktusbog* [The new cactus book]. Copenhagen: Gyldendal.

Kohout, M. D. 2020. "Big Bend Ranch State Park." *Handbook of Texas Online*, Texas State Historical Association.

Kurz, D. R. 1979. "Cacti of Big Bend National Park, Texas." Master's thesis. Southern Illinois University, Carbondale.

Laudermilk, J. 1945. "Desert Midget and Jungle Giant." *Desert Magazine* 8, no. 8: 9–10.

L'Héritier, C. 1789. In *Hortus Kewensis; or, A Catalogue of the Plants Cultivated in the Royal Botanic Garden at Kew*, by W. Aiton. London: Longman, Hurst, Rees, Orme, and Brown.

Liebmann, F. 1869. *Chênes de l'Amérique Tropicale*. Leipzig: Léopold Voss.

Lietava, J. 1992. Medicinal Plants in a Middle Paleolithic Grave Shanidar IV? *Journal of Ethnopharmacology* 35, no. 3: 263–66.

Linnaeus, C. 1737a. *Genera Plantarum*. Lugduni Batavorum: C. Wishoff.

———. 1737b. *Hortus Cliffortianus*. Amsterdam.

———. 1753. *Species Plantarum*. Holmiae: Impensis Laurentii Salvii.

———. 1756. *Centuria II. Plantarum . . .* Uppsala: Exc. L. M. Höjer.

———. 1759. *Systema Naturae, Editio Decima* 2. Holmiae: Impensis Direct. Laurentii Salvii.

Lira, R., and J. Caballero. 2002. "Ethnobotany of the Wild Mexican Cucurbitaceae 1." *Economic Botany* 56, no. 4: 380–98.

Little, E. L. 1980. *The Audubon Society Field Guide to North American Trees: Western Region*. New York: Alfred A. Knopf.

Lloyd, F. E. 1911. *Guayule (Parthenium argentatum Gray), a Rubber-Plant of the Chihuahuan Desert*. Washington, DC: Carnegie Institution.

———. 1942. *The Carnivorous Plants*. Waltham: Chronica Botanica.

l'Obel, M. de. 1581. *Plantarum seu stirpium icones*. Antwerp: Christophori Plantini.

Loden, S. 2012. "Fort Leaton Comes to Life." *El Solitario*. Texas Parks and Wildlife Department, Big Bend Ranch State Park.

Löfling, P., and C. Linnaeus. 1758. *Iter Hispanicum*. Stockholm: Ttryckt på Direct. Lars Salvii Kostnad.

Loureiro, J., de. 1790. *Flora Cochinchinensis*. Ulyssipone: Typis et expensis Academicis.

MacLeod, W. 2008. *River Road Vistas: A Journey along the River Road*. Alpine: Texas Geological Press.

Marsigli, L. F. 1725. *Histoire Physique de la Mer*. Amsterdam: Aux De'Pens De La Compagnie.

Matthews, J. F., D. W. Ketron, and S. F. Zane. 1994. "The Seed Surface Morphology and Cytology of Six Species of *Portulaca* (Portulacaceae)." *Castanea* 59, no. 4: 331–37.

Mattiza, D. B. 1993. *One Hundred Texas Wildflowers*. Tucson: Southwest Parks and Monuments Association.

Maxwell, Ross A. 1967. "The Big Bend of the Rio Grande: A Guide to the Rocks, Geologic History, and Settlers of the Area of Big Bend National Park." Texas Bureau of Economic Geology.

McDougall, W. B., and O. E. Sperry. 1951. *Plants of Big Bend National Park*. Washington, DC: US Government Printing Office.

McKann, M. H. 1975. "The Recreation Potential of Chorro Canyon, Presidio County, Texas." A thesis in Park Administration. Lubbock: Texas Tech University.

Metcalfe, O. B. 1903. "The Flora of the Mesilla Valley." Undergraduate thesis. New Mexico College of Agriculture and Mechanic Arts, Mesilla Park.

Metcalfe, S. E. 2006. "Late Quaternary Environments of

the Northern Deserts and Central Transvolcanic Belt of Mexico." *Annals of the Missouri Botanical Garden* 93, no. 2: 258–73.

Mickel, J. T., and A. R. Smith. 2004. *The Pteridophytes of Mexico*. New York: New York Botanical Garden Press.

Miller, P. 1754. *The Gardeners Dictionary*. London: John and James Rivington.

Missouri Botanical Garden. 2019. *Tropicos*. tropicos.org/Home.aspx.

Moerman, D. E. 2009. *Native American Medicinal Plants: An Ethnobotanical Dictionary*. Portland, OR: Timber Press.

Morey, R. 2008. *Little Big Bend: Common, Uncommon, and Rare Plants of Big Bend National Park*. Lubbock: Texas Tech University Press.

Mühlenpfordt, F. "Neue Cacteen." 1846. *Allgemeine Gartenzeitung* 47: 371.

Muller, C. H. 1951. *The Oaks of Texas*. Renner: Texas Research Foundation.

NatureServe Explorer. *An Online Encyclopedia of Life*. http://explorer.natureserve.org/.

New York Botanical Garden. 2019. *C. V. Starr Virtual Herbarium*. sweetgum.nybg.org/science/vh/.

Nilsson, K. B. 1994. *A Wild Flower by Any Other Name: Sketches of Pioneer Naturalists Who Named Our Western Plants*. San Francisco: Yosemite Association.

Nored, G. 2014a. *Big Bend Ranch State Park Road Guide*. Texas Parks and Wildlife Department.

———. 2014b. *Campsites of the Big Bend Ranch State Park*. Texas Parks and Wildlife Department.

Nuttall, T. 1818. *The Genera of North American Plants, and a Catalogue of the Species, to the Year 1817*. Philadelphia: D. Heartt.

———. 1821. *A Journal of Travels into the Arkansas Territory during the Year 1819*. Philadelphia: T. H. Palmer.

———. 1840. In *A Flora of North America: Containing Abridged Descriptions of All the Known Indigenous and Naturalized Plants Growing North of Mexico; Arranged According to the Natural System*, by J. Torrey and A. Gray. London: Wiley and Putnam.

Ohl, A. J., and W. A. Cloud. 2001. *Archeological Survey of Select Boundary and Power Line Segments: Big Bend Ranch State Park, Presidio County, Texas*. Center for Big Bend Studies, Sul Ross State University, and Texas Parks and Wildlife Department.

Parkinson, J. 1640. *Theatrum Botanicum: The Theater of Plants; or, An Herball of Large Extent*. London: Thomas Cotes.

Pavón, J. A., and H. Ruiz López. 1799. *Flora Peruviana et Chilensis*. Madrid: Typis Gabrielis de Sancha.

Pennington, C. W. 1963. *The Tarahumara of Mexico*. Salt Lake City: University of Utah Press.

Perry, M. 2008. "Big Bend Ranch Is Ready to Show Off." *Alpine Avalanche*, May 1.

Persoon, C. H. 1807. "Synopsis Plantarum, seu enchiridium botanicum." Paris: Treuttel and Wurtz.

Phillips, B., D. Sholly, K. H. Blizzard, and C. Beckham (maps). 2010. *Big Bend Ranch Biking Guide: The Other Side of Nowhere*. Texas Parks and Wildlife Department.

Pliny the Elder. 77 CE. *Natural History*. English online translation by J. Bostock and H. T. Riley (1855). perseus.tufts.edu/hopper/text?doc=Plin.+Nat.+toc.

Plukenet, L. 1691. *Phytographia. pars altera*. London: Sumptibus autoris.

———. 1692. *Phytographia. pars tertia*. London: Sumptibus autoris.

———. 1696. *Almagestum Botanicum*. London: Sumptibus autoris.

Plumier, C. 1703. *Nova Plantarum Americanarum Genera*. Paris: J. Boudot.

———. 1756. In *Plantarum Americanarum . . . t. 81(1)*, by J. Burman. Amstelædami: S. Schouten; Lugd. Batav.: G. Potuliet and T. Haak.

Poole, J. M., W. R. Carr, D. M. Price, and J. R. Singhurst. 2007. *Rare Plants of Texas*. College Station: Texas A&M University Press.

Powell, A. M. 1998. *Trees and Shrubs of the Trans-Pecos and Adjacent Areas*. Austin: University of Texas Press.

Powell, A. M., and P. R. Manning, illustrator. 1994. *Grasses of the Trans-Pecos and Adjacent Areas*. Austin: University of Texas Press.

Powell, A. M., and J. F. Weedin. 2004. *Cacti of the Trans-Pecos and Adjacent Areas*. Lubbock: Texas Tech University Press.

Powell, A. M., and R. D. Worthington. 2018. *Flowering Plants of Trans-Pecos Texas and Adjacent Areas. Sida, Botanical Miscellany* 49. Fort Worth: Botanical Research Institute of Texas.

Pursh, F. T. 1814. *Flora Americae Septentrionalis; or, A Systematic Arrangement and Description of the Plants of North America*. London: White, Cochrane.

Rafinesque, C. S. 1836–38. *New Flora and Botany of North America*. Philadelphia.

Rafinesque, C. S., ed. 1832–. *Atlantic Journal, and Friend of Knowledge*. Philadelphia.

Ralston, B. E., and R. A. Hilsenbeck. 1989. "Taxonomy of the *Opuntia schottii* Complex (Cactaceae) in Texas." *Madroño* 36, no. 4: 221–31.

Richardson, A. T. 1977. "Monograph of the Genus *Tiquilia* (*Coldenia*, sensu lato), Boraginaceae: Ehretioideae." *Rhodora* 79, no. 820: 467–572.

Riskind, D. H. 1993. Personal communication.

———. 2007. "A Sense of Place: El Solitario; What's in a

Name?" *El Solitario* (Fall). Texas Parks and Wildlife Department, Big Bend Ranch State Park.

———. 2019. Personal communication.

———. 2021. Personal communication.

Riskind, David, and Dan Sholly. 2009. *Roads to Nowhere.* Texas Parks and Wildlife.

Rivinus, A. Q. 1690. *Introductio generalis in rem herbariam.* Leipzig: Christoph. Guntheri.

Robbins, W. W., J. P. Harrington, and B. Freire-Marreco. 1916. *Ethnobotany of the Tewa Indians.* US Bureau of American Ethnology, Bulletin 55. Washington, DC: Smithsonian Institution.

Roberts, T., T. Gibbs, and J. Gibbs. 2017. *Trails through Time: Archeological Reconnaissance of Selected Trail Corridors, Big Bend Ranch State Park, Presidio and Brewster Counties, Texas, 2004–2010.* Texas Parks and Wildlife, State Parks Division, Cultural Resources Program.

Royen, A., van. 1740. *Florae Leydensis prodromus.* Leiden: Samuelem Luchtmans.

Ruhlman, J., L. Gass, and B. Middleton. 2012. "Chihuahuan Deserts Ecoregion." In *Status and Trends of Land Change in the Western United States—1973 to 2000,* edited by B. M. Sleeter, T. S. Wilson, and W. Acevedo. US Geological Survey Professional Paper 1794-A.

Rydberg, P. A. 1917. *Flora of the Rocky Mountains and Adjacent Plains . . .* New York: Self-published.

Salm-Reifferscheidt-Dyck, J. 1850. In *Plantae Lindheimerianae: An Enumeration of F. Lindheimer's Collection of Texan Plants . . . ,* by G. Engelmann and A. Gray. Boston: Freeman and Bolles.

Sanchez, J. M. 1999. *Archeological Reconnaissance of Upper Fresno Canyon Rim, Big Bend Ranch State Park, Texas.* Alpine, TX: Center for Big Bend Studies, Sul Ross State University.

Saunders, B. D. 1976. "A Brief Historical Survey of the Big Bend Area." In *Bofecillos Mountains, A Natural Area Survey, No. 12.* Lyndon B. Johnson School of Public Affairs. University of Texas at Austin.

Scheele, G. H. A. 1848a. *Cassia lindheimeriana* Scheele. *Linnaea* 21: 457.

———. 1848b. *Desmanthus velutinus* Scheele. *Linnaea* 21: 455.

———. 1849a. *Diospyros texana* Scheele (family Ebenaceae). *Linnaea* 22: 145.

———. 1849b. *Euphorbia villifera* Scheele. *Linnaea* 22: 153.

Scheinvar, E., N. Gámez, G. Castellanos-Morales, E. Aguirre-Planter, and L. E. Eguiarte. 2016. "Neogene and Pleistocene History of *Agave lechuguilla* in the Chihuahuan Desert." *Journal of Biogeography* 44: 322–34.

Schultes, R. E., A. Hofmann, and C. Ratsch. 1998. *Plants of the Gods: Their Sacred, Healing, and Hallucinogenic Powers.* Rochester, VT: Healing Arts Press.

Seemann, B. C. 1856. *The Botany of the Voyage of H.M.S. Herald: Under the Command of Captain Henry Kellett, R.N., C.B., during the Years 1845–51.* London: Lovell Reeve.

SEINet Portal Network. 2019. swbiodiversity.org/seinet/index.php.

Sessé, M., and J. M. Mociño. 1887–91. *Plantae Novae Hispaniae.* Mexico City: Ignacio Escalante.

———. 1891–97. *Flora Mexicana.* Mexico City: Ignacio Escalante.

Shaw, E. A. 1987. *Charles Wright on the Boundary, 1849–1852; or, Plantae Wrightianae Revisited.* Westport, CT: Meckler Publishing.

Shreve, F., and I. L. Wiggins. 1964. *Vegetation and Flora of the Sonoran Desert.* Stanford, CA: Stanford University Press.

Simpson, B. B. 2013. "*Krameria bicolor,* the Correct Name for *Krameria grayi* (Krameriaceae)." *Phytoneuron* 62: 1.

Singh, A. 2008. "Review of Ethnomedicinal Uses and Pharmacology of *Evolvulus alsinoides* Linn." *Ethnobotanical Leaflets* 12, no. 1: 734–40.

Sloane, H. 1707. *A Voyage to the Islands Madera, Barbados, Nieves, S. Christophers, and Jamaica . . .* London: Printed by B. M. for the author.

Smith-Savage, S. 1996a. "Culture History: Late Historic Period to Present." In *Archeological Reconnaissance on Big Bend Ranch State Park, Presidio and Brewster Counties, Texas, 1988–1994.* Alpine, TX: Center for Big Bend Studies, Sul Ross State University.

———. 1996b. "Appendix A: Historic Sites." In *Archeological Reconnaissance on Big Bend Ranch State Park, Presidio and Brewster Counties, Texas, 1988–1994.* Alpine, TX: Center for Big Bend Studies, Sul Ross State University.

Sowerby, J., et al. 1863–72. *Sowerby's English Botany.* London: R. Hardwicke.

Sprengel, K. P. J. 1807–8. *Historia Rei Herbariae.* 2 vols. Amsterdam: Tabernae librariae et artium.

Standley, P. C. 1920–26. *Trees and Shrubs of Mexico.* Smithsonian Institution, Contributions from the United States National Herbarium, vol. 23. Washington, DC: Government Printing Office.

———. 1937. *Flora of Costa Rica.* Chicago: Field Museum of Natural History.

Standley, P.C., and J. A. Steyermark. 1946–58. *Flora of Guatemala.* Fieldiana: Botany. Chicago: Chicago Natural History Museum.

Steyermark, J. A. 1963. *Flora of Missouri.* Ames: Iowa State University Press.

Subils, R. 1984. "Una nueva especie de *Euphorbia* Sect. *Poinsettia* (Euphorbiaceae)." *Kurtziana* 17: 125–30.

Swank, G. R. 1932. "The Ethnobotany of the Acoma and Laguna Indians." Master's thesis, Department of Biology. University of New Mexico, Albuquerque.

Sweet, R. 1827. *Sweet's Hortus Britannicus; or, A Catalogue of Plants Cultivated in the Gardens of Great Britain . . .* London: J. Ridgway.

Taylor, L. 2011. "Ninety-Six Miles in Ninety-Six Hours: The Marfa-Terlingua Freight Road, 1899–1960." *Journal of Big Bend Studies* 23: 7–20.

Taylor, N. C., and M. Terry. 2015. "*Euphorbia abramsiana* (Euphorbiaceae): New to Texas." *Phytoneuron* 24: 1–7.

Texas beyond History. 2004. *Wax, Men, and Money: Candelilla Wax Camps along the Rio Grande.* texasbeyondhistory.net/waxcamps/index.html.

Thornton, R. J. 1812. *The British Flora; or, Genera and Species of British Plants Arranged after the Reformed Sexual System and Illustrated by Numerous Tables and Dissections.* London: J. Whiting.

Tidestrom, I. 1935. "New Arizona Plant Names." *Proceedings of the Biological Society of Washington* 48: 39–44.

Tidestrom, I., and T. Kittell. 1941. *A Flora of Arizona and New Mexico.* Washington, DC: Catholic University of America Press.

Tinkham, E. R. 1941. "Biological and Faunistic Notes on the Cicadidae of the Big Bend Region of Trans-Pecos Texas." *Journal of the New York Entomological Society* 49: 165–83.

———. 1944. "Faunistic Notes on the Diurnal Lepidoptera of the Big Bend Region of Trans-Pecos, Texas, with the Description of a New *Melitaea*." *Canadian Entomologist* 76: 11–18.

Torrey, J. 1848. In *Notes of a Military Reconnaissance from Fort Leavenworth, in Missouri, to San Diego, in California, Including Parts of the Arkansas, Del Norte, and Gila Rivers*, **by** W. H. Emory. US Army Corps of Topographical Engineers. 30th Cong., 1st sess. [House] Ex. doc.; — no. 41. Washington, DC: Wendell and Van Benthuysen.

———. 1853. In *Report of an Expedition down the Zuni and Colorado Rivers*, by L. Sitgreaves. U.S Army. Corps of Topographical Engineers. Senate Executive Document No. 59, 32d Congress, 2d sess. Washington, DC: R. Armstrong.

Torrey, J., and A. Gray. 1840. *A Flora of North America: Containing Abridged Descriptions of All the Known Indigenous and Naturalized Plants Growing North of Mexico; Arranged According to the Natural System.* London: Wiley and Putnam.

———. 1855. In *Pacific Railroad Surveys: Reports of Explorations and Surveys, to Ascertain the Most Practicable and Economical Route for a Railroad from the Mississippi River to the Pacific Ocean.* Washington, DC: Beverley Tucker.

TPWD. 2009. *The Crawford-Smith Ranch of Fresno Canyon.* Big Bend Ranch State Park, interpretive brochure, April.

———. 2011. "Interpretive Guide to Big Bend Ranch State Park."

———. 2012. *Sauceda Historic District: History and Walking Guide*, Big Bend Ranch State Park.

———. 2019. *Rare, Threatened, and Endangered Species of Texas by County, July 17, 2019.* tpwd.texas.gov/gis/rtest/.

Traditional Herbal Blogspot, 2011. *Traditional Herbal Medicine Herbs.* traditionalherbl.blogspot.com/2011/04/inmortal-asclepias-asperula.html.

Turner, B. L. 1959. *The Legumes of Texas.* Austin: University of Texas Press.

———. 2005. "*Stemodia coahuilensis* (Scrophulariaceae), a New Record for the United States." *SIDA, Contributions to Botany* 21, no. 3. Botanical Research Institute of Texas.

Turner, B. L., H. Nichols, G. Denny, and O. Doron. 2003. *Atlas of the Vascular Plants of Texas.* Fort Worth: Botanical Research Institute of Texas.

United States Department of Agriculture, Natural Resources Conservation Service, USDA PLANTS Database. https://plants.sc.egov.usda.gov/java/.

Vail, A. M. 1898. "Studies in the Asclepiadaceae III." *Bulletin of the Torrey Botanical Club* 25, no. 4:171–82.

Valdés-Reyna, J. 2003. "*Dasyochloa* Willd. ex Rydb." In *Flora of North America*, vol. 25, *Poaceae*, part 2, pp. 45, 47–48, edited by Mary E. Barkworth, Kathleen M. Capels, Sandy Long, and Michael B. Piep. New York: Oxford University Press.

Van Devender, T. R., C. E. Freeman, and R. D. Worthington. 1978. "Full-Glacial and Recent Vegetation of Livingston Hills, Presidio County, Texas." *Southwestern Naturalist* 23, no. 2: 289–302.

Van Devender, T. R., and W. G. Spaulding. "Development of Vegetation and Climate in the Southwestern United States." Science 204 (4394): 701–10.

Vines, R. A. 1960. *Trees, Shrubs, and Woody Vines of the Southwest.* Austin: University of Texas Press.

Von Reis, S., and F. J. Lipp Jr. 1982. *New Plant Sources for Drugs and Foods from the New York Botanical Garden Herbarium.* Cambridge, MA: Harvard University Press.

Walter, T. 1788. *Flora Caroliniana.* London: J. Fraser.

Wang, X., and Zheng, S. 2010. "Whole Fossil Plants of *Ephedra* and Their Implications on the Morphology, Ecology, and Evolution of Ephedraceae (Gnetales)." *Chinese Science Bulletin* 55: 675–83.

Warnock, B. H. 1960. "*Cirsium turnerae* (Compositae), New Species from Trans-Pecos Texas." *Southwestern Naturalist* 5, no. 2: 101–2.

———. 1970. *Wildflowers of the Big Bend Country, Texas.* Alpine, TX: Sul Ross State University.

———. 1974. *Wildflowers of the Guadalupe Mountains and the Sand Dune Country, Texas.* Alpine, TX: Sul Ross State University.

———. 1977. *Wildflowers of the Davis Mountains and Marathon Basin, Texas*. Alpine, TX: Sul Ross State University.

Warnock, B. H., and M. C. Johnston. 1960. "New Combinations in Texas *Euphorbias* of Subgenus *Chamaesyce*." *Southwestern Naturalist* 5, no. 3: 170.

Watahomigie, L. J. 1982. "Hualapai Ethnobotany." Peach Springs: Hualapai Bilingual Program, Peach Springs School District 8: 25.

Wayne's Word: An Online Textbook of Natural History. "Poison Oak: More Than Just Scratching The Surface." www2.palomar.edu/users/warmstrong/ww0802.htm.

Weiser, K. 2019. "Redford, Texas, and the Lost Mission of El Polvo." In *Legends of America*. May. www.legendsofamerica.com/tx-redford/.

Wells, P. V. 1966. "Late Pleistocene Vegetation and Degree of Pluvial Climatic Change in the Chihuahuan Desert." *Science* 153: 970–75.

———. 1977. "Postglacial Origin of the Chihuahuan Desert Less Than 11,500 Years Ago." In *Trans. symp. vol. on the Biological Resources of the Chihuahuan Desert, US and Mexico*, edited by R. H. Wauer and D. H. Riskind. National Park Service, Washington, DC.

Whitehouse, E. 1936. *Texas Wildflowers in Natural Colors, Including Many Common Plants of the Southwest*. Austin: Privately published, distributed by Texas Book Store.

Whiting, A. F. 1939. "Ethnobotany of the Hopi." *Museum of Northern Arizona University, Bulletin* 15: 98.

Wilde, M., and S. Platt. 2011. "A Life-Giving Trail: Documenting the Environmental History of Alamito Creek." *Journal of Big Bend Studies* 23: 39–62.

Williams, L. O. 1965. *The Orchidaceae of Mexico*. Tegucigalpa: Escuela Agricola Panamericana.

Wislizenus, A. 1848. *Memoir of a Tour to Northern Mexico, Connected with Col. Doniphan's Expedition, in 1846 and 1847*. Washington, DC: Tippin and Streeter.

Wooton, E. O., and P. C. Standley. 1915. *Flora of New Mexico*. Washington, DC: US Government Printing Office.

Worthington, R. D. 1995. *A Floral Inventory of the Big Bend Ranch State Natural Area, Presidio and Brewster Counties, Texas*. University of Texas at El Paso, Department of Biological Sciences.

Wrede, J. 1997. *Texans Love Their Land: A Guide to 76 Native Texas Hill Country Woody Plants*. San Antonio: Watercress Press.

Yancey, F. D. 1997. *The Mammals of Big Bend Ranch State Park, Texas*. Special Publications, Museum of Texas Tech University, No. 39, October.

Yarborough, S. C., and A. M. Powell. 2002. *Ferns and Fern Allies of the Trans-Pecos and Adjacent Areas*. Lubbock: Texas Tech University Press.

Yates, R. G., and G. A. Thompson. 1959. *Geology and Quicksilver Deposits of the Terlingua District, Texas*. Geological Survey Professional Paper 312. Washington, DC: US Government Printing Office.

INDEX

Abert's wild buckwheat, 47
Abrams' spurge, 17, 92, 214, 387
Abutilon
    malacum, 17
    parvulum, 50
    wrightii, 51
acacia(s), 4, 7, 52, 371
Acaciella angustissima, 54, 86
Acalypha phleoides, 51, 85
Acanthaceae, 89, 97–98
acanthus (family), 61, 89, 97–98
Acebuches Canyon, 20, 102, 403
Acleisanthes
    angustifolia, 94, 281
    chenopodioides, 94, 282
    longiflora, 94, 283
    parvifolia, 81, 88, 379, 386
Acourtia
    nana, 89, 128
    wrightii, 89, 129
Adiantum capillus-veneris, 95, 322
African rue, 29, 94, 280, 389
Agarito, 7, 34, 44, 51, 53, 90, 148, 403, 405
Agave
    havardiana, 89, 122
    lechuguilla, 89, 123, 404, 417
agave(s), 4–5, 43, 122, 123
Ageratina wrightii, 62–63, 85
Agua Adentro
    Mountain, 33–34, 328, 382
    Pens, 32–33, 35–36, 39, 45
    Spring, 33, 35–36, 105, 113–14, 131, 138, 177, 188–89, 197–98, 207, 223, 235, 255–56, 274, 279, 309–10, 320, 335, 346, 349, 355, 368–69, 373, 375, 403
Aguirre, Ramon, 86
Aguja de Tascate, 32, 48, 54
Aizoaceae, 89, 99–100
Alabama lip fern, 49
Alamito
    community, 31, 403

    Creek, 3, 11, 28–30, 83–86, 101, 236, 246, 354, 403, 419
    Creek, Upper, 83, 85, 149, 189, 209, 338, 340, 346, 348, 357
Alamito Creek-Terneros Creek Lowlands, v, 28–31, 33, 277
Alamito Dam, 33, 40
Agua Adentro Mountain, 33–34, 328, 382
Agua Adentro Pens, 32–33, 35–36, 39, 45
Alamo de Cesario Creek, 59, 69, 235, 403
Alamo de Cesario Creek, South Fork, 59, 69, 130–31, 242, 254, 345–46, 349
Alamo Seco Creek, 29, 31
Alamo Springs Road, 83
alfalfa, 22
Alismataceae, 89, 101
alkali heliotrope, 74
Allionia incarnata, 78
Allowissadula holosericea, 51
allthorn, 4, 7, 51, 60, 78, 93, 256
allthorn (family), 93, 256
Aloysia gratissima, 25, 35, 47, 50, 60, 74, 96, 368
Aloysia wrightii, 35, 45, 96, 369, 405
amaranth (family), 89, 102–5
Amaranthaceae, 89, 102–5
Amaranthus palmeri, 34
Amaryllidaceae, 89
amaryllis (family), 89
Ambrosia monogyra, 8, 20, 25, 89, 130
Amsonia
    longiflora, 53, 81
    palmeri, 25, 55, 85, 89, 109
Anacampserotaceae, 89, 107
Anacardiaceae, 89, 108
Anderson, Robert O., vii, 4
Andropogon glomeratus, 45
angel trumpets, 46, 53, 94, 283
annual bastard cabbage, 16, 390
antelope horns, 85, 89, 110

Anulocaulis
    eriosolenus, 15, 71, 75, 79, 94, 284, 386
    leiosolenus var. *lasianthus*, 7, 81, 94, 285, 379
Apache plume, 7, 53, 83, 95, 340
Apache(s), 3, 22, 27–28, 53, 126, 340, 403–4, 412
Aphanostephus ramosissimus var. *humilis*, 16
Apocynaceae, 89, 109–18
Apodanthaceae, 89, 119
Apodanthera undulata, 52, 79, 92, 204, 405, 412
Aquilegia chrysantha var. *chrysantha*, 7–8, 43–44, 76, 95, 334
archaeological, xii–xiii, 2, 10, 109, 407–8
archaeological periods
    Archaic, Early, 2
    Archaic, Late, 3
    Archaic, Middle, 2
    Cretaceous, 2, 78, 210
    Paleo-American, 2
    Paleozoic, 2, 57, 61
    Pleistocene, 5, 417, 419
    Prehistoric, 2, 411
    Protohistoric, 3
Arcytophyllum fasciculatum, 63–64, 71
Arecaceae, 89, 120
Arenosa, Arenosa Campground (Group Camping Area), 12, 25, 169, 182, 195, 302, 366, 403
Argemone chisosensis, 15, 66, 94, 294
Argyrochosma microphylla, 52, 66
Argythamnia serrata, 85
Ariocarpus fissuratus var. *fissuratus*, 62, 71, 78, 90, 162
Aristolochia wrightii, 17, 44, 55, 85, 89, 121
Aristolochiaceae, 89, 121
Arizona
    carlowrightia, 22
    centaury, 8, 43, 85

cockroach, 81, 89, 116, 387
cottonwood, 34–36, 41, 45–46, 53, 69, 74, 83, 85, 95
goldenaster, 17, 69, 387
snakecotton, 42, 44, 85, 89, 102
spikemoss, 66
spurge, 35, 75
Armstrong, Bob, vii, 4, 51
arrowleaf mallow, 52, 93, 272, 388
arroyo fameflower, 35, 50, 54, 89, 107
arroyo fameflower (family), 89, 107
Arroyo
de los Ojitos, 50
Mexicano, 17, 33, 42, 48, 52–54, 76, 102–3, 108, 111, 115, 129, 140–41, 148, 156, 188, 195, 197, 201, 204, 207–8, 212, 228, 238, 240, 251, 254, 257–58, 263, 278, 298, 306, 308–9, 314, 335, 338, 341, 357, 360, 369, 373–74
Mexicano, lower, 44, 105, 243, 282, 286–87, 310
Mexicano, upper, 42, 121, 169, 198, 218, 279, 292
Primero, 32–33, 42–43, 76, 143, 188, 211, 290, 334
Primero Trail, 42, 76
Arundo donax, 6, 94, 305, 403
Asclepias
asperula, 89, 110, 418
oenotheroides, 89, 111
scaposa, 89, 112, 386
ash, 7–8, 44, 66, 74, 85–86, 94, 242, 286, 372, 404
ash, volcanic, 2, 16, 35, 393, 400
ash flow, 20, 25, 27, 35, 42
ashy sandmat, 62, 67, 92, 217
ashy-leaf bahia, 31
Asparagaceae, 89, 122–27
asparagus (family), 89, 122–27
Aspicarpa hyssopifolia, 93, 269
Asteraceae, 89, 128–47, 394–95, 397–98
Astragalus, 228
Astragalus
terlinguensis, 92, 227
wootonii, 92, 228
Astrolepis
cochisensis, 36, 50
integerrima, 36
sinuata, 36
athel, 8, 15, 34, 96, 365
Atriplex, 191–93
Atriplex
acanthocarpa ssp. *acanthocarpa*, 91, 191

canescens, 77
elegans var. *elegans*, 91, 192
lentiformis ssp. *lentiformis*, 91, 193, 386
August Santleben, 3
Auras Canyon, 27, 36, 121, 184
Auras Creek, 27, 36,
autumn sage, 64, 71, 77, 93, 260
Ayenia, 271
Ayenia
filiformis, barberry (family), 63
microphylla, 64
pilosa, 93, 271, 386

baby jump-up, 50
Baccharis salicifolia, 89, 131
baccharisleaf penstemon, 60, 71
Bandidos Creek, 36
Baños de Leyva, 46, 403
Barbados cherry (family), 93
barberry (family), 90, 148
Barton Warnock Visitor Center, xiii, 10, 13, 80, 82
Basin and Range faulting, 25, 29, 33
Basin and Range Province, 2, 4
Battle of Rancherias, 22
BBNP, 1, 13–14, 25, 43, 71, 73, 76, 117, 120, 150, 158, 163, 182, 214, 221, 229, 260, 288, 297, 338, 382
BBR, 1–8, 10, 13–16, 22, 25, 28–29, 31, 33–34, 36, 39, 42–46, 49, 52, 54–55, 59–61, 65, 67, 75, 83, 85–87, 112–13, 116, 118, 145, 158, 170–71, 174, 183, 214, 241–42, 253, 260, 263, 271, 277, 291, 294, 297, 317, 319, 324, 329–30, 376, 381–83, 391, 397, 407–9
bean (family), 92, 227–39
Bear Mountain milkweed, 61, 65, 89, 112, 386
bearded dalea, 64
bearded flatsedge, 50, 79, 92, 207
beargrass, 7–8, 60, 78, 89, 125, 301
bebedero(s), 6, 43, 46, 403
beebrush, 25, 35, 50, 74, 96, 368
beech (family), 93, 240–43
bellflower (family), 91, 188
Berberidaceae, 90, 148
Berberis trifoliolata, 90, 148
Berlandier's flax, 61
Berlandier's wolfberry, 20
Bernardia obovata, 44, 64
Bidens leptocephala, 35
Big Bend
bluebonnet, 17, 46–47
croton, 75, 386

purplish prickly pear, 182
ringstem, 15, 71, 75, 79, 94, 284, 386
silverleaf, 35, 42, 60, 64, 81, 96, 350
Big Bend National Park, xi, xiii-xiv, 1, 409, 415–16
Big Bend Ranch State Natural Area, vii, 1, 4, 87, 409, 419
Big Bend Ranch State Park, iii, v, vii, ix, xi–xiii, 1, 4–5, 9–11, 407–8, 411–19
Big Hill, 13, 17–20, 22, 173, 183, 185, 266, 382
Big Hill prickly pear, 20, 91, 183, 388
big saltbush, 8, 14–16, 22, 91, 193, 386
big sandbur, 25
bigbract verbena, 55
bighorn sheep, 17, 20
bigleaf mistletoe, 85, 96, 372, 389
big-needle pincushion cactus, 22, 386
Bignoniaceae, 90, 149
biking, 1, 51, 61, 78, 407, 416
birthwort (family), 85, 89, 121
bittersweet (family), 91, 190
bitterweed, 52
Black Gap WMA, 5, 171, 382
black grama, 45, 94, 308
Black Hills
(range), 11, 29–31, 125, 140, 148, 157, 169, 210, 226, 244, 257, 274, 313, 361, 377
Creek, 28–29, 31, 251
Road, 47
bladder mallow, 51
blind prickly pear, 7, 17, 22, 39, 71, 75
blue milkwort, 85, 95, 314
Blue Range, 8, 56, 59, 61, 66, 71
bluntscale bahia, 34, 79
Boerhavia
gracillima, 15, 35
linearifolia, 60
triquetra var. *intermedia*, 22
Bofecillos
Canyon, 13, 20, 22, 27, 33, 35–36, 223, 231, 236, 252, 295, 346, 373
Creek, 27–28, 35–36, 348
Mountains, ix, 2, 10–11, 13, 17, 25, 29, 31–35, 39, 43, 46, 48, 51, 73, 76, 78, 83, 161, 295, 317, 407, 412, 417
Mountains: Highlands, v, 10, 33–47
Mountains: Llano, v, 10, 48–55
Peak, 33, 36
Plateau, ix, 6
Road, 12, 27–29, 139, 163, 168, 182
Spring, 33, 35–36
Volcano, 2, 27, 35, 48, 52

Bogel, W.W., Gus, Gallie, Graves, 4
Boke's button cactus, 81, 88, 172, 379
Bolboschoenus maritimus subsp. *paludosus*, 92, 206, 386
Bommeria hispida, 95, 323
Bonaire lip fern, 36, 55, 95, 325
Bonamia repens, 62, 71, 88, 379, 386
Boquillas (Formation, limestone), 14, 64, 73, 76–77, 80, 82, 404
Boquillas lizardtail, 15, 94, 288, 379
borage (family), 75, 90, 150–55
Boraginaceae, 90, 150–55, 267, 416
Botella
    (residence), 31–34, 47, 69, 365, 403
    Junction, 11, 28, 32, 34
    Ranch, 34
    Road, 34
    Spring, 34
Botella-Black Hills Road, 31, 47
Bouchea linifolia, 96, 370
boundary ephedra, 81
Bouteloua, 31, 306–7
Bouteloua
    aristidoides, 42
    curtipendula, 94, 306
    erecta, 94, 307
    eriopoda, 94, 308
    hirsuta, 17
Bouvardia ternifolia, 54
bracted bedstraw, 49
Brassicaceae, 90, 156–60
Brewster County, 1, 5, 11, 13, 98, 102, 104, 112, 171, 229, 239, 285, 382
Brickellia
    coulteri, 35
    eupatorioides var. *chlorolepis*, 90, 132
    laciniata, 50, 69
    lemmonii var. *lemmonii*, 62, 78, 386
bristlecup sandmat, 61
bristly nama, 31, 77, 93, 251
Bromeliaceae, 90, 161
bromeliad (family), 90, 161
broom milkwort, 42
broomrape (family), 94, 292–93
brownfoot, 42, 44, 53, 85, 89, 129
Buckley's centaury, 61, 64, 93, 247, 390
buckthorn (family), 95, 338–39
buckwheat (family), 95, 316–18
Buda (Formation), 63
Buddleja
    marrubiifolia, 34, 71
    scordioides, 69
Buena Suerte, 73, 75, 403
Buena Suerte Trail, 71–75, 78, 80–82, 124, 147, 154, 166, 168, 177, 180, 187, 224, 229, 239, 244, 257, 266, 285, 306, 316–17, 338, 350–51, 364
buffalo gourd, 35,
buffalobur nightshade, 22, 25, 96, 362, 390
bunkhouse, xiv, 4, 51
Burnt Camp, 62–63
Burnt Camp Trailhead, 59, 61–62, 65
bush croton, 86
bushy bluestem, 45
bushy wild buckwheat, 62–63, 69, 80, 95, 317, 379
buttercup (family), 76, 95, 334–36
buttonbush, 8, 50–51, 69, 74, 95, 345

Cabeza de Vaca, Álvar Núñez, 3, 14, 28
Cactaceae, 90, 162–87, 395, 411–13, 416
cactus (family), 17, 28, 39, 78, 81, 90–91, 162–87, 379, 386–87, 394, 397–98, 411, 413–15
Calibrachoa parviflora, 74
California
    caltrop, 20
    loosestrife, 81, 93, 268
    trixis, 49, 86, 90, 144
Calliandra iselyi, 92, 229, 386
caltrop (family), 96, 374–78
camel (caravan, expedition), 3, 31
camels, 3
Camino Viejo Trail, 80, 403
Camp Polvo, 27
Campanulaceae, 91, 188
Campsite Guide, xii, 10
candelilla, 4, 7, 60, 73, 76, 78, 81–82, 85, 92, 216, 403, 418
candle cholla, 90, 164
Cannabaceae, 91, 189
Cañon de los Bandidos, 36
Cañon de Rancho Viejo, 35
canyon grape, 8, 17, 35–36, 42, 44–45, 96, 373
cardinal flower, 7–8, 35–36, 43–44, 91, 188
Carlowrightia
    arizonica, 22
    linearifolia, 89, 97
    serpyllifolia, 81
Carrasco, Guadalupe, 41, 404
Carrasco, Mateo, 39
carrizo, 5–6, 305, 403
Casa Piedra
    (community), 403
    Road, 12, 28–31, 34, 83–84, 86, 182, 280, 355
    Trailhead, East, 30, 86
    Trailhead, West, 30, 83
Casa Ramon, 83–84, 86
Casa Reza, 20, 33
Castilleja
    lanata ssp. *lanata*, 94, 292
    sessiliflora, 94, 293
Cat Spring, 86
catalpa (family), 90, 149
catchfly prairie gentian, 8, 22, 93, 246
catclaw
    acacia, 31, 51, 60, 74, 83
    cactus, 74–75, 81, 91, 178
    mimosa, 49, 74, 85
catnip noseburn, 45
cattail (family), 15, 22, 36, 44, 96, 367, 414
CD, 1–2, 4–7, 13, 16, 123–24, 174–75, 177, 210, 256, 303, 338,
Celastraceae, 91, 190
Celtis
    pallida, 25
    reticulata, 91, 189
Cenchrus myosuroides, 25
central basin, 57, 59, 61–63, 65–67
Cephalanthus occidentalis, 95, 345
Cerro Boludo, 48–50
Cerro Chilicote, 48, 53
Cerro Chilicote Loop, 48, 52–53, 107, 319
Cerro de las Burras, 27
Cerro de las Burras Loop, 27, 114, 149, 167, 210, 226, 282, 374
Cerro Elephante, 33, 35–36, 39
Cerro Prieto Windmill, 30–31
Cerro Redondo, 11, 29–30
Cerros Prietos, 31
Cevallia sinuata, 71
Chamaesaracha
    coniodes, 63
    coronopus, 96, 353, 386
    pallida, 63, 69
    sordida, 31
    villosa, 25
cheeseweed, 8, 20, 25, 89, 130
Chenopodiaceae, 91, 191–94
Chenopodiastrum murale, 91, 194
Chenopodium incanum, 83
chicken-thief stickleaf, 267
Chihuahua Trail(s), 3, 28–29, 85, 155, 185–86, 217
Chihuahuan Desert, ix, xiv, 1, 3–4, 172, 408–9, 414–15, 417, 419
Chihuahuan Desert scrub, 60, 78
Chihuahuan ringstem, 7, 81, 94, 285, 379

Chillicothe, 404
Chillicothe Ranch, 3
Chillicothe-Saucita Ranch, 3
Chilopsis linearis, 90, 149
Chinati Mts., 29, 83, 85, 122, 133, 295, 316, 330, 382, 404
chiqueras, 40, 42, 404
Chisos Mountain false Indian mallow, 51
Chisos pricklypoppy, 15, 66, 94, 294
Chloracantha spinosa, 20
cholla, 7, 15, 22, 27–28, 35, 52, 75–76, 78, 82, 90, 163–65, 394
Chorro Canyon, 33, 43–44, 76, 102, 108, 123, 176, 188, 198, 211, 219, 240, 242, 270, 286, 290, 326, 328, 332, 334, 341, 346, 367, 373, 382, 407–8, 415
Chorro Vista Campsite, 43, 58
Chorro Vista Trailhead, 43
Christmas mistletoe, 42, 82, 85, 96, 371–72
Chrysactinia mexicana, 60, 64, 404
Chupadero Ranch, 86
Chupadero Spring, 86, 206
Cibolo Creek, 6, 85, 404
Cienega
    (area), v, 11, 83–86
    4WD roads, 86
    Camp, 85
    Creek, 6, 83, 85–86, 100, 132, 157, 164, 172–73, 192, 208, 210, 216, 222, 224, 232–33, 236, 245–46, 255, 257, 262, 268, 286–87, 289, 299, 304, 306, 314, 323, 339, 343, 346, 351, 354, 356, 371–72
    Creek 4WD Loop, 86
    Gorge, 83–86, 109, 156, 176, 225, 304, 364, 372
    Mountain(s), 5, 8, 29, 83, 85–86, 102, 110–11, 129, 140–41, 144, 153, 189, 212, 222–23, 226, 231, 240, 261, 314, 332, 360–61
    Mountains Rhyolite, 85
    park residence, 83, 85, 100, 192
    Trail, 84, 225
Cinco Tinajas, 11, 32, 46, 48–51, 107, 109, 132, 136, 138, 144–45, 157, 196, 202, 207, 242, 252, 270, 283, 292, 296, 298, 303, 310, 314, 318, 326–27, 332, 345, 351, 357, 361, 363, 404
Cirsium
    turneri, 90, 133, 379, 386
    undulatum var. *undulatum*, 90, 134
Cissus incisa, 49

Citrullus lanatus, 16, 386
Clematis
    drummondii, 95, 335
    pitcheri var. *dictyota*, 95, 336, 386
Cleomaceae, 91, 195–96
cliff waxwort, 50
climbing wartclub, 25, 34
Closed Canyon, 13, 20–21, 114, 146, 152, 155, 169, 182, 226, 233, 281, 377
Closed Canyon Creek, 22
clumped dog cholla, 15, 28, 76, 78, 82, 90, 163
cluster flaveria, 17, 387
Coahuila blazingstar, 15, 93, 267, 381–82
Comanche
    chief, 17, 406
    Creek, West Fork, 73, 80, 82, 104, 160, 191, 371
    Crossing, 239
    prickly pear, 27, 35, 184, 389
    Trail, 14
Comanche(s), 3, 28, 34, 205, 415
Commelina erecta var. *angustifolia*, 46
Commelinaceae, 91
Commicarpus scandens, 25, 34
common
    button cactus, 62, 81, 91, 172
    cob cactus, 176
    purslane, 44, 53
    reed, 6, 8, 15, 20, 25
    threesquare, 20, 390
comparison of BBR and BBNP, 1
Condalia
    ericoides, 60
    warnockii var. *warnockii*, 95, 338
Connelly, Henry, 3
Conoclinium dissectum, 50
Contrabandista Spur Road, 80
Contrabandista Spur Trail, 71–72, 81, 116, 144, 219, 370
Contrabando
    Canyon, 15, 72, 81, 214, 234, 303, 305
    Creek, 13, 15–16, 61, 73, 78–81, 97, 126, 163, 178, 204, 207, 216, 218–19, 273, 309, 375, 378
    Dome, 73–74, 78–80, 317
    Lowlands, 2, 15–16, 43, 59, 61, 70–71, 73, 78–79, 175, 376–77, 382
    Mountain, 72, 81
    Movie Set, 15, 378
    Trail, Dome, 73, 79–81, 122, 180, 316

    Trail, East, 73, 80, 153
    Trail, Multiuse Trail, 16
    Trail, Waterhole, 80, 224
    Trail, West, 78–79, 81, 147, 162–63, 168, 180, 187, 221, 226, 239, 256, 281, 316, 356
    Trailhead, East, 73
    Trailhead, West, 16, 73, 78
    Waterhole, 11, 72, 81, 97, 216, 268, 370
Convolvulaceae, 91, 197–200
Convolvulus equitans, 36
copper zephyrlily, 52, 89, 106
Corydalis aurea, 93, 245
Corynopuntia
    aggeria, 90, 163
    grahamii, 35, 76
Coryphantha
    echinus var. *echinus*, 71, 81, 386
    macromeris var. *macromeris*, 22, 386
cosmopolitan bulrush, 25, 92, 206, 386
cottonwood(s), 5–8, 20, 22, 31, 34–36, 44, 46–47, 50, 77, 85–86, 242, 372, 403
Cottsia gracilis, 93, 270
Coues, Elliott, US Army surgeon-naturalist, 320, 373, 412
Coulter's brickellbush, 35
Crassulaceae, 91, 201–2
Crawford-Smith House, 44, 73, 75–76
creeping cliff brake, 35, 389
creeping rockvine, 62, 71, 88, 379, 386
creosote
    (shrub), 4, 5, 7, 31, 52, 60, 78, 96, 117, 173, 180, 377
    scrub, creosote, 128, 153, 163, 210, 224, 256–57, 267, 285, 292, 302–3, 309, 311, 313, 317, 356
    scrub, creosote-mesquite, 316
crestrib morning glory, 35, 45
cretaceous, 2, 57, 59–60, 62, 64, 66–67, 71
Cretaceous Period, 2, 78, 210
Cretaceous Shutup Conglomerate, 66
crossosoma (family), 92, 203
Crossosomataceae, 92, 203
Croton
    bigbendensis, 75, 386
    fruticulosis, 86
    pottsii var. *pottsii*, 92, 212
    sancti-lazari, 92, 213
Crow Town, 12, 26–27
crowpoison, 50
Cryptantha mexicana, 75
Crystal Trail, 80

cucumber (family), 92, 204
Cucurbita foetidissima, 35, 404
Cucurbitaceae, 204, 397, 415
Cuesta de los Mexicanos, 43
Cuesta Primo, 36
Cueva Larga, 34, 45, 404
Cuevas Amarillas, 404
Cuevas Amarillas Tuff, 35–36
Cupressaceae, 92, 205
curlytop knotweed, 25
Cuscuta
    indecora, 31
    umbellata, 91, 197
Cylindropuntia
    imbricata var. *imbricata*, 22, 27, 52, 75, 78, 164
    kleiniae, 90, 164
    leptocaulis, 90, 164–65
Cyperaceae, 92, 206–8
Cyperus
    esculentus var. *macrostachyus*, 75, 386
    odoratus, 75
    squarrosus, 92, 207
Cyphomeris gypsophiloides, 46
cypress (family), 92, 365

Dagger Flat Sandstone, 61
Dalea
    bicolor, 54
    formosa, 42, 44, 49, 52, 119
    frutescens, 49, 119
    glaberrima, 15
    lachnostachys, 69
    neomexicana, 31, 81
    pogonathera, 64
    wrightii, 44
damianita, 60, 64, 404
Dark Canyon, 13, 17–18, 22
Dasylirion leiophyllum, 89, 124
Dasyochloa pulchella, 94, 309
date palm, 67, 89, 120, 389
Datura
    quercifolia, 96, 354
    wrightii, 96, 355
David's spurge, 92, 122, 218
Davis Mountains mock vervain, 52, 387
Davis, John, 31
Davis, William T., 34
dayflower, 46
Dayia havardii, 28, 60, 379
deergrass, 53
Del Carmen (Formation), 60, 63–64, 85
dense ayenia, 64
Dermatophyllum secundiflorum, 42, 44
Descurainia pinnata, 25

desert
    Christmas cholla, 7, 52, 90, 164–65
    evening primrose, 28
    hackberry, 25
    holly, 52, 89, 128
    Indianwheat, 25, 94, 302, 389
    marigold, 31, 35
    myrtlecroton, 44, 64
    night-blooming cereus, 39, 91, 186, 379
    olive, 7–8, 50, 60, 85, 403
    pincushion cactus, 22, 36, 91, 173–74, 379
    rosemallow, 36, 39, 45, 86
Desmanthus velutinus, 92, 230, 417
Desmodium neomexicanum, 50, 386
devil's claw, 28–29, 277–78
Devil's Graveyard, 59, 69
Diamond A Cattle Company, 4
Diamond A Ranch, 36, 39
Dichondra argentea, 54
Diospyros texana, 92, 209, 417
Diploid purple prickly pear, 25, 28–29, 31, 52, 91, 182, 388
dog cholla, 7, 15, 28, 35, 76, 78, 82, 90, 163
Dog Cholla Trail, 80
dogbane (family), 89, 109–18
Dos Pilas (Spur), 61, 404
doubleclaw, 17, 94, 278
downy paintbrush, 61, 64, 94, 293
downy wolfberry, 64, 71, 74, 78, 85, 96, 356
drainage from Sauceda to Cinco Tinajas, 51, 109, 196, 252, 327, 345, 363
Drymaria pachyphylla, 79, 386
Dryopetalon auriculatum, 15, 386
Duncan's pincushion cactus, 91, 173–74, 379, 387
Durango yellowcress, 90, 158, 380, 390
Dutchman's breeches, 46
dwarf
    ayenia, 64, 93, 271, 386
    false pennyroyal, 20
    Indian mallow, 50
Dyschoriste linearis var. *decumbens*, 54

Eagle Mt, 59
eagle-claw cactus, 69, 81, 90, 166, 178
earlobe mustard, 15, 386
Early Archaic Period, 2
Eaton's lip fern, 49
Ebenaceae, 92, 209, 417
Echeveria strictiflora, 42, 46, 85, 91, 201

Echinocactus horizonthalonius var. *horizonthalonius*, 90, 166
Echinocereus
    coccineus var. *paucispinus*, 90, 167, 379, 413
    dasyacanthus, 91, 168
    enneacanthus var. *enneacanthus*, 22
    stramineus var. *stramineus*, 91, 169
    viridiflorus var. *canus*, 91, 170, 379
    x *roetteri* var. *neomexicanus*, 61
Echinomastus
    mariposensis, 91, 171, 379, 386, 414
    warnockii, 31, 78–79, 386
Echols, William, 3, 31
Eclipta prostrata, 16
El Camino del Rio, 13, 408, 411
El Despoblado, 2, 404
El Oso Ranch, 39
El Padre al Altar, 16, 404
El Peñasco, 17
Eleocharis montevidensis, 92, 208
Emory, William, US Army Major, ix, 3, 130, 162, 215, 286, 308, 374, 418
Emory Peak, 1
Emory's mimosa, 60, 85, 388
Encino Loop Trail, 51, 140, 244, 282, 377
Encino Nature Trail, 51
Engelmann's prickly pear, 7, 17, 22, 52, 71, 91, 184–85
Entradas, 5, 404
Ephedra
    aspera, 81
    trifurca, 92, 210
ephedra, 7, 210, 216, 418
Ephedraceae, 92, 210, 418
Epipactis gigantea, 94, 290
Epithelantha
    bokei, 81, 88, 172, 379
    micromeris, 91, 172
Epixiphium wislizeni, 94, 300, 387
Equisetaceae, 92, 211
Equisetum xferrissii, 92, 211, 387
Eremothera chamaenerioides, 94, 287
Eriogonum
    abertianum, 47
    havardii, 95, 316, 387
    jamesii, 45
    rotundifolium, 42
    suffruticosum, ix, 95, 317, 379
    tenellum, 47
    wrightii var. *wrightii*, 46
Erodium
    cicutarium, 93, 248
    texanum, 93, 249, 387

Erythranthe
    chinatiensis, 94, 295, 379, 387
    inamoena, 94, 296, 387
Escobaria
    dasyacantha, 91, 173, 379
    duncanii, 91, 173, 174, 379, 387
    sneedii var. *albicolumnaria*, 91, 175, 379
    tuberculosa var. *varicolor*, 91, 176
escobilla butterflybush, 69
Espejo, Antonio de, 3
Eucnide bartonioides, 93, 264
Euphorbia
    abramsiana, 92, 214, 387, 418
    acuta, 92, 215, 387
    albomarginata, 78
    antisyphilitica, 92, 216, 403
    arizonica, 35, 75
    chaetocalyx var. *chaetocalyx*, 61
    cinerascens, 92, 217
    davidii, 92, 218
    eriantha, 74
    exstipulata, 92, 219
    hyssopifolia, 36
    indivisa, 92, 220, 387
    perennans, 92, 221, 379
    revoluta, 92, 222
    serpillifolia, 17
    serrula, 52
    setiloba, 92, 223
    simulans, 20, 88, 379
    spurca, 67, 387
    stictospora, 92, 224
    theriaca, 67, 75, 387
    villifera, 92, 225, 387, 417
Euphorbiaceae, 92, 212–26, 297, 394, 417–18
Euploca
    greggii, 52
    powelliorum, 90, 150, 387
Eustoma exaltatum, 93, 246
evening primrose (family), 28, 63, 69, 77, 94, 287–89
evergreen sumac, 8, 36, 44, 60, 85
Evolvulus alsinoides, 91, 198, 417

F Mountain, 85
Fabaceae, 92, 227–39, 396
Fagaceae, 93, 240–43
fairy swords, 50, 55
Fallugia paradoxa, 95, 340
false broomweed, 78
false grama, 25, 94, 307
fameflower (family), 96, 363–64
fascicled bluet, 63–64, 71

Faver, Milton, 6, 85
feather dalea, 42, 44, 49, 52, 119
Fendlera linearis, 93, 250, 379
Fendler's bladderpod, 31
Fendler's lip fern, 66
fern acacia, 54, 86
Ferocactus hamatacanthus var. *hamatacanthus*, 91, 177
Ferriss's scouring rush, 43–44, 74, 76, 92, 211, 387
fewflower beggarticks, 35
fiddle dock, 8, 16, 95, 318
fig marigold (family), 22, 89, 99–100
figwort (family), 96, 350
flatglobe dodder, 52, 91, 197
Flatirons, 57–58, 63, 76
flatspine stickseed, 77
Flaveria trinervia, 17, 387
flax (family), 93, 263
fleshy tidestromia, 15, 82, 89, 104, 380, 390
Flourensia cernua, 49
flower of stone, 8, 42, 49, 96, 351
fluffgrass, 7, 79, 94, 309
FM 170, 12–13, 15, 25, 27–30, 32, 72, 78, 229, 408, 411
foreign cloak fern, 95, 328, 381–82, 388
Forestiera angustifolia, 50, 60, 85, 403
Fort Davis, xiv, 29, 85, 323
Fort Leaton, 28, 277, 415
Fort Leaton State Historic Site, xii, 10, 28
Fouquieria splendens, 93, 244, 405
Fouquieriaceae, 93, 244
four-o'clock (family), 46, 94, 281–85
fourwing saltbush, 8, 77,
Fowlkes brothers, Edwin H., J. M. (Mannie), wife Frankie, 4, 34, 39–40, 50, 63, 69
Fowlkes era, 25, 33, 39
fragrant flatsedge, 75
fragrant heliotrope, 52
Fraxinus
    greggii, 66
    velutina, 94, 286
Fresno
    Canyon, v, ix, 16, 33, 42–44, 48, 52–53, 57, 59, 62–64, 67, 73–79, 81, 108, 111, 117, 130, 152, 159–60, 199, 204, 211, 227, 235, 251, 254, 260, 264, 268, 274, 281, 285, 296, 335, 345–46, 358, 366, 368, 375, 407, 412, 417–18
    Creek, 16, 59, 67, 73–74, 76–77, 79, 104, 178, 180, 232, 262, 284, 286–87

    Creek, lower, 178, 232, 287
    Creek Bridge, 16
    Creek Trailhead, 16, 73
    Divide Trail, 16, 78–79, 284
    Formation, 17
    Mine, 11, 56, 71, 75–76, 81
    Peak, 11, 56–58, 62–63, 250, 301, 317
    Ranch, 16, 73
    Rim Trail, 52–53, 283, 292
    Volcano, 27
    West Rim Map, 43
Fresno-Terlingua Monocline, 2, 69, 78
fringed monkeyflower, 94, 295, 379, 387
fringed puccoon, 50, 90, 151
Froelichia arizonica, 89, 102
Fuirena simplex, 43
Fumariaceae, 93, 245
fumitory (family), 93, 245
Funastrum
    crispum, 89, 113
    hartwegii, 89, 114
    torreyi, 89, 115

Gaillardia pinnatifida, 52, 387
Galium microphyllum, 49
gentian (family), 93, 102, 246–47
Gentianaceae, 93, 246–47
geological, v, ix, xii, 5, 8, 10, 27, 59, 63–64, 247, 279, 393–401, 407–8, 412, 415, 417, 419
geological history, 2, 59, 78
geology, xii–xiii, 2, 7, 10, 27, 59, 260, 407, 412, 414–15, 419
Geraniaceae, 93, 248–49
geranium (family), 93, 248–49
giant
    fishhook cactus, 35–36, 81, 91, 177,
    helleborine, 43, 94, 290
    reed, 8, 15, 25, 94, 305, 403
Giliastrum
    acerosum, 95, 311
    stewartii, 44
glandleaf milkwort, 71, 79
Glandularia pubera, 52, 387
Glandulicactus uncinatus var. *wrightii*, 91, 178
Glen Rose (Formation), 60, 62, 64, 83, 85, 297
glory of Texas cactus, 15, 78, 91, 187
Glossopetalon spinescens var. *spinescens*, 92, 203
goathead, 79, 96, 378
golden columbine, 7–8, 43–44, 76, 95, 334

golden crownbeard, 60
golf ball cactus, 78, 80
Gomphrena nitida, 69
Goodding's willow, 83, 96, 348
goosefoot (family), 91, 191–94
goosefoot moonpod, 27, 44, 52, 94, 282
Graham's dog cholla, 35, 76
grama, 5, 7, 17, 25, 31, 42, 45, 94, 150, 306–8
grape (family), 96, 373, 413
grass (family), 305–10, 409, 414, 416
grass(es), ix, 4–7, 20, 25, 31, 42, 45, 53, 60, 79, 94, 150, 206, 305–10, 393–94, 409, 414, 416
Grassy Banks (Campground), 11, 14, 16–17, 135, 137, 153, 193, 305, 318
gray five eyes, 63,
gray oak, 8, 44–46, 50, 54, 60, 62, 71, 85, 93, 240, 242
graybeard cactus, 91, 170, 379
Gray's cloak fern, 95, 330, 381–82
Great Wall of Chihuahua, 16
greenleaf five eyes, 96, 353, 386
Gregg, Josiah, US Army guide and interpreter, 106, 118, 128, 130, 155, 158, 186, 191, 194, 217, 260, 267, 283–84, 331, 360
Gregg's ash, 66
Gregg's cloak fern, 65, 71, 95, 331, 381–82, 388
Grindelia oxylepis, 90, 135, 381, 387, 414–15
groovestem bouchea, 69, 81, 96, 370
Guaiacum angustifolium, 34, 85, 404
Guajillo, 67, 390
Guale
    1 (campsite), 40, 252, 338, 352
    2 (campsite), 40–41
    Mesa, 13, 20, 25, 39–41, 107, 166, 179, 198, 213, 233, 270, 281, 308, 332, 341, 350, 404
    Mesa Road, 25, 33, 39, 40–41, 102, 151, 292
    Mesa Trailhead, 25, 40
guayacán, 34, 85, 404
guayule, 60, 181, 293, 404, 415
Guilleminea densa, 51
Gutierrezia microcephala, 45
Gymnosperma glutinosum, 42
gyp daisy, 78, 81, 90, 147, 380
gypsum tansy aster, 7, 22, 90, 146, 381–82, 390

Habranthus longifolius, 89, 106
hairy
    bommer, 50, 95, 323
    crinklemat, 7, 81
    five eyes, 31
    grama, 17
    spurge, 85, 92, 225, 387
    tubetongue, 55
Haploesthes greggii var. *texana*, 78
Haplophyton cimicidum var. *crooksii*, 89, 116, 387
Hardy (Pittman), Jean, 60–61, 87, 291, 391, 409, 414
Harper, S. S. (Ted), 86
Hartweg's evening primrose, 69,
Hartweg's twinevine, 25, 27, 35, 89, 114
Havard, Valery, US Army surgeon, ix, 122, 124–25, 161–62, 307, 316–17, 330, 405, 414
Havard's
    agave, xiv, 22, 43, 80, 89, 122
    dayia, 28, 60, 379
    nama, 69
    plum, 7, 35, 42, 44, 50, 54, 88, 380
    wild buckwheat, 64, 80–81, 95, 316, 387
heartleaf groundcherry, 96, 360
heartleaf rockdaisy, 47
heath carlowrightia, 50, 54, 69, 89, 97
heath cliffrose, 60, 71, 85, 95, 343
Hebecarpa barbeyana, 95, 314
Hechtia texensis, 90, 161
Hedeoma nana, 20
helecho de la Llave, 87, 95, 324, 381, 388, 391, 404
Helenium amphibolum, 50, 387
Heliomeris longifolia, 50, 387
Heliotropium curassavicum, 74
hemp (family), 91, 189
Herissantia crispa, 51
Hesperidanthus linearifolius, 49
Heteropogon contortus, 94, 310
Heterotheca arizonica, 17, 69, 387
Hexalectris warnockii, 94, 291, 379
Hibiscus
    coulteri, 36, 39, 45, 86
    denudatus, 39
Highlands,
Highway 170, 1, 13, 29, 78, 82, 146, 152, 337
hiking, xi, xiii-xiv, 1, 10, 16, 39, 42–43, 50–52, 54, 61, 71, 78, 86, 119, 124, 382, 407
Hill, Robert T., 14
hillside vervain, 64,
Hinckley's oak, 93, 241, 297, 380
historic past, 3

Hoban, Father Joseph, 34, 43
Hoffmannseggia glauca, 52
hog potato, 52
Holland, Clay T., 63, 69
Hondo Oil and Gas Company, 4
hooded arrowhead, 15, 89, 101, 390
hoodoos, 16, 27
Hoodoos, 11–12, 25, 114, 139, 206, 234, 305–7, 310, 362
hooked water clover, 52, 94, 276, 388
Hopi tea greenthread, 67, 90, 143
horse purslane, 35, 85, 89, 100
horseback riding, 1, 51, 78, 407
horsetail (family), 92, 211
Horsetrap
    Bike and Hiking Trail, 50, 54
    Spring, 32, 48, 50, 97, 150, 189, 217, 348, 368
    Spring Loop Trail, 48, 50, 189
Houstonia acerosa var. *acerosa*, 60
Howard, George, 3
Howard's Ranch, 3, 6, 48, 52–53
huisache, 25, 390
human habitation, 2, 35
human prehistory, 2
Humphris, John, 31, 34
hybrid cloak fern, 36
hydrangea (family), 93, 250
Hydrangeaceae, 93, 250
Hydrophyllaceae, 93, 251–53
Hymenothrix pedata, 34, 79
Hymenoxys odorata, 52
hyssopleaf asphead, 60, 64–65, 93, 269
hyssopleaf sandmat, 36

Ibervillea tenuisecta, 44, 62
Inner Loop (Trail), 62, 162, 172
International Dark Sky Park, 1
invasive
    (plants, species, exotics), 5–8, 15, 25, 193, 248, 280, 305, 362, 365–67, 378, 396
    geranium, 248
    perennial, 305
    shrub, 193
Ipomoea
    cordatotriloba var. *torreyana*, 91, 199, 388
    costellata, 35, 45
    cristulata, 45, 50
    hederacea, 91, 200, 388
Ipomopsis
    laxiflora, 95, 312, 388
    longiflora var. *neomexicana*, 95, 313, 388

Iresine leptoclada, 89, 103
Iron ipomopsis, 52, 95, 312, 388
Isely's feather duster, 61, 69, 92, 229, 386
Isocoma pluriflora, 74
ivy treebine, 49
ivyleaf morning-glory, 34, 91, 388

Jackson Pasture, 59
Jackson Pens, 56, 59–60, 118, 140, 256, 357
Jackson Pens Trail, 60, 69
Jackson, Pearl Andrew, 59–60
James's monkeyflower, 22, 25, 47, 94, 296, 387
James's nailwort, 78
James's wild buckwheat, 45
Jatropha dioica var. *graminea*, 92, 226
Javelin
    Campsite, 42
    Pens, 33, 41–42
    Road, 17, 33, 41–43, 50, 52–54, 126
    Trailhead, 17, 42
javelina bush, 60
Jefea brevifolia, 69
jimmyfern, 36, 50
jimmyweed, 74
Juglandaceae, 93, 254
Juglans microcarpa, 93, 254
Julia's goldenrod, 36, 44
Jumanos, 3
Juncaceae, 93, 255
Juncus nodosus, 93, 255
juniper, 5, 8, 60, 190, 205, 366, 406
juniper globemallow, 63–64, 71, 94, 275
juniper roughland, 7–8, 167, 243, 329
juniper woodland, 5
Juniperus pinchotii, 92, 205
Justicia
    pilosella, 55
    warnockii, 61–62, 71, 89, 98, 379

Kallstroemia
    californica, 20,
    grandiflora, 96, 374
    parviflora, 96, 375
    perennans, 96, 376, 379
knotted rush, 22, 85, 93, 255
Koanophyllon solidaginifolium, 90, 136
Koeberliniaceae, 93, 256
Koeberlinia spinosa var. *spinosa*, 93, 256
Krameria
    bicolor, 93 258, 417
    erecta, 93, 257

Krameriaceae, 93, 257–58, 417
Kunth's evening primrose, 77

La Cienega, 85
La Cuesta, 12, 20, 130, 183, 234
La Guitarra, 20, 40, 404
La Iglesia, 33, 40, 222, 325
La Junta, 3, 404, 415
La Monilla, 17
La Monilla Campsite, 43
La Morita, 85
La Mota
    (mountain), 11, 32–33, 46–47, 49–51, 54, 103, 195, 404
    1 (campsite), 46
    2 (campsite), 46,
    Road, 33, 45–46, 48, 51, 167
lacy tansyaster, 45
Lajitas
    (community), xi, xiii, 1, 3, 11, 13–14, 16–17, 20, 31, 61, 72, 78, 81, 100, 233–34, 239, 374, 381, 391, 403–5, 408
    1 mile west of, 15, 117, 192, 194, 214, 304, 318, 359, 365–66
    1 ¼ miles west of, 15, 163, 187, 193, 294, 300, 302, 313
    Boat Launch, 14–15, 193
    Golf Resort, 15
    Mesa, 13, 16, 73, 79
    Mesa, North, 13, 80–82
    Mesa, South, 14–15, 79, 104, 117, 142, 160, 187, 191, 252, 267, 284, 287–88, 351, 374
    miles from, 15–17, 20, 22, 25, 27–28
    Resort, 4
    Stables, 15
    Trading Post, 14
Lamiaceae, 93, 259–62
lanceleaf cottonwood, 50
lanceleaf groundcherry, 8, 96, 359, 389
land use changes, 5
Lantana achyranthifolia, 64, 405
Lappula occidentalis, 77
Larrea tridentata, 96, 377, 414
Las Burras
    1 (campsite), 39
    2 (campsite), 39
    3 (campsite), 39
    Canyon, 27, 39
    Creek, 27,
    Road, 25, 27, 33, 36, 39, 123, 136, 167
    Trailhead, 27, 39

Las Cuevas, 11, 32, 35–36, 100, 107, 115, 118, 124, 131, 140, 195, 205, 232, 244–46, 267, 279, 294, 314, 348, 350, 373–74, 404
Late Archaic Period, 3
Late Pleistocene, 5, 419
Late Prehistoric Period, 3
leaf flower (family), 94, 297
leafy rosemary-mint, 64, 66, 71, 93, 259
leather flower, 45, 95, 336, 386
leatherstem, 7, 27, 31, 42, 53, 85, 92, 226
leatherweed, 62, 92, 212
Leaton Trading Post, 3
Leaton, Ben, 3, 28
lechuguilla, 2, 4–5, 7, 31, 52, 60, 89, 98, 123, 150, 161, 180, 297, 404, 417
Lefthand Shutup, 56, 59–61, 65–66, 71, 97, 110, 114–15, 118, 133, 150, 190, 203, 209, 227, 250, 258–59, 263, 292, 294, 311, 327
Lefthand Shutup Creek, 59, 61, 65
Lemmon's brickellbush, 62, 78, 386
Leucophyllum minus, 96, 350
Leyva
    Canyon, 33, 45–49, 51–52, 202, 242, 283, 319, 327, 348, 405
    Creek, 33, 45–46, 48, 52, 231, 238, 254, 346, 349, 354–55, 357
    Dome, 31
    Escondido (Spring), 33, 45–46, 49, 201
    Trailhead, 46–47
limerock brookweed, 77
Limpia blacksenna, 86
Linaceae, 93, 263
Lincoln Land and Cattle Company, 4
Lindheimer's copperleaf, 51, 85
Lindheimer's senna, 52, 92, 237, 404
linear-leaf four o'clock, 50–51
Linum
    berlandieri var. *filifolium*, 61
    rupestre, 93, 263
Lithospermum incisum, 90, 151
little mallow, 15, 388
little walnut, 8, 46, 69, 74, 93, 254
littleleaf moonpod, 81, 88, 379, 386
liverwort, 50
living rock cactus, 62, 71, 78, 90, 162
lizardtail gaura, 22, 94, 289
Llano
    (area), v, 6, 10–11, 32–33, 46, 48–55, 78, 330, 405
    Dome, 52, 167, 185, 320, 333
    Loop, 48, 52–53, 105–6, 122–24,

126, 128, 140, 165, 167, 182, 197, 204, 224, 228, 235, 253, 256, 262, 272, 276, 306, 312, 353, 364, 377–78
   Loop, West, 52
   plain, 48, 51
   plateau, 2, 48
Llavea cordifolia, 95, 324, 381, 388, 391
Lloyd's hedgehog cactus, 61
Loasaceae, 93, 264–67
Lobelia cardinalis, 91, 188
London rocket, 25
longcapsule suncup, 15, 17, 44, 74, 94, 287
longleaf false goldeneye, 50, 387
longleaf jointfir, 27, 81, 92, 210
longpetal echeveria, 42, 46, 85, 91, 201
longstalk greenthread, 60
loosestrife (family), 93, 268
Lopez, Fray Nicolas, 3
lopseed (family), 94, 295–96
Los Alamos
   (site, area), 11, 56, 59–61, 69, 81, 97, 132, 166, 224, 227–28, 248, 251, 313, 317, 335, 407
   4WD Loop, 59–61, 69, 78, 229
   Campsite, 59–60, 69
   powerline feeder route, 51
   Road, 60, 69
   turnoff, 60, 69, 73, 128
Los Cuates, 32–33, 40–41, 405
Los Hermanos, 43, 405
Los Ojitos, 48, 50, 97, 151, 157, 173, 176, 232, 248
Los Portales, 59, 62–63, 72, 77, 121, 133, 250, 275, 315-   316, 326–27, 329, 405, 407, 414
lotebush, 85, 95, 339
Louisiana vetch, 20
Lower Shutup
   (drainage), 56, 59, 66–67, 69, 76, 115, 120, 132, 134, 157, 189, 205, 209, 243, 250, 260, 268, 315, 327, 335, 343, 360, 370
   Creek, 59, 62, 67, 73, 76
   Trailhead, 59, 62–63, 65, 67
Lupinus havardii, 17, 46–47
Lycium
   berlandieri, 20
   puberulum var. *puberulum*, 96, 356
lyreleaf parthenium, 60
lyreleaf twistflower, 25, 88, 380
Lythraceae, 93, 268
Lythrum californicum, 93, 268

Machaeranthera tanacetifolia, 90, 137
Maclura pomifera, 46, 66
Macroptilium atropupureum, 92, 231
madder (family), 95, 345
Madera Canyon Campground, 16–17, 131, 274, 287, 358, 366
Madrid
   Falls, 11, 32–33, 43–44, 107, 264, 307, 322, 334, 339, 369
   Falls Road, 42–43, 122
   Falls Trail, 157, 198, 201, 368
   House, 33, 43, 76, 156, 191, 411, 413
   Springs, 44
Madrid, Andres, 43–44
Madrid, Eusebia, 43
maidenhair fern, 8, 36, 44, 95, 322
maidenhair fern (family), 95, 322–33
Malacomeles pringlei, 95, 341, 388
mallow (family), 95, 271–75
Malpighiaceae, 93, 269–70, 411
Malvaceae, 93–94, 271–75, 394
Malvella
   parvifolia, 15
   sagittifolia, 93, 272, 388
Mammillaria
   lasiacantha, 78, 80
   meiacantha, 91, 179
   pottsii, 91, 180
Mandevilla macrosiphon, 62
manybract groundsel, 46, 90, 138
many-flowered stickleaf, 266
Map
   Discovery Topographic, xii, 10
   Exploration, xii, 10
   Interactive, xii, 10
Maravillas (Formation), 62, 66
Marfa-Terlingua Road, 14, 73, 76, 418
Margaranthus solanceus, 96, 357
margined rockdaisy, 25
mariola, 31, 34–35, 44, 52, 60, 85, 90, 140
mariposa cactus, 91, 171, 379, 386
Marsilea vestita, 94, 276, 388
Marsileaceae, 94, 276
Martyniaceae, 94, 277–78
Masada Ridge, 11, 56–57, 59, 61, 71, 150, 181, 205, 209, 243, 259–60
mat nama, 45
mat rock spiraea, 65, 95, 342
McGuirk's Tanks, 56, 59, 61, 103, 274
McGuirk's Tanks Campsite, 59, 61
mealy goosefoot, 83
Mecardonia procumbens, 50
Mecheranero, Colonel, 34

Medicago sativa, 22
mejorana, 64, 405
Melampodium leucanthum, 53
Melochia pyramidata, 93, 273, 388
melón loco, 52, 79, 92, 204, 405
Mendoza, Dominguez de, 3
Menodora
   longiflora, 50
   scabra, 34–35
Mentzelia
   asperula, 93, 265
   longiloba, 266
   longiloba var. *chihuahuensis*, 20
   mexicana, 93, 266, 388
   oligosperma, 267
   pachyrhiza, 93, 267
Meriwether, G. C. (Gay), 63, 69
Mesa de Anguila, 1, 14–15, 382, 405
mesa dropseed, 31
mesa greggia, 31
mesquite, 4–7, 31, 43, 60, 85, 118, 193, 301, 371, 403, 411
mesquite scrub, 116, 173, 262, 316, 329
Mexican
   blue oak, 44, 93, 242
   buckeye, 17, 36, 44, 81, 83, 405
   clammyweed, 25, 47, 91, 195
   cryptantha, 75
   devilweed, 20
   gumweed, 15–16, 90, 135, 381, 387
   navelwort, 22, 90, 152, 388
   skullcap, 22, 93, 261
   stickleaf, 17, 81, 93, 266, 388
   tiquilia, 79, 81
Mexican Revolution, 3, 27, 29, 34, 81
Mexicano
   1 (campsite), 43
   2 (campsite), 43
   Falls, 11, 32–33, 42–44, 322, 334
   Falls Trailhead, 43–44
Middle Archaic Period, 2
mignonette (family), 95, 337
milkwort (family), 95, 314–15
Mills, Primo, 36
mimicking matspurge, 20, 88, 379
mimosa, 4, 7, 232
Mimosa
   aculeaticarpa var. *biuncifera*, 49, 74, 85
   emoryana, 60, 85, 388
   texana, 92, 232, 388
   turneri, 92, 233
mine(s), 2, 11, 14, 56, 59, 67, 71, 73, 75–76, 80–81, 85, 171

mining, 6, 59, 66–67, 73, 75–76, 171, 241
mint (family), 93, 259–62
Mirabilis linearis, 51
Mischer, Walter, 4, 14
mistletoe (family), 96, 371–72
Monilla Canyon, 17, 42, 121, 137, 141, 155, 157, 199, 212, 237, 252, 262, 273, 294, 337, 361, 373
Monilla Creek, 17, 42, 238
Moraceae, 94, 279
Morita, 405
Morita Creek, 83, 85
Morita Ranch (Formation), 83
Mormon tea (family), 92, 210
morning-glory (family), 91, 197–200, 388
Mortonia scabrella, 91, 190
Morus microphylla, 94, 279
mountain mustard, 49
movie set, 12, 15–16, 27, 101, 135, 366–67, 378
Muhlenbergia rigens, 53
mulberry (family), 94, 279
muleear spurge, 55, 92, 220, 387
mustard (family), 90, 156–60
Myriopteris
　alabamensis, 49
　aurea, 95, 325
　fendleri, 66
　gracilis, 71
　lindheimeri, 50, 55
　rufa, 49
　scabra, 95, 326
　windhamii, 95, 327, 414
　wrightii, 49–50

naked brittlestem, 28
Nama
　havardii, 69
　hispidum, 93, 251
　torynophyllum, 45
　undulatum, 69
narrowleaf
　fendlerbush, 63, 93, 250, 379
　globemallow, 75, 93, 274
　moonpod, 22, 74, 78, 94, 281
　spiderling, 60,
Nasturtium officinale, 90, 156, 415
needle grama, 42
Needle Peak, 56–57, 62–63, 144
needleleaf bluet, 60
needleleaf gilia, 61, 95, 311
Neolloydia conoidea var. *conoidea*, 91, 181
Nerisyrenia camporum, 31

netleaf hackberry, 7, 35–36, 50, 53, 83, 91, 189
netted globecherry, 46, 83, 96, 357
nettleleaf goosefoot, 15, 91, 194
New Mexico
　cliff fern, 86
　dalea, 31, 81
　silverbush, 85
　tickflower, 50, 386
Nicotiana
　glauca, 96, 358
　obtusifolia, 20
nightshade (family), 96, 353–62
nipple cactus, 62, 91, 179
Nitrariaceae, 94, 280
nitre bush (family), 94, 280
Nolina erumpens, 89, 125
Nopalera (Trailhead), 25, 36, 39, 405
Norsworthy, Jeanne, 16
North Lajitas Mesa, 13, 80–82
northern dewberry, 36, 95, 344
Notholaena
　aliena, 95, 328, 381, 388
　aschenborniana, 95, 329, 381, 388
　grayi ssp. *grayi*, 95, 330, 381
　greggii, 95, 331, 381, 388
　standleyi, 95, 332
Nothoscordum bivalve, 50
novaculite, caballos, 61–63, 66–67, 112, 169–70, 176, 201, 212–13, 311, 397
Nuttallanthus texanus, 49, 388
Nyctaginaceae, 94, 281–85, 393, 413

oak, 5–6, 45–46, 60, 138, 157, 177, 240–42, 291, 293, 404, 416
oak mistletoe, 46
oak-juniper communities, 190
oak-juniper roughland, 8, 167, 243, 329
oakleaf datura, 46, 83, 85, 96, 354
ocotillo, 4, 6–8, 31, 34–35, 51–52, 54, 60, 81, 93, 244, 405
ocotillo (family), 93, 244
Oenothera
　boquillensis, 94, 288, 379
　brachycarpa, 63, 77
　curtiflora, 94, 289
　hartwegii ssp. *hartwegii*, 69
　primiveris, 28
　suffrutescens, 76
Ojinaga, 13, 28
Ojito Adentro, 11, 27, 32–33, 35–36, 108, 159, 173, 177, 188–89, 245, 252, 262, 270, 322, 325, 332, 344, 358, 367, 373, 405
Ojo

　Blanco, 43
　Chilicote, 11, 32, 48, 52–53, 127, 129, 148, 189, 218, 226, 237, 283, 292, 340, 346, 360
　de León, 20, 32–33
　de Papalote Alto, 45
　Escondido, 32, 46, 329
　Escondido Campsite, 46
　Escondido Pens, 45–46, 48
　Escondido Spring, 46, 48
　Mexicano, 32–33, 42, 48, 279
ojo de vibora, 44, 91, 198, 405
Old Entrance Road Trail, 34–36, 45
old man's beard, 75, 95, 335
Oleaceae, 94, 286
Oligomeris linifolia, 95, 337
olive (family), 94, 286
Omphalodes alienoides, 90, 152, 388
Onagraceae, 94, 287–89
Opuntia
　azurea var. *diplopupurea*, 91, 182–83, 388
　azurea var. *discolor*, 91, 183, 388
　azurea var. *parva*, 182
　camanchica, 27, 35, 184, 389
　dulcis, 91, 184, 389
　engelmannii var. *engelmannii*, 91, 184–85
　rufida, 17, 22, 39, 71, 75
orange caltrop, 27, 34, 96, 374
orange flameflower, 49
orchid(s), 1, 7, 43, 376, 396
orchid (family), 94, 290–91
Orchidaceae, 94, 290–91, 376, 394, 411, 419
oreganillo, 35, 45, 96, 369, 405
oreja de perro, 85, 90, 153–54, 405
Oreocarya palmeri, 61, 80, 389
Organ Mountain blazingstar, 45, 93, 265
Orobanchaceae, 94, 292–93
osage orange, 46, 66
Oso
　Creek, 39, 201
　Loop, 33, 35–36, 39–40
　Mountain, 1–2, 11, 32–33, 39–40, 49, 83
　Spring, 25, 39, 201
Ospital, Sanson, 34
Outer Loop (Trail), 61–63, 67, 77, 181, 212, 217, 222, 309

Packera millelobata, 90, 1, 26138
pack rat midden analysis, 5
Palafoxia riograndensis, 90, 139, 389
pale false nightshade, 63, 69

paleface rosemallow, 39
paleflower ipomopsis, 15, 69, 95, 313, 388
Paleo-American period, 2
Paleozoic, 2, 57, 61
Paleozoic period, 2
palm (family), 89, 120
Palmer's bluestar, 25, 55, 85, 89, 109
Palmer's cat's eye, 61, 80, 389
palmleaf thoroughwort, 50
Palo Amarillo Creek, 27–29, 33–35, 111, 140, 148, 200, 346, 355, 374–75
Palo Amarillo Spring, 34
Panther
    Canyon, 20, 33, 141, 143, 146, 149, 208, 218, 246, 251, 255, 289, 322, 326, 351, 374
    Creek, 20
    Mountain, 33, 40, 42
    Spring, 33
Panther Canyon (in Solitario), 59, 69
Papalote
    Colorado (Campsite), 33, 40, 45
    de la Sierra, 86
    Encino (Campsite), 48, 51–52
    Llano (Campsite), 48, 52
    Llano Nuevo (Campsite), 52, 320
    Ramon, 59, 62–63
    Severo, 36
Papalotito Colorado, 32–33, 40
Papaveraceae, 94, 294
park history, 3, 407
park plant communities, 7
Parkinsonia aculeata, 92, 234, 406
Paronychia jamesii, 78
Parry's holdback, 71, 389
Parthenium
    argentatum, 60, 404, 415
    confertum, 60
    incanum, 90, 140
Paso al Solitario, 56, 59–61, 65, 69, 222, 275, 311
Paso al Solitario 4WD Road, 166, 215
Paso Lajitas, 14, 405
Patarabueyes, 3
pearly globe amaranth, 69
Pectis prostrata, 55
Peganum harmala, 94, 280, 389
Pellaea
    intermedia, 35, 389
    wrightiana, 95, 333
Peniocereus greggii, 91, 186, 379
Penstemon
    baccharifolius, 60, 71
    dasyphyllus, 94, 301

Perdiz Conglomerate, 29, 83, 85
perennial caltrop, 96, 376, 379, 380
perennial sandmat, 79, 81, 92, 221, 379
Perityle
    dissecta, 86, 380
    parryi, 47
    vaseyi, 25
Pershing, General John "Black Jack," 14
Persicaria lapathifolia, 25
persimmon (family), 66, 71, 83, 92, 209
Peruvian spikemoss, 50, 55, 96, 352
Petrophytum caespitosum, 95, 342
Phacelia
    infundibuliformis, 93, 252, 389
    popei, 93, 253
    rupestris, 20, 389
Phaseolus
    angustissimus, 77, 389
    filiformis, 43
phlox (family), 95, 311–13
Phoenix dactylifera, 89, 120, 389
Phoradendron
    leucarpum ssp. *tomentosum*, 96, 371
    macrophyllum, 96, 372, 389
Phragmites australis, 15, 20, 25
Phrymaceae, 94, 295–96
Phyla nodiflora, 22
Phyllanthaceae, 94, 297
Phyllanthopsis arida, 94, 297, 380
Physalis
    angulata var. *lanceifolia*, 96, 359, 389
    hederifolia, 96, 360
Physaria
    fendleri, 31
    purpurea, 90, 157
Physiographic
    Region 1, v, xiv, 13
    Region 2, v, 28–29
    Region 3, v, 27–28, 33, 48
    Region 4, v, 46, 48,
    Region 5, v, 57
    Region 6, v, 16, 67, 71, 73
    Region 7, v, 28, 83
Phytolaccaceae, 94, 298
Pico de Fierro, 62
Pico de las Aguilas, 42, 48, 54, 369
Picradeniopsis absinthifolia, 31
pigweed, 34, 197
Pila
    de Gato, 48, 405
    de los Muchachos, 43
    de Tascate, 48, 54
    Montoya, 59
    Montoya 3, 77
    Montoya Road, 72, 77, 262

    Montoya Trail, 73, 76–78, 111, 121, 125, 134, 190, 349
    Montoya Trailhead, 76–77, 129, 260
Pilostyles thurberi, 89, 119, 389
pink baby's-breath, 51, 55, 96, 363, 390
pink windmills, 78
piñon, 5, 406
piñon-juniper, 5
Pitcher, Zina, US Army surgeon, 336
Plagiochasma rupestre, 50
plains dozedaisy, 16
plains ironweed, 50, 90, 145
plant communities, xi, 7
Plantaginaceae, 94, 299–304
Plantago
    hookeriana, 15
    ovata, 94, 302, 389
    rhodosperma, 49, 389,
plantain (family), 94, 299–304
plateau rocktrumpet, 62
Pleistocene, 5, 417, 419
plume tiquilia, 60, 90, 155
Poaceae, 94, 305–10, 418
pointed sandmat, 69, 92, 215, 387
poison ivy, 42, 44–45, 89, 108, 406
pokeweed (family), 94, 298
Polanisia uniglandulosa, 91, 195
Polemoniaceae, 95, 311–13
Poliomintha glabrescens, 93, 259
Polvo, El, 27, 39, 43–44, 405, 419
Polygala
    alba, 64
    macradenia, 71, 79
    scoparioides, 42
Polygalaceae, 95, 314–15
Polygonaceae, 95, 316–18
Polygonum
    argyrocoleon, 16, 389
    aviculare, 15, 389
Pomaria melanosticta, 71, 389
Pool, John Sr., 85
Pope, John, US Army Captain, 253
Pope's phacelia, 52, 93, 253
poppy (family), 94, 294
Populus
    fremontii subsp. *mesetae*, 95, 346–47
    x *acuminata*, 50
Porophyllum scoparium, 90, 141
Portulaca
    halimoides, 95, 319
    oleracea, 44, 53
    pilosa, 44,
    suffrutescens, 95, 320
    umbraticola var. *lanceolata*, 95, 321, 389

Portulacaceae, 95, 319–21, 415
Potts' mammillaria cactus, 71, 74–75, 78, 81, 91, 180
Powell Herbarium, xi, xiii, 386–91
Powell, A. Michael, 150, 170, 175, 182–83, 409
Powell, Shirley, 175, 183
Powell's heliotrope, 71, 90, 150, 387
Powerline Trail, 51–52
prehistoric past, 2
prehistory, 2, 411
Presidio
    (City), 1, 3, 13, 15, 28, 31, 101, 307, 365, 381, 408, 411
    Basin, 13
    bolson, 22
    de la Junta de los Rios, 3
    del Norte, ix, 3, 34, 124, 184, 281, 343
pretty dodder, 31
prickleleaf dogweed, 54
prickly lip fern, 44, 63, 95, 326
prickly pear, 4, 7, 163–64, 182, 398, 405
Primero Trailhead, 43
Pringle's swallowwort, 39
Proboscidea
    althaeifolia, 94, 277
    fragrans, 29
    parviflora, 94, 278
propellerbush, 25, 93, 270
Prosopis
    glandulosa var. *torreyana*, 92, 235
    pubescens, 92, 236, 389
prostrate knotweed, 15, 389
Protohistoric Period, 3
Prunus havardii, 35, 42, 44, 50, 54, 88, 380
Psathyrotopsis scaposa, 28
Pseudognaphalium luteoalbum, 16, 389
Pteridaceae, 95, 322–33, 394, 414
Puerta Chilicote Trail junction, 52
Puerta Chilicote Trailhead, 48, 53
purple groundcherry, 71
Purpus' tumblemustard, 36, 90, 159
Purshia ericifolia, 95, 343
purslane (family), 95, 197, 319–21

Quercus
    hinckleyi, 93, 241, 380
    oblongifolia, 93, 242
    pungens, 93, 243
Quincula lobata, 71

Rábago y Terán, Pedro de, 31
Rancherias
    4WD (Loop) Road, 40–41, 151
    Campsite, 41
    Canyon, 20, 22, 33, 39–41, 114, 173, 185, 208, 232, 246, 255, 261, 264, 268, 326, 337, 367, 371, 373, 377
    Canyon, lower, 367
    Canyon, upper, 41, 373
    Canyon Trailhead, 22
    Creek, 22, 40, 146, 152, 382
    Dome, 20
    Falls, 20, 22, 32, 41
    Spring(s), 20, 33, 40–41, 201, 223, 226, 310, 346, 350–51
    Spring Trailhead, 41
    Trail, East, 326, 375
    Trail, East Trailhead, 20
    Trail, Loop, 20, 22, 40–42
    Trail, Loop Trailhead, West, 22
    Trail, West, 40
ranches, 2–3, 6, 39, 406
ranching, 3, 6, 31, 33, 39–40, 42, 48, 51, 73
Rancho Moreno, 25, 27
Rancho Viejo, 11, 32–35, 100, 223, 406
Rancho Viejo Trail, 34–35
range ratany, 31, 60, 93, 257
Ranunculaceae, 95, 334–36
Rapistrum rugosum, 16, 390
rat midden analysis, 5
ratany (family), 93, 257–58
Rawls (Formation), 42, 52
Rawls, Thomas, 59
red berry juniper, 8, 35, 54, 60, 71, 92, 205
red cyphomeris, 46
red dome blanketflower, 52, 387
Redford, 1, 3, 11–13, 15, 20, 22, 27, 100, 365, 381, 406, 419
Redford Basin, 2, 13, 27
redseed plantain, 49, 389
redstem stork's bill, 50, 69, 93, 248
red-tip rabbit tobacco, 16, 389
Resedaceae, 95, 337
resurrection fern, 8, 42, 351
retama, 8, 15, 17, 20, 25, 92, 193, 234, 406
Rhamnaceae, 95, 338–39
Rhinotropis
    lindheimeri var. *parvifolia*, 22
    nudata, 95, 315
Rhus
    microphylla, 44, 46, 50, 53, 69, 85
    virens var. *virens*, 36, 44, 60, 85
Rhynchosia senna var. *texana*, 45
Rhynchosida physocalyx, 67
Righthand Shutup
    (Trail), 59, 61–64, 76–77, 106, 113, 123, 132, 213, 243, 259–60, 271, 341, 407, 414
    hills, slopes above, near, north of, 112, 162, 172, 177, 213
    interior slopes north of, above, 65, 269
    Outer Loop Trail beyond, 222
    slopes southeast of, 216
    Solitario rim above, 64, 117, 247, 269, 275, 293, 315- 316, 329, 331, 350, 356
    Trailhead, 72, 77
river dalea, 15
River Road, 13, 15, 17, 28, 135, 139, 234, 264, 358, 407–8, 411, 415
River Road (BBNP), 158
Rivina humilis, 94, 298
Road to Nowhere, 66–67, 105, 219
Roads to Nowhere, xii, 10, 407, 417
rock art, xiii, 2, 8, 46, 49, 73, 77, 414
rock
    flax, 66, 93, 263
    milkwort, 22
    phacelia, 20, 389
Rock Quarry Trail, 80
rock-shelter(s), 2, 20, 34, 45, 76
Rockcrusher Road, 81, 191
Rocky Mountain sage, 52, 390
Roemer's acacia, 8, 25, 35–36, 60, 71
Rorippa ramosa, 90, 158, 380, 390
Rosaceae, 95, 340–44, 398
rose (family), 95, 340–44
rose bladderpod, 69, 85, 90, 157
rouge plant, 49, 94, 298
rough menodora, 34–35
roundleaf wild buckwheat, 42
Rubiaceae, 95, 345
Rubus flagellaris, 95, 344
Rumex pulcher, 95, 318
rush (family), 93, 255
Russian thistle, 15

sacred datura, 29, 34, 96, 355
Sagittaria montevidensis ssp. *calycina*, 89, 101, 390
Salicaceae, 95, 346–48
Salix gooddingii, 96, 348
Salsola tragus, 15
salt cedar, 5–6, 8, 15, 17, 25, 96, 365, 366, 406
Saltgrass Draw, 56, 59, 70–71, 203, 209, 240, 243, 259–60
Salvia
    greggii, 93, 260
    reflexa, 52, 390

Samolus ebracteatus var. *cuneatus*, 77
San Carlos
    (canyon), 239
    Caldera, 17
    Tuff, 17
sand spikerush, 47, 92, 208
sandpaper oak, 44, 64, 66, 93, 243
sandpaper vervain, 44
Santa Elena
    (Formation), 60, 64, 297
    Canyon, 14, 158, 408, 411
    limestone, 29, 57, 60, 63, 71, 78
Santana
    Caldera, 17
    Mesa, 13, 17, 20, 32, 183, 406
    Tuff, 20
Sapindus saponaria, 96, 349
Sauceda (site)
    xiii–xiv, 1, 3, 11, 17, 28–29, 32, 34, 42–43, 46, 48, 50–52, 54, 59, 103, 131, 134, 143, 148–49, 179, 207, 228, 238, 272, 339, 354–55, 378, 391, 406–7, 418
    drainage from Sauceda to Cinco Tinajas, 51, 109, 196, 252, 327, 345, 363
    ranch house, 51
    Ranger Station, xii, 10, 51
    Road, 12, 27–36, 39, 45, 48, 51–52, 56, 59, 61, 69, 72, 77–78, 100, 163, 204, 214, 225, 272, 282
    Volcano, 48
Saucita (Ranch), 3, 6, 51, 406
Saucita Spring, 51
sawtooth spurge, 52
scaled cloak fern, 46, 63, 65, 95, 329, 381–82, 388
scarlet beeblossom, 76
scarlet bouvardia, 54
scarlet morning glory, 45, 50
Schoenoplectus pungens, 20, 390
Schott's acacia, 78, 81, 93, 239
scrambled eggs, 47, 85, 93, 245
screwbean mesquite, 8, 31, 36, 45, 83, 85, 92, 236, 389
Scrophulariaceae, 96, 350, 418
Scutellaria potosina var. *tessellata*, 93, 261
sea urchin cactus, 71, 81, 386
seaside petunia, 8, 74
sedge (family), 92, 206–8
Sedum wrightii, 91, 202
Seep Spring, 20
seepwillow, 8, 17, 35, 45, 51, 69, 89, 131, 193
Selaginella
    arizonica, 66
    lepidophylla, 96, 351
    peruviana, 96, 352
    wrightii, 50
Selaginellaceae, 96, 351–52
Senecio flaccidus var. *flaccidus*, 50
Senegalia
    berlandieri, 67, 390
    greggii, 31, 51, 60, 74, 83
    roemeriana, 25, 35–36, 60, 71
Senna
    lindheimeriana, 92, 237, 404
    pilosior, 34, 75
    wislizeni, 92, 238
Sentinel Rock, 35
Sesuvium verrucosum, 89, 99, 390
Seymeria scabra, 86
Shafter, 83, 85, 241
shaggy
    false nightshade, 25
    portulaca, 44
    stenandrium, 64, 79
shortfruit evening primrose, 63, 77
shorthorn jefea, 69
showy menodora, 50
shrubby
    purslane, 52, 95, 320
    senna, 17, 25, 46, 54, 92, 238
    umbrella thoroughwort, 49, 90, 136
Sida abutifolia, 78
sidecluster milkweed, 34, 42, 44, 85, 89, 111
sideoats grama, 7, 25, 94, 306
Sidneya tenuifolia, 51, 67, 85
Sierra
    Blanca (Dome), 83, 85, 190, 201
    El Mulato, 13, 27
    Rica, 2, 5, 13, 17, 20, 34, 40, 43, 79, 406
silver dalea, 54
silver ponyfoot, 54
silverlace cactus, 61, 71, 91, 175, 379
silverleaf nightshade, 85
silversheath knotweed, 16, 389
sinkerleaf purslane, 46, 50, 53, 95, 319
siratro, 36, 45–46, 92, 231
Sisymbrium irio, 25
Skeet Canyon, 48–50, 55, 107, 124, 134, 136, 140, 145, 208, 240, 287, 303, 319, 321, 323, 325, 333, 352
skeletonleaf goldeneye, 51, 67, 85
slender lip fern, 71
slimjim bean, 43
slimleaf bean, 77, 389
slimlobe globeberry, 44, 62
slimlobe rockdaisy, 86, 380
slimseed spurge, 81, 92, 224
slimstalk spiderling, 15, 35
small leaf silver fern, 52, 66
smallflower milkwort, 63–64, 67, 95, 315
Smith Canyon, 44, 159, 290, 367
Smith Spring, 44, 76, 322, 334
smooth sotol, 7, 35, 50, 52, 54, 60, 81, 89, 124
smooth spiderwort, 51, 91, 196
snapdragon, 50
soapberry (family), 96, 349
soaptree yucca, 7, 10, 13, 52, 78, 89, 126
soft twinevine, 35, 45, 89, 115
Solanaceae, 96, 353–62
Solanum
    citrullifolium, 96, 361
    elaeagnifolium, 85
    ptycanthum, 79, 390
    rostratum, 96, 362, 390
Solidago juliae, 36, 44
Solitario
    (area), v, 2, 5, 11, 16, 43, 49–50, 53–54, 56–71, 73, 76, 87, 112, 170, 174–75, 178, 216, 222, 241, 260, 291, 297, 308, 340, 391, 398, 406–7, 412–14, 416–17
    central basin, 57, 59, 61–63, 65–67
    Dome, 62, 87, 409, 414
    Inner Loop, 162
    Overlook, 59
    Peak, 11, 56–57, 62, 98, 103, 169, 172, 179, 213, 240, 258, 265, 317, 341, 405
    rim, 2, 63–64, 77, 110, 117, 269, 275, 293, 315–16, 329, 331, 350, 356, 382
    rim, east, 8, 59
    rim, north, 122–23, 126, 177, 180, 237
    rim, northwest, 60, 98, 112, 119, 124–25, 149, 154–55, 161, 168, 181, 187, 190, 205, 213, 229, 235, 240, 244, 247, 249, 256–58, 269, 275, 283, 293, 301, 311, 343, 350
    rim, west, 62
    Road, 59, 61, 65, 69, 73, 78, 118, 128
    Road East, 59, 61, 65
    Road West, 61, 128
sotol, 2, 4–8, 78, 124–25, 297, 301, 406
sotol-yucca, 203, 257, 316
South Lajitas Mesa, 14–15, 79, 104, 117, 142, 187, 191, 252, 267, 284, 287–88, 351, 374

South Leyva Campground, 51, 244, 256
South Orient Railway, 85–86
southern annual saltmarsh aster, 74
southern cattail, 15, 22, 36, 44, 96, 367
Spanish culture, 3
Spanish dagger, 7, 13, 17, 35, 53, 60, 89, 127
Spanish expedition(s), 3
Spanish occupation, 3
spearleaf, 15, 64, 75, 89, 117
spearleaf sida, 67
Sphaeralcea
    angustifolia, 93, 274
    digitata, 94, 275
spiderflower (family), 91, 195
spiderwort (family), 91, 196
spike burgrass, 45
spikemoss (family), 96, 351–52
spiny greasebush, 66, 71, 92, 203
spinyleaf zinnia, 60
splitleaf brickellbush, 50, 69
Sporobolis flexuosus, 31
Sprague, Julian, 4
spreading
    chinchweed, 55
    sida, 78
    snakeherb, 54
    spiderling, 22
spurca spurge, 67, 387
spurge (family), 92, 212–26
squareseed spurge, 67, 92, 219
Standley's cloak fern, 8, 36, 49, 95, 332
stargazing, 1
state park rules, 8
stemmy four-nerved daisy, 60, 390
Stemodia coahuilensis, 94, 303, 381, 390, 418
stemsucker (family), 89, 119
Stenandrium barbatum, 64, 79
Stephanomeria pauciflora, 90, 142
Stewart's gilia, 44
stickleaf (family), 93, 264–67
stinging cevallia, 71
stinkweed, 71
stonecrop (family), 91, 201–2
strawberry cactus, 7, 22, 25, 31, 42, 62, 71, 91, 169
strawberry hedgehog cactus, 22
Streptanthus carinatus ssp. *carinatus*, 25, 88, 380
Sue Peaks (Formation), 60
sumac (family), 89, 108
sunflower (family), 89–90, 128–47
sweet prickly pear, 27, 39, 91, 184, 389

swollenstalk sneezeweed, 50, 387
Symphyotrichum divaricatum, 74

Tahoka daisy, 90, 137
talayote, 39, 386
Talinaceae, 96, 363–64
Talinopsis frutescens, 89, 107
Talinum
    aurantiacum, 49, 390
    paniculatum, 96, 363, 390
    polygaloides, 96, 364, 390
tall wild buckwheat, 47
tallow weed, 15
Tamaricaceae, 96, 365–66
tamarisk (family), 96, 365–66
Tamarix
    aphylla, 96, 365
    chinensis, 96, 366
tanglehead, 7, 25, 42, 94, 310
Tanque
    Blanco, 73, 78
    Lara, 39–40
    Talique, 48
tansy mustard, 25
Tapacolmes, 27
Tapado Canyon, 13, 20, 25, 39–41, 109, 114, 130, 194–95, 238, 246, 266, 270, 296, 368
tarbush, 7, 49
Tascate
    1 (campsite), 48, 54
    2 (campsite), 54
    Hills, 6, 8, 48, 50, 52–55, 97, 102, 107, 109, 121, 124, 136, 167, 205, 213, 220, 238, 240, 244, 276, 320–21, 325, 333, 341, 352, 363–64
tatalencho, 42
Tecoma stans, 34, 50
Teepees Roadside Park, 17, 127
Terlingua
    (community), xi, xiii, 13, 73, 160, 171, 174, 227, 229, 285, 406, 413
    Creek, 59, 65, 71, 160, 227, 406, 408, 411
    (Mercury) District, 412, 419
    quicksilver (mercury) mines, 14, 76
    Ranch, 59, 71
    Uplift, 2, 11, 16, 56–57, 59, 61, 69–71, 73, 78–79, 81, 98, 133, 150, 153–54, 160–62, 168–69, 171, 174–75, 180, 185, 215, 229–30, 284, 331, 343, 356, 382
Terlingua milkvetch, 75, 92, 227
Terlingua spurge, 67, 75

Terneros
    Creek, 3, 28–29, 31, 33–34, 45–47, 69, 151, 182, 208, 277
    Creek, upper, 47, 245, 264, 296, 368
    Ranch, 31, 47
Terneros/Black Hills Creek Junction, 28
Tesnus (Formation), 63, 66–67
Tetraclea coulteri, 71
Tetraneuris scaposa, 60, 390
Teucrium depressum, 93, 262
Texas
    bindweed, 36
    claret-cup (cactus), 27, 39, 46, 52, 55, 61, 90, 167–68, 379
    cone cactus, 61–62, 71, 91, 181
    crinklemat, 7, 60, 71, 90, 154
    false agave, 7, 60, 71, 90, 161
    milkvine, 60, 89, 118
    mimosa, 50, 74, 92, 232, 388
    mountain laurel, 42, 44
    mulberry, 42, 94, 279
    persimmon, 66, 71, 83, 92, 209
    purplespike, 94, 291, 379
    rainbow (cactus), 7, 28, 61, 71, 78, 81, 91, 168
    shrub, 47, 89, 103
    snoutbean, 45
    stork's bill, 93, 249, 387
    thelypody, 64, 90, 160, 380
    toadflax, 49, 388
Texas Geological Survey, 247, 279. 419
Thamnosma texana, 46
Thelesperma
    longipes, 60
    megapotamicum, 90, 143
Thelocactus bicolor var. *bicolor*, 91, 187
Thelypodiopsis purpusii, 90, 159
Thelypodium texanum, 90, 160, 380
thickleaf drymary, 79, 386
Thompson's yucca, 7, 10, 13, 50–52, 54, 60, 78
threadleaf groundsel, 50
threadleaf snakeweed, 45
threadstem sandmat, 62, 92, 222
Three Dike Hill, 25–27
threeflower goldenweed, 7, 69, 81, 390
Thurber's stemsucker, 61, 89, 119, 389
Thymophylla acerosa, 54
Tidestromia
    carnosa, 89, 104, 380, 390
    lanuginosa, 89, 105
tinaja(s), 49, 59, 63, 67, 400, 406
Tiquilia
    canescens, 90, 153, 405
    gossypina, 90, 154

greggii, 90, 155
  mexicana, 79, 81
Tomostima cuneifolia, 75
toothed serviceberry, 8, 42, 44, 54, 64, 95, 341, 388
Torrey's tievine, 17, 45, 77, 91, 199, 388
tortolo, 406
Tortolo Creek, 83
Toxicodendron radicans, 89, 108
Tradescantia leiandra, 91, 196
Tragia ramosa, 45
Tragus berteronianus, 45
Trans-Pecos
  ayenia, 63
  carlowrightia, 81
  croton, 54, 60, 64, 92, 213
  false boneset, 49, 69, 90, 132
  maidenbush, 62, 94, 297, 380
  poreleaf, 86, 90, 141
  senna, 34, 75
tree cholla, 7, 22, 27, 52, 75, 78, 164
tree tobacco, 8, 15, 17, 75, 96, 358
Tres Papalotes, 11, 56, 59, 61, 65–67, 102, 128, 143, 178, 217, 230, 314, 364, 406
Trianthema portulacastrum, 89, 100
Tribulus terrestris, 96, 378
Trixis californica, 90, 144
trumpetflower, 34, 50
tubercled saltbush, 15, 91, 191
tubular slimpod, 53, 81
tufted cottonflower, 51
turkey tangle fogfruit, 22
Turner's mimosa, 15, 92, 233
Turner's thistle, 63, 66, 90, 133, 379, 386
Typha dominguensis, 96, 367
Typhaceae, 96, 367

US
  Army, 3, 27–29, 73, 76, 253, 418
  Highway 67, 83
  -Mexico Boundary Survey, 27, 107, 115, 119, 148, 157, 162, 169, 196, 239, 243, 262, 281, 286, 330, 334, 374
Ugalde, Juan de, 3, 22
Ungnadia speciosa, 17, 36, 44, 81, 83, 405
unicorn plant (family), 94, 277–78

Vachellia
  constricta, 78
  farnesiana, 25, 390
  schottii, 93, 239
  vernicosa, 75, 78

Vado Colorado, 27, 406
varicolor cob cactus, 44, 55, 91, 176
Vasquez House, 30–31, 34
Vasquez, Natividad Jr., 31
vegetation, 5–7, 22, 29, 35–39, 52, 59–60, 62, 66, 73, 76, 78, 87, 124, 221, 241, 266, 399, 407, 409, 414, 417–19
velvet ash, 8, 44, 74, 85–86, 94, 286
velvet bundleflower, 64, 67, 92, 230
Verbena
  bracteata, 55
  hirtella, 64
  scabra, 44
Verbenaceae, 96, 368–70
Verbesina encelioides, 60
Vernonia marginata, 90, 145
Veronica anagallis-aquatica, 94, 304, 390
vervain (family), 96, 368–70
Vicia ludoviciana, 20
Villa, Francisco "Pancho," 3, 14, 73
villous lip fern, 63, 66, 95, 327
Viscaceae, 96, 371–72
viscid acacia, 75, 78
Vista de Bofecillos, 32, 34, 105, 184, 223, 244, 374
Vista del Chisos, 43
Vitaceae, 96, 373
Vitis arizonica, 96, 373

walnut (family), 93, 254
Warnock, Barton, 13, 98, 124, 133, 221, 224, 229, 285, 297, 324, 338, 381, 391, 408–9
Warnock's
  cactus, 31, 78–79, 386,
  justicia, 61–62, 71, 89, 98, 379
  snakewood, 81, 83, 95, 338
warty caltrop, 79, 96, 375
water clover (family), 94, 276
water hyssop, 22, 85, 94, 299
water plantain (family), 89, 101
water speedwell, 85, 94, 304, 390
watercress, 85, 90, 156
waterfall(s), vii, 2, 20, 33, 35–38, 44, 81, 290, 295–96, 322, 334
waterleaf (family), 93, 251–53
watermelon, 16, 386
watermelon nightshade, 49, 96, 361
wavyleaf
  cloak fern, 8, 36
  thistle, 90, 134
  twinevine, 62, 89, 113
wax-making, 2, 73, 76
West Contrabando — Fresno Creek Trailhead, 73–74

West Indian nightshade, 79, 390
western
  honey mesquite, 7–8, 52, 60, 69, 92, 235
  sea purslane, 22, 89, 99, 390
  soapberry, 35, 46, 69, 96, 349
  umbrella-sedge, 43
wheelscale saltbush, 15, 91, 192
white milkwort, 64
white ratany, 60, 66, 93, 258
whitemargin spurge, 78
whitethorn, 78
whitewhisker fiddleleaf, 69
whitlowwort, 75
Whit-Roy Mine, 75–76, 81
willow, 5, 7–8, 22, 31, 36, 44, 50–51, 348, 372, 403, 406
willow (family), 95–96, 346–48
winged spurge, 17
wingpod purslane, 50, 55, 95, 321, 389
wire lettuce, 90, 142
Woodsia neomexicana, 86
woolly
  butterflybush, 34, 71
  paintbrush, 42, 53, 94, 292
  tidestromia, 34–35, 89, 105
woollyflower spurge, 74
Wooton's loco, 44, 51, 92, 228
Worthington, Richard, ix, 5, 8, 87, 168, 175, 382, 390–91, 408–9, 416, 418–19
Wright's
  cliff brake, 52, 55, 95, 333
  dalea, 44
  Dutchman's pipe, 17, 44, 55, 85, 89, 121
  Indian mallow, 51
  lip fern, 49–50
  snakeroot, 62–63, 85
  spikemoss, 50
  stonecrop, 46, 49, 91, 202
  wild buckwheat, 46

Xanthisma
  gypsophilum, 90, 146, 381, 390
  spinulosum, 45
Xylorhiza wrightii, 90, 147, 380
Xylothamia triantha, 69, 81, 390

Yedra
  1 (campsite), 45
  2 (campsite), 45
  Canyon, 40, 45, 105, 108, 131, 199, 223, 231, 236, 240, 265, 308, 336, 346, 360–61, 369, 373

Canyon Trail, 34
Creek, 46
Road, 45
Trailhead, 35, 45, 115, 336
yellow
    flameflower, 52, 67, 85, 96, 364, 390
    Indian mallow, 17
    nutgrass, 75, 386
    rocknettle, 15, 22, 47, 93, 264

yerba de tago, 16
yucca(s), 4–5, 7, 54, 127, 203, 257, 297, 301, 316
Yucca elata, 89, 126, 413
Yucca Formation, 60, 62, 64
yucca moth, 126–27
Yucca
    thompsoniana, ix, 10, 13, 50–52, 54, 60, 78
    torreyi, 89, 127
Yuma spurge, 34, 36, 92, 223

Zeltnera
    arizonica, 43, 85
    calycosa, 93, 247, 390
Zinnia acerosa, 60
Ziziphus obtusifolia, 95, 339
Zygophyllaceae, 96, 374–78

Other titles in the Kathie and Ed Cox Jr. Books on Conservation Leadership, sponsored by The Meadows Center for Water and the Environment, Texas State University

Politics and Parks: People, Places, Politics, Parks
George L. Bristol

Money for the Cause: A Complete Guide to Event Fundraising
Rudolph A. Rosen

Green in Gridlock: Common Goals, Common Ground, and Compromise
Paul W. Hansen

Hillingdon Ranch: Four Seasons, Six Generations
David K. Langford and Lorie Woodward Cantu

Heads above Water: The Inside Story of the Edwards Aquifer Recovery Implementation Program
Robert L. Gulley

Border Sanctuary: The Conservation Legacy of the Santa Ana Land Grant
M. J. Morgan

Fog at Hillingdon
David K. Langford

Texas Landscape Project: Nature and People
David Todd and Jonathan Ogden

Discovering Westcave: The Natural and Human History of a Hill Country Nature Preserve
Christopher S. Caran and Elaine Davenport

Rise of Climate Science: A Memoir
Gerald R. North

Wild Lives of Reptiles and Amphibians: A Young Herpetologist's Guide
Michael A. Smith

Wild Focus: Twenty-five Years of Texas Parks & Wildlife Photography
Earl Nottingham

The Art of Texas State Parks: A Centennial Celebration, 1923–2023
Andrew Sansom and Linda J. Reaves

Duck Walk: A Birder's Improbable Path to Hunting as Conservation
Margie Crisp